COSMOLOGICAL EFFECTS OF SCATTERING IN THE INTERGALACTIC MEDIUM

To order additional copies or download an electronic version, visit
http://www.lulu.com/content/7198406

*Cover image illustrates the density of galaxies in the Hubble Ultra Deep
Field in which only a couple of the objects are stars in the Milky Way.
The image was obtained from STScI. It was prepared for NASA
under Contract NAS5-26555. The image is available on the internet at:
http://hubblesite.org/gallery/album/the_universe/pr2004007m/large_web/*

The Ancient Mariner Books
www.theancientmarinerbooks.com
Vaughan Publishing

*This book is dedicated to my wife and children who have tolerated my abstraction of thought over the years that these and related concepts were a major preoccupation of my daytime and sleepless nighttime hours. These wonderful people deserved much more than my full attention and too often received very much less. Anyway...this was frequently what I was thinking about when they'd ask and I was too preoccupied to say more than, "Oh, **nothing**." In a way that was a reasonable approximation to the **something** whose density is at most a very few electrons and protons per cubic meter. But inappropriate nonetheless! I hope this is in some however-slight way an acceptable, although certainly inadequate, excuse.*

CONTENTS

List of Figures   *xi*
Preface   *xxi*
Forward   *xxiii*
Reader's Guide   *xxvii*
Chapter 1:   Introduction   1
Chapter 2:   Physical Characteristics of Objects Observed at Cosmological
             Distances   15
             a.   the sun and its position among stars   17
             b.   our local environment   21
             c.   cepheid variable stars   24
             d.   supernovae   27
             e.   quasi-stellar objects   30
             f.   gamma ray bursts   32
             g.   galaxies   34
             h.   galactic development versus 'evolution' of the universe   36
             i.   spectroscopic categorization of galaxies   39
             j.   clusters of galaxies and their cosmological setting   43
             k.   other objects of cosmological concern   43
             l.   cosmological observation methods   44
Chapter 3:   Characterization of the Intergalactic Medium   47
             a.   qualitative features   47
             b.   quantifying the intergalactic plasma density ranges   49
             c.   quantifying the intergalactic plasma temperature ranges   54
             d.   statistical properties of the plasma   57
             e.   applicability of relativistic analyses   60
             f.   typical misconceptions concerning properties
                  of the intergalactic medium   61
Chapter 4:   Interactions between Radiation and Electric Charges in the
             Substances through Which It Propagates   65
             a.   the electromagnetic wave equations   65
             b.   energy contained in electromagnetic fields   67
             c.   luminous flux   68
             d.   forces imposed on individual material charges   71
             e.   dispersion formulas   75
             f.   the Lorentz-Lorenz formula   78
             g.   composite Lorentz-Lorenz formulas   81

	h.	application of formulas to the intergalactic plasma	84
	i.	application of formulas to neutral hydrogen 'clouds'	89
Chapter 5:		Refraction of Radiation in Dispersive Media and Its Implications to Astronomy and Cosmology	91
	a.	the speed of light in a medium and its implications	92
	b.	Snell's law and some of its ramifications	94
	c.	common electromagnetic refraction phenomena	96
	d.	refractive properties of earth's atmosphere	97
	e.	effects of refraction in the solar chromosphere	102
	f.	expected refraction effects in the intergalactic medium	107
Chapter 6:		Absorption of Radiation in Dispersive Media and What Is to Be Expected in a Redshifting Intergalactic Plasma	109
	a.	Lambert's law	110
	b.	optical depth and related scattering concepts	111
	c.	broadband absorption in a plasma medium	113
	d.	effects of redshift on plasma absorption	116
	e.	deriving plasma properties from absorption data	122
	f.	absorption in neutral hydrogen	125
Chapter 7:		Absorption by Neutral Hydrogen in the Lyman-α Forests	129
	a.	distinctive spectra of quasi-stellar objects (QSOs)	129
	b.	quantum phenomena involved in absorption by neutral hydrogen clouds	130
	c.	the line width of atomic spectra	132
	d.	comparing emission-absorption profile data	135
	e.	anomalous effects of redshift on Lyman-α absorption	140
	f.	redshift distribution of neutral hydrogen clouds	141
Chapter 8:		The Unique Constraints of Forward Scattering	149
	a.	characterization of the scattered field	149
	b.	coherence of scattered radiation from separated events	151
	c.	concept of 'extinction' in a scattering medium	156
	d.	integrating coherent scattering effects	158
	e.	determining the extinction interval and base angle	159
	f.	characterization and limitation on spectral invariance	161
Chapter 9:		Relativistic Effects in High Temperature Plasma	167
	a.	significance of the transverse component of electron velocity	168
	b.	concepts of 'simultaneous' wave planes and coincident observation	169
	c.	locally stationary reference frame and local time	171
	d.	applying relativity to coherent forward scattering	173
	e.	Lorentz transformation equations	174
	f.	relativistic aberration equation	174
	g.	application of Lorentz equations to forward scattering	176
	h.	reversing the perspective of aberration	178
	i.	combining effects for individual electrons	181

      j.   geometrical considerations   184

      k.   relativistic time constraints   185

      l.   angular convergence of the detected radiation   186

      m. wavelength characteristics of the detected radiation 190

      n.   wavelength characteristics of the emanating radiation 192

      o.   convergence of the emanating radiation   193

Chapter 10: Implications of Conservation Laws to Scattering by Plasma
Electrons   197

      a.   Compton scattering formulas   197

      b.   applying conservation formulas to forward scattering 199

      c.   assessing the magnitude of wavelength change at each
extinction   203

Chapter 11: Derivation of a Distance-Redshift Relation   209

      a.   extending energy conservation to a redshift relation 209

      b.   exploring the range of applicability   212

      c.   matching Hubble's constant   215

      d.   some background on the distance-redshift relation 218

      e.   other examples of redshift occurring in ionized plasma 224

      f.   arguments supporting a logarithmic relationship 226

Chapter 12: Cosmographic (Metric) Predictions of the Standard and
Scattering Models   231

      g.   a variety of distance metrics, each with problems 231

      a.   cosmographic underpinnings of the standard model 235

      b.   comoving line-of-sight distance   238

      c.   comoving transverse (proper motion) distance 240

      d.   angular diameter distance   242

      e.   apparent angular size   244

      f.   luminosity distance   248

      g.   comoving volume   249

      h.   'lookback' time and distance   251

      i.   confusions on assignments of distance   253

      j.   the Tolman test of surface brightness   255

      k.   issue of model flexibility versus scientific refutability 256

Chapter 13: Comparing Predictions of the Scattering and Standard
Models against Observations   261

      a.   measurements of angular separation   261

      b.   luminosity vs. redshift of standard candles   264

      c.   details of SN1A data   267

      d.   gamma ray burst data   269

      e.   surface brightness data   270

Chapter 14: Claims of Acceleration of Expansion and Time Dilation
in SN1A Data   275

      a.   rationale for inferring an acceleration of expansion 276

      b.   background on time dilation claims   278

      c.   SN1A decay rate considerations   279

	d.	mimicking of time dilation by a Malmquist bias	282
	e.	more recent claim of time dilation in SN1A data	285
Chapter 15:	Comoving Number Densities of Galaxies		289
	a.	cautions in predicting comoving number densities	289
	b.	the Schecter distribution of galaxies	297
	c.	estimating galaxy counts versus redshift	303
	d.	assessing luminosity effects on observations	304
	e.	wrap up on comoving number density analyses	310
	f.	claimed evolution of galaxy types with redshift	314
Chapter 16:	Techniques and Estimates Used in Evaluating the Density of the Universe		317
	a.	mass-luminosity relationships	317
	b.	Newton's laws and Keplerian mechanics	320
	c.	the virial theorem	321
	d.	gravitational lensing	326
	e.	mass measurement results	329
	f.	calculating the mass of galaxy clusters	330
	g.	characteristics of 'rich' galaxy clusters	333
	h.	determining the mass of individual galaxies	339
	i.	calculating contributions to galaxy rotation	342
		1. central massive bulge – radius R_b and density ρ_b	342
		2. uniform thickness disk – thickness Δ_d and density ρ_d	343
		3. uniform density spherical halo – density ρ_h	343
Chapter 17:	Profound Implications of the Plasma Redshifting in Rich Cluster Cores That Produces Apparent Velocity Scatter		345
	a.	partitioning properties of the intergalactic medium	345
	b.	determining static redshift scatter of cluster galaxies	348
	c.	follow-on activities with regard to galaxy clusters	358
Chapter 18:	Assessing Requirements for, and an Alternative Explanation of 'Dark Matter'		363
	a.	current rationale for 'dark matter'	363
	b.	What acts as dark matter and where does it occur?	366
	c.	a different explanation of the velocity dispersion	367
	d.	intracluster gas contribution to cosmological redshift	370
	e.	observations of redshift clumping	374
	f.	the rationale for 'dark matter' in rotating galaxies	378
	g.	additional data on the 'dark matter halos' of galaxies	379
	h.	resolution of 'dark matter' in the scattering model	380
Chapter 19:	The Measured Background Radiation Spectrum		381
	a.	high temperature thermal spectrum	382
	b.	the microwave background spectrum	386
	c.	the precise determination of the spectrum	388
Chapter 20:	Theoretical Considerations Involving the Redshifting of a Blackbody Spectrum		395

a. characterization of the blackbody spectrum 395
b. temperature dependence of the blackbody spectrum 399
d. dependence on the density of matter 403
e. effect of redshift on a blackbody spectrum 406
e. Wien's law 408
f. considerations of an expanding thermal surface 410
g. alternative consideration of a stationary state 412

Chapter 21: The Relevance of Olbers' Paradox 415
a. description of the paradox 415
b. the proper resolution of the paradox 418
c. thermalization effectiveness of various forms of matter 419
d. quantifying 'sky cover' for uniform density particles 420
e. sky cover as the closure criterion for a 'cavity surface' 422
f. applying the analysis to thermodynamic electrons 425
g. parameter averaging appropriate to H_0 and η 428

Chapter 22: Standard Model Explanation of Cosmological Effects 433
a. the standard model scenario 433
b. the standard model calculations 440
c. counter argument discussion 442

Chapter 23: The Scattering Model Explanation of Cosmological Effects 447
a. hydrogen-helium ratio – coincidence or explanation 450
b. determining the extent of the agreement 453
c. the thermalization process 455
d. implied kinetic temperature in a redshifting medium 458
e. implications of variability of dynamic pressure 460
f. arguments for perpetuation 463
g. summarizing the scattering model 465

Chapter 24: Thermonuclear Reactions in the Intergalactic Plasma 467
a. hypotheses with regard to the hydrogen-helium ratio 467
b. significance of primordial hydrogenous plasma 469
c. equilibrium equations for an hydrogenous plasma 471
d. proceeding as far as helium 477
e. determining the elemental abundance for two models 481
f. the other possibility – massive object explosions 483

Chapter 25: Matters of Gravity 487
a. Einstein's "greatest error" 488
b. the 'cosmological equation' 492
c. the effects of pressure 493
d. uniformity of matter in the universe 494
e. expanding universe hypothesis 495
f. the 'critical density' 498
g. the missing matter 500
h. inherent problems in the theory 501

Chapter 26: Cosmogony and Other Flights of Fancy 503

	a.	theories of everything and other flawed logic	503
	b.	mixing mythology and theology with science	507
	c.	exploitation of the 'anthropic principle'	508
	d.	more mundane errors in reasoning	517

Chapter Z+1: Conclusions — 519

    a. brief summary of results of the scattering model — 520

    b. extending analytical results — 524

        1. Absorption effects applicable to the dispersion associated with the propagation of electromagnetic radiation through a plasma — 524

        2. Convergent diffraction effect associated with forward scattering processes through a plasma medium — 525

        3. Implications of redshift occurring throughout a scattering medium on the thermalization process — 526

    c. deeper questions — 527

    d. looking back at strongly held misinterpretations — 530

Appendix A: Electromagnetic Theory of Radiation — 535

Appendix B: Refraction in Spherically Symmetric Electron Densities — 543

Appendix C: Reopening the Book on Black Holes — 547

Appendix D: Frequently Encountered Constants — 557

    a. universal physical constants — 557

    b. constant properties of objects in the universes — 557

    c. unit conversion constants — 557

Bibliography — 559

Index of (First-Appearing) Authors — 575

LIST OF FIGURES

	View through a field of galaxies	xxvi
Figure 1:	Hertzsprung-Russell diagram of stellar characteristics	18
Figure 2:	Facts associated with the electromagnetic spectrum and spectroscopy	19
Figure 3:	Our location in the Milky Way galaxy	21
Figure 4:	View toward the center of the Milky Way in infrared	22
Figure 5:	The "Local Group" of galaxies with 49 galaxies	22
Figure 6:	The closest 200 clusters of galaxies containing 53,000 galaxies	23
Figure 7:	The closest 100 superclusters with 63,000,000 galaxies	24
Figure 8:	The seven trillion closest galaxies in ten million superclusters	25
Figure 9:	Cepheid variable stellar luminosity decline profile	26
Figure 10:	The two cepheid variable types period - luminosity relationships	26
Figure 11:	Edwin Hubble's initial tentative conjecture	27
Figure 12:	Supernova luminosity decline profile	28
Figure 13:	Redshift versus distance indicator and required SN1A data correction	29
Figure 14:	A composite spectra of QSOs in their rest frame	31
Figure 15:	A couple of examples of QSO spectra	31
Figure 16:	Light-curves for a single-peaked gamma ray burst event with a good signal-to-noise ratio	32
Figure 17:	The distribution of detected GRBs	34
Figure 18:	Galaxy developmental stages sometimes mistakenly referred to as 'galaxy evolution'	35
Figure 19:	Images of several galaxy types	37
Figure 20:	Normalized spectral intensity for a b_j=19.2 emission-line galaxy at a redshift of z=0.067 and a b_j=19.3 absorption-line galaxy at z=2.46	39
Figure 21:	The mean spectra corresponding to the five 2dF spectral types	40
Figure 22:	Elliptical galaxy spectra	41
Figure 23:	Spiral SO and SBO galaxy spectra	41
Figure 24:	A set of spiral Sb galaxy spectra	40
Figure 25:	Nearby galaxy M81 as observed at various frequencies	43
Figure 26:	The UBVRI (or 'Johnson system') of color filters	44
Figure 27:	K-correction as a function of redshift and galaxy type	45
Figure 28:	Constituents of intergalactic space	50
Figure 29:	Illustration of line of sight calculations	51
Figure 30:	Product of intracluster plasma parameters	53
Figure 31:	Light element ionization properties	55
Figure 32:	Classical versus relativistic representation of dependence of total energy on electron velocity	58
Figure 33:	Relativity implication on high temperature plasma electron velocity	59

Figure 34: Non-normalized Maxwell-Boltzmann distribution of electron
            velocities for various temperatures at equilibrium conditions   61
Figure 35: Field vector, propagation, and coordinate direction assumptions   67
Figure 36: The creation of a scattered field by displacement of electrons   73
Figure 37: Relationship between resonant frequency and absorption constant   75
Figure 38: Real and imaginary components of complex index of refraction   79
Figure 39: Range of variation in complex index of refraction with changing γ 80
Figure 40: The allowed dispersion domains involving ω, ω_o, and γ 81
Figure 41.a: The various parameter domains affecting $|Re(n) - 1|$ 82
Figure 41.b: The various parameter domains affecting $Im(n)$ 82
Figure 42: The emerging form of the plasma absorption factor   87
Figure 43: Snell's law of refraction with stratified indexes of refraction   94
Figure 44: Apparent compression of solar disk at sunset   98
Figure 45: Angles to actual and observed solar disk   98
Figure 46: Atmospheric density and temperature with altitude data   99
Figure 47: Atmospheric index of refraction for various wavelengths   100
Figure 48: Atmospheric deflection due to refraction at various viewing angles   101
Figure 49: Lensing effects earth's atmosphere would have from deep space   102
Figure 50: Deflection of starlight during eclipse of 1922   103
Figure 51: Electron density measures for the solar atmosphere   104
Figure 52: Index of refraction and associated deflection angles of observations
            made through various minimum altitudes in earth's atmosphere   105
Figure 53: Illustration of a few scattering medium concepts   112
Figure 54.a: Dependence of total absorption on wavelength for various
            values of propagation distance   115
Figure 54.b: Dependence of total absorption on propagation distance
            for various values of wavelength   115
Figure 55.a: Luminous flux loss in propagating through intergalactic plasma   117
Figure 55.b: Luminous flux loss in plasma as a function of redshift   117
Figure 56.a Dependence of total absorption on *observed* wavelength
            for various values of redshift   120
Figure 56.b Inset for small values of total absorption on *observed*
            wavelength for various values of redshift   120
Figure 57.a Dependence of total absorption on redshift
            for various values of *observed* wavelengths   121
Figure 57.b Inset for small values of total absorption on redshift
            for various values of *observed* wavelengths   121
Figure 58.a: Range of values of absorption factor in the exponent of
            Lambert's law applied to the Lyman-α absorption line for
            various values of γ 127
Figure 58.b: The affect of γ on the peak Lyman-α absorption   127
Figure 59: Lyman-alpha forest associated with QSO HE 2217-2818 at z=2.4   130
Figure 60: Hydrogen emission/absorption line series   131
Figure 61: Absorption factor showing the natural line width of the hydrogen
            Lyman-α line   133

Figure 62: Spectral line broadening effects   134
Figure 63: The operative phenomena of Lyman–α clouds, show the effect of
            HI clouds on the observed spectra.   136
Figure 64: The measured redshift of a young galaxy in a cluster near
            TN J1338-1942 which is at z = 4.1   137
Figure 65: Depiction of operative phenomena of observations over distances
            for which redshifting is appreciable with intervening HI clouds   138
Figure 66: The Gunn-Petersen trough as it affects QSO spectra   139
Figure 67: Typical Lyman-break galaxy spectra vs. redshift, with
            IRAC GOODS Legacy team survey limits indicated   140
Figure 68: Form of predicted absorption characteristics by neutral hydrogen
            on both sides of the Lyman-alpha absorption line   144
Figure 69: The Gunn-Petersen trough of neutral hydrogen absorption as
            implied by the Lorentz-Lorenz formula and Lambert's law   145
Figure 70: Symbolized effect of the predicted absorption characteristics
            of neutral hydrogen shown in figure 68 and 69 on uniform spectra
            on both sides of the Lyman-alpha absorption line showing a very
            basic agreement with the observations shown in figure 67   147
Figure 71: Illustrating conditions for coherent forward scattering   152
Figure 72: Illumination from two separated scattering events   153
Figure 73: Geometrical considerations for integrating effects of forward
            scattering   155
Figure 74: Coherency domain in a scattering medium   157
Figure 75: Illustrating conditions for distributed source coherence   164
Figure 76: Spectral coherence applicable to thermal sources   164
Figure 77: Distances traveled by electrons after scattering event   167
Figure 78: Scattering planes of two electrons from their perspectives
            at the tips of respective coherency domains   170
Figure 79: Situation with the various electrons that 'coincide' at P_o at time $t_o=0$ 174
Figure 80: Temporal relationships of events on wave plane surfaces of incident
            radiation resulting from Lorentz transformation   175
Figure 81: Geometrical relationship between coherency domain in L_o and
            its Lorentz-transformed counterpart in L_β 177
Figure 82: Geometrical and temporal relation between scattering events
            for relatively moving electrons   179
Figure 83: Geometrical relationship between the coherency domain in
            L_o and its Lorentz-transformed counterpart in L_β 180
Figure 84: Suggestion of a composite view of the various scattering domains
            for the variously moving electrons   182
Figure 85: Depiction of unique coherency domains applicable to relativistic
            electrons   183
Figure 86: Geometrical considerations that allow a spherical wave front to be
            treated as approximately a planar wave front   185
Figure 87: Frequency of electrons obtaining forward scattering intensity from
            various angles for $T = 4 \times 10^7$ K   187

Figure 88:   Slight angular preference of individual electron scattering   194
Figure 89:   Angular dependence of continuing forward-scattered radiation   195
Figure 90:   Effective coherency domain about (and scattering from) P_o in L_o 196
Figure 91:   Depiction of Compton scattering relations   198
Figure 92:   Geometry of Compton-like scattering effect on an individual plasma
             electron involved in forward-scattering   199
Figure 93:   The continual refocusing of electromagnetic energy through
             extinction via high energy plasma electrons   201
Figure 94:   Distance versus number of extinction intervals (longer wavelengths)   206
Figure 95:   Distance versus number of extinction intervals (shorter wavelengths) 207
Figure 96:   The logarithmic form of the relationship between distance and
             number of extinctions   208
Figure 97:   Form of relationship for 'redshift' versus number of extinctions   211
Figure 98:   The emergence of a distance versus redshift relation independent
             of wavelength over a broad range   212
Figure 99:   The broad range of applicability of the linear phase of distance-
             redshift relation and the tremendous disparity at extremely
             short wavelengths   213
Figure 100: Distances and wavelengths at which meaningful distance-redshift
             relations begin   214
Figure 101: Predicted logarithmic relationship between distance and redshift   217
Figure 102: History of efforts to refine measurements of H_o 220
Figure 103: Several plasma redshift application domains   227
Figure 104: Illustration of the significance of the logarithmic relationship
             between distance and redshift   228
Figure 105: Illustration of the need for testing the logarithmic relationship
             between distance and redshift in a nonlinear spacetime   229
Figure 106: Alternative conceptions of spacetime metrics   233
Figure 107: Cutaway view of distance metrics used by standard models using
             the analogy of curvature in three-space: Line-of-sight comoving
             distance D_C, comoving transverse distance D_M, angular diameter
             distance D_A for assessing angular size, and lookback distance D_{LB} 235
Figure 108: Line-of-sight comoving distance D_C versus redshift predictions for
             the *standard model* with various density parameter values (but with
             $\Omega_R = 0.0$ and $\Omega_M + \Omega_\Lambda = 1.0$) as well as for the *scattering model*   240
Figure 109: Comoving transverse (proper motion) distance D_{CM} versus
             redshift predictions for the *standard model* with various
             density parameter values as well as the *scattering model*   242
Figure 110: Angular diameter distance D_A versus redshift predictions for the
             standard model with various density parameter values. The
             analogy for the *scattering model* is also shown.   243
Figure 111: Apparent angular size predictions of the *Standard Model* with
             various percentages of baryonic mass, along with the curve
             for the scattering model   245

Figure 112: Luminosity distance (D_L in centimeters) *standard model* with
various density parameter values as well as for the *scattering model* 247

Figure 113: Luminosity distance modulus (D_M in relative magnitude)
predictions for the *standard model* with various density
parameter values as well as the *scattering model* 248

Figure 114: The dimensionless comoving volume element with several
density parameter variations for the standard model 251

Figure 115: Circumscribed volume out to a given redshift (panels a and b) 252

Figure 116: Lookback time/distance predictions for the *standard model* with
various density parameter values as well as the *scattering model* 253

Figure 117: Standard model flexibility versus scattering model stability 257

Figure 118: The angular size data predictions for the various models with
plots of median angular sizes of samples of galaxies (open
circles), galaxies in narrow luminosity range (crosses), and
quasars (filled circles) 262

Figure 119: The angular distance relation is plotted for three different standard
model cosmologies, all assuming H_0 = 60 km per sec per Mpc;
scattering model added 263

Figure 120: Redshift-magnitude relation for radio galaxies 265

Figure 121: The SNIA supernova data (log plot) 266

Figure 122: The SNIA supernova data superimposed on model predictions 267

Figure 123: The SNIA supernova individual data sets and set residuals 268

Figure 124: GRB luminosity equivalent 270

Figure 125: The basis for proclamations of acceleration of the universe 277

Figure 126: SN1A decay data for events of differing brightness 281

Figure 127: SN1A brightness decay profiles 283

Figure 128: Confused inferences of time dilation in SN1A profiles 284

Figure 129: Day-by-day spectra of 13 SN1A supernova. The standard and
maximum deviations are shaded light and dark respectively
with the mean indicated by the dark line 286

Figure 130: SN1A aging data interpreted as time dilation 287

Figure 131: Breakpoints labeled (Lf) of number of galaxies of a given inherent
luminosity that will be observed with a given instrument 293

Figure 132: Predicted constraint from the various models 294

Figure 133: Distribution of absolute (inherent galaxy luminosity) with redshift 295

Figure 134: Simplified predictions of comoving density 296

Figure 135: Surveys indicating the dramatic increases in numbers of observed
galaxies as one increases the resolving power of instrumentation 297

Figure 136: Rest-frame K-band luminosity function in four redshift bins 298

Figure 137: Schecter luminosity distribution function and its color dependence 299

Figure 138: Observed K-band galaxy number-luminosity function 300

Figure 139: Galaxy differential number counts dN(m) from a number of
surveys are plotted to completeness limits in five color bands 302

Figure 140: Rest-frame K-band luminosity function – current and redshifted 302

Figure 141: 2dF galaxy redshift survey (2dFGRS) data   303
Figure 142: Luminosity functions of 3000 galaxies partitioned into the five
   identified 2dF types   305
Figure 143: The effect of redshift on the galaxy luminosity function $\Phi(L,z)$ 306
Figure 144: The differential number of galaxies to be encountered of a given
   inherent luminosity in a solid angle and redshift interval   308
Figure 145: The total galaxy count in a three order of magnitude interval of
   inherent luminosity as a function of redshift for various models   309
Figure 146: The differential galaxy count as a function of redshift for the
   standard and scattering models   310
Figure 147: The differential galaxy count for various instrument capabilities
   as predicted by the scattering model   311
Figure 148: Distribution of photometric redshifts and a best-fit analytic
   description as well as the distribution of those entries for which
   spectroscopic redshifts were available   311
Figure 149: The uncertainty in photometric versus spectroscopic redshifts   312
Figure 150: Number of galaxies versus observed luminosity out to the
   various redshift limits ($\alpha = 0.65$)   313
Figure 151: Elbow in the number of galaxies versus observed luminosity
   curves for various values of α 313
Figure 152: Magnitude of galaxy number counts in the B and K passbands   314
Figure 153: Kepler's and Newton's contributions to orbital mechanics   321
Figure 154: Escape criterion for gravitationally bound systems   325
Figure 155: Velocity dispersion in the Coma cluster   327
Figure 156: Geometry of gravitational lensing   328
Figure 157: Observation using a gravitational lens   329
Figure 158: CFA Redshift Survey data   331
Figure 159: Temperature in intracluster core regions of PKS 0745-191   335
Figure 160: X-ray intensity of cluster gases in PKS 0745-191   336
Figure 161: Estimated temperature of cluster gases in PKS 0745-191   336
Figure 162: Cluster radial velocity dispersion (sigma-r) vs. gas temperature
   for 41 clusters   337
Figure 163: Implied baryonic mass density for various clustering assumptions   338
Figure 164: The lower panel shows a composite rotation curve produced
   by combining the CO result and HI from the top panel data for
   the outer regions   340
Figure 165: Rotation curves of spiral galaxies obtained by combining CO data for
   the central regions, optical for disks, and HI for outer disk and halo   341
Figure 166: Rotation rates plotted on a log scale show the bulge rotation is
   Keplerian, in the case of the Milky Way galaxy, from the center
   out to 0.001 of its total radius   341
Figure 167: Rotation rates vs. distance from the center of the galaxy M31   342
Figure 162: Electron density measures in intracluster plasma   346
Figure 169: Electron temperature measures in intracluster plasma   346

Figure 170: Dynamic pressure product of intracluster plasma parameters   348
Figure 171: Determination of redshift 'velocity' of galaxy #1 in cluster plasma   349
Figure 172: Redshift 'velocity' of galaxies at various angles and radii in cluster   351
Figure 173: Insert from previous figure   352
Figure 174: Redshift 'velocity' of galaxies at various angles and radii in cluster   353
Figure 175: Plasma redshifting at various angles through a galactic cluster   354
Figure 176: Simplifying geometrical symmetry of galaxy cluster distribution   355
Figure 177: Plasma redshift scatter at various angles through a galactic cluster   356
Figure 178: Determination of average line-of-sight separation of galaxy clusters   360
Figure 179: Increased plasma redshift effect through the core of galaxy clusters   369
Figure 180: Cosmological impact of redshift through the core of galaxy clusters   371
Figure 181: Punctuated expression of Hubble's cosmological redshift constant   372
Figure 182: Mapping process using Hubble's cosmological constant   373
Figure 183: Grouping of galaxies that results using Hubble mapping   374
Figure 184: Apparent compression of field galaxy redshift spacing   375
Figure 185: Density ripples in the SDSS redshift survey from $0 < Dec < 6$ 376
Figure 186: Peaks and valleys in the SDSS redshift profile   377
Figure 187: The background spectrum modeled at 40 keV   383
Figure 188: The measured X-ray background spectrum   384
Figure 189: Spectrum of the extra-galactic sky – energy from the radio band
       to gamma-rays   387
Figure 190: Measured microwave background radiation   389
Figure 191: "Theoretical" blackbody radiation for 2.725 K temperature   390
Figure 192: Tolerances on NASA Microwave background data   390
Figure 193: FIRAS data with exaggerated error bars included   391
Figure 194: Blackbody radiation for 2.725 K temperature   391
Figure 195: Measured variations in microwave background radiation intensity   392
Figure 196: "Processed" variations in background radiation intensity   392
Figure 197: WMAP cosmic microwave background data   393
Figure 198: Apparatus for observing 'cavity' radiation   396
Figure 199: Blackbody spectrum and its first derivative   397
Figure 200: Blackbody spectrum plotted on log scale   398
Figure 201: The effect of temperature on spectral distribution   399
Figure 202: The effect of temperature and redshift on spectral distribution   400
Figure 203: 'Cavity' embedded in an ideal gas at a fixed temperature   401
Figure 204: Radiation situation in an ideal gas   402
Figure 205: thermodynamic gas density constraints   405
Figure 206: The effect of redshift on blackbody spectral distribution   407
Figure 207: The contributions to be integrated to obtain the intensity of
       radiation realized for a given wavelength in an extended
       redshifting medium   413
Figure 208: Cancellation of distance related relationships in observation   416
Figure 209: View through a slice of randomly distributed objects of uniform size   417
Figure 210: Surface brightness dominated by closer of uniformly sized objects   417
Figure 211: Sky coverage (curve #2) of uniformly sized objects   422

Figure 212: Distance and redshift to half sky cover of plasma electrons   424

Figure 213: Log of the ratio of the average kinetic and radiational temperatures versus the average electron density of an extended thermal medium   429

Figure 214: Reconstructed figure from Misner (1973) depicting key elements of the standard cosmological model   434

Figure 215: The period and impact of Alan Guth's inflation   436

Figure 216: Diagram taken from Wagoner et. al. (1967) depicting key elements of the standard cosmological model   437

Figure 217: A schematic outline of cosmic history – the standard model   440

Figure 218: The universe as conceived according to the scattering model   448

Figure 219: The universal average and typical intracluster gas densities   457

Figure 220: Kinetic temperature versus electron density of the medium as implied by the observed 2.725 K background radiation   459

Figure 221: Cross section of D-T, D-D, and D-^{3}He reactions   476

Figure 222: Proton fusion process reaction rates   477

Figure 223: Deuterium (D-X) reaction rate data   478

Figure 224: Interrelationships of thermonuclear reactions leading to ^{4}He   479

Figure 225: Thermonuclear stability diagram – excluding elements beyond helium   480

Figure 226: Nuclear abundance percentages as a function of temperature and time for the big bang model   482

Figure 227: Elemental abundances that would be realized as a function of density in a large high temperature exploding object   483

Figure 228: Elemental abundances that would be realized as functions of temperature in any large high temperature exploding object   485

Figure 229: Applying Gauss's integral theorem to embedded 'Hawking spheres'   491

Figure 230: Thermodynamic forces active in the structure- producing processes of the universe   493

Figure 231: The manifest uniformity of the universe at large enough scales   496

Figure 232: A visualization of Einstein's conception of a four-space universe   497

Figure 233: Theories of everything tend to divert attention away from science   504

Figure A1: The right-hand vector cross product rule   536

Figure A2: Plane and circularly polarized solutions of Maxwell's homogeneous differential equations   540

Figure A3: Minkowski spacetime diagram showing the propagation of advanced and retarded waves from an emission locus at $(x,t)=(0,0)$ 541

Figure B1: Geometry of refraction in a spherically symmetry situation   543

Figure B2: Detail of geometry for recursive triangle   544

Figure B3: Determination of integration interval   546

Figure C1: Nuclear charge densities   551

Figure C2: Significance of fuzziness in the mass distribution in a 'fermion gas' of neutrons as would be realized in a collapsed neutron star   554

Preface

The objective throughout the development of this cosmological model based on scattering in the intergalactic medium has been to rely exclusively on established physical concepts. As a consequence the resulting approach to accounting for the Hubble relation, luminosity-redshift data, background radiation, and other phenomena that have become classed as 'cosmological' does not depend on speculative physical concepts or theories. It most certainly does not require an expanding universe emerging from a primeval explosion of however many dimensions, demanding unknown physical concepts to justify an inflationary period, evolving physical constants, or mysterious assortment of parameter values to satisfy various tests of theory, etc..

While some physical and mathematical concepts must be understood as a prerequisite for understanding the effects of scattering of electromagnetic radiation by an intergalactic medium, there is nothing with which anyone familiar with the most basic concepts of physical theory should have much difficulty. All concepts employed by these hypotheses are described in detail so that anyone with a keen mind should be able to follow the discussions whether the details of the mathematical equations are fully understood or not. Nor should any reader work through derivations for which conclusions seem obvious or are already understood by him or her.

The author has attempted to graphically illustrate those concepts that seem most difficult to understand and/or most essential to his conclusions. He hopes that the dutiful reader will expend only as much effort as he or she will consider well spent in the pursuit of an increased understanding of this vast universe we glimpse so magnificently with the help of the intergalactic medium through this so-short a window in spacetime.

Ray Bonn, 2009

Forward

This volume discusses a series of discoveries with regard to the nature and behavior of the vast regions of space between galaxies that explain phenomena that have previously been aggregated using quite complex theoretical conjectures into what is commonly referred to as a 'standard cosmological model'. This independent basis of explanation therefore becomes an 'alternative' to that 'standard'.

It is well known that intergalactic space not totally devoid of matter, nor is it 'cold'. In fact, it is filled with hydrogenous plasma whose temperature and density, and therefore its ionization level and free electron density, vary over a broad range with distance from the centers of galaxy clusters. Electromagnetic energy transmitted through such a plasma is redshifted by a relativistic transverse Doppler mechanism in an amount commensurable with the dynamic pressure of the plasma. This pressure is proportional to the product of the temperature and density of the plasma. Thus, the total accumulated redshift of light transmitted over appreciable distances depends upon the average of the pressure along the transmission path of the radiation. Since galaxies are typically grouped into galaxy clusters that are uniformly distributed throughout the universe, the amount of observed redshift over cosmological distances is naturally the same in all directions.

In addition to Lyman-alpha forests where neutral hydrogen virtually obliterates high frequency radiation from great distances by normal molecular absorption processes, there is a plasma absorption effect that results in a broad band redshift-dependent attenuation of radiation at all frequencies. This aspect of scattering phenomena produces what is equivalent to the relativistic effect of time dilation just as if there were recessional velocities of the sources of the radiation.

Together these plasma effects result in the observed redshift-distance and luminosity-redshift relationships without requiring a universal expansion. The predicted effects in a uniform distribution of galaxy clusters result in observed comoving number densities of galaxies with redshift. This and other predictions agree closely with those of the standard cosmological model, but are in fact more precise than any version of that model according to virtually every metric. Supernova observations at great distances require modification of the standard model to include a conjectured acceleration in order to obtain agreement with observations. The scattering model also excels with regard to predicting surface brightness with redshift, subtended angle-redshift dependence of distant objects, etc..

The enigma of cold (2.725 K) microwave background radiation in a universe, all of whose components exhibit temperatures that are many orders of magnitude greater, is resolved by the same redshifting effects in an extended stationary state thermal medium. It is shown that the effects of scattering that is the cause of thermalization and which establishes and maintains the equilibrium in any medium, results in a separation of the observed radiation spectrum temperature from the temperature of the medium itself whenever there is redshifting of the associated radiation. The amount of this separation is again a function of the temperature and density of the medium, but in this case rather than the average of the product of these parameters, it is the product of the separately averaged parameter values that is involved. Thus, the more dense and intensely hot regions in galaxy cluster cores contribute more substantially to observed cosmological redshift, while the much more vast sparse regions at somewhat lower temperatures that exist between clusters provide the primary contribution to the thermalization of the observed background radiation.

The extreme effectiveness of the plasma redshifting mechanism in the extremely hot and relatively dense regions about the centers of galaxy clusters results in the so-called 'fingers of god' phenomenon observed in redshift surveys. It is the extreme 'spectral velocity' gradients across these gravitationally bound clusters that necessitated incorporating the assumption of 'dark matter' as a conceived means of producing the observed redshift scatter if that were to have been caused by the gravitationally-induced orbital velocities. However, if these were actual velocities as assumed by the standard cosmological model, they would tear the structures apart.

The large predicted plasma reshifts encountered across these structures average out at cosmological distances to effect the observed Hubble constant, predicting also the easily observed major fluctuations in the numbers of galaxies at increasing redshift in redshift surveys. Similarly the rotation rates of large spiral galaxies that extend further than their luminous mass would suggest result from increased density of the plasma halos in the vicinity of the more massive galaxies rather than from mysterious dark matter.

Finally, one must account for the relative abundances of the light elements. The universal relative abundance of hydrogen and helium, as well as deuterium, requires developments other than those that occur in stars. It requires much more extreme high-temperature explosive forces. The seminal work done as the primary support for the rationale assumed by the standard model is compatible with less than a 'big bang' of the entire universe. In fact the temperature and density requirements are not that dissimilar to what is experienced in the hot plasma at the centers of the larger galaxy clusters. The characteristic reactions certainly occur at temperatures and densities associated with conflagrations called 'gamma ray bursts' that are observed throughout our universe. Whatever their cause, these extreme explosions recycle matter back into intergalactic regions in a primordial hydrogenous plasma state tending to maintain the abundance ratios and a stationary state of the universe itself.

This volume, although concentrating on the individual effects of scattering in a plasma medium, provides in toto a comprehensive cosmological model that is based exclusively on these electromagnetic effects.

Reader's Guide

This volume will differ considerably from what the reader may have expected of a book whose title announces only that it discusses cosmological effects of a given cause. This is not, as one might have supposed, just a discussion, far less a consensus, of current thinking with regard to that topic. It is rather a treatise documenting investigations into the nature of the physical processes of dispersion in the intergalactic medium whose profound ramifications are manifest in observations at cosmological distances. No doubt because of the dearth of material in intergalactic space and a prolonged lack of understanding of what little there is, the more subtle aspects of what would otherwise be merely mundane processes of dispersion in a scattering medium have not previously been investigated to a sufficient depth. Nor, therefore, have these effects been acknowledged as significant to our view of the distant universe. As the reader will learn, this lack of understanding of the contribution of scattering to observation processes has resulted in a totally inaccurate perception of the universe that is observed at cosmological distances.

In one sense this volume should be more exciting than a mere popularization discussing the state of cosmology but only addressing the resolved thinking of others concerning antinomies within its scope. This book strives to go beyond what is currently *understood*. It resolves the paradox of an apparent expansion of the universe without necessitating an *actual* expansion. It rejects the introduction of a big bang that must be assisted by an unaccountable inflationary period thereafter, rejecting also the resurrection of Einstein's acknowledged "greatest error" in order to account for what is observed. It does this without introducing any *exciting new physics* or exotic matter, or attempting to revoke Copernicus's insightful *cosmological principle* that has kept science to its task as disenthralled observer. No; there is nothing like that to which the reader may have become accustomed. Here hee or she will just find traditional disciplines of physics applied to plasma properties of the intergalactic medium. That is all.

Nonetheless, there is considerable complexity. This forward is directed toward helping the reader deal with that complexity and laying out what are sometimes arduous lines of reasoning to show him or her where each essential piece to this cosmological puzzle can be found and how they all fit together. That might otherwise prove difficult.

Of course this volume is not intended as a text to be employed in a curriculum for a university degree in astrophysics. However, with regard to those physical principles on which the effects depend, it may occasionally appear so. The material has been presented in this way so as to provide a complete and self-contained exposition. The reader who understands electromagnetic theory should probably scan rapidly through chapters that discuss traditional absorption and forward scattering theory except as noted below. Anyone who is skeptical of the possibility of a 'tired light' theory ever being able to resolve the associated notorious deflection problem should probably pore over those chapters that merge previously unrelated effects. The aberration effect of special relativity in hot plasma produces the necessary deflection as an integral part of forward scattering. The application of conservation of energy and momentum equations similar to what is applied to Compton scattering determines associated energy and momentum losses that ultimately produce a Doppler-like redshift. There is a detailed explanation of how coherent forward scattering by which we view everything we see, including cosmological objects, is altered accordingly by scattering through the intergalactic medium.

The various parts of this book specifically address the following topics: the cosmological observations themselves that provide the basis for refutation testing so essential to the process of science, the effects of dispersion that affect our view of the universe, and the accepted rationale for predictions made by proponents of the standard model. Progress in explaining the author's scattering model is not continuous. There are topics that need to be properly understood as a basis for understanding subsequent aspects of the model. For example, after coming to terms with the physics of electromagnetic scattering, one must understand the aberration and transverse Doppler effects of special relativity before one can understand the affect that hot plasma has on the forward scattering process. One must understand the plasma redshifting process to appreciate the affect irregularities in plasma parameters will have on the resulting cosmological redshift.

The author is not at all sure that the organization he has chosen for presenting this material is ideal. However, he is confident that the information has been sufficiently well covered so that the intelligent reader, willing to uncover it, can do so.

What follows is a reader's guide to what to expect in each chapter of the book, what key results accrue from each. This should assist the reader in skipping through it at a reasonably rapid pace to accommodate his level of interest in this new cosmological model.

Background Information

CHAPTER 1:

Here the reader will find a brief overview of the author's thesis, an overview comparison of this thesis with the standard cosmological model, and discussion of the approach the author will take to present his thesis.

CHAPTER 2:

Cosmology involves the observation of certain categories of astronomical objects that become the 'standard candles' for assessing extreme distances. A range of observation technologies and expertise is employed in this process. This chapter provides an informative catalog of the objects and observation techniques involved in these so-essential observational aspects of cosmology. Figures are provided that illustrate the nature of these objects and viewing technologies.

CHAPTER 3:

The effects to be described in this volume involve the unique characteristics of physical processes associated with the transmission of light through the intergalactic medium. That any view of the distant universe will inevitably be through a field of intermediate galaxies is obvious when one inspects images such as the Hubble field presented in the figure on the following page obtained from the Hubble gallery. In the figure all but a very few of the objects are galaxies, each containing billions of stars. This chapter describes what is known of the pervasive medium in and among these myriad galaxies that so affects our view of the even more distant universe.

Electromagnetic interactions of radiation and matter

CHAPTER 4:

Since it is in all cases the effects on the transmission of electromagnetic radiation that is at issue, this chapter provides a basis for understanding that radiation and the nature of its interactions with the charges present in any non-vacuous medium. A complete vacuum is never realized. The distance that light must travel through a medium determines whether a medium of a given density can be treated as a vacuum or not. Electromagnetic field expressions and the physics of scattering as embodied in the Lorentz-Lorenz formulas determine this.

http://hubblesite.org/gallery/album/the_universe/pr2004007m/large_web/

CHAPTER 5:

The velocity of light in a scattering medium does not comply with the universal restriction on the speed of light in a vacuum for every observer that is the second postulate of the special theory of relativity. The velocity, in fact, depends upon the electron density encountered in the medium. This results in dispersion and refraction wherever there are variations in the electron density as there is in the earth's atmosphere. Refraction encountered in earth's atmosphere is explored and compared with the effects of gravitational bending of light applicable to gravitational lenses employed in cosmology.

CHAPTER 6:

Although not directly involved in cosmological redshift, absorption plays a key role in observations that are essential to testing any hypothesis concerning cosmological effects. The attenuation properties of the medium including its optical depth and absorption are discussed. In particular, it is shown that in a redshifting medium such as the intergalactic plasma, there is broadband absorption that reduces luminosity by a factor of $1/(z+1)$, where z denotes redshift. This is in addition to the $1/(z+1)$ quantum effect due to the reduced energy in redshifted photons, and the classical effect involving the inverse square of distance. This broadband absorption is equivalent in effect to what standard models have attributed to time dilation of a receding source.

CHAPTER 7:

A hallmark of the various arguments for 'evolution of the universe' and a primary criticisms of any attempt at a 'tired light' explanation of redshift, has been the neutral hydrogen absorption profiles that intensify with increasing redshift in what are denominated "Lyman-alpha forests". This chapter shows that a uniform density of neutral hydrogen clouds in the intergalactic medium would produce precisely the observed narrow resonance band absorption effects when there is a redshifting medium throughout the regions between neutral hydrogen clouds.

Coherent forward scattering

CHAPTER 8:

A significant aspect of the scattering of light is that by which a dispersive medium effectively replaces photons as a part of a coherent forward scattering process. This process allows imaging of distant objects. The topic is not generally discussed in much detail because it has long been presumed that the wavelength of transmitted radiation is not affected by this process. So the effects through which photons are continually replaced have not seemed all that important to astronomy other than in observations made specifically to refine constraints on the constancy of the velocity of light propagated through a vacuum. Nonetheless, we will describe this forward scattering process in some detail as preparation for its primary role in the redshifting process. A determination is made of what is referred to as the 'extinction interval'. This distance after which incident photons are replaced by 'cloned' versions of themselves is inversely proportional to the wavelength of the radiation.

This preparation is a precursor to an altered determination of the forward scattering process applicable to cases where the involved scattering electrons have relativistic speeds. Diagrams of how the forward scattering process works and the scope of distances involved in effecting the process of extinction are presented.

CHAPTER 9:

A brief discussion of Einstein's special theory of relativity pertinent to its effects on light transmission is provided in this chapter. The transverse Doppler effect and relativistic aberration more generally are shown to significantly alter the geometrical relations involved in coherent forward scattering. Previous analyses of forward scattering by Born, Wolf, and others specifically omitted cases involving electrons with relativistic speeds from their investigations of secondary radiation. In hot plasmas these restrictions cannot be made to apply. Clearly, for the intergalactic, and particularly intracluster plasma, an altered form of the process is what applies. The concepts associated with the effects of relativistic aberration are clearly illustrated in this chapter.

CHAPTER 10:

The ramifications of the altered forward scattering process due to relativistic effects described in the previous chapter are pursued here. There is an implied lengthening of wavelength that occurs because of a Compton-like effect at each forward scattering extinction interval. It is shown how the amount of this lengthening of wavelengths at each extinction interval is related to the electron temperature of the plasma. In every conceivable case it is extremely minute. However, just as for transverse Doppler generally, the effect is unilaterally to increase wavelength at each extinction interval so that these, however 'tiny', effects will not 'average out' over distance but will instead accumulate proportionately with the number of extinction intervals through which the radiation propagates. The interactions associated with this process are characterized by conservation of energy and momentum. How these concepts apply are clearly illustrated in this chapter.

The emergence of a distance redshift relation

CHAPTER 11:
The lengthening of wavelength at each forward scattering event covered in the previous chapter will result in the transfer of momentum (and energy) to the medium. This in itself does not constitute 'redshifting' of radiation. However, in combination with the inverse proportionality of the extinction interval with wavelength, this being the distance between the wavelength-lengthening events, a redshift results. It is inversely proportional to the wavelength of the incident radiation as it must be, and is, therefore, indistinguishable from a recessional Doppler redshift traditionally associated with an expanding universe. The average dynamic pressure of a plasma medium determines the amount of the combined effect over a given distance.

What this means is that it is the average product of the electron temperature and density that determines the amount of redshift to be incurred over that given distance. Properties of the plasma required to match Hubble's constant – that has previously been used to measure what was considered to have been the rate of expansion of the universe – are assessed for comparison with what was presented in chapter 3 and will be elaborated in much more detail in chapter 23. Other instances where a plasma redshifting effect can be observed are discussed as a means for refutation testing. The significance of this product will be pursued in later chapters in the context of refuting inferences of 'dark matter' as well as pursuing implications of a thermalization process that has resulted in the microwave background radiation.

Cosmographic predictions and observations

CHAPTER 12:
Predictions of this scattering model of redshifting must be compared with those of the currently accepted versions of the standard cosmological model in order to establish differences that can be used as discriminators in determining which of the various models is correct. 'Cosmographic' predictions, i. e., quantified observable effects on the

observations of objects at cosmological distances, must be defined. For a meaningful comparison one must understand the basis of the predictions. So there is a brief discussion of the various versions of the standard cosmological model with their various selections of density parameters that each produce quite different predictions.

Despite the hullabaloo about a distance-redshift relation at extreme distances, 'distance' *per se* is not something that can be directly measured. Several distance-related parameters *are* observable although each is a somewhat theory-laden concept. There are considerable differences in predictions with regard to the associated observations; this is especially true even for those model variations within the umbrella of the 'standard cosmological model'. Perhaps the most useful of the metric predictions for comparison with actual data is the luminosity distance modulus. In this useful case, the author's scattering model predictions are very close to those of what is currently the most acceptable 'concordant' version of the standard cosmological model.

We introduce Tolman's surface brightness test of expansion.

Pros and cons of 'flexibility' exhibited by the standard model and the refutable inflexibility of the scattering model are discussed.

CHAPTER 13:

In this chapter we present observations of cosmographic data together with predictions of the various models. The scattering model predicts most cosmographic parameter values indistinguishably from the emerging consensus version of the standard cosmological model. The exception is in observed object diameter data for which all of the standard model versions predict strange angular distortions including distant objects appearing larger than the same object at a closer range. No such phenomena has been observed. The scattering model predicts expected Euclidean relationships that are closest to what is observed. On the Luminosity-distance relationship and the distance modulus, the scattering model predictions are better than those of any of the standard model versions. With the SN1A super nova sightings which have extended the distance to which observations can be made, the scattering model fits the data precisely whereas the best fit of the standard model versions requires adjustment of Hubble's constant to accommodate a perceived acceleration followed by a deceleration.

Detailed discussion of Lubin and Sandage's (2001) effort to perform the Tolman surface brightness test of expansion is provided.

Again, observational data overwhelmingly favor the scattering model predictions over any of the alternative standard model versions.

CHAPTER 14:

The rationale for the hypothesis that the universe's expansion rate is accelerating is examined. This conjecture is totally unnecessary in the scattering model. This extraordinary claim spawns supporting conjectures of "mysterious negative energy", "eternal inflation", and similar pithy phrases for which meaningful explanation has not been forthcoming. After a decade of wrestling with such illusive concepts there is now a contingent willing to abandon the cosmological principle altogether rather than abandon the standard cosmological model.

Time dilation associated with excessive recessional velocities has been a major factor in several aspects of the standard cosmological model. Of course the notion of time dilation derives from Einstein's special theory of relativity. Luminosity profiles of SN1A events show an aging process to which the concept of time dilation should certainly apply if expansion is indeed the mechanism of cosmological redshift. However, there is an inevitable Malmquist bias that filters observations in favor of the more luminous events, and more luminous SN1A events are associated with a much lengthier aging process that mimics a time dilation effect. Arguments for and against confirming the occurrence of time dilation are analyzed in this chapter.

CHAPTER 15:

Galaxy surveys that provide the number of galaxies observed at each redshift are considered in this chapter using predicted volumetrics appropriate to the new scattering, and various versions of the standard, models. There are literally trillions of galaxies as discussed in chapter 2. The issue of whether there is an 'evolution' in the number or type of galaxies that existed billions of years ago compared to the current epoch is major to the cosmology community. The scattering model – in part, perhaps, because it is based solely on the physics applicable in our current vicinity – is more or less based on an extended (if not actually infinite) uniformly filled universe very much like what we observe all around us. The standard models all assume an evolution from a big bang, out of which the universe as we see it today has 'evolved'. So there should be gross differences in the numbers and types of galaxies observed at unit redshift and beyond compared to what we see in the immediate environs of the Milky Way. However, galaxies in advanced stages of development observed in the distant past make this presumption difficult to accept. Even the determined age of stars in our own Milky Way galaxy makes it hard to believe. But it is still a matter for science to resolve by observation and comparison with predictions of the various models. That comparison is complicated by the facts of there being insufficient high resolution data from high enough redshift surveys due to current limitations of instrumentation. It is also complicated by the fact that predicted numbers according to the scattering model and the concordance standard model are extremely similar. This chapter provides data on the numbers of galaxies actually observed in various *windows* in various wavelength bands.

Conjectures of evolution of galaxy type by those who concede the possibility of numerical stability in comoving number densities are investigated as well. Their evidence is not convincing.

The distribution of matter and radiation in the universe

CHAPTER 16:

Much of cosmology from Einstein's first musings onward, and even more so at the current time, has been about the density of the universe. According to Einstein's general theory and gravitational model there is a critical density above which the universe would eventually collapse back upon itself. Less than that and it would expand forever. All standard models subscribe to one or the other of those basic agenda options; differences involve how that density is made up. Is it just that mundane stuff to which physics and chemistry has traditionally applied or is the universe comprised of weirder forms of matter never encountered on planet earth?

For these and other reasons, much effort has gone into assessing the amount of observable mass distributed throughout the universe. These efforts have involved what we know of the luminosity-mass ratio of stars and galaxies derived from our own neighborhood in and around the Milky Way. When this total mass of a large representative volume of the universe is taken into account, it is found to be insufficient to 'close' the universe, i. e., the numbers indicate that the universe should expand forever according to Einstein's formulation.

However, the 'virial theorem', derivative to Kepler's law of planetary motions and Newton's theories of motion and gravitation can also be used to assess the total mass of gravitationally bound systems from the peculiar motions of their constituents that can be measured spectroscopically. Thus, the masses of huge clusters of galaxies can be assessed by two independent methods. Spectacularly, the results of the two assessments do not agree. If one accepts redshift as being caused exclusively by recessional velocities, it follows that there must be considerably more 'gravitational mass' than 'luminous mass' in the universe. This has given rise to a resurrection of what Einstein considered to have been his 'greatest error' and the introduction of 'dark matter'. In the currently most accepted of the many versions of the standard cosmological model, that resurrected error accounts for about seventy percent of the mass in the universe.

All the confusion arises because the *presumed* Doppler redshift dispersion of galaxies in these clusters is much too high in most cases for the involved galaxies to even be considered gravitationally 'bound'.

Similarly, the observed rotation rates of spiral galaxies are quite excessive, continuing to spiral more rapidly than their luminous mass would suggest, requiring 'halos' that interpenetrate and extend to the outer reaches of these structures. This too is analyzed and resolved as due to plasma mass and redshift effects.

CHAPTER 17:

The author explores the data with regard to the intense plasmas known to exist throughout the cores of galaxy clusters. The electron densities and kinetic temperatures in these regions induce large redshifts – on the order of a hundred times greater than the universal average. The increased plasma densities and temperatures are then shown to account for the additional redshift based on the same redshifting process that produces the lesser cosmological redshift with a lesser average dynamic pressure throughout intergalactic regions.

CHAPTER 18:

In this scattering model explanation a new prediction begins to emerge, which is that galaxies in redshift surveys should be much more bunched than predicted by any standard model. This 'bunching' of redshifts has been documented by other researchers and is readily apparent in all redshift surveys. This fact is accounted in the scattering model as a result of increased redshifting in intracluster regions.

CHAPTER 19:

The observed thermal background radiation spectrum is described in this chapter. This includes in addition to the well-known microwave background radiation, the uniform background of ultraviolet and X-ray radiation. Sources of this radiation have still not been fully identified even though this uniform background was first detected even before the microwave background radiation. It is now clear that at least most of this high temperature thermal ultraviolet and X-ray radiation comes from among the clustered galaxies like those included in the figure provided on page xxvi.

Details of the microwave background radiation observations are discussed, including the degree to which it can be shown to be 'blackbody' radiation at the precise temperature of 2.725 K.

CHAPTER 20:

A blackbody radiation spectrum exhibits features that provide clues to its origin. In this chapter we discuss the nature of this 'cavity' radiation, what is implied by the Planck distribution, how the thermalization of photons occurs, and how energy is redistributed by being scattered by interactions with matter until it achieves the blackbody form. Alternative theoretical notions of how spectra are differently affected by redshift caused by one mechanism rather than another are discussed. A redshift mechanism effective throughout a cavity in thermal equilibrium would reduce the temperature of the radiation relative to kinetic temperature of the material particles involved in its thermalization. Even though at a different temperature the radiation would still maintain its blackbody form.

CHAPTER 21:

An analysis of Olbers' paradox, although it is certainly no longer a legitimate 'paradox' in any mysterious sense, is an instructive exercise. It is instructive because too many people who should know better believe that this supposed paradox is entirely resolved by redshift in a finite universe. That is not the case. Resolution involves densities of objects and the distances between them in a uniform distribution to effect complete enclosure as appropriate to 'cavity' radiation. This completely eliminates all but the intergalactic medium from culpability in this process. Stars would effect an insignificant fraction of the total sky cover even in an infinite universe.

The impact of redshift on thermalization of radiation is then discussed in the context of Olbers' paradox. This is where its most profound ramifications are manifest. Including clues to why the background radiation temperature is so much less than the temperature of everything else in the universe.

Cosmological explanations of the microwave background

CHAPTER 22:

In this chapter a more comprehensive view of how the standard cosmological model accounts for various observed phenomena is presented. Most notably we are concerned here with the supposed origin of the microwave background radiation. The discussion involves a detailed timeline scenario of the happenings from the time in the past at which the big bang is supposed to have occurred down to the present time. The description proceeds from Guth's inflationary period, the major purpose of which is to spline together emergence from a single point in spacetime into a sizeable enough region that what we see today could with some plausibility be expected to have developed.

After creation of basic particles, the explanation is that there was an annihilation of a billion times as many particles as are left in the universe today by matter and anti-matter pairing off into oblivion with only a slight numerical majority of matter prevailing. The annihilation process converts the mass of two particles into high-energy gamma radiation and non-interactive neutrinos. That is the explanation given by standard cosmological model adherents for the origin of the microwave background radiation.

Shortly thereafter a resulting hydrogenous plasma is envisioned to have converted 24% of its remaining mass into helium via thermonuclear reactions and thus to have produced the universally observed light element percentages that are pervasive in the current universe. Deuterium and mere traces of other light elements are explained as having been created at that time. The rest of what we observe of heavier elements has had to be created by the various on-going thermonuclear processes in stars. In this explanation it is perceived as merely a coincidence that there is such precise agreement between energy released in creating universal light element abundances and the energy present in the current microwave background radiation.

It is treated as a very coincidental fact applicable only to our particular moment of existence in the accepted timeline of events.

All the while these spectacular events are thought to have been occurring, universal expansion is envisioned to have continued unabated with the gamma radiations from annihilation continuing to be scattered by the material particles, a process continuously thermalizing radiation to maintain a blackbody form. Since expansion is conceived as 'adiabatic', electrons, nuclear particles, and radiation would all have been maintained at the *same* temperature even as temperatures continued to fall.

At a temperature somewhere around 3,400 K charged particles would have begun to unite by chemical bonding to create neutral hydrogen and helium. Scattering is perceived as no longer taking place, with (as one researcher puts it) "the universe continuing to expand beneath the photons". The photons proceed from this 'last scattering' as though from a solid wall on an expanding cavity at the 'decoupling' temperature. Photons continue to be redshifted as though from that receding surface. Finally the observed 2.725 K blackbody radiation is realized at the current moment and it continues to cool.

CHAPTER 23:

In this chapter a similar exposition is undertaken to summarize how the scattering model accounts for various observed cosmological phenomena, again most notably the microwave background. There is no timeline scenario of happenings in this case. The universe is envisioned as existing in a *stationary state* with a perpetuated energy balance, which is in no way to state that it is perceived as 'stationary' – far from it. The scattering model addresses all facets of the cosmos as having arisen naturally as a part of the day-to-day operations of the universe.

The facts to be accounted are the same: cosmological redshift characterized by Hubble's constant of 7.14×10^{-29} cm^{-1}, a universal helium percentage of 24% by mass, an extremely uniform 2.725 K blackbody background radiation spectrum with 4.176×10^{-13} ergs cm^{-3} of energy coming in from all directions. It varies by less than one part in 10^{-4} in all directions. In addition to that we know that the density of luminous (baryonic) mass in the universe is between 10^{-31} and 10^{-30} gm cm^{-3}. We also know that redshift is spread in passing through clusters of galaxies, that the overall redshift distribution of galaxies has a lumpy structure, and that there is a uniformly distributed background of ultraviolet and X-ray radiation without specific sources.

All these facts are employed as parts of the self-consistent explanation of the scattering model. The energy from conversion of hydrogen to helium is exactly the amount of energy that appears as blackbody radiation of 2.725 K after it has been thermalized by interacting with matter – which is what redshifting *is* in the scattering model. Unlike the final phase of the standard cosmological model explanation, the redshift must be treated as occurring throughout, and associated with, the medium. Energy taken from redshifted radiation becomes particle kinetic energy to be re-emitted as thermal radiation.

In the scattering model the blackbody radiation temperature will not be identical to the kinetic temperature of the material partition of the universe because of the redshifting that occurs throughout the universe. The density and temperature of the material partition of the universe can be derived from the facts of universal abundances of the elements and the energy vested in the microwave background radiation. These values are derived and accord well with what is observed rather than pretending that the material universe is actually at 2.725 K despite all measurements to the contrary. These results allocate redshifting phenomena primarily to intracluster gases through which most of the universe is seen with large gaps of lower redshifting between.

Where does this on-going conversion of hydrogen to helium take place? Some might occur in the hottest intracluster gas regions perhaps, but probably the majority is from the irregularly occurring gamma ray bursts that provide an on-going environment very similar to that theorized by the one-time big bang of the standard model.

Theoretical underpinnings of cosmological models

CHAPTER 24:

Thermonuclear reactions in hydrogenous plasma are the origin of the light elements in both cosmological models. In this chapter we explore this phenomenon from both perspectives. The seminal paper by Wagoner, Fowler, and Hoyle (1967) laid out the scheme for light element production that was in direct contradiction to much of Gammow's flawed original conception. Their analyses apply to more or less any sufficiently hot hydrogenous plasma. Furthermore, their approach assumes an equilibrium at each succeeding temperature; expansion is not an integral part of the determinations. So the approach is easily adapted to either an expanding or stationary state model like that of the scattering model for which equilibrium conditions are more persistent even if diminutive versions of the big bang are sporadically interjected.

CHAPTER 25:

Certainly Einstein's development of the general theory of relativity produced a burst of activities relating directly to cosmology. Einstein directed much of his own efforts during the rest of his life to problems in this realm. It was the urge of solving the ultimate physical puzzles, no doubt, that drove him.

Even the fact of gravitation being weaker than electromagnetic forces by forty orders of magnitude does not diminish the fact that it is gravitation that plays such a key role in the motions of astronomical objects. Like electrostatics, Newton's classical gravitational theory involves an inverse square law of force. A large body of theory has been developed over the centuries since Newton, including the work of Poisson whose noteworthy equation that related inverse square law forces to density. Einstein considered Poisson's equation key to the development of his theory of cosmology. However, applying the equation to a uniform distribution of matter in an infinite universe seemed to him to produce problems. His modeled universe could not be kept from collapsing. So he altered the time-honored equation such that collapse could be avoided by inserting a fudge factor, Λ. Then, when he became aware of Hubble's discoveries that seemed to imply an expanding universe that would have to have had some initial impetus to have gotten it going, Einstein decried Λ as being his "greatest error".

As an arrow in flight can not be called back, cosmologists now herald Einstein's acknowledged "greatest error" as a god-send that accounts for four times as much matter to the universe as the regular old matter with which we are all familiar, and that to which our university physics and chemistry classes have pertained. But wait, in the last decade, in addition to 'dark matter' they have introduced a new form of 'dark energy' that they consider to comprise 82% of the entire mass of the universe, dwarfing even the supposed preponderance of 'dark matter'. Those are but the latest products of a disproportionate urge to theorize with a willingness to fudge predictions if necessary to defend and confirm a preferred notion, to all of which Einstein's greatest acknowledged error seems to have given license.

A re-evaluation of the mind set and reasoning that went into producing that original error is in order. The author contends that the fudge factor Einstein inserted into Poisson's equation addressed what was not actually a problem in the first place. The symmetry Einstein assumed as pertinent to solving the dynamics of the universe as a whole is brought into question by this author.

CHAPTER 26:

Cosmogony is a temptation for cosmologists as well as everyone else. The big questions that border on religion seem to beckon practitioners in this field and goad them into very flawed reasoning. Knowing what answers one wants before doing the analyses, performing rote procedures to effect an appearance of statistical validity, reading ancient myths into current realities all seem to have affected researchers who should have known better.

Strings, and multiverses are all attached to the current standard cosmological model so that some at least minimal discussion of the issues involved in these theoretical excursions seems appropriate. The standard model involves some fifteen universal constants – so far – to account for observations. Yet the theoretical foundations provide no help in determining *a priori* what values to assign to these constants. It is even proposed that the values of these 'constants' might be determined willy-nilly at the time a universe is created as though we know more about universes in general than we know about our own. Some universes in the currently popular 'multiverse' (only one of which could we ever observe) might have one value of the gravitation constant, electronic charge-to-mass ratio, etc., while others could have their own unique values. Like professor Pangloss of Voltaire's *Candide*, we just happen to live in the best of all possible worlds where apples fall downward, like charges repel each other, and the study of physics can be something other than a schizophrenic nightmare.

These later theories have predicted nothing. There are some claims that the microwave background radiation was predicted by the standard cosmological model. It wasn't. All the predictions were way off. When it was discovered by accident, those who discovered it believed it to be the result of pigeon feces in their antenna. That is how well it was predicted. They received a Nobel prize for finding it.

Sometimes in addition to proclaiming elegance, string theorists jest about having discovered gravitation. They jest. It is in much the same sense that the current author discovered rain just this morning.

The cosmological principle has been the bulwark of science since Copernicus's day. Now the anthropic principle is being used to supplant it as a means to justify otherwise incompatible theoretical positions. There is even an emerging group who believe we exist in a very peculiar, albeit large scale, section of the universe, very different from other regions.

Conclusions

CHAPTER Z+1:

The author summarizes the successes of his scattering model of cosmology, citing comparisons of its predictions with those of various versions of the standard cosmological model to demonstrate its superiority both observationally and philosophically. Not least of this success is that scattering in the intergalactic medium accounts for such a broad a scope of phenomena that it unifies cosmological observations.

Three key innovative methods employed in this investigation are discussed with regard to the appropriateness of their application to the analysis of standard model hypotheses. It is suggested that there is a serious lack of thoroughness in their not having previously been applied. The extent to which plasma absorption pertains to the standard model must be determined; if in that model it plays the similar role, that would cast doubt on the assumed time dilation factor in luminosity-redshift relationships. The mechanism of plasma redshifting applies also in the context of the standard model throughout the various phases of that scenario, significantly altering what is claimed for that model. Finally, the analysis of the impact of redshift on thermalization of energy infused into a diffuse medium must be addressed in all models. No one seems to have previously addressed any of these significant mechanisms. Whatever conclusions have been drawn without having performed these critical analyses are invalidated by their omission.

The scattering model makes a major break with tradition. There is no grand unified theory that pontificates that this is the way the universe *is*. This model merely accepts what is observed with a disenthralled interest in mechanisms are that make it appear so. It is the author's position that any properly performed scientific study must proceed in that way.

Finally, we must consider what is implied by the scattering model. On the one hand it seems to imply that the universe is, if not infinite, at least much more extensive than we have so far observed. It implies that rather than a running down from a mysterious infusion of energy to initiate an entire universe, its natural processes produce an ongoing equilibrated balance that is compatible with conservation laws. Replenishment of primordial neutrons is envisioned as the result of gamma ray bursts from erupting black holes where mass-energy had been trapped; this maintains light element abundances at the observed levels by the interaction of freed particles with the ever present intergalactic hydrogenous plasma. Thus, helium is created anew by thermonuclear reactions in the exploding effluvium. The plasma ejected from such gamma ray burts cools as it expands, hydrogen clouds result, gravity eventually takes over as the dominant applicable force, protogalaxies, galaxies, and galactic cluster structures form, eventually black holes develop in an ineluctable, never-ending cycle. That is the 'big picture' envisioned as the backdrop for this model.

And, of course, we must wonder what it all 'means' even if what it *means* must remain largely subjective. Certainly from this narrow slit in the total scheme of things from which we have been granted this grand opportunity to view the cosmos and question its workings, it is tempting to consider the gift, our promontory, and abilities that we apply to this endeavor as more significant than they are. That reaction may be a natural healthy response to what we observe – but something to guard against in scientific endeavors nonetheless.

Chapter 1

Introduction

"It is sensible and prudent that people should continue to think about alternatives to the standard model, because evidence is not all that abundant. The examples presented here do show that the observational constraints are far from negligible, however... The moral is that the invention of a credible alternative to the standard model would require consultation of a considerable suite of evidence. It is equally essential that the standard model be subject to scrutiny at a still closer level than the alternatives, for it takes only one well-established failure to rule out a model, but many successes to make a convincing case that cosmology really is on the right track." Peebles (1993), p. 226.

Certainly among the most spectacular phenomena of the cosmos are those whereby we are enabled to look back so extremely far into the distant past – distances of virtually unfathomable magnitude. It is indeed amazing that intergalactic space is so wondrously transparent. With proper equipment one can observe to distances for which the finest resolution of the unassisted human eye – which can only see about 6,000 of the closest stars – blurs images of many billions of galaxies so like our own Milky Way. With its many billion of stars of which our own sun is but a modest example, the Milky Way galaxy is itself rather typical of large spiral galaxies of which billions exist.

To see further and more than anyone has ever seen continues to be merely a technological challenge and *not* an essential limitation of our species or the universe. The awe-inspiring aspects of this privilege and associated challenge derive directly from the uniquely agreeable properties of the intergalactic medium.

1

Understanding the properties of this medium that account for our universe being so extremely available to our investigations is one of the most exciting challenges of our time. The intergalactic medium provides a most startling demonstration of the wisdom in the adage: *One must sometimes look into how we see in order to understand more fully the nature of what we see.* Cosmology is in general an area where this is most certainly true.

There is obviously a tremendous current awareness of the *facts*. We will review many of these facts pertinent to what has been learned concerning the objects of which this universe is comprised in the next chapter. We will then change our focus briefly to consider the nature of the intergalactic medium. Next we will consider the processes that affect observations made through it. Those are areas the author feels cosmologists would do well to more fully appreciate. Characteristics of intergalactic plasma and its associated processes accommodate, but also ineluctably *alter*, observations of apparent characteristics of the objects.

It is not surprising that any reasonable attempt to explain the cosmos, or put forward hypotheses aimed at explaining its observed characteristics, would begin by reviewing astronomical observations. It must then, of course, call into account the appropriate electromagnetic, relativistic, thermodynamic, and quantum mechanical effects to account for the various aspects of the observations. Thus, known effects in these various fields of endeavor are used here to explain how we see the distant regions of our universe – the impact of absorption and the modes of transmitting images via coherent forward scattering through the intergalactic medium on the observed characteristics of the images themselves. The properties of the intergalactic plasma appreciably alter these processes, producing effects that are quite different from those of similar processes recreated in earth-bound laboratories or that take place naturally through our atmosphere. These processes are also very different than those that would take place if space were a *complete* vacuum. These essential differences make a clear distinction between what are otherwise very familiar physical processes of observation and what occurs in the intergalactic medium. This difference has ultimately produced what has too often been perceived as, and attributed to, physical differences in the observed objects themselves.

A 'scattering model' based on these plasma scattering effects will gradually emerge from these pages with predictions presented for comparison with the predictions of the 'standard cosmological model' currently favored among cosmologists. To do this, the anticipations of the most tenable of the many variations of the standard model will be elaborated to accommodate a fuller and completely honest comparison.

For most of a century Hubble's law of expansion of the universe has been accepted as the only reasonable inference from an observed relationship between the redshift and distance of galaxies observed on a cosmological scale. Redshift, according to Hubble's hypothesis, is a Doppler effect caused by the recessional velocities of the objects themselves. Distant objects have thereby been assumed to be receding at extreme speeds with an as yet inaccurately determined deceleration attributed to gravitation, or (as more recent investigations suggest) an *acceleration* of unknown origin, and to experience dramatic (although not yet completely theorized) evolutionary developments as well. It is typically thought that when all effects are taken into account, the observed peculiar dependency of angular dimensions, luminosity, and comoving number density of distant galaxies on redshift will result.

The uniform microwave background radiation has been fairly successfully modeled, although using awkwardly constructed ad hoc assumptions. The model definitely did not accurately predict the temperature of the microwave background radiation. Other than acknowledging that there would have to have been a redshifted remnant of radiation produced during such an explosive early phase of the universe, it predicted very little in that regard. Nor is that explanation without its considerable problems.

So there are indeed inconsistencies with observations, including the requirement for an otherwise-unaccountable inflationary expansion phase. There is a too early emergence of giant elliptical galaxies and a large-scale structure of the universe which seems incompatible with the smooth background radiation, elaborate epiphanies notwithstanding. There is the anomaly of stars within globular clusters that orbit our own galaxy whose ages, based on their metalicity, may actually exceed the supposed age of the universe. The angle-redshift relation of distant galaxies continues to baffle modelers. An insufficient mass density of luminous matter to account for perceived gravitational effects in accounting for intracluster galactic motions and galactic rotational energy distributions seems to demand mysterious unobserved 'dark matter', etc.. But, by and large, the model has accounted sufficiently well for observations that its demise is not typically anticipated. There is no major tendency of the scientific community to abandon this standard model because of these inconsistencies; instead, theoretical investigations continue into variations on tried approaches that might ultimately account for them. Multiple universes of unimaginable numbers and qualities are hypothesized to mitigate the strangeness hypothesized about the observable universe – attempts to place our universe into a tenable range of *possible* universes.

The current investigation breaks with the tradition of apologies for those awkward aspects of the standard model. It illustrates instead that ostensible characteristics of distant regions of the cosmos might better be described as merely the effects associated with observation through an indefinitely extended intergalactic plasma. It proposes that a statistically uniform distribution of astronomical phenomena in a flat extended space-time universe would appear to be expanding and to possess nonlinear comoving number densities of galaxies as well as producing apparent exceptions to the virial theorem in denser intracluster plasma. Many of the inconsistencies currently encountered by all of the propounded variations of the accepted standard model are thus avoided by the scattering model presented here.

Of course any viable cosmological model must also account for microwave background blackbody radiation. That is treated as a separate but related issue here, somewhat as it has been with the standard models except that a solution fits much more directly into the scattering model without ad hoc assumptions. Long ago Hoyle and others suggested that the microwave background might have resulted from the thermalization of energy released in conversion of hydrogen to helium in attaining its universally encountered proportions.

The extremely high energy density (even if low characteristic temperature) inherent in the observed microwave background radiation provides an essential clue that its origin differs in very essential ways from the inferences typically made with regard to it based on the standard model. The requirement for a different hypothesis is evident in the data itself. We will show that the 2.725 K temperature of the microwave background blackbody radiation can be derived directly from nucleosynthesis of helium with the observed baryon density and a universal average temperature of about a thousand kelvin.

However, a much higher average dynamic pressure involving the product of temperature and density is realized throughout the intergalactic regions because of the extremely high values that occur in untracluster plasmas at the cores of clusters of galaxies through which

4

we observe the universe at large. This average determines a value that accommodates the realization of the Hubble constant for cosmological redshift. The much higher value within rich clusters produces extreme redshift dispersion in the clusters themselves that has given rise to claims that over 70% of the mass in the universe is 'dark matter'.

Unlike variations on the standard model theme, the approach to accounting for cosmological appearances described in this volume neither employs unconventional physics nor posits new (or varying) physical constants. Both of these artifices have been used repeatedly in shoring up the standard cosmological model. However, several unique analytical techniques are employed including a non-Doppler redshifting mechanism, a broadband absorption appropriate to redshifting in hot plasma, and an approach to determining blackbody characteristics in a redshifting environment. None of these analytical tools that are solidly based on accepted physical principles has been applied to cosmological studies before, and each should be applied in every such study.

The previously unanticipated basis for a distance-redshift relation propounded here is shown to be inherently unavoidable in observations made through high temperature plasma. The relationship results as a combined effect of wavelength dependence of 'extinction distance'* for forward scattering, previously undocumented special relativistic aberration effects that alter the constraints on any forward scattering involving high-speed plasma electrons, and Compton-like wavelength alteration. Energy and momentum of the radiation are transferred to the medium via a non-dispersive absorption effect that unilaterally diminishes the luminosity of redshifted objects. This results functionally, although not physically, in an effect that is directly analogous to the relativistic time dilation effect on luminosity attributed to recessional velocities (universal expansion) in the standard model.

The scattering model's unique redshifting effect avoids the previously anticipated blurring of images that was thought to be the necessary concomitant of any such 'tired light' model. The predicted

---

* The term "extinction" will be used throughout this volume in a sense that is other than the usual parlance of astronomers for whom it often refers simply to absorption. As used in this volume the term pertains to optical phenomena associated with forward scattering for which absorption *per se* is not involved. The *extinction distance* or *extinction interval* refers to how far light must travel through a medium before a photon of its electromagnetic radiation will be replaced by a virtually identical successor photon. This process of extinction is essential to observation via forward-scattering through *any* non-vacuous medium. Extinction distance is inversely proportional to the electron density of the medium as we will see.

proportionality is equivalent to (without adopting) the radial Doppler interpretation of the observed cosmological redshift relation. Matching Hubble's constant depends on a specific average value of the *dynamic pressure* of the intergalactic plasma medium. The average value involves the product of the kinetic electron temperature and density over the distance to an observed object; it must be approximately 4.13×10^3 K cm^{-3}.

The temperature and density parameters that determine the dynamic pressure can be estimated indirectly from various related measurements. Then there is the observed ratio of distances of lines of sight through intergalactic and intracluster regions that determines a universal average of the combined product of temperature and density. Finally there are the more direct measurements of the spectra of intracluster gases from which estimates of density and temperature have been derived. All these measurements are compatible with this average of 4.13×10^3 K cm^{-3} required by the mechanism that effects Hubble's constant in the scattering model.

This dynamic pressure is somewhat greater than anticipated for intergalactic regions, but it is orders of magnitude less than that found within galaxy clusters. It will be shown that Hubble's constant results as a weighted average of these products along any line of sight to observed cosmological objects. The extremely obvious regular peaks and valleys in galaxy densities in redshift surveys recommend such a resolution. This conjecture that forward scattering in plasma is the cause of the observed cosmological and more extreme intracluster spectroscopic redshift warrants serious consideration. Investigation and analysis of redshift surveys provides ample data with which to discriminate between alternative cosmological models. Comparisons are presented here that overwhelmingly support the scattering model.

With regard to arguments against the viability of redshift resulting from forward-scattering in a plasma medium, we note that the well-established theoretical invariance of spectra of electromagnetic radiation propagated through a scattering medium will be demonstrated to apply only to a media model invalidated by high temperature plasmas. Conservation formulas applicable to forward scattering constraints in a medium involving plasma electrons with relativistic thermal velocities, will be shown to impose a very minor lengthening of wavelengths at each extinction interval. (These formulas are those applied by Compton to X-ray scattering in more restrictive situations.)

6

The resulting forward scattering process is such that it will *not* thereby produce a net angular blurring of the forward-scattered image as had formerly been thought to be an unavoidable obstacle of any such explanation. The process acts instead to continuously refocus the imaging process.

The tiny 'reddening' effect caused by the coherent scattering of single photons is not in itself a 'redshift', since such wavelength increases would not be proportional to wavelength and could therefore clearly be distinguished from a Doppler redshift if extinction were to occur at regular intervals. However, the wavelength proportionality of increases in wavelength is imposed upon this reddening process through an inverse dependency on wavelength of *extinction distance* in any medium. Thus a distance-redshift relation does, in fact, result that is indistinguishable from a radial Doppler redshift so that the observed relation between redshift and distance results. The absorption that applies to the intergalactic plasma produces a redshift-luminosity relation that is similar to the most viable of standard model versions. So, in general the plasma scattering model predictions are extremely similar to many of the predictions obtained from the currently accepted 'concordance' variation of the standard model. Thus, predictions to match observations are obtained without acquiescing to hypotheses that Hubble himself had only reluctantly accepted. Redshift, luminosity characteristics, and other cosmological effects result as by-products of the processes of observation through high temperature plasma rather than as pertaining directly to the observed astronomical phenomena.

Universal values of temperature and density are determined by compatibility with current microwave background radiation data. This determination assumes that the microwave background has resulted from the thermalization of the byproduct radiation from fusion of 24% (by mass) of the universe's hydrogen into helium. Notice, however, that a product of these separately averaged universal values is naturally very different than the average value of the product of these parameters.

Collaboration with regard to many apparent 'cosmological' effects also being attributable to scattering by the intergalactic medium is readily available. For example, luminosity, surface brightness, and angular relations, although some are accurately predicted by one or another version of the standard cosmological model have proven to be problematic in requiring unique parameter values to best account for each effect. The luminosity-redshift relation predicted by the scattering model agrees with observations, differing only slightly from the most accurate predictions of the standard model. It provides better agreement with actual data than can be obtained using any other theoretically justified model at the present time. In fact, the data for which Riess, et al. (1998) observed a recent *'acceleration'* is in complete and precise agreement with the scattering model with no acceleration implied – or indeed any expansion whatsoever. To obtain as good an agreement Riess et al. had to propose not only acceleration, but a 'jerk' (rate of change in acceleration), with a deceleration phase all propelled by a mysterious "dark energy" source.

Of course an ionizing flux of X-ray radiation as observed from intracluster gases is required – and implied – in effecting the high dynamic pressure. The very uniform background distribution of X-rays was discovered even before the microwave background radiation. An explanation has only recently been forthcoming for standard models and this is incomplete and unsatisfying. The spectra does not seem to be attributable to any or any number of specific individual sources although primarily associated with rich plasma cores of galaxy clusters.

We will demonstrate that in a stationary state situation of an extensive universe, homogeneous at the highest level, thermalization of gamma radiation from nuclear fusion of helium to ultimately achieve 2.725 K blackbody background radiation occurs naturally when redshifting of scattered radiation involved in the process occurs. This results directly from thermodynamic considerations involving a diffuse medium with redshifting of associated scattered radiation.

The abundances of the elements and the microwave background radiation are often cited as key evidence in favor of the standard model cosmologies even though neither the temperature of the background radiation nor the elemental abundances were ever accurately predicted. These were merely 'embraced' by the theory after the fact of their discovery. As stated by Arp, Burbidge, and Hoyle (1990):

"...a particular value for the baryon-to-photon ratio needs to be assumed *ad hoc* to obtain the required abundances. A theory in which results are obtained only through *ad hoc* assumptions can hardly be considered to acquire much merit thereby.

The relative 'primordial' abundances of deuterium, helium and lithium are often cited as strong evidence in favor of the standard cosmological model, but again as stated by Arp, et al.:

"According to our point of view, on the other hand, the required abundances follow inescapably from the density-temperature relation that holds in all bodies with masses large compared with that of the Sun."

In fact much earlier – in the widely accepted seminal papers which derived expected elemental abundances – it was stated that there is no need to assume an initial *ad hoc* ratio of matter to anti-matter. The amount of energy distributed throughout our universe as background radiation is very closely matched by the amount of energy that has been released by the production of the known percentage of helium from the universally accepted primordial hydrogenous plasma. This energy will be 'thermalized' by scattering in the intergalactic medium, indeed from essentially all the matter in the universe.

In addition to determining the microwave background energy density and temperature, we will show that the hydrogen-helium ratio determines an average baryonic (and thereby an electron) density and temperature for the material aspects of the universe as well. All of these findings are in complete agreement with the scattering model.

Thus, there are three directly measurable universal numbers – the percentage abundance of helium, the energy density of background radiation, and Hubble's constant. These values are all uniquely compatible with features and predictions of the plasma scattering model. That phenomenal agreement would need to be regarded as mere coincidence if the *ad hoc* assumptions of the standard cosmological model referred to by Arp, et al. were to be valid. See, for example, Wagoner, Fowler, and Hoyle (1967, pp. 23-24) whose seminal paper established the expectation for the percentage abundances of the elements. The following is quoted from that paper:

"...Because in a cosmological expansion baryon density decreases as R^{-3} while the radiation density decreases as R^{-4}, the coincidence is an accident if the $3°$ K is a relic of a cosmological fireball. On this view the expansion factor R has increased since the fireball by a factor of 10^9 so that no such coincidence could have obtained over most of the expansion. It would be an accident of the present epoch. This is not the case if the observed radiation results from the thermalization of energy from the ... hydrogen to helium conversion..."

Although there are slight adjustments that must be made to the observations after more than forty years, this argument remains completely valid. In this way, a full accounting for the 2.725 K background blackbody radiation without *ad hoc* presumptions has been shown to be completely compatible with thermalization in a redshifting environment with scattered electromagnetic radiation propagated through intergalactic plasma. The intracluster plasma gas densities and temperatures that accommodate the observed cosmological redshift may contribute at some low level to the thermonuclear recycling that maintains both elemental abundances and background radiation, but in very much the spirit of those seminal studies to which all standard cosmological models defer, that function is more likely to have been provided exclusively by the massive explosions heralded as 'gamma ray bursts'. The temperatures associated with the energy release of these tremendous explosions map directly to that required for light element production – in fact, very like that conjectured for the big bang by the standard model.

Apparent problems with applying the virial theorem*, which addresses intracluster dynamics, are circumvented by the scattering model. Extreme variations in redshift denominated "fingers of God" have been major considerations with regard to the hypothesis that the universe must be comprised of a large percentage of 'dark matter'. Although there had been theoretical expectations that more mass would be found to 'close' the universe according to Einstein's general relativity, dark matter does not fulfill even that perceived expectation.

The increased redshift across galactic clusters results as an obvious corollary for the scattering model. It is handled as a direct consequence of the same mechanism that produces 'cosmological' redshift phenomena. In this case the increased redshift across such rich clusters is associated with, and results from, the observed increased plasma density and temperature within these structures and not, as has been supposed, primarily due to increased orbital velocities requiring appreciably more than the observed luminous mass. This prediction does not require *ad hoc* hypotheses of mysterious increases in mass. The resolution requires no more than the observed baryonic matter including that observed in intracluster gases.

---

* The virial theorem is used to determine the mass of clusters of galaxies based on observed radial velocity variations of individual galaxies within the cluster. Redshift variations within individual galaxies and that of galaxies within galaxy clusters take on more extreme values than their luminous mass would justify. This is a primary source of conjectures with regard to 'dark matter.' This will be discussed in detail in a later chapter.

The measured parallax-distance relation also serves as a direct assurance that dispersion by intergalactic plasma is responsible for the obvious appearance. Significance of many parallax measurements have been disavowed by *ad hoc* conjectures of evolutionary developments to accommodate variations of the standard cosmological model for which different deceleration constant and evolutionary developments would be required here than in forging agreement with other data. The scattering model in a universe characterized by Euclidean geometry leads us to expect nothing but the relationship that has been observed. Note, of course, that distance is *not* a directly observed quantity and is a somewhat different quantity when one *backs out* the assumptions that have in some cases been invasively integrated into 'distance' data as typically presented.

'Lyman-alpha forests', the denomination given the absorption profiles of very distant objects such as quasars and 'Lyman break galaxies', have suggested the possibility of an increasing density of intermediate neutral hydrogen clouds as one observes objects at greater and greater redshifts. The profiles are often used as evidence to support arguments that the universe is, in fact, undergoing evolutionary change. At first blush the data seems quite persuasive. However, in looking into implications of attenuation by neutral hydrogen in these intermittent 'clouds', one finds a more obvious explanation. There is an absorption domain beneath the Lyman-alpha line, for which a uniform column density of neutral hydrogen implicit in a uniform density of HI clouds in conjunction with a combined effect of the short emission but redshifted wavelengths produce an exponential increase in absorption with redshift. Of course, 'volume increment' metrics differ between the various standard models and that for the scattering model such that a uniform distribution for the scattering model correlates with an increasing density for the standard model.

Other evolutionary arguments result in part from the parameter assignments of the various standard models and an underestimation of a theory-laden 'lookback distance' parameter that is not otherwise directly amenable to measurement.

Finally, it should be noted that in the rubric of cosmological models the currently proposed hypothesis falls under the general heading of a 'tired light' theory. That alternative to Edwin Hubble's and Georges LeMaître's bold hypotheses was introduced by Fritz Zwicky in 1929, but seems never to have actually been seriously entertained by anyone else of note. It is mentioned fairly often as a 'straw man' against which to defend standard models, however. This one-sided argument has been tolerated because no one had been able to discover a mechanism to effect a reduction in the momentum (inversely proportional to the wavelength of electromagnetic radiation) of photons without an associated deflection angle. Naturally, blurring by repeated angles of diffraction would preclude photons proceeding onward in a straight line from the source of the radiation to the observer.

But as a matter of mere heuristic plausibility it must have struck Zwicky and evidently Hubble from time-to-time as well as per Assis et al. (2008), that it was strange to assume a photon of light could interact with matter billions of times with no energy losses. Interactions via the extinction phenomenon are an inevitable part of the coherent forward scattering process by which we view distant objects. Rejecting such inherently plausible arguments requires a similar gullibility to believing there could be perpetual motion, or that there might be no wear and tear on tires meeting the road on cross-continental journeys. For there to be *no* associated loss of energy or momentum defies reason. It was the absurdity of such a presumption that was responsible for the author's preoccupation with discovering the viable mechanism of momentum and energy transfer associated with forward scattering phenomena.

For a perspective of views on the possibility of a *tired light* theory ultimately proving successful, see for example Geller and Peebles (1972, pp. 1-5), who state in introducing their "Test of the Expanding Universe Postulate":

"The most obvious alternative, a tired-light model, is not mentioned in many reviews of cosmology, given short shrift in others (Withrow 1959; Zel'dovich 1963; Alfvén 1966). Not least of the reasons for the widespread acceptance of the expansion hypothesis is the lack of a reasonable alternative basis for redshift...We ask whether a tired-light cosmology can do as well. If it cannot, the result may be interpreted as evidence for expansion. Of course, the evidence would not be conclusive, for in the absence of even a tentative physical basis for the tired-light effect we are free to add as many embellishments as necessary to secure agreement with observation. Therefore, adequate establishment of the case is a matter of judgement."

This is logic gone amok! Failure of an alternative is *not* tantamount to success in any truly scientific arena. And yet... this logic has continued to hold sway. Although to the author's knowledge no viable 'tired light' model has previously been put forward, after eighty years cosmologists still argue in papers co-authored by huge numbers of researchers against the silent 'tired light' contingent – no doubt because of its obvious inherent appeal. See for example Blondin et al. (2008).

In any case more than thirty-five years after Geller and Peeble's quoted statement above, two things are now certain: standard models have indeed *continued to be embellished* and *a tired-light model has finally been given a physical basis that accounts for observations* at least as well as any "embellished" version of the standard cosmological model.

Cosmological redshift no longer justifies retention of Hubble's recessional velocity, or otherwise characterized expansion, hypothesis. The luminosity-redshift data is precisely in agreement with the scattering model. Nor does the ad hoc assumption that billions of times the mass of our entire universe was required to produce the 2.725 K microwave background radiation make any sense at all. It is too easily accounted as the byproduct of the fusion required to produce the 24% helium from the ubiquitous primordial hydrogenous plasma, which accords well with the observed temperature and density of the universe around us. So there seems to be no residual unaccounted phenomena to support the standard model although the list of claims to the contrary against *any* and *all* tired light alternatives continues unabated.

There would seem to be too few reasons to retain the more radical theoretical accouterments of the standard cosmological model, including emergence of our entire universe from nothing and a still physically unjustifiable inflationary phase that defies all physical laws to get it kick started. There is no need to resurrect Einstein's "greatest error" that was incorrect for more reasons than the one of which he became aware, just to account for 'dark matter'. The sordid history of that artifact is one of intrigue in fudging formulas to fit data. And the ever so mysterious 'dark energy' that is supposed to propel a phased 'acceleration' was only hypothesized because none of the standard models could precisely match the magnitude-redshift plots of distant SN1A supernova data which is fit precisely by the scattering model. Absolutely no valid rationale can be given for the conjectured mysterious energy. All these co-dependent props are engaged in defending an indefensible model that defies and mocks the very physics from which valid models derive. Proponents argue based on anthropic principles with regard to a 'multiverse' of universes somewhat like our own but with totally different physical laws and universal constants.

All these professor Panglosses profess that "we live in the best of all these possible worlds" where physics happens to *work*, when in fact, it seems quite certain that we live in the *only* possible universe.

The basic theory behind the standard cosmological model – as against Hubble's observations and the Doppler redshift interpretation of that data – is, of course, Einstein's general relativity. Beyond what Einstein considered to have been his "greatest error" in introducing a fudge factor to avoid a situation where his mathematical representation of the universe failed to predict observations, was an egregious flaw in the symmetry used in framing his argument against the possibility of an indefinitely extended space-time. The proper symmetry is *not* with regard to an undefinable center of the universe but with regard to the position of *any* test particle (observer) that must be considered as at the center of its own universe in such an analysis. We cannot situate ourselves outside the universe looking *in*. When the proper symmetry is used the fudge factor is not required in Poisson's equation in order to avoid collapse of an indefinitely-extended stationary state universe.

At the outset of this introduction the author quoted Peebles' open challenge concerning "the invention of a credible alternative to the standard model" requiring analysis of a "considerable suite of evidence". That is certainly true. However the almost insurmountable difficulty encountered by anyone accepting that challenge is not so much the scope of the evidence or subtlety of arguments supporting the established standard, but the sheer number and zealotry of its 'standard bearers'. Each year they generate a tremendous number of technical papers, many just cluttering the intellectual landscape with conclusions of narrow scope. Co-authored by entire departments and inter-institutional working groups, each affirms some minor inference of that *standard*, repeating a mantra of common arguments, but nonetheless demanding individual refutation. It is not clear how Peebles would envision a single researcher attacking such formidable defenses. Clearly he did not envision the possibility of success.

It will require a generation of independent thinkers to break through that "considerable suite" of obfuscations. This is the problem typically encountered by any alternative scientific paradigm, of course, as was so eloquently described by Kuhn (1962). Therefore, with no illusions of grandeur we ignore that problem and proceed directly to the *scientific* issues in that "considerable suite of evidence" pertinent to invention of a valid alternative to the standard cosmological model.

14

Chapter 2

Physical Characteristics of Objects
Observed at Cosmological Distances

Certainly most of what we know of distant regions of the universe is inferred from what we have observed in our own more immediate environs. The 'universal laws' of physics have largely been discovered in earthbound laboratories. Newton's gravitation theory pushed this envelope in matching up with Kepler's observations of the 'heavens'. "Universal", "laws", "heavens", what is meant by the terms?

Let us acknowledge right off that the *laws* of nature are in no way proscriptions of the gods; their *universality* pertains merely to their observed invariance. Facts associated with their invariably holding true with regard to the observed phenomena provide them with a degree of *universality*; that is all. *Heavens*? Don't read too much into it. These initial caveats are merely to warn the reader that everyone who enters this arena brings a certain amount of baggage with them. Recognition of this fact is helpful in reminding us that we must travel lighter. But don't throw everything overboard. This author will not abandon time-honored physical constants or the discovered laws of physics without hard evidence warranting such changes. But adherents of the standard model have done that whenever that model seems to be in jeopardy.

By and large the science we have learned from residence on this modest planet that circles a rather modest star about two thirds of the way out on one of the spiral arms of a fairly respectable galaxy is valid and useful. What we know of the cosmos has been built on that.

Initially parallax estimates using the earth's orbit provided the distance to a few nearby stars. Determination of absolute luminosities associated with various types of stars defined classes of stars that could be identified at great distances. Determination of invariant patterns in certain of these astronomical objects enabled defining what has become known as 'standard candles' whose apparent luminosities imply a distance. A gradual accumulation of information on the invariance of characteristics of objects has improved, and with it, the assessment of distance. With the use of standard-candles distance estimates have become available to objects that had previously been indeterminably *remote*. Continued discovery of telltale characteristics of increasingly luminous objects increases the depth of our assessments of 'distance'.

Determination of the period-luminosity relationship of cepheid variables – including the discovery that there are two types of cepheid variable stars that at one point required modification of the Hubble constant by about a factor of two – was perhaps the most significant of the early discoveries. These standard candles allowed the first tentative determination that observed nebulous blotches were well beyond the furthest reaches of our own Milky Way galaxy, were in fact galaxies in their own right. From such discoveries Hubble made his first tentative conjectures using a linear distance-redshift relationship. Employing the Doppler recessional velocity interpretation of redshift suggested to Hubble the more audacious hypothesis that these recessional velocities unilaterally increase with distance to the observed galaxies.

This distance-redshift relation greatly extended the distance to which a 'distance' could be assigned. Inferred invariance of luminosity by galaxy type extended the range of the redshift-luminosity relation well beyond the resolution of individual stars. The extreme luminosity of supernova and the established invariances in the temporal profile of these events further extended the reach of astronomical observations. Ultimately, with the advent of gamma ray burst observations involving almost unimaginable luminosities, the sky seems to be the limit. The redshift of a gamma ray burst has already been assessed to be in excess of six in one particular case, with greater distances certainly to follow.

Of course there are also the quasi-stellar objects (QSO) we will discuss in some detail with regard to absorption. Absorption, in the plasma scattering model, plays a very significant role in the observed functionality of luminosity with redshift. Emission and absorption anomalies associated with QSO spectra have made them less amenable for use as standard candles, however.

Since all these astronomical objects are the subjects of interest in cosmological research, we had better learn more about them.

a. the sun and its position among stars

Stars can be categorized in many ways. Of course, size, mass, luminosity, and internal and surface temperatures are major among these. There is a direct relationship between size and mass to other properties. We will discuss the mass-luminosity relationship later.

Using these ostensible features apparent in the observations of stars, the Hertzsprung-Russell diagram provides indications of where stars fit among their brethren. Figure 1 illustrates a typical example of such a diagram with the sun shown as a central figure on the main sequence of stars with rough indications of size superimposed as the dashed lines labeled in terms of the sun's radius, R. The sun is ten times larger and over a thousand times brighter than Barnard's Star, for example, but many thousands of times dimmer than Deneb.

Naturally, the stars whose names appear on the diagram are neighbors of ours. But there is no reason to believe that stars everywhere in our galaxy and other galaxies are much different than those included in this ensemble. Mira, for example, is a long period cepheid variable – a type that we will see as having been very useful in determining how far away our neighboring galaxies are. Barnard's Star has very little mass. It is a 'red dwarf' that is approximately 6 light-years away from us, which is why we can even see it at all. Betelgeuse is a 'red supergiant' that has an immense outer atmosphere that would swell beyond the bounds of Neptune's orbit in our own solar system. Sirius A, because of its close proximity to us, appears as the brightest star in our sky. Sirius B is a 'white dwarf'.

After the work of von Fraunhofer in the mid-1800s in which he explored lines in the solar spectra, it was soon realized that the solar spectrum was typical of stars generally. Most stellar spectra were, in fact, continua with just such absorption lines. (See figure 2.) Many similarities were detected, but there was also broad variation in the relative strengths of these lines from one star to another. In the growing archives of complicated stellar spectra a pattern emerged when the spectra were sorted according to the relative strength of hydrogen

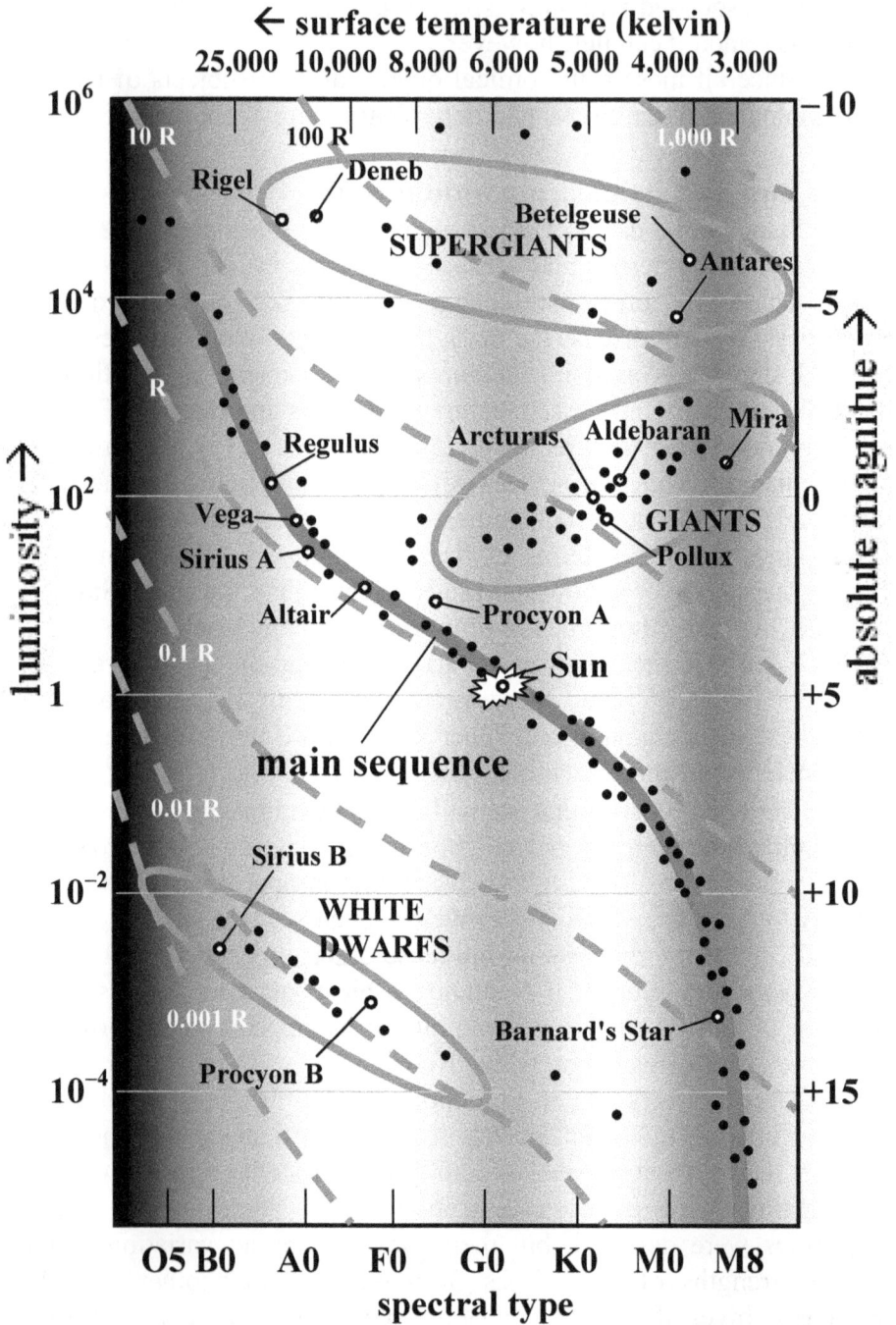

Figure 1: Hertzsprung-Russell diagram of stellar characteristics

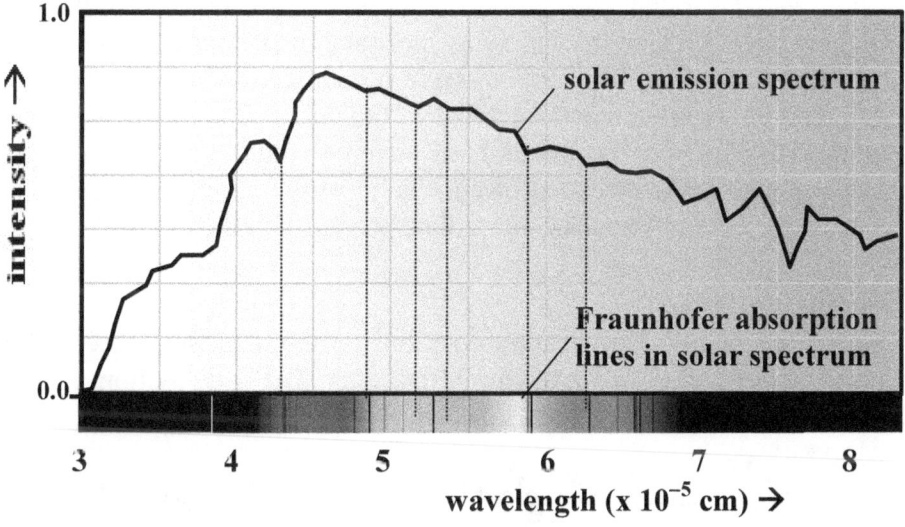

Figure 2: Facts associated with the electromagnetic spectrum and spectroscopy

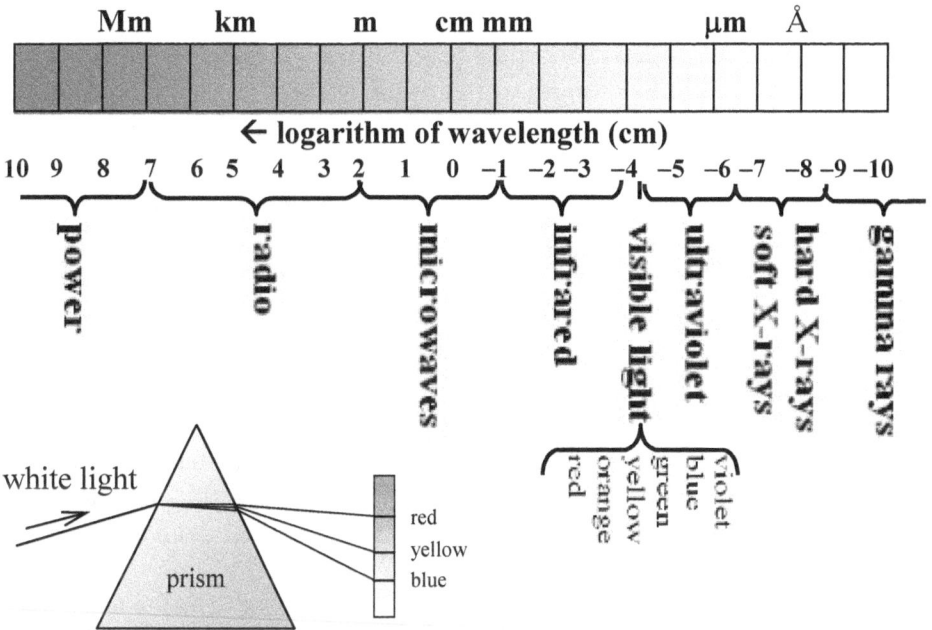

absorption. Those lines appeared in nearly all of the available high-quality spectra. Stars were then assigned letters between A and P indicating the strength of their hydrogen absorption lines. Those with the strongest hydrogen absorption became type A, the weakest type P.

However, the strength of hydrogen absorption features in stellar spectra does not directly correlate with many physical properties of the star, although it does correlate somewhat with surface temperature. In the late 1800s many of the assigned letter categories were dropped as inconsequential, and the rest were re-organized to align with stellar temperature, from the hottest to the coolest as follows: O, B, A, F, G, K, and M. Numbers were added to further divide this sequence, so that within any letter category there would be ten subtypes running from 0 to 9, so that an A0 star is hotter than an A1 star, but cooler than a B9 star. In this classification scheme the Sun is G2.

The reason that the lettering scheme is somewhat altered from its original sequence is because the hydrogen absorption line features on which the scheme was based do not increase and decrease directly with temperature. For example, the most prominent Balmer series lines are found in neither the hottest nor the coolest stars. (Refer to chapter 7.) The extremely hot O stars are so hot (T ~ 25,000 K) that the hydrogen in their atmospheres is completely ionized, so there aren't any Balmer lines. The coolest stars, those in category M, are so cool (T ~ 3,000 K) that virtually all their hydrogen exists in its ground state. There aren't enough high energy photons to transition hydrogen to a higher state. Thus, in this case also there are no Balmer line features.

Two new spectral classes have been added more recently to the cool end of the sequence (L and T). It turns out that these stars are very common, but they were not discovered until the 1990s when technological advances allowed their observation. They have extremely low luminosities and radiate primarily in the infrared.

main sequence
About 80% of all stars occupy positions on the main sequence. More massive larger stars are at the top left, lighter stars at the bottom right.

giant stars
There are two categories of giant stars. All are very massive; the mass is greater for hotter supergiants on the left; radii are greater on the right.

white dwarf stars

These stars are very hot and have densities of up to 10^8 that of the sun.

b. our local environment

In figure 3 we have reproduced an artist's rendition of the Milky Way galaxy and the sun's position within this structure. Its full extent is about 100,000 light years across with a thickness of about 2,000 light years. It contains over 200 billion stars. Our sun and the constellations that can be observed from our location are shown. This image and the five that follow, as well as many more such informative images, can be found at *http://www.atlasoftheuniverse.com/galaxy.html*. The site is provided by Richard Powell.

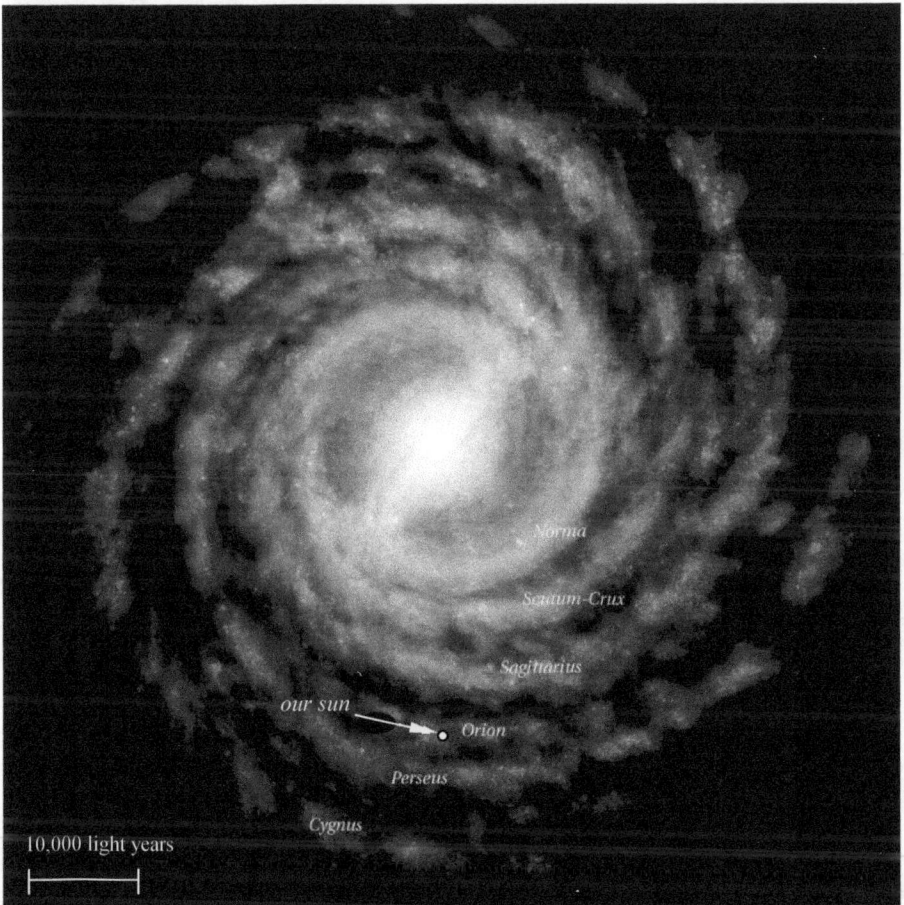

Figure 3: Our location in the Milky Way galaxy

21

Naturally we cannot see much through the disk of the Milky Way. Most of the known cosmos must be observed at higher galactic latitudes. The Milky Way itself is more visible in the infrared whose wavelengths are longer than those of optically visible light. This is illustrated in figure 4 that is available at the same source as figure 3.

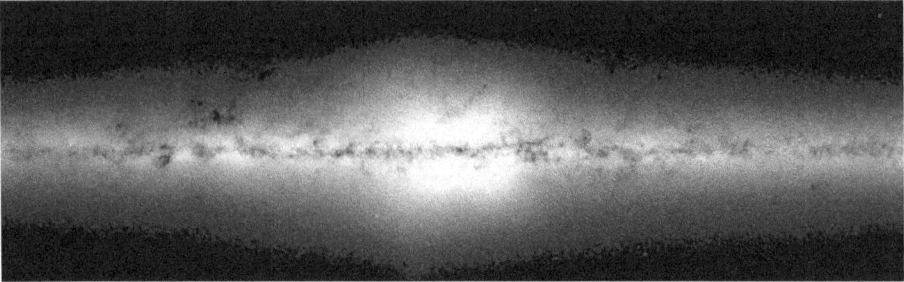

Figure 4: View toward the center of the Milky Way in infrared

The Milky Way is a prominent member of the "Local Group" of galaxies shown in figure 5. This group is central to figure 6; the Milky Way is at the center of all diagrams from the same source as the above.

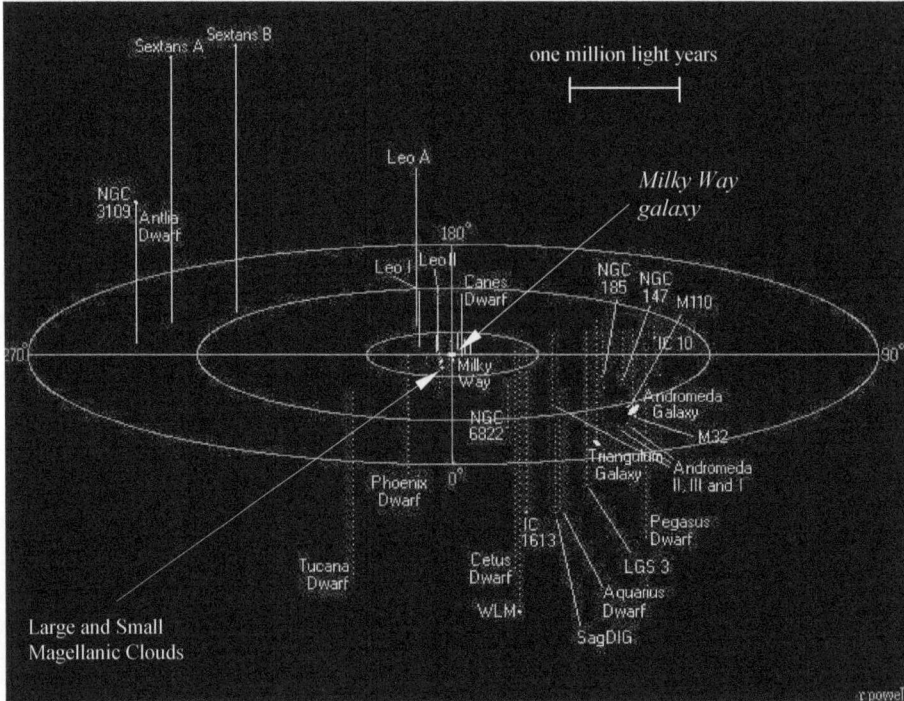

Figure 5: The "Local Group" of galaxies with 49 galaxies

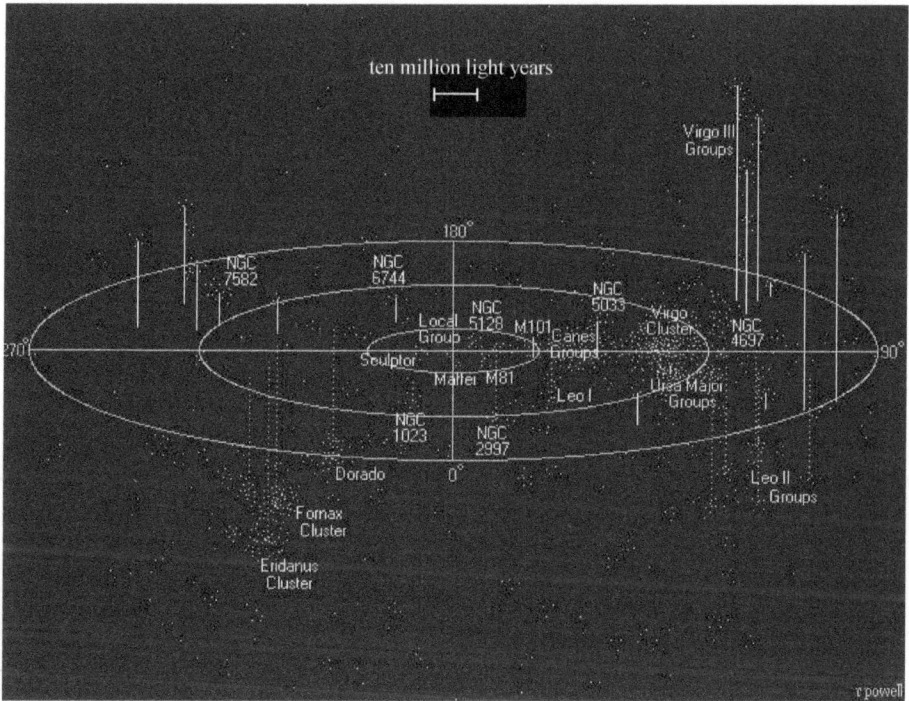

Figure 6: The closest 200 clusters of galaxies containing 53,000 galaxies

Cosmology is concerned with distances that dwarf our Milky Way galaxy, the Local Group, and neighboring clusters. In figure 7 obtained from the same source, we begin to see what can properly be considered cosmological structures. In this diagram there are 100 superclusters within a billion light years of the Milky Way, each may contain millions of galaxies. There are 60 million galaxies within the purview of this diagram. But that is only the beginning. Within ten billion light years there are seven trillion galaxies in 10 million superclusters, of 25 billion Groups, with 350 billion large galaxies like our own Milky Way. This is shown in figure 8 from the same source.

Our species can see only about 6,000 of the brightest stars with the naked eye, and yet we now know the positions and features of literally trillions of galaxies with many billions of stars each. This awe inspiring feat was accomplished incrementally, one step at a time using 'standard candles' and a host of technologies from Gallileo's primitive telescopes up through the orbiting Hubble telescope and multifarious methods employing the full electromagnetic spectrum.

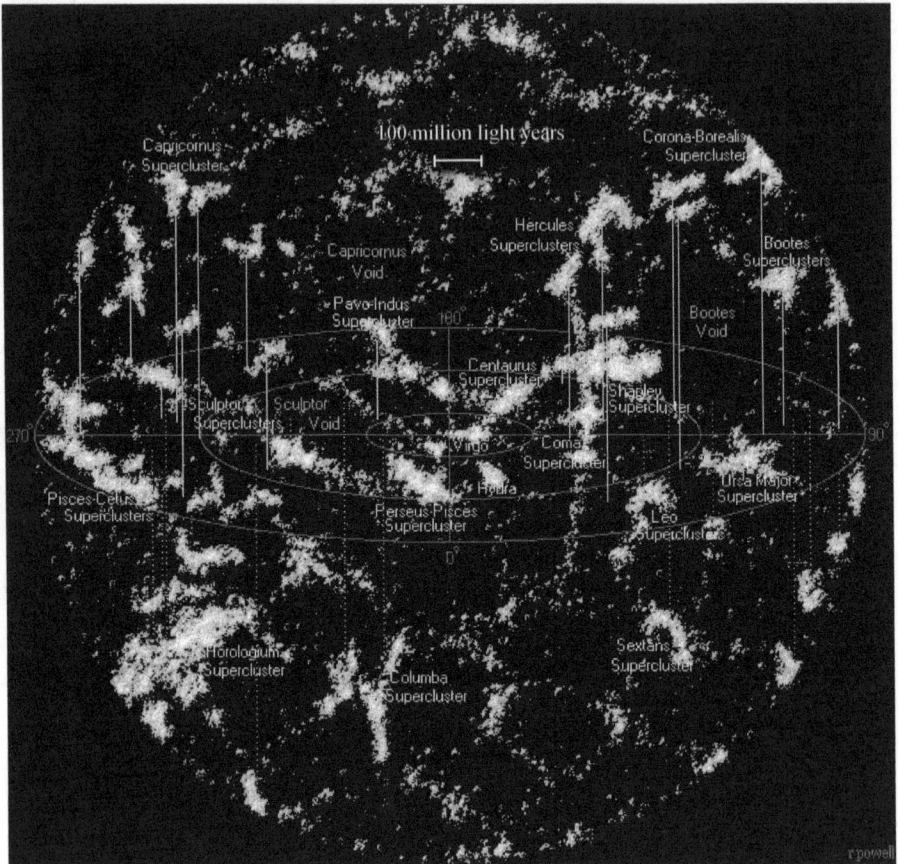

Figure 7: The closest 100 superclusters with 63,000,000 galaxies

c. cepheid variable stars

Cepheid variables are a class of stars whose luminosities vary with a very predictable periodicity. A dramatically rapid rise followed by a gradual decline in luminosity is characteristic of their extremely repeatable phenomenological histories. These objects have light curves like that shown in figure 9 of a cepheid in the Large Magellanic Cloud.

Henrietta Leavitt discovered that the luminosity and periodicity were linked by a linear relationship – the more luminous the star, the longer the period. This relationship is what provided the impetus to Hubble's endeavor to establish reliable distances to objects outside our own Milky Way galaxy. The period - luminosity relationship shown in figure 10 was the key to that effort. From this initial bootstrap inference galaxies in the general neighborhood of the Milky Way could be assigned distances. From the redshift of host galaxies an inferred distance-velocity relation was established as the next big hurdle.

From these discoveries Hubble was able to show that the spectra of galaxies had strong similarities, but that they were 'shifted' with a predominance of fainter galaxies showing a lengthening of wavelength in their spectra. The luminosity data for the cepheids were a ready means of assigning a distance; then by attributing the redshift to velocity by interpreting the redshift as a Doppler effect he was able to come up with plots like the one shown in figure 11. With the Hubble telescope individual cepheid variables could now be used to extend this distance by a factor of ten, but now we have brighter 'candles'.

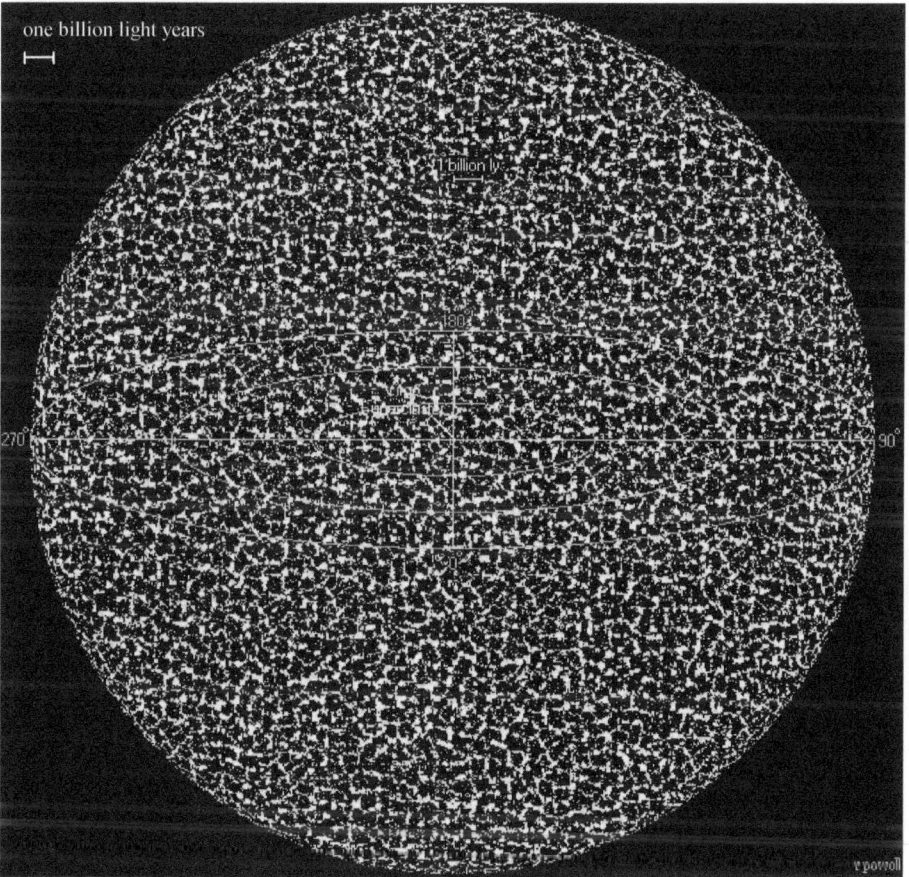

Figure 8: The seven trillion closest galaxies in ten million superclusters

Hubble found other distance indicators such as the luminosity of certain classes of galaxies et cetera that then became the 'standard candles' of choice in extending his diagrams.

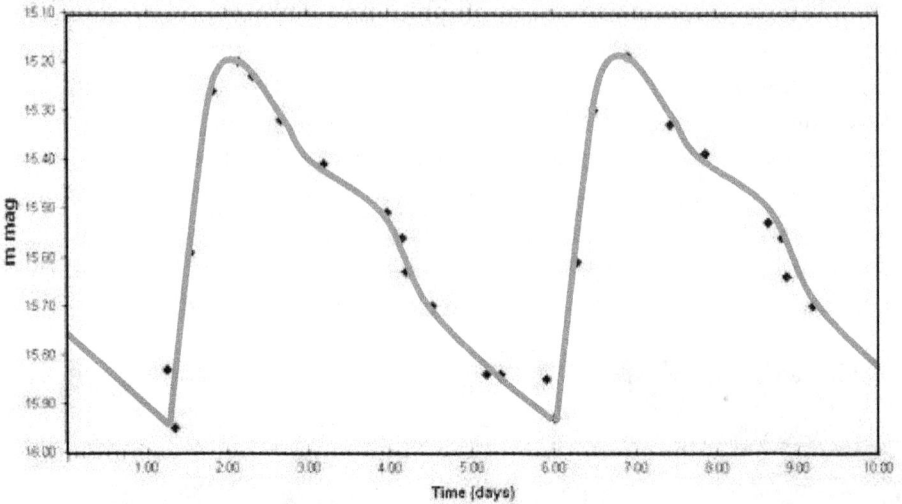

Figure 9: Cepheid variable stellar luminosity decline profile

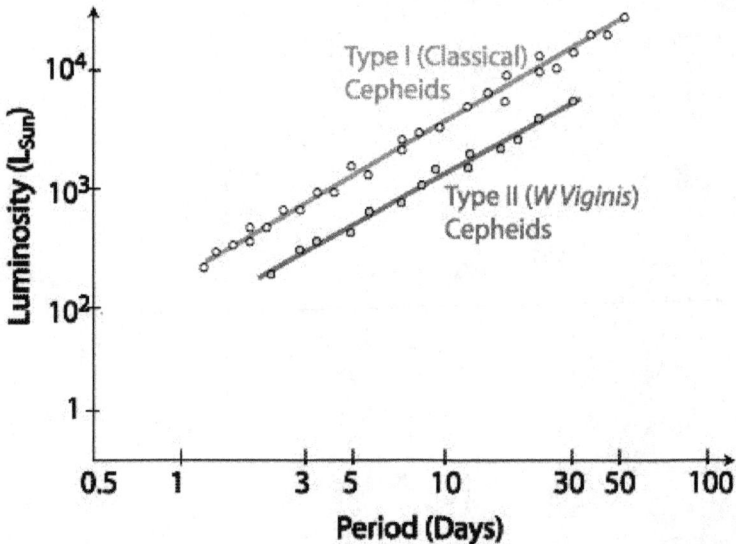

Figure 10: The two cepheid variable types period - luminosity relationships

Eventually it was noticed that the dramatic explosion of stars called "supernova" occur in distant galaxies just as they do irregularly in our own. These soon became standard candles of choice for gauging the distances to more and more remote galaxies for which apparent velocities continued to increase approximately linearly although at very large distances it is more of a logarithmic relationship as we will see.

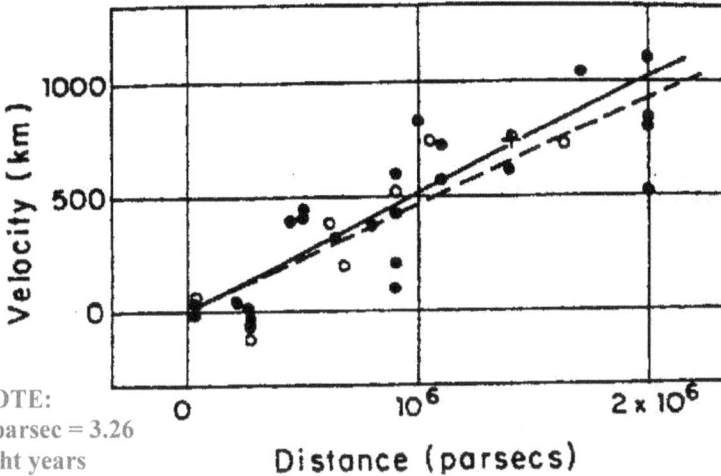

Figure 11: Edwin Hubble's initial tentative conjecture

d. supernovae

Two basic types of stellar explosions called supernovae occur. Type Ia, denominated 'SN1A' here, result as the explosive demise of white dwarf stars. These explosions are particularly amenable to use as standard candles. They are increasingly important to cosmological research both because of extreme luminosities that afford observation opportunities at great distances and because they exhibit characteristic features that accommodate their use as standard candles.

To a first order approximation, light curves of all SN1A type supernovas observed in the Blue-band look the same. See figure 12 for a characterization. Earlier phases of supernovae eruptions are inevitably unobserved as a faint dwarf star emerges from obscurity to become as bright as the rest of its galaxy in some cases. The initial very rapid increase in luminosity involves brightness increasing by up to 3 magnitudes in 15 days. This phase ends with a maximum in luminosity. Thereafter the luminosity begins a fairly rapid decline in brightness (on the order of a tenth of a magnitude per day) for the next 3 to 4 weeks. About a month or two after the maximum, the rate of decline is reduced to a steady one percent of a magnitude per day. Finally luminosities will have declined sufficiently so that the original obscurity of the source is restored. However, this characteristic feature of observation is largely apparent when observations are made in the blue region of the electromagnetic spectrum; when observed in longer wavelength bands the peak in not nearly so pronounced. This is illustrated in the inset of figure 12 of profiles observed in other bands.

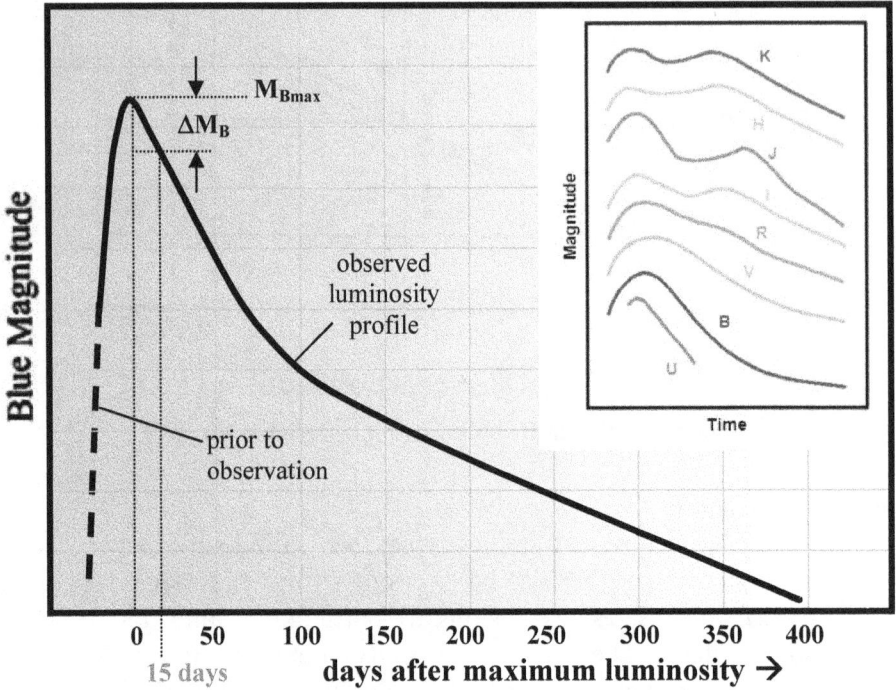

Figure 12: Supernova luminosity decline profile

SN1A luminosity curves in the blue band exhibit very distinct variations with regard to absolute magnitude and period of decline. Importantly, there is a direct correlation between the decline rate of the luminosity and the width (duration) of the maximum, as well as the peak magnitude of the supernova as shown in figure 13 that gives rise to their current use as 'standard candles'.

However, the problem with observations of very distant objects is that there is selection involved through which a subtle 'Malmquist bias' can be introduced such that only the brightest objects and events of a given class are observed at great distances. Naturally the data will have preferentially selected extremely energetic objects/events of a given type. See Plionis (2002) for a characterization of the effect. It applies in particular to SN1A data. There is also some confusion about

28

whether extremely powerful events such as supernova 2006gy are even properly classified as SN1A rather than Type II or other category supernovae that do not qualify as 'standard candles'.

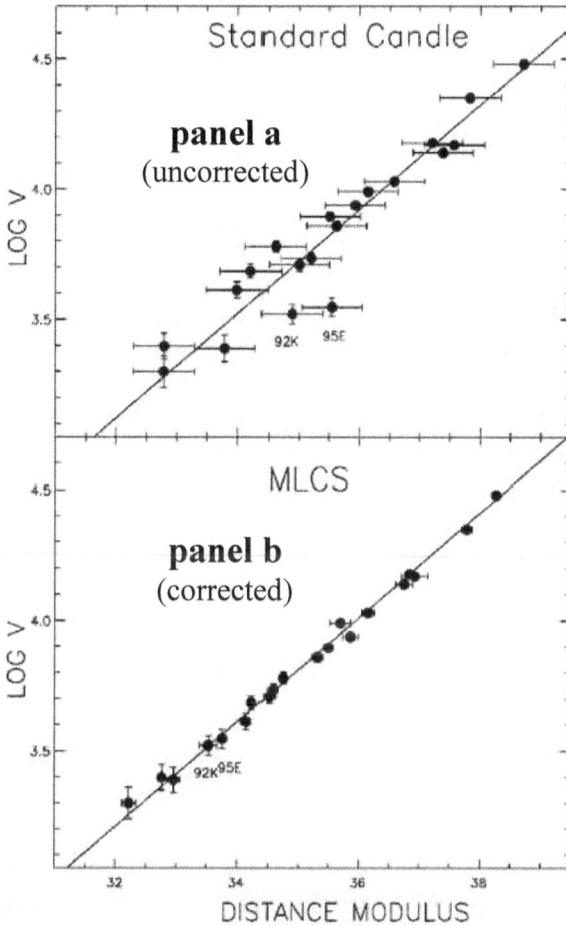

Figure 13: **Redshift versus distance indicator and required SN1A data correction.** [data from Ruiz-LaPuente (2004)]

Initially it was presumed that all SN1A events had identical characteristics to justify heralding them as the powerful new standard candles of choice. This is reminiscent of early application of cepheid variable stars independent of type. SN1A decay profiles exhibit lengthier decay intervals for the more powerful supernovae. Ruiz-LaPuente (2004) indicates that as early as 1977, it was realized that there was a required correlation between the brightness at maximum and the rate of decline of the light curve. Nonetheless, it was the mid-nineties before corrections were formulated to legitimize their 'standard candle' status as shown by Ruiz-LaPuente in panel b of figure 13.

The longer decay period is an indication of an internal process difference, although it has been taken to suggest that time dilation may occur to confirm extremely rapid recession of the more distant objects implied by the standard cosmological model. However, since only the brightest supernovae can be observed at these extreme distances and they inherently decay more slowly, the increased decay time naturally disguises an operative difference other than time dilation.

e. quasi-stellar objects

Continuing our exhibition of objects that appear within galaxies, quasi-stellar objects (as they were initially denominated because of confusion with whether they were stars within our own galaxy), QSOs as they came to be acronymed, or just 'quasars', have a very distinctive spectra with extremely broad hydrogen Lyman-α line emission. It is now fairly certain that these are associated with active galactic nuclei. In figure 14 an extended spectrum is illustrated for a typical quasar. A couple of examples of specific quasar spectra at very different distances are shown over a lesser extent of their spectra in the panels of figure 15.

In addition to other anomalous absorption and emission lines, the extremely jagged absorption lines on the short wavelength side of the Lyman-α emission peak become increasingly apparent at higher redshifts. This distinctive characteristic of distant quasars has been denominated the 'Lyman-α forests' for obvious reasons. The panels in figure 15 suggest that very short wavelength emissions from the quasar source are absorbed prior to observation. A primary explanation has been absorption by 'clouds' of neutral hydrogen during the passage of radiation through regions between the quasar and observation. Whether these clouds have been ejected from the quasar itself or are characteristic of intergalactic space has been debated as an issue pertaining to whether the universe is evolving.

If these clouds are associated with the quasar itself, then more distant quasars and their environs are very different than those, which are closer. If they are associated more generally with intergalactic space, the short wavelength radiations are assumed to have become redshifted to the same wavelength as the Lyman-α resonance frequency of neutral hydrogen. In either case, individual photons of the radiation are assumed to have been absorbed by the process of transitioning hydrogen atoms to their next higher energy levels and there are strong presumptions with regard to implications of evolution of the universe.

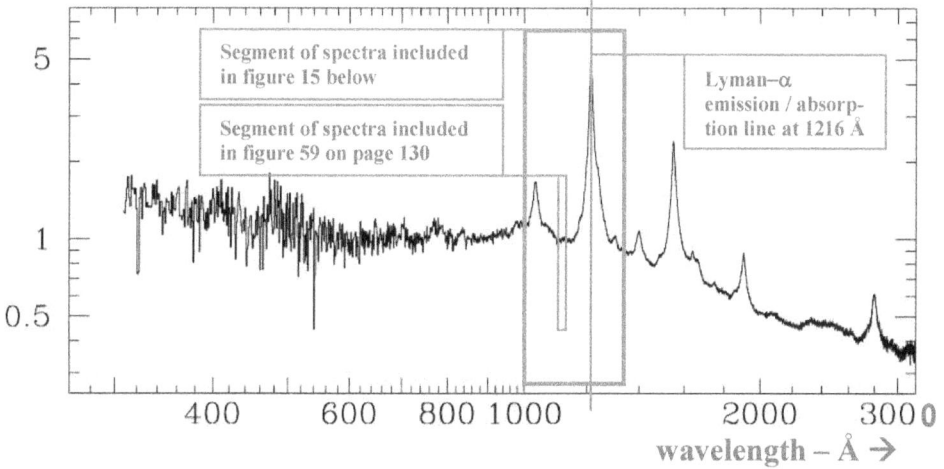

Figure 14: A composite spectra of QSOs in their rest frame
(http://www.journals.uchicago.edu/ApJ/journal/issues/ApJ/v565n2/54470/54470.
html?erFrom=2226131762376180548Guest)

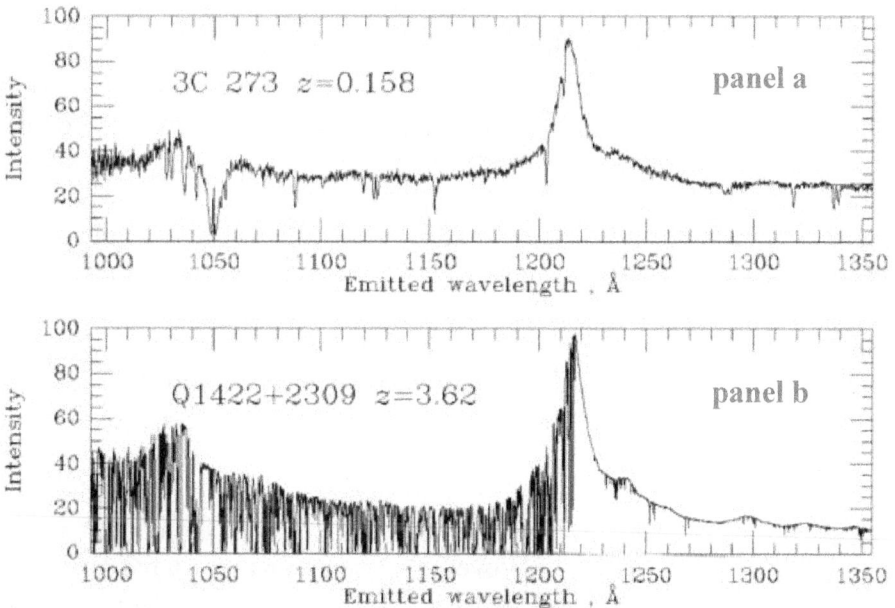

Figure 15: A couple of examples of QSO spectra

f. gamma ray bursts

Gamma ray bursts (GRB) are the latest in a long line of astronomical phenomena to be applied to cosmological research. Radiations from these GRBs have the furthest reach and they are undoubtedly the most mysterious of all cosmological objects. These tremendous explosions radiate in high energy, short wavelength regions of the electromagnetic spectrum, assuring that they can be seen at extreme redshifts. They are characterized by emission of tremendous bursts of hard X-ray and gamma ray photons with a total of up to 10^{54} ergs of energy released in a very few seconds. See figure 16.

These objects can sometimes be identified with a galaxy for which an accurate independent redshift assessment can be obtained. It is thought that they might be associated with the explosion of a black hole central to the galaxy or some otherwise characterized compact object as recent theorists suggest. Refer to Barceló et al. (2009) and Smolin (2007), or Appendix C, for a related discussion. In figure 17 from Ghirlanda et al. (2006) it can be seen that to date objects have been sighted whose redshifts are in excess of 6. Recently a relationship has been established that may extend our depth perception even beyond that. GRB appear throughout the sky with no particular preference with regard to the galactic plane as illustrated in the inset.

An important consequence of applying GRBs as standard candles through the recently discovered E_{peak}, E_γ correlations or the empirical 3D E_{iso}- E_{peak} - t_{break} correlation (Ghirlanda et al., 2006) is that the data can be combined with that of SNIA. GRBs already extend the observational redshift range well beyond the present (and furthest potential) SNIA predicted redshift limit of $z \sim 1.7$. See Aldering et al., (2002) for a more detailed rationale. The fact that GRBs employ the gamma ray band of the electromagnetic spectrum assures immunity to the dust that imposes limitations on longer wavelength transmissions. Since the availability of large numbers of SNIA supernova at very low redshifts allows more accurate calibration, GRBs will probably remain a supplement to SNIA observations, used primarily because of their extreme redshift limit.

GRBs did not initially seem to provide a viable standard candle capability in their own right without unambiguous association with a galaxy of known redshift. However, correlations that have now been established among the several observed quantities accommodate an accurate assessment of the total energy and the peak luminosity emitted by a specific burst. Using these correlations, GRBs may indeed become extremely useful as standard candles.

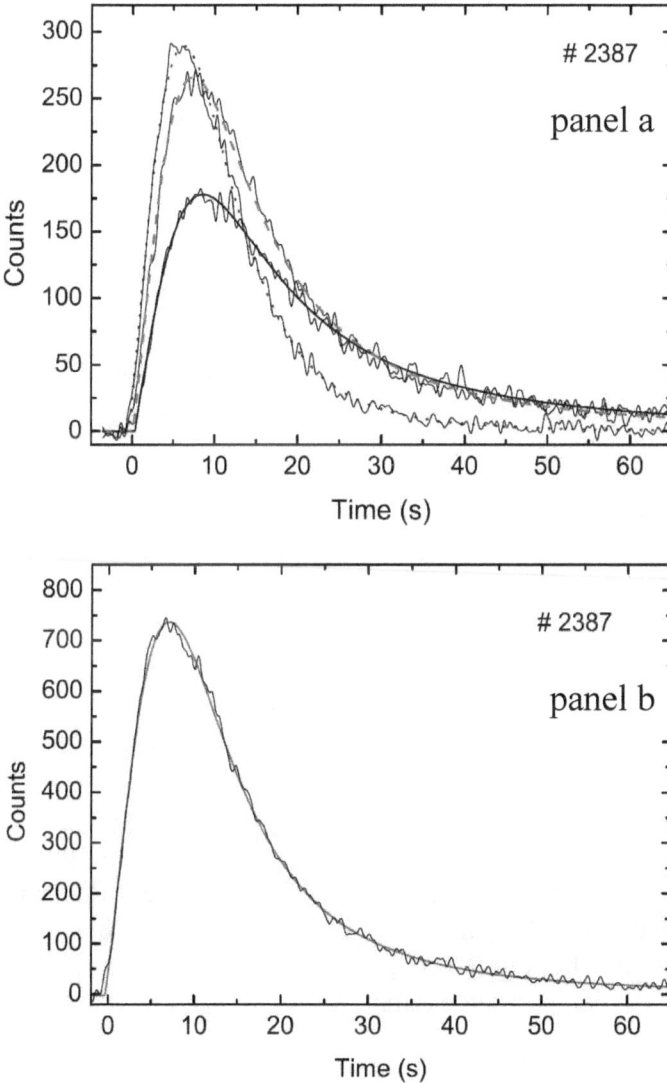

Figure 16: Light-curves for a single-peaked gamma ray burst event with a good signal-to-noise ratio (GRB 930612 #2387). Panel a shows the best fit to the observations in three channels, 1 (25–55 keV, solid line), 2 (55–110 keV, dashed line), and 3 (110–320 keV, dotted line). Panel b shows the best-fit "bolometric" light-curves resulting from all channels.

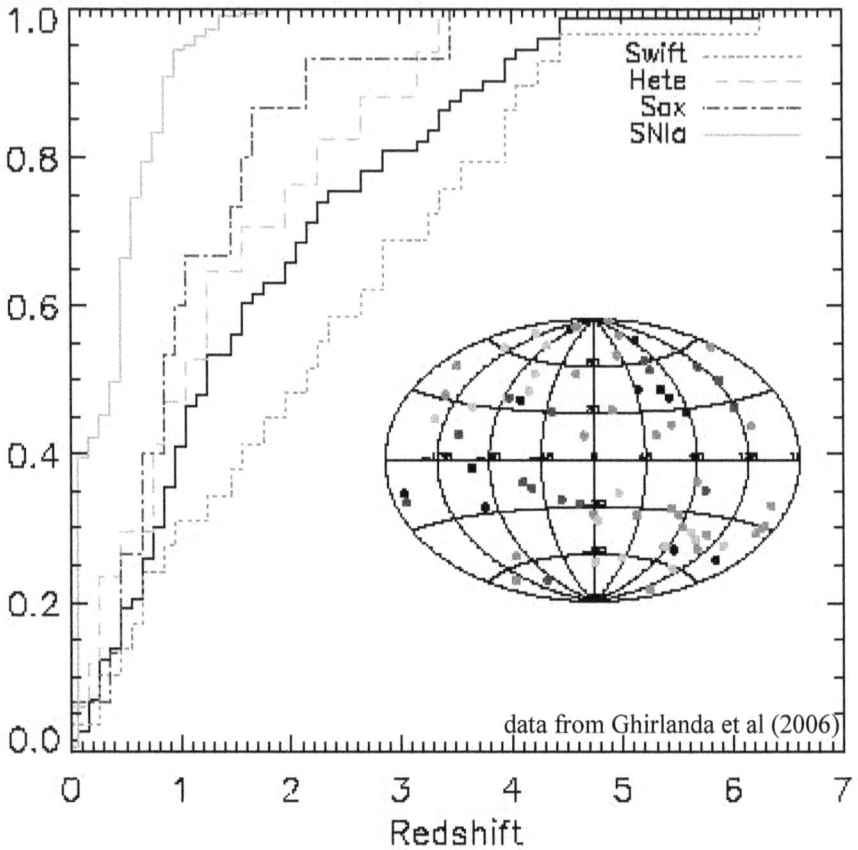

Figure 17: **The distribution of detected GRBs**

A pertinent correlation is that between the rest frame peak spectral energy, E_{peak} and a properly corrected total energy emitted in γ-rays, E_γ. Another recent discovery relates the total GRB luminosity, its peak spectral energy, and a characteristic time scale related to the variability of the rapid emission. The latter correlation is model independent, being based exclusively on emission timing properties.

g. galaxies

We come finally to galaxies that are the primary individual objects of interest in cosmology. Galaxies come in various shapes and sizes. Hubble devised the following sequence of categories of galactic structures that has remained very much in use over the years since its introduction.

Galaxy Properties and the Hubble Sequence

E	→	S0	→	Sa	→	Sb	→	Sc	→	Sd/Irr

Pressure supported	→	**Rotation supported**
Passive	→	**Actively star forming**
red colors	→	**blue colors**
hot gas	→	**cold gas and dust**
old	→	**still forming**
high luminosity density	→	**low luminosity density**

This Hubble sequence has proven to be surprisingly robust with many galaxy properties aligning as tabulated above, although by no means all of the physical properties of galaxies correlate very precisely with this morphological classification. Masses, luminosities, sizes, etc. do not correlate with it, for example. For every Hubble type there is a large spread in all of the fundamental physical properties.

The sequence involves two separate branches as shown in figure 18. This diagram has been denominated the 'Hubble tuning fork'. Hubble originally believed that galaxy development proceeded from left to right. This notion has changed; the reversed order seems now to have given rise to the associated developments. The developmental stages of galaxies proceed from massive hydrogen clouds that eventually begin to spiral due to gravitational forces, compacting as they develop, on through convergence into massive elliptical galaxies.

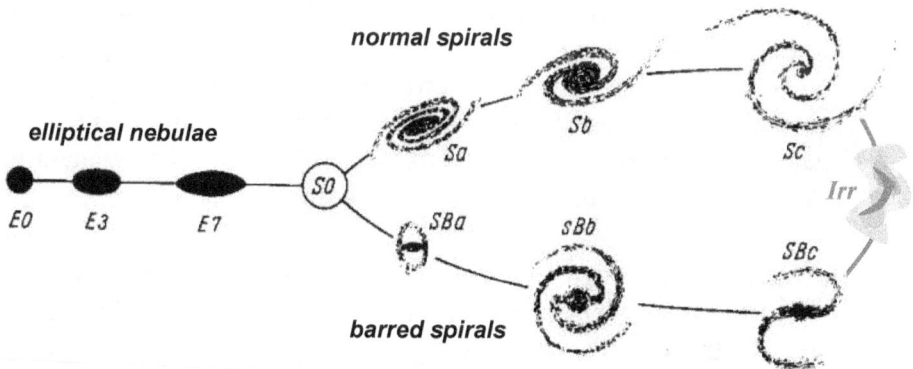

Figure 18: Galaxy developmental stages sometimes mistakenly referred to as 'galaxy evolution'

Round elliptical galaxies are classified as E0, the most elongated ellipticals being classified as type E7. The Irr is a category of amorphous blobs of stars including the two Magellanic clouds in the immediate neighborhood of our own Milky Way galaxy. The term spiral galaxy derives from the bright arms that spiral outward. These are comprised of bright O and B stars, both evidence of recent star formation. Also dust lanes may occur between the arms that obscure the transmission of light. Figure 19 provides several example galaxies. There are two sequences of spiral galaxies that have been categorized:

Sa, Sb, Sc, and Sd
In these spirals, the central bulge is less important with the disk itself becoming the more important feature. The spiral arms are more open and may in some cases be less well defined than in the other category of spirals.

SBa, SBb, SBc, and SBd
These galaxies are distinguished from the first category of spirals by having a central, linear bar through the center that joins the spiral arms.

SO as well as Irr
The class of transition galaxies between the ellipticals and spirals are the S0 galaxies, also called "lenticulars". S0 galaxies have a rotating disk in addition to a central elliptical bulge, but the disk does not include spiral arms or prominent dust lanes. Lenticulars may sometimes exhibit a central bar, in which case they are categorized as SB0. These are believed to have probably derived from spirals as a part of the developmental process from spirals to ellipticals. The irregulars are viewed as early forms of galaxies more generally associated with the first category of spirals than the second.

h. galactic development versus 'evolution' of the universe
Galactic developments do not constitute an 'evolution' (or even a change) in the universe itself, unless perchance we find in looking back to increasing distances that the later developmental phases have not yet appeared upon the stage. Then, of course, the issue of *evolution* does enter the picture, although 'evolution' is an ill-chosen word for

giant elliptical – M87 (NGC 4486)

spiral – NGC 628

barred spiral – NGC 1300

Figure 19:   Images of several galaxy types

what is to be assessed in this regard.* Furthermore, there is mounting evidence that late period galaxies *do* occur even at extreme redshifts.

Nevertheless, it is meaningful to assess whatever differences there seem to be between the universe as it exists today and as it appeared in the distant past. Such data is increasingly becoming available to modern astronomy using its high powered tools. If indeed there are differences of kind that suggest that there are trends from a primordial state to what we see today, then that would be significant. There is much effort being expended in that direction. We will later specifically discuss the efforts of Conselice (2004) and Melbourne (2007), for example, and question their fairly typical conclusions.

The confusion in the term 'evolution' involves the stages of development of galaxies from mere hydrogen clouds up through huge spirals not unlike our own Milky Way and onward to giant ellipticals. This counter development is characterized in figure 18. This galactic development trend has been presumed to correlate well with a general scenario envisioned for the universe as a whole according to standard cosmological models. Early phases of the universe's development according to all such models involve initial dense plasma conditions that eventually 'evolve' to phases characterized by cooler temperatures for which neutral matter (primarily in the form of hydrogen clouds) is precipitated to collapse into larger structures.

However, should the universe prove to be much more extensive and older than most cosmologists typically suppose, it could, for all intents and purposes be modeled as an infinite universe. In this case it would be reasonable to suppose also that it is homogeneous, and isotropic well beyond what we currently observe. In that case – except for problems of observation selection effects – one would be justified in assuming that populations of these naturally occurring objects would be evenly distributed throughout space, with variations due only to random fluctuations in local conditions. Furthermore, one would have to suppose that galaxies of a given type would have been the same then as they are now.

---

* The word itself is borrowed from one of the most outstanding scientific success stories. But one can only improperly refer to mere changes that occur over time in individuals of a current population as 'evolution'. The very meaning of Darwinian evolution is intimately tied to reproduction and other biological phenomena that, by application to cosmology, stretch metaphors beyond useful credibility.

i. spectroscopic categorization of galaxies

In addition to, or supplementing, morphological galaxy types as described above, galaxies have been categorized by electromagnetic emission and absorption characteristics. These characteristics derive, of course, from the stars and gases contained in the galaxies. Younger galaxies will have predominantly younger, bluer stars. These telltale emission and absorption lines in their spectra are essential to accurate determination of the redshift of the galaxy. In figure 20 Colless (1999) illustrates a distinction between 'emission line' and 'absorption line' galaxies. Notice that these two galaxies were observed at different redshifts such that the various emission and absorption lines – for example H_α, H_β, H_γ, et cetera occur at different observed wavelengths even though their emission wavelengths had to have been identical.

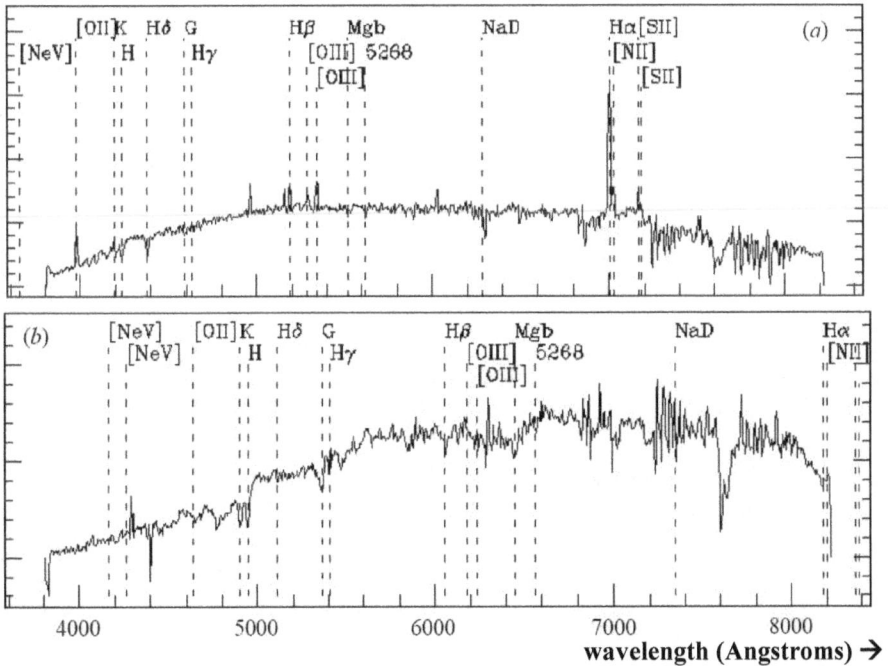

Figure 20: Normalized spectral intensity for a b_j=19.2 emission-line galaxy at a redshift of z=0.067 and a b_j=19.3 absorption-line galaxy at z=2.46

By identifying the emission (and absorption) lines in the two panels of the figure one can easily determine the redshift of the two galaxies. One can also thereby identify the common emission and absorption features in the infra red region that suggests a local cause.

From this kind of analyses five unique spectral types were identified for use in the 2dF Redshift Survey to be discussed in more detail further on in this volume. The partitioning of 3,000 individual galaxy spectra into five categories, whose redshift-corrected means are illustrated in figure 21, gave rise to the category types used by the 2dF analyses.

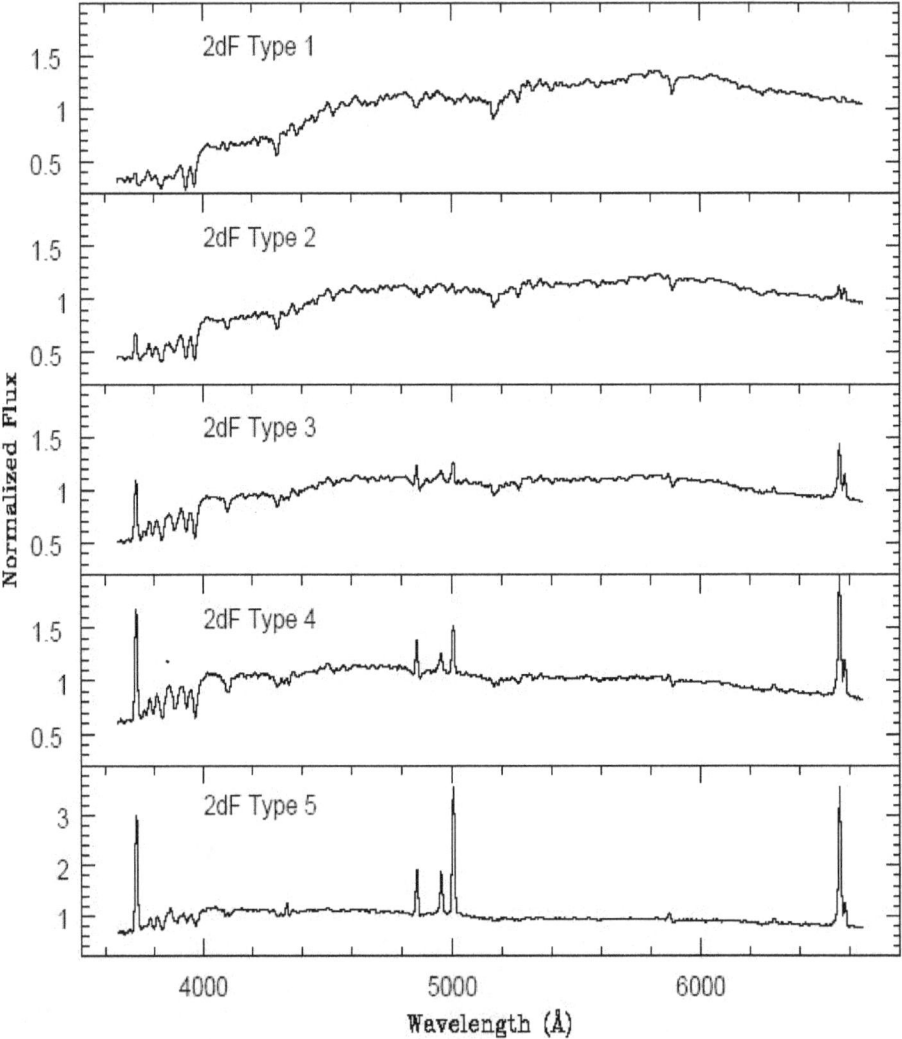

Figure 21: The mean spectra corresponding to the five 2dF spectral types

Of course there is a strong correlation between morphological and spectrographic galaxy types. Figures 22 through 24 illustrate the differences in spectra of various morphological galaxy types. Clearly, the 2dF and Hubble categorizations can be matched up fairly closely.

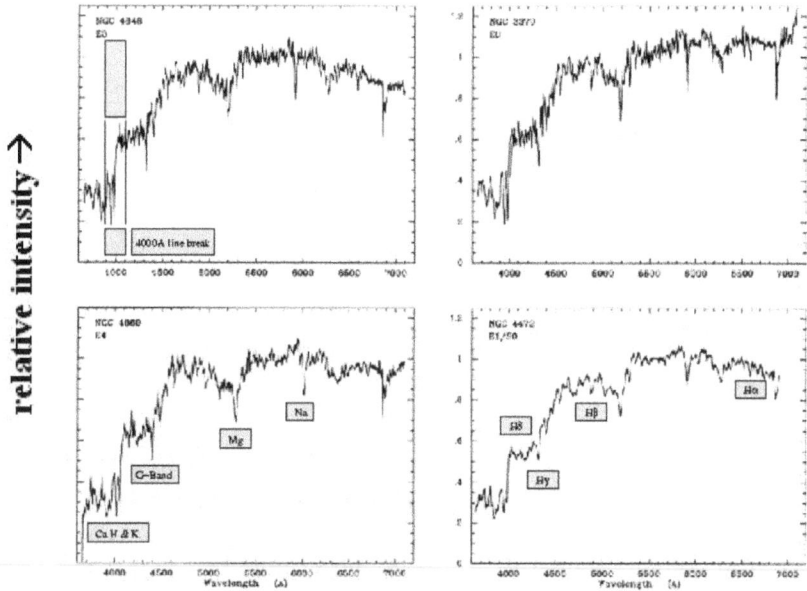

Figure 22:   Elliptical galaxy spectra

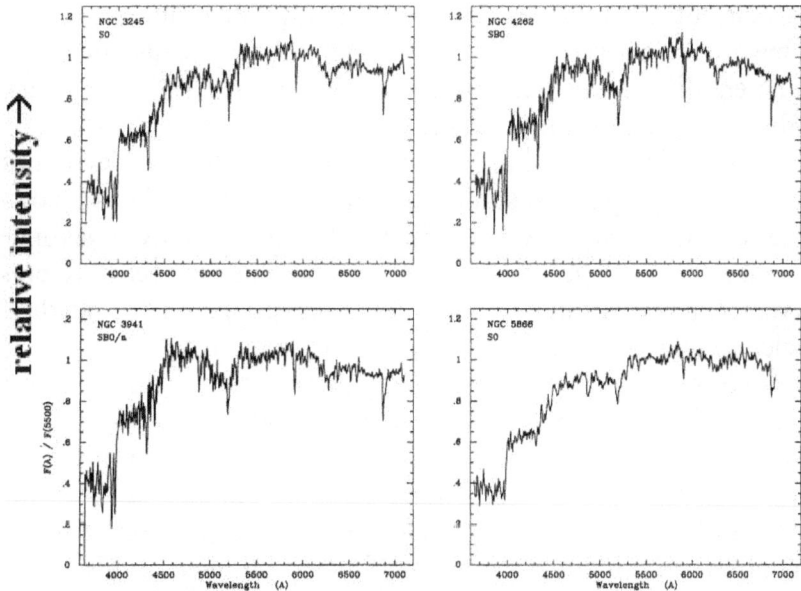

Figure 23: Spiral SO and SBO galaxy spectra

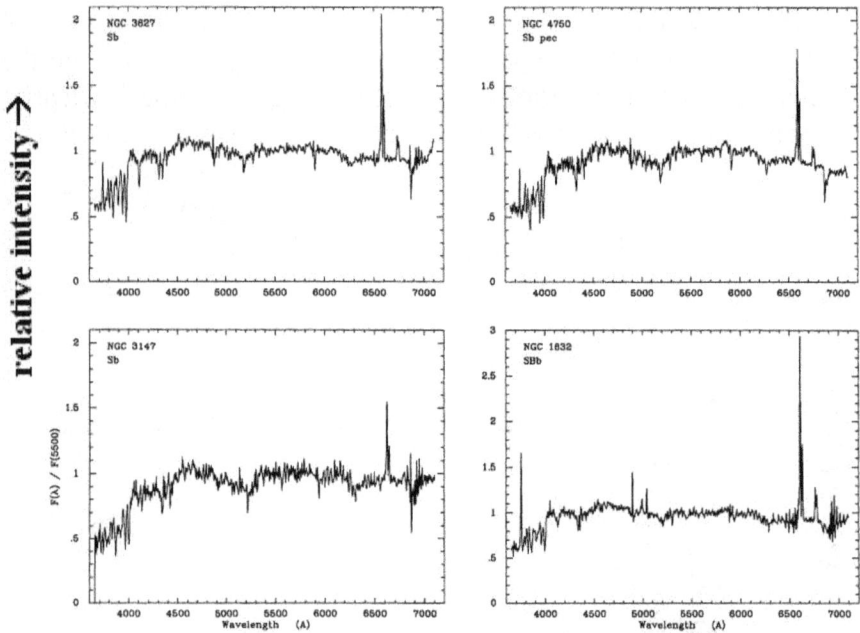

Figure 24: A set of spiral Sb galaxy spectra

But when redshift is taken into account there are further apparent changes in the spectra of distant galaxies. These changes are largely due to observational effects rather than inherent differences in the spectra of the galaxies observed, although there are some changes that have been classed as 'evolutionary' in origin because they seem to be inherent differences from spectra of similar local ($z \ll 1.0$) galaxies. But these conjectures too are quite problematical, as we will see.

Much of what might otherwise seem to be *real* 'evolutionary' differences in galaxies of similar type at different epochs is recognized as due to what is called a "K-correction". This acknowledges inherent differences in the spectra of galaxy types and that redshift presents different segments of these spectra for observations made with different wavelength bands. That is what requires correction.

So with all this dependence on the spectra of galaxies and in the measured flux density of telescopic observations it is particularly important that we understand a little of what is involved in these tasks that are the daily bread of astronomers. Without clearly disentangling effects of observation we would have a distorted perception of what is being observed. In quite another, but associated, area of this arena, this author also has addressed effects of observation on what is observed.

j. clusters of galaxies and their cosmological setting

Rich clusters of galaxies are believed to be the largest 'objects' in the Universe. These galaxies are bound together by gravitation in much the same way as stars are bound within a galaxy and atoms within stars. Rich clusters involve concentrations of hundreds (or even thousands) of galaxies within a region spanning many millions of light years. They are extremely bright sources of X-rays. Here hot intracluster plasma gases fill the space between galaxies with an average electron density, $\rho_e \approx 10^{-3}$ cm^{-3} and electron temperatures ranging from 10^7 to well over 10^8 K. The high temperatures are inferred to derive from gravitational causes associated with their large masses of from 10^{14} to 2×10^{15} times the solar mass. Larger clusters possess a high percentage of early-type morphology galaxies. Galaxies in smaller galaxy groups and 'field galaxies' are more typically older, i.e., those found at the left of Hubble's 'tuning fork'.

The massive clusters of galaxies shown in figures 7 and 8 will warrant a considerable amount of our attention. We will address the associated suppositions of 'dark matter' that resides (if at all), primarily in and about such clusters.

k. other objects of cosmological concern

We have concentrated in this chapter on an overview of what we know about galaxies and how we know it.

Other objects of cosmological concern than those we have mentioned certainly exist. Among these are a multifarious variety of stars that comprise the luminous functions of galaxies. It seems certain now that black holes, or at least "black stars" as Barcelo et al. (2009) insist as alternative, do exist. But we can know very little about them because of their very nature. Planets, of which there must be a tremendous number throughout the universe, many of which are very probably similar to our earth, must certainly exist in large numbers as well. However, they have little to do with discussions of cosmology because even at five billion years of age our planet, for example, is generations removed from its cosmological origins. The substances in and on earth have had to be cycled through the life cycles of innumerable stars within our own Milky Way galaxy.

Of course, the microwave background radiation is not an 'object' in any traditional sense, although it seems at times to be treated as though it were. Attempts to associate its tiny variations with major galactic cluster structural variations has indeed become an obsession of cosmological research. We reserve detailed discussion of that topic as well as the relative abundances of the elements to later chapters.

l. cosmological observation methods

Brightness and the general structural appearance of galaxies are strongly dependent on the wavelength at which observation takes place. This includes major differences in morphological shape in telescopic images of galaxies obtained in different electromagnetic frequency domains from X-ray and ultraviolet (UV), up through visible optical colors to the near infrared, microwave, and radio frequencies. Galaxies appear `lumpy' in X-ray and UV, with increasingly smoother images resulting when viewed in visible light and longer wavelengths. See figure 25.

images from P. Armitage

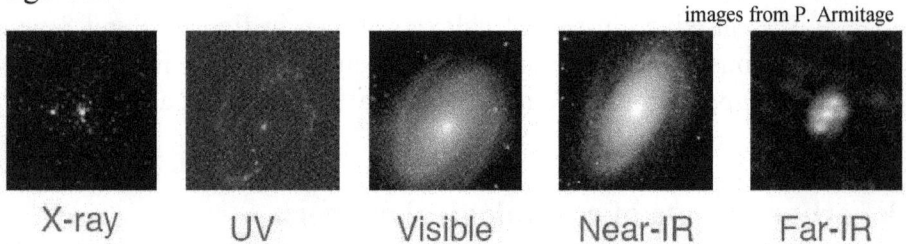

| X-ray | UV | Visible | Near-IR | Far-IR |

Figure 25: Nearby galaxy M81 as observed at various frequencies

Specific wavelength 'bands' are employed in the measurement of luminosity of an observed object. Bolometric flux on the other hand includes all the radiant energy emitted by the object across the entire spectrum. Integrating (summing) all this data is the only way to assess the total energy radiated by an object. So to estimate this total flux density, the intensity of radiation must be measured at multiple wavelengths using various filters such as those shown in figure 26 with the results from each added together. Much of the data used in cosmological analyses is slanted toward assessments of whether 'evolution' has occurred or not, and the rest frame wavelength at which the object is observed can easily suggest changes that may not have occurred at all. Use of wavelength filters narrows the search to a specific domain of the spectrum to which an investigation pertains.

The sensitivity of individual filters used to estimate the total amount of flux in, for example, 'U', 'B', 'V', 'R', and 'I' that refer respectively to 'ultraviolet', 'blue', 'violet', 'red', and 'infra-red' light fluxes, are included in figure 26. The first three of these comprise what is denominated, "three-color apparent '*bolometric*' flux" – UBV. This 'Johnson system' of filters accommodates much current cosmological observation.

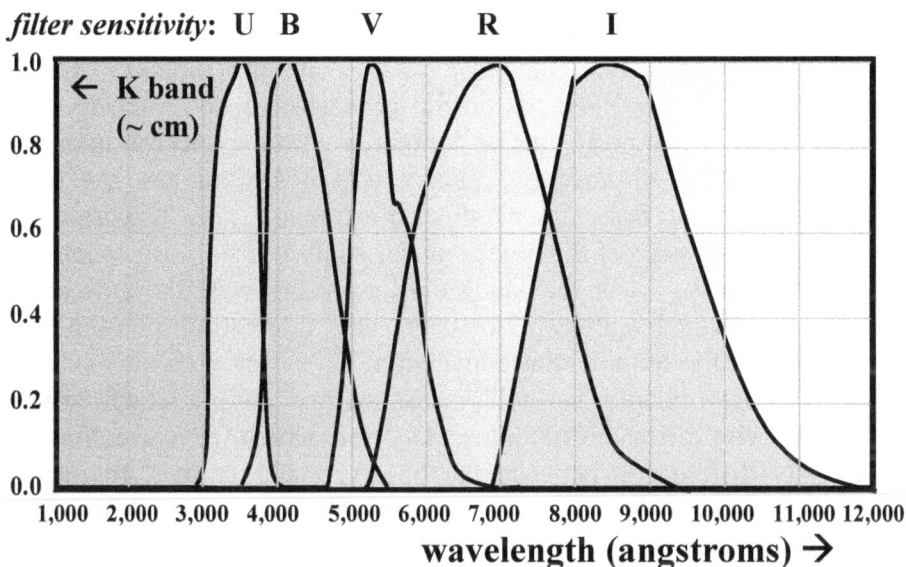

Figure 26: The UBVRI (or 'Johnson system') of color filters with the average effective wavelengths of the ultraviolet filter at 3,600 Å, the blue filter at 4,400 Å, the violet filter at 5,500 Å, the red filter at 7,000 Å, and the infrared filter at 8,800 Å. The 'K band is far to the left (~ 1 cm)

K correction of luminous flux adjusts an astronomical object's magnitude (log of luminous flux) to convert a measurement of the intensity of an object at redshift z to that of a measurement made in the 'rest frame' of the object (i. e., at z = 0). If one could measure all electromagnetic radiation emitted from an object (that denominated 'bolometric flux', as clarified above), K correction would not be required. But since it *is* typically required, we must try to understand it. Ellis (1997) discusses in some detail the luminosity measures for the various types of galaxy. Blain et al. (2002) shows associated strange behavior at considerably longer wavelengths.

If the concern is (and it typically is) differential luminosity at specific frequencies, $f = c / \lambda$, where λ is the wavelength of the radiation, then K-correction is required. A frequency-dependent factor is multiplied times the measured luminosity, because the redshifted object emitted its flux of radiation in a different wavelength band with a different inherent intensity than the one in which it is being observed. The K-correction takes into account general intensity functions of the spectra of the object in question.

See for example the spectra of galaxies shown in figures 20 through 24. Each type of galaxy has a unique intensity versus wavelength profile. When the profile is redshifted, the intensity at a given frequency can no longer be compared with the spectral intensity at that frequency of a similar galaxy whose spectra has not been redshifted without first making this adjustment. This K correction would be unnecessary if the objects under study had uniform spectra, i. e., a 'flat' intensity level for which the luminosity was the same at all frequencies (wavelengths). However, any interesting non-uniform luminosity profile requires this conversion.

In a later chapter we will discuss metrics, among which will be the luminosity distance modulus, D_L, for which K correction of luminosity differences between the observed and emitted frequency bands is required. K corrections are typically incorporated into data used by any cosmological model. We will not attempt to pre-empt expertise in this area, and discuss it here only to provide an awareness of what has been involved in obtaining the data we use.

At extreme distances luminosity characteristics of entire galaxies have tended to be used as the standard candles of choice for classes of galaxies that are easily identified by their morphology – giant ellipticals, giant spirals, radio galaxies, etc.. Since the value of the luminosity distance parameters are cosmological model-dependent, the expected differential flux may also be somewhat model-dependent.

Of course in this summary chapter we have just scratched the surface of the data required to realize success on this intellectual journey upon which we have embarked.

Chapter 3

Characterization of the Intergalactic Medium

What is known of the objects discussed in the preceding chapter has been obtained by observations through a most unique medium. The impact of associated electromagnetic scattering effects are significant. This volume is more or less dedicated to an exploration of those effects.

Considerable investigation has been directed over the past several decades toward an improved understanding of the composition and structure of this intergalactic medium although typically from a very different perspective than our current focus. See Barcons, et al. (1991) and more recent studies. From such research has arisen a wealth of information and an improved understanding which, although still incomplete, reveals much about this virtually transparent substance.

a. qualitative features

The vast regions between galaxies are known to be comprised primarily of high temperature, low density, ionized plasma because only such a medium could exhibit the virtually complete transparency that allows observation of objects to distances beyond ten billion light years. Nonetheless, even such an unobtrusive medium must scatter a certain amount of electromagnetic radiation. Such dispersion (even though minimal) must introduce effects that would allow its presence to be inferred even if not directly detected. As will be shown, however, restrictions on the ranges of temperature and density of such a plasma medium reduce the calculated magnitudes of usual dispersion effects below ready observability according to the applicable Lorentz-Lorenz formulas. Attenuation by absorption is considered negligible by most investigators. We will show farther on, however, an observable absorption effect that accounts for radiation energy lost due to a plasma redshift mechanism. A luminosity diminution factor attributed directly to this phenomenon mimics a previously supposed time dilation.

It had been theoretically anticipated that 'forward' scattering of radiation would not, nor had it ever been demonstrated to, produce an alteration of the spectra of radiation propagated through such a medium. According to extensions of Wolf's scaling law, for example, forward scattered radiation would in all such cases exhibit spectral invariance (Wolf, 1986, 1989). Therefore, other than through indirect null observations and misinterpretations that establish constraints on its density and temperature, the illusive intergalactic medium has been thought to have little, if any, affect on our empirical view of the cosmos. This misguided anticipation will be re-evaluated here.

Observations of the extra galactic universe have been shown to be compatible with limited possibilities of composition, density and temperature of this medium (Barcons, 1991; Silk, 1976; Sunyaev, 1969; Gunn, 1965). Unless temperatures were extreme, a redshifting 'Lyman alpha trough' (assuming a preponderance of hydrogen in the composition of the medium, and other absorption lines otherwise) would obliterate the spectra of distant objects beneath the Lyman (or other analogous) continuum limit. The Lyman-α trough is, in fact, observed in the observations of very distant quasars and Lyman break galaxies, which helps determine temperature and density constraints.

In addition, it has been believed that if the density were too great and/or the temperature too low, then constituents of the medium would collapse into galactic clusters generating characteristic X-ray emissions that are not observed in particular (Gunn, 1972). If the medium were too diffuse, it would be incompatible with the X-ray spectrum that *is* observed from intracluster gases, which would more quickly disseminate out into intergalactic regions (Sarazin, 1986; Misner, 1973). Together, these features, to the extent that they are in fact based upon actual observations, serve to quantify constraints on the characteristics of the intergalactic medium.

Largely, however, it is from the facts associated with the degree to which it is *not* being directly observed, that we know that the

48

intergalactic medium must exist in a state which is highly unlikely (perhaps even *least* likely) to be observed. This could only be the case if its properties were to satisfy the following approximate quantitative conditions on density and temperature:

b. quantifying the intergalactic medium density ranges

In particular according to the references cited above and others, these qualitative conditions would only result if the average density of intergalactic plasma were to satisfy the conditions on density:

$$10^{-32} \leq \rho_{ig} \leq 10^{-29},$$

where ρ_{ig} is the mass density of the intergalactic medium measured in units of grams per cubic centimeter. A value near the high end of this range has sometimes been justified on theoretical grounds associated with the general theory of relativity as providing a major contribution to the 'critical mass' anticipated by Einstein's general theory.* The low end is based on the amount of matter it has been possible to observe by its various luminous effects. Additional constraints imply that if the density were to be at the high end of its range, then the temperature must also be at the high end of *its* range as well in order to maintain the observed degree of ionization. The same would apply for low densities, which could similarly be linked to lower temperatures.

If we assume the mass density of the universe to be comprised primarily of baryons which is by no means a currently popular position, then the implied baryon density is given approximately as follows:

$$\rho_B = \rho_{ig} / m_B$$

where m_B is the mass of a baryon (primarily nucleon, i. e., proton, p^+ or neutron, denoted n). The approximation $m_B = 1.67 \times 10^{-24}$ gram applies to a hydrogenous plasma that is generally assumed to pertain.

---

* See for example Silk (1980) where a value of the density of the universe is inferred from Einstein's theory in terms of the gravitational constant and Hubble's constant. With current estimates of Hubble's constant this would suggest a density of from 3 to 5 x 10^{-30} gm cm^{-3}. However, refer to a later chapter where the validity of this sort of theoretical inference is questioned in some detail.

There seems to be no reason according to any observation or viable theory to assume composition of the intergalactic medium would appreciably differ from the universally observed elemental composition of baryonic matter percentages. This universally observed composition is about 76 percent hydrogen, 24 percent helium by mass with mere traces of isotopes of other elements. Figure 28 illustrates this approximate distribution of protons and ^{4}He in four representative partitions of space each with four baryons. In reference to this figure, the associated table illustrates the relative abundance of particles by volume with temperature-related aspects to be discussed later. For comparison, at standard temperature and pressure in our atmosphere a cubic meter of a similar substance (hydrogen gas with the appropriate fraction of helium) would contain Avagadro's number (i. e., 6.0×10^{23}) of baryons and approximately equal numbers of associated electrons.

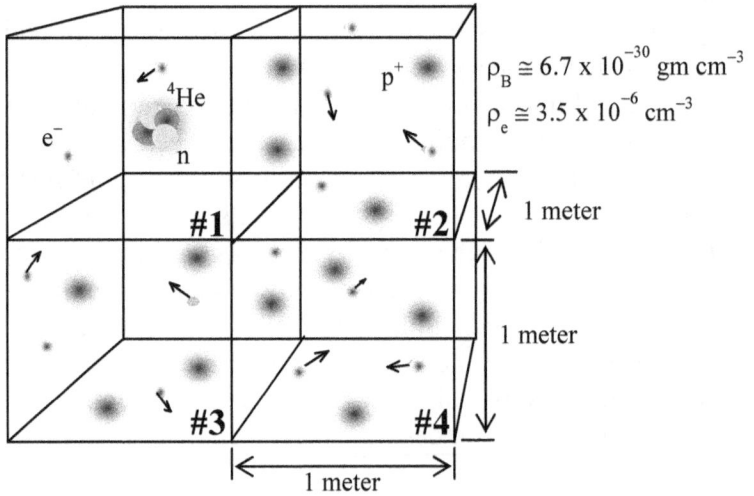

$\rho_B \cong 6.7 \times 10^{-30} \text{ gm cm}^{-3}$

$\rho_e \cong 3.5 \times 10^{-6} \text{ cm}^{-3}$

partition		Contained baryons	contained electrons
#1	4	1 ^{4}He nuclei (2 protons plus two neutrons)	2 electrons
#2 – #4	12	12 protons	12 electrons
#1 – #4 (sum)	16	14 protons, 2 neutrons	14 electrons

Figure 28: Constituents of intergalactic space

Thus, the mass density range indicated above for the universe as a whole implies the approximate range of electron density as follows:

$$10^{-8} \leq \rho_e < 10^{-5} \text{ electrons cm}^{-3}$$

An associated distance to which a line of sight must extend to terminate at the 'surface' of an electron in intergalactic space is implied. The electron cross section is 6.65×10^{-25} cm^2. It is larger than would be obtained using the *classical electron radius* of $r_e = e^2 / m_e c^2 \cong 2.82 \times 10^{-13}$ cm, where $e = 4.80 \times 10^{-10}$ statcoulombs represents electronic charge, $m_e = 9.109 \times 10^{-28}$ gm the electron rest mass, and $c = 2.9979 \times 10^{10}$ cm/sec is the speed of light in a vacuum.

However, even near the high end of the electron density range, the average distance, d to anything like a material surface applicable to a 'surface brightness' appropriate to thermal blackbody radiation from the intergalactic medium, would exceed 10^{30} cm. This is determined as illustrated in panel a of figure 29 using the formula:

$$d = 1 / (\rho_e \times 6.65 \times 10^{-25}) \cong 1.5 \times 10^{30} \text{ cm.}$$

The value at right applies to an electron density of 10^{-6} electrons cm^{-3}.

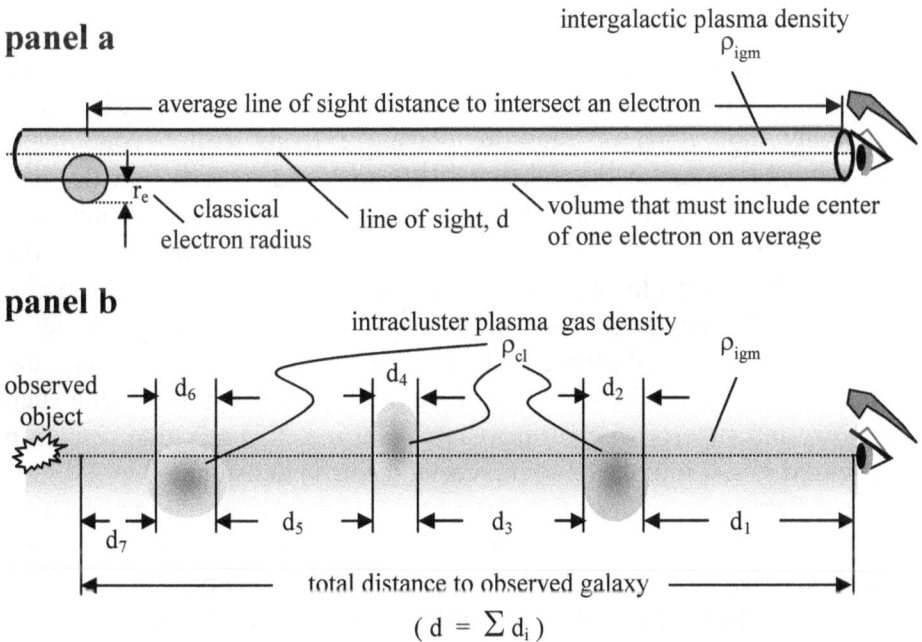

panel a

intergalactic plasma density ρ_{igm}

average line of sight distance to intersect an electron

r_e

classical electron radius

line of sight, d

volume that must include center of one electron on average

panel b

intracluster plasma gas density ρ_{cl}

ρ_{igm}

observed object

d_6 d_4 d_2

d_7 d_5 d_3 d_1

total distance to observed galaxy

$(d = \Sigma d_i)$

Figure 29: Illustration of line of sight calculations

Of course irregularities in electron density are encountered along the line of sight in observing objects at extreme distances through intergalactic space. There are neutral hydrogen clouds along the way that contribute substantially to short wavelength (sub Lyman-α line) absorption, but otherwise do not affect observations very much.

There are also very dense regions of intracluster plasma as discussed by Cowie and Perrenod (1978), Bahcall (1999), Reiprich (2006), Hicks et al. (2002), Loewenstein (2003), and many others. In panel b of figure 29 we broach an important aspect of observations made by light propagating through these gases encountered in galactic clusters intermediate between the observer and an observed object.

'Cluster core' regions make substantial contributions to the weighted average of the dynamic pressure encountered along any line of sight. It is this dynamic pressure that imposes the most severe of the impacts of the intergalactic medium on observations at cosmological distances. Plasma gas density and temperature determine this pressure and are each known to be several orders of magnitude higher in cluster cores than corresponding values for the intergalactic medium between clusters generally. These most extreme regions of galactic cluster cores associated with d_2, d_4, and d_6 shown in panel b of figure 29, are typically only a few Mpc across. (Mpc refers to a 'megaparsec', i. e., 3.26×10^6 light years, which is roughly 3.1×10^{24} cm.) In fact, the most intense regions are assessed as having radii less than an Mpc. But relatively high dynamic pressure plasma extends well beyond these most intense cores at the centers of clusters and merges into the less dense medium that pervades all of intergalactic space. The material in these regions is basically of the same 'primordial' substance as the medium between clusters. Loewenstein (2003) states, for example, "...results demonstrated the origin of cluster X-ray emission as a thermal primordial plasma enriched by material processed in stars and ejected in galactic winds." Certainly explosive stellar phenomena such as nova and supernova eject heavier nuclei into this mix that results in increased X-ray intensity and increased electron densities.

Separations between clusters tend to be more than an order of magnitude greater than cluster core dimensions. Ignoring for the moment extremely low densities between clusters, the baryonic mass of a typical galactic clusters is the order of 10^{47} to 10^{48} grams within a radius of 5 to 10 Mpc – much of this being plasma gas. Thus, if the density of the universe as a whole is less than Einstein's critical density, ρ_0, as appears to be the case, average separation of typical clusters must be on the order of 50 to 100 Mpc. Figure 30 shows representative density data found in galactic clusters and the surrounding regions.

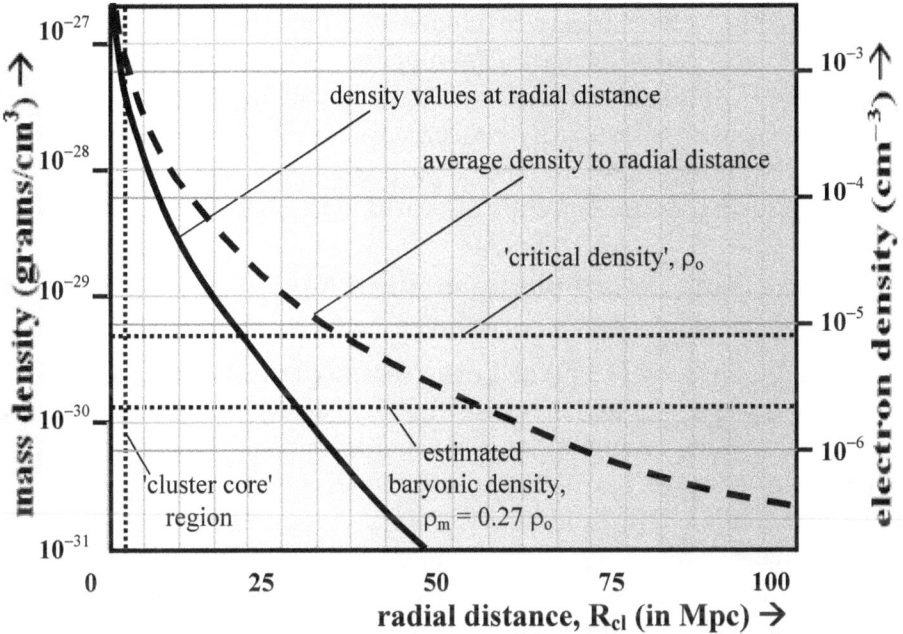

Figure 30:   Example baryonic density at various radii from center of a cluster

In our investigations average dynamic pressures encountered on any line of sight observation will be what impacts the nature of what is observed. Thus, it will be the average of the product of electron density and temperature with which we will be most concerned. So the effect of the relatively small distances through intermediate clusters will contribute in a major way to cosmological redshift because of the so extreme values of these parameters. Observations of intracluster gas density imply a lower range of intergalactic medium density as given above. We will find that this data is in general agreement with all observed baryonic mass densities of the universe. So if we include the variations we should considerably increase the range of densities to be encountered in any line of sight involved in cosmological observations.

Einstein's field equations suggest the 'critical mass density', ρ_0 for compelling theoretical reasons associated with the general theory of relativity and the standard cosmological models. Estimates based on observation suggest a dearth of luminous baryonic matter in the universe to meet this objective. So estimates based on measurements have tended toward the lower end of the range provided above. We will find similar demands in accounting for the microwave background. Measurements have largely excluded the unobserved intergalactic medium. Evidence increasingly suggests that it contains a large percentage of the total baryonic matter of the universe. In our current endeavor we will be primarily concerned with electron density, and in particular with more intense plasma properties in cluster gases, thereby resolving several issues currently allocated to mysterious 'dark matter'.

c. quantifying intergalactic medium temperature ranges

An average temperature of $10^3 \leq T_{ig} \leq 10^6$ kelvins at $z < 2$ was accepted by Sunyaev (1969) and more recently by Lehner et al. (2007). Their theory-laden approaches insist that any characteristic temperature specify an *epoch*, beginning at what is denominated 'decoupling' – a time when, according to the standard model, atoms are conceived to have first combined and ultimately gravitated into stars and galaxies, to finally have *reheated* intergalactic regions to their current temperatures.

However, this range seems low for the following reasons: First of all, a gas with only traces of elements beyond hydrogen will only be weakly ionized. Stripping of electrons from *all* constituent atoms must be quite complete except in isolated 'clouds' if we include intracluster gases. Figure 31 shows the abundance of the various ionization levels of hydrogen and helium as functions of kinetic temperature, each shown for one (solid line) and ten (dotted line) dynes dynamic pressure, with the considerably lower pressures reducing temperatures further.

Secondly, lines of sight passing through galaxy clusters that 'cosmological' observation entails, involve orders of magnitude more extreme conditions as indicated by Cowie and Perrenod (1978), and more recent investigators, e. g., Bahcall (1999) and Hicks et al. (2002).

The range of anticipated temperatures for the intergalactic medium is reminiscent of temperature ranges in the Hertzsprung-Russell diagram provided as figure 1 on page 18. There is clearly a real sense in which the intergalactic medium is part and parcel of the same universe as its objects, hardly something to which a temperature of 2.725 K would seem legitimately to apply.

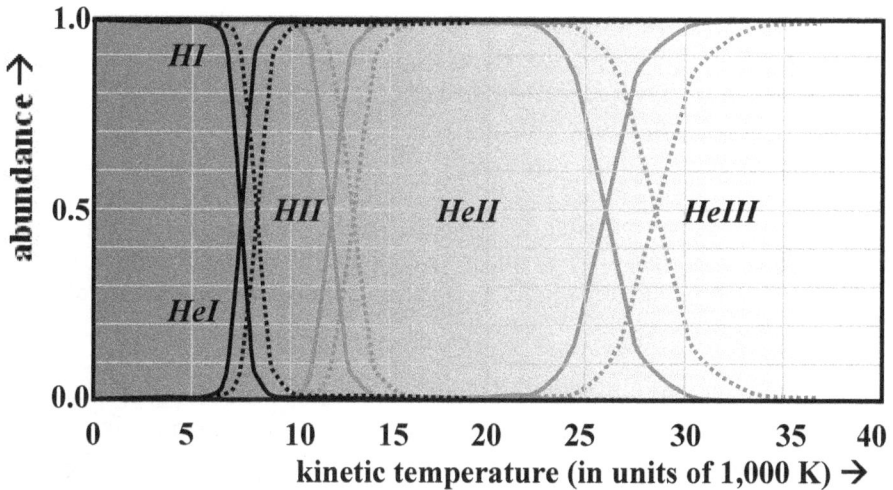

Figure 31: Light element ionization properties

Although there seems to be little in the way of an observed ultraviolet background flux as cited by Sunyaev (1969) and Bothus (1998), a considerable density of shorter wavelength radiation is well confirmed, so there is indeed extragalactic data to suggest higher temperatures. Such high-energy radiation ionizes neutral hydrogen halos of galaxies at a sharp edge (Maloney and Leiden, 1991) although absorption by neutral hydrogen within our own galaxy obscures such observations for the Milky Way galaxy itself. That there is an extremely high level of X-radiation background that is not associated with our own galaxy and cannot be attributed completely to distinct sources has been known for some time. See, for example, Silk (1980) and Sunyaev (1976). This radiation shows an approximate power law distribution of photons with energies consistent with thermal origins within the medium itself. All of this suggests considerable evidence for the existence of intergalactic thermal radiation from an exceedingly hot plasma at temperatures *greatly* in excess of 10^6 K. See for example, Burrows and Medenhall (1991), Kahn (1991), and Bahcall (1999).

The following should be noted in this context also, particularly by any who might underestimate the weight of such evidence. A natural confusion arises because the intensities of this radiation are lower than for a blackbody in this temperature range. The intensity of thermal radiation from a diffuse plasma (which is our only means of accurately determining the temperature of the intergalactic medium, and is in turn its only means of cooling) is typically very much less than would be expected from Planck's formula for blackbody radiation. In particular, we have according to Post (1993),

"...the fortunate fact that a tenuous plasma is optically very 'thin' over almost all of its emission spectrum means that, as might be expected from Kirchhoff's law, radiation is greatly reduced with respect to the Planck value, so that under the proper circumstances a plasma with a kinetic energy of 10^8 K might radiate at a rate equivalent to the radiation rate from a blackbody at radiation temperatures of only a few hundred kelvins."

Also to be considered in assessing its temperature are the various types of radiation mechanisms that come into play in plasma. See for example, Post (1993) for more detailed explanations. There are three such mechanisms:

1)   The generation of X-rays – bremsstrahlung radiation – that involves deflections of electrons by the heavier ions.

2)   The similar generation of radiation involving the interactions between the electrons themselves. This phenomena occurs primarily for extremely high energy relativistic electrons for which the radiation energy is greater than 500 keV (thousand electron volts) or wavelength of about 10^{-10} cm well into the gamma ray spectrum characteristic of temperatures exceeding 5 x 10^9 K.

3)   The excitation of bound electrons attached to the heavier ions. Naturally this involves plasmas for which the temperatures are insufficient to fully ionize all of the atoms, and therefore would only apply for temperatures less than about 10^5 K.

Clearly, the first type (and possibly some of the third type) of radiation is what one should expect as most prevalent from the stripping implications of the minimal atomic absorption. The existence of isolated clouds in the Lyman-α forests and protogalaxies certainly does not significantly nullify this inference for intergalactic regions in general as we will see. Apparently the X-ray background to be inferred from these facts is what is observed to the extent that the low density and our obscured observation post allow such observations at all. Temperatures required to ionize even the lighter elements, to so completely eliminate emission lines, must be at about 10^4 kelvins or greater. To smoothly transition with intracluster plasmas of higher electron densities and extreme temperatures, the temperature between clusters must be fairly appreciable as well. These conditions exist *now* as well as in epochs past. Although, if the temperature were to exceed 10^9 by very much, there would be a continuous gamma ray spectrum rather than the isolated bursts that are observed.

Implied high averages of temperatures persist into intergalactic regions or absorption would have *re-entered* the picture in our 'epoch'. It definitely has not. So it seems reasonable to substantially increase estimates of the upper bounds of the temperature range from earlier estimates to include the transitions from intracluster gases as follows:

$$10^3 \leq T_{ig} \leq 10^9$$

where again T_{ig} is the electron temperature of the intergalactic medium measured in kelvins. More recent investigations suggest that the upper end of this range is directly applicable to the hot plasma that has been observed within 'rich clusters' of galaxies. Refer to Bahcall (1999) for a detailed discussion of conditions within such clusters.

d. statistical properties of the plasma

Sarazin (1986) stated that if a homogeneous plasma were not in thermal equilibrium, it would be restored to that condition in a period of time less than, t_{eq}, where:

$$t_{eq} \approx 3.3 \times 10^5 \, [\, T_{ig} \times 10^{-8} \,]^{3/2} \, [\rho_e \times 10^3 \,]^{-1} \leq 2 \times 10^9 \text{ years.}$$

Note that the upper limit specified here involves opposite extremes of the temperature and density ranges. Even in that case the time required to restore an altered equilibrium is considerably less than the Hubble time or 'age of the universe' according to the standard model. The expected (as against maximum) period of time is much shorter than the duration of observed processes that would appreciably have altered this equilibrium. The large-scale structure of the universe suggests that at a reasonable estimate of peculiar velocities of galaxies, i. e., up to 600 km/s, it would take 1.6×10^{11} years for a galaxy to travel the 100 Mpc distance to cross typical voids. Furthermore, the model pursued in this investigation is associated with a stationary state of the intergalactic medium. So in this study it is reasonable to assume that there has been ample time to achieve an isotropic Maxwellian distribution of electron energies.

At the high end of the temperature range speeds of many of the electrons would be appreciable relative to that of light, so we must determine the extent to which relativistic treatment is required. Key to that determination is the extent to which the classical kinetic energy formula approximates the relativistic formula, as follows:

$$E_R = m_e c^2 [(1 - (v/c)^2)^{-\frac{1}{2}} - 1] \overset{?}{\approx} E_C = \frac{1}{2} m_e v^2$$

Here c is the speed of light in vacuum, m_e the electron mass. The two formulas are plotted in figure 32, where it can be seen that the classical approximation, E_C works quite well below half the speed of light.

Under equilibrium conditions the average energy (per degree of freedom) is $\frac{1}{2}$ k T, where k = 1.380 x 10^{-16} ergs/degree kelvin is Boltzmann's constant. So for the three degrees of freedom that apply to the electron velocities with which we will primarily be concerned in forward scattering investigations, average energy must be:

$<E> = 3/2$ k T_e ergs

The average electron velocity divided by the speed of light, <v>/c, to which we will refer as β_e, of non-relativistic individual electrons in a Maxwellian distribution will be given by:

$$\beta_e \cong [3 k T_e / m_e c^2]^{\frac{1}{2}}$$

$$= 1.83 \times 10^{-5} T_e^{\frac{1}{2}} \text{ cm/sec}$$

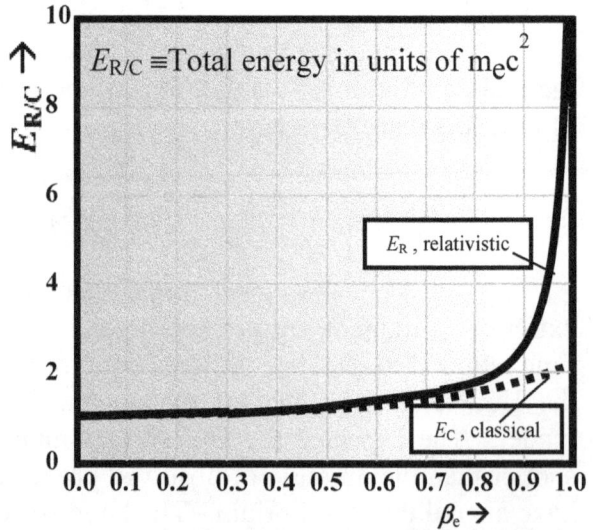

Figure 32: Classical versus relativistic representation of dependence of the total energy on electron velocity

according to the classical kinetic energy formula. However, for temperatures appreciably greater than 10^8 K, this non-relativistic formula does not provide a good approximation as shown in figure 33 below. The appropriate formula in this domain is:

$$\beta_e = \left[\, 1 - \frac{1}{\left(\dfrac{k T_e}{m_e c^2} + 1\right)^2}\, \right]^{\frac{1}{2}}$$

The properties of the intergalactic medium suggested by earlier discussions are those of an extremely diffuse and (particularly in intracluster gases) thermodynamically hot plasma medium. Only about one in four cubic meters of space as depicted in figure 28 would contain a helium nucleus (alpha particle). The velocities of each particle type will be distributed uniquely according to their own Maxwellian distributions associated with their respective masses as shown in the table below. The distances covered by the respective root mean square (rms) velocities in one nanosecond (10^{-9} second) are also illustrated in the table. During this time interval light would proceed 30 centimeters.

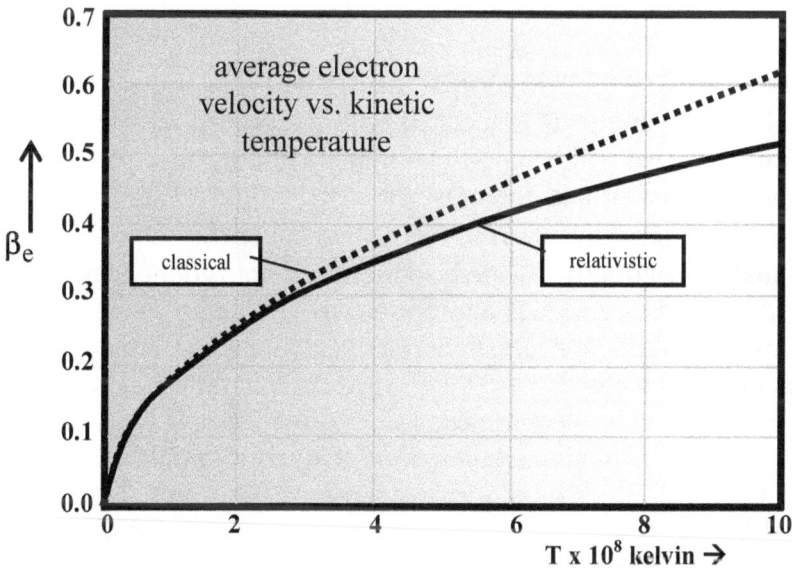

Figure 33: Relativity implication on high temperature plasma electron velocity

If δ_m is the mean free path of an electron prior to colliding with another ion in a plasma (including more distant Coulomb interactions as well as direct collisions), it has been shown by Sarazin (1986) that:

$$\delta_m = 3^{1/2} \, [k \, T_e]^2 \, / \, [\, 4 \, \pi^{1/2} \, \rho_e \, e^4 \, \ln \Lambda \,],$$

particle type	rms velocity (at 10^8 K)	distance traveled per nanosecond
photon of light	3.0×10^{10} cm/sec	30 cm
electron	5.4×10^9 cm/sec	5.4 cm
proton	1.3×10^8 cm/sec	0.13 cm
alpha particle (He^4)	6.3×10^7 cm/sec	0.063 cm

where $\ln \Lambda \cong 40$ for values of T_e and ρ_e throughout the admissible ranges for these parameters as discussed above. This yields a mean free path of on the order of 10^7 light years. This value collaborates low expected absorption properties from this cause in the intergalactic medium, since it is such collisions during the passage of a photon that effects traditional absorption. This also corroborates that even though there is a high temperature, bremsstrahlung radiation will not be all that appreciable, but some minimal absorption is to be expected.

e. applicability of relativistic analyses

Relativistic effects associated with the ratio β have a profound effect on much that will be presented in this volume. Since Maxwellian distributions are first and foremost the distributions of the energies of particles – *not* velocities derived from classical energy formulas in any case – that are thermally affected, so the treatment is quite natural. The range of temperature values appropriate to producing effects that are being investigated here necessitate relativistic considerations whenever the velocity parameter is involved. However, it should be noted that the more accurately determined value provided by the relativistic velocity formula is in all cases *less than* that which would be predicted by the classical formula as was shown in figure 33.

Refer to figure 34, which illustrates the implied differences in Maxwellian distributions of these electron velocities for the various temperatures. Distribution differences are shown for a range of temperature values up to $T_e = 4 \times 10^8$ K. Clearly, for temperatures exceeding 10^7 K the difference becomes appreciable, but it is always less than what would be predicted classically.

The magnitude of β_e (when it is not raised to a power greater than unity) should never be disregarded without consideration wherever it appears as a factor in the formulas of physical phenomena associated with electrons in the intergalactic medium. In the case of forward scattering, such relativistic effects will be shown to introduce a major qualitative difference in the observed phenomena, the derivation of which is a primary objective of this investigation.

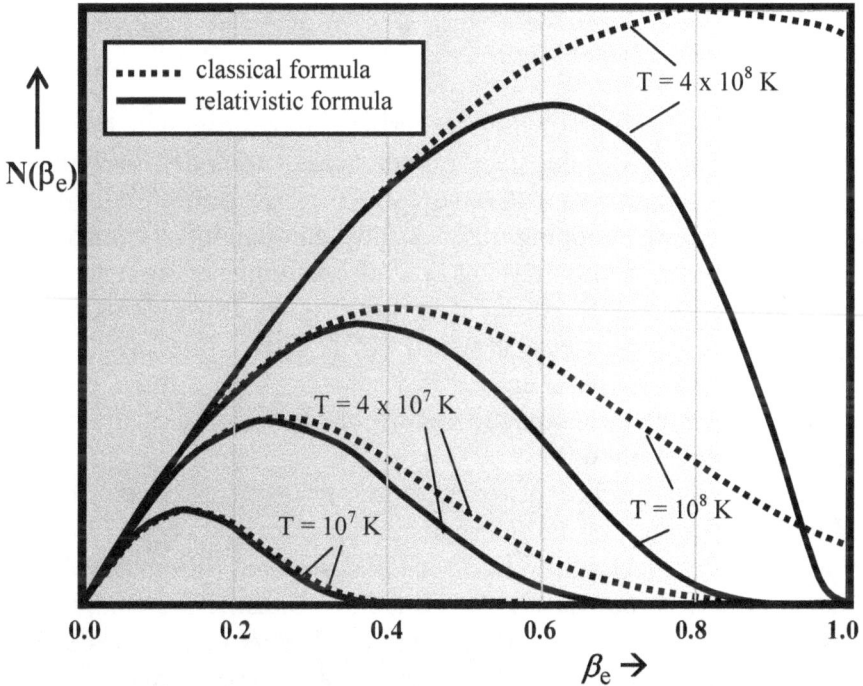

Figure 34: Non-normalized Maxwell-Boltzmann distribution of plasma electron velocities for various temperatures at equilibrium conditions

f. typical misconceptions concerning the properties of the intergalactic medium

A frequent misconception concerning the intergalactic medium is that *there's nothing out there*. Well there *is* something out there! A major part of the universe exists in that nebula-less domain between the objects that we actually observe only because of its unique properties.

But that is not the biggest surprise for most of us. It seems that virtually everyone (a consensus that amazingly is not limited to uninformed laymen) has the misguided impression that the intergalactic medium in the *present epoch* must be extremely cool. That, in particular, its temperature must correspond in some way to that of the uniform microwave background radiation that is less than 3 kelvins. Riess (1998), whose scholarship on other accounts seems quite reliable, states for example that,

"...there are huge contrasts between the stars with their blazing surfaces (and still hotter centers) and the sky between them, which is almost at the 'absolute zero' of temperature – not quite, of course, because it is warmed to 2.7 degrees by the microwave 'echoes' from the big bang."

This misconception obviously derives from promoted theoretical notions associated with the standard cosmological model for which the universe as a whole is said to have cooled from a primordial 'fireball' with temperatures dropping dramatically throughout its subsequent history. It is depicted as having cooled continuously from well over 10^{10} K at one second, to 10^9 K, after one minute. It is claimed to have reached 10^7 K at one week, down to 10^4 K after a brief ten thousand years, and finally to this mere 2.725 K that is inferred from one small segment of the electromagnetic spectrum we observe as microwave background radiation today.

The usual inference from these projections, i. e., that the ambient temperature of the intergalactic medium must be 2.725 kelvin is incorrect – not just a little in error mind you, but astronomically and most definitely astrophysically so. It is hot. Galaxies are comprised of stars – they are *hot*. The plasma within clusters of galaxies is even hotter – on the order of from 10^7 to 10^8 K, and in some case as high as 10^9 K. As we have seen, the temperature of the intergalactic medium that merges into this milieu must also be appreciable to account for observations. Those are all the major observed constituents of the universe other than a smattering of dust and black holes that radiate with spectacular effect only because of the matter falling into them. The universe is *not*, therefore, very *cool* in our epoch! It is hot – *very* hot. Period – or exclamation point as appropriate..

So to what does Riess refer as "the sky between" that "is almost at 'absolute zero'? A vacuum between the various charged particles that populate the vast regions between what we *see* in our universe? No; the 'temperature' of a vacuum has no meaning. Only by pre-empting observation by a questionable theory could the vast regions between galaxies be considered cold.

But since the deduction from presumptions has seemed valid to so many for so long, we should probably reconsider the extent to which premises on which they are based are themselves false. The conjecture that somehow a natural descent into a vast cold expanse that supposedly only seems to have been halted by the emergence of stars whose brilliance in turn re-heated the intervening medium does not in reality fit any of the data. The only justification for such claims is the existence of a 2.725 K blackbody radiation spectrum that is retrofitted for compatibility with that conjecture.

As we will show, it seems evident that it is radiation from scattered gamma ray bursts that exist throughout the depths of the observable universe whose extremely *high* temperatures produce on-going thermonuclear reactions that generate gamma rays, all of which when thermalized become the microwave background radiation. Such reactions not only produce the current light elemental abundances from primordial hydrogenous plasma, but also the background radiational energy as a residue. That radiation – that would not initially have been blackbody radiation, but once thermalized by having been scattered by matter – predominantly the intergalactic medium as we will show – eventually becomes blackbody microwave radiation. This is *not* by virtue of adiabatic expansion followed by a naïvely conceived recession of a 'wall of last scattering', but because initial and final energy densities must be equal in a universe for which the conservation of energy applies.

The higher electron and baryonic kinetic temperatures that contrast with the very much cooler radiation residue is an artifact of a thermalization process necessitated by a redshifting environment. This much higher average plasma kinetic temperature contributes to the much maligned and misinterpreted redshift, absorbing energy thereby lost by the radiation. All these features have been incorrectly inferred to indicate an expanding, currently cold, universe that seems to be oblivious to the conservation of energy. The more viable alternative is a universe that merely produces the *appearance* of such nonsense.

We seem in actuality to be living in the midst of on-going events that most theorists have attributed to the first few minutes after the big bang. So in contradistinction to Eric Lerner's thesis, that "the big bang never happened" (1991), this author believes that it *did* happen, or rather, that it *does*. The big bang is in essence happening *now* – all around us. It always has been and always will be. Plasma conditions like those that would have been pursuant to a big bang naively conceived as having happened at a single point in spacetime seem instead to be continuing prerequisites for the current universe as we know it. In deference to Lerner, this author is also of the opinion that such a singular instant never occurred. What is happening throughout the universe now is what may well always have happened and we have no scientifically legitimate reasons to presume that it will not continue indefinitely. However, the intergalactic medium accommodates our hardly having to notice it. Happily for now, we live in a very protected hollow sheltered from this on-going conflagration.

For all meaningful discussion the intergalactic medium is an endless *expanse* rather than a result of universal *expansion*. That it happens to effect *appearances* of a finite and expanding universe is, of course, a very interesting fact we will be pursuing.

Chapter 4

Interactions between Radiation and the Electronic Charges in Substances through Which It Propagates

Electromagnetic radiation throughout its broad spectrum is, of course, the product of interactions of electrically charged particles – in general one material entity interacting with another by way of electromagnetic transmissions. The form of radiations that are emitted and detected are in all cases described by Maxwell's equations of electrodynamics, which, despite the many subsequent conceptual reevaluations with quantum and relativity theories, still describe that behavior most effectively and very adequately in most cases.

We will not go into exhaustive explanation of electrodynamics here, although a more complete summary is provided as Appendix A for the interested reader. Readers are expected to appreciate, however, that all radiations by which we learn to know about distant regions of the cosmos are propagated in accordance with those laws. And furthermore, the scattering phenomena that are more specifically the subject matter of the current investigation must be explained in terms of electromagnetic fields associated with such material interactions.

a. the electromagnetic wave equations

The use of *plane polarized* waves as a basis for our discussions is merely an acknowledgement that the most general solutions of Maxwell's equations can be represented as linear combinations of these simpler forms. Appendix A discusses solution of Maxwell's equations for readers wanting more information and provides a ready reference for justifying related assumptions made in the body of this treatise.

As electromagnetic radiation propagates to greater and greater distances from a relatively localized source, its behavior can be approximated by that of a *plane wave* with ever increasing accuracy because limited surfaces on very large spheres are nearly planar. This usual assumption is justified to fairly high precision for radiation that propagates to a distance that is considerably greater than the average *extinction distance*.[*] The validity of results based on this assumption will later be shown to apply even for analyses of forward scattering in a very high temperature medium for which the assumption might not seem rigorously to apply for geometrical reasons to be discussed.

Even though not quite so easily explained with spherical wave fronts as will be seen, the forward scattering analyses can be shown to remain valid except as specifically noted. So, since plane wave representation is valid and easiest to explain and understand, we will discuss the incident electromagnetic vector fields originating in distant regions of the universe using this simplified functionality of plane polarized electromagnetic waves.

We will also assume the incident waves to be monochromatic, i. e., of a single wavelength. Again, this is to simplify considerations, but again it is completely valid for broader generalization since all photons of radiation can be represented as linear summations of just such monochromatic waves. The propagation of these – thus-simplified waveforms – along the direction of an arbitrarily chosen positive z axis may be specified using the following expressions:

$$\mathbf{E}(z,t) = \hat{i}\, E_0\, e^{\,i(\,\omega t - 2\pi\, \mathbf{n}\, z/\lambda\,)} \text{ and}$$

$$\mathbf{H}(z,t) = \hat{j}\, H_0\, e^{\,i(\,\omega t - 2\pi\, \mathbf{n}\, z/\lambda\,)},$$

where the field vectors $\mathbf{E}(z,t)$ and $\mathbf{H}(z,t)$ are assumed to be polarized such that the electric field $\mathbf{E}$ is aligned always with the x axis, whose unit vector is $\hat{i}$ and the magnetic field $\mathbf{H}$ always along the y axis, whose

---

[*] This is the distance after which the properties of the medium determine propagation characteristics. The concept of 'extinction' and the term 'extinction distance' will be explained in considerably more detail further on.

unit vector is $\hat{j}$. The angular frequency of the radiation is given by ω, its wavelength by λ. The index of refraction **n** possesses a 'real' value of unity for propagation through a vacuum implying that there is no 'imaginary' part to contribute to absorption in our initial treatment; its definition and functional dependence on properties of the medium will be discussed further on. The symbol e is the base of the natural logarithms; it is a number, $e \cong 2.718$. Of course i is the 'imaginary' square root of minus one. The scalar quantities E_0 and H_0 are assumed to be constant over considerable ranges of space and time for plane waves. Angular frequency and wavelength are related by the formula:

$$\omega = 2\pi\, c\, /\, \lambda$$

The implied coordinate directions and perspectives employed in analyses to be performed and the figures used to illustrate concepts described throughout this volume are as illustrated in figure 35.

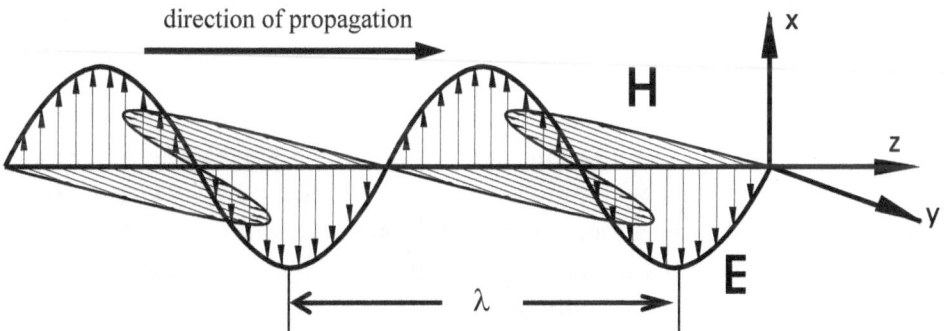

Figure 35: Field vector, propagation, and coordinate direction assumptions

b. energy contained in electromagnetic fields

The energy intensity, $E(z,t)$, of these electromagnetic fields is given by the following expression:

$$E(z,t) \;=\; 1/2\,[\; \mathbf{E}(z,t)^{*} \bullet \mathbf{E}(z,t) + \mathbf{H}(z,t)^{*} \bullet \mathbf{H}(z,t)\;]$$

where the symbol ($\bullet$) is the 'dot product' defined along with a discussion of this and other equations in Appendix A. The superscripted asterisk (*) implies *complex conjugation* (i. e., sign reversal of the *imaginary* components of these solutions to Maxwell's field equations), so that for monochromatic radiation propagating in a medium with a *real* index of refraction, we have:

$$E(z,t) = 1/2 \, (E^2{}_0 + H^2{}_0)$$

Actually, of course, all electromagnetic radiations (including what are sometimes considered to be *monochromatic* photons) possess energy that is spread throughout a range of wavelengths of the electromagnetic spectrum as mentioned above. So that in general there is a required integration of monochromatic fields over all included wavelengths in order to assess the net field intensities:

$$\mathbf{E}(z,t) = \int_0^\infty \mathbf{E}(z,t,\kappa) \, d\kappa$$

and

$$\mathbf{H}(z,t) = \int_0^\infty \mathbf{H}(z,t,\kappa) \, d\kappa$$

where $\kappa \equiv 2\pi / \lambda$ is known as the *wave number*.

The monochromatic field wave functions $\mathbf{E}(z,t,\kappa)$ and $\mathbf{H}(z,t,\kappa)$ can be represented by complex vector quantities like those shown earlier. The representation of the total associated *energy* contained in the component fields of the radiation is given by the integral over all the monochromatic wave functions:

$$E(z,t) = 1/2 \int_0^\infty \{ \, \mathbf{E}(z,t,\kappa)^* \bullet \mathbf{E}(z,t,\kappa) + \mathbf{H}(z,t,\kappa)^* \bullet \mathbf{H}(z,t,\kappa) \, \} \, d\kappa$$

c. luminous flux

A typical astronomical observable that is directly measurable is the quantity of energy contained in observed electromagnetic radiation. It is called *luminous flux*, which is the amount of energy passing through a unit area in a unit of time as follows:

$$f(z) = \int_0^1 E(z,t) \, dt$$

This is the quantity that is typically *the* observable of record in astronomical observations of *standard candles* that must be taken into account in comparing predictions of any cosmological model, whether *the scattering model* to be described here, versions of the *standard model*, or any other conceivable model that attempts to explain cosmological phenomena.

There are many terms associated with various quantitative measures of different aspects of radiant energy. *Luminosity* is a quantity, which we will usually represent by L. We will use the term rather loosely on occasion as the rate at which a star or galaxy, for example, radiates energy in all directions. Sometimes we will state this as *inherent* or *absolute* luminosity, using L_0. This refers to the innate ability of a source at given distance from an observer to generate a measured luminous flux. In classical physics luminous flux was considered to be exclusively proportional to the inherent luminosity of the object and the inverse square of distance, r, as follows:

$$f(r) = \Omega \, L_0 \, / \, 4 \, \pi \, r^2$$

The constant of proportionality, $\Omega / 4 \, \pi$, is determined by the amount of solid angle subtended by the aperture of one's viewing instrument, Ω. However, when measuring luminosity characteristics of cosmological objects at a distance, r with a redshift, Z, the applicable formulas differ considerably from that classical anticipation as follows:

$$f(Z) = \Omega \, L_0 \, / \, 4 \, \pi \, [\, 1 + Z \,]^2 \, r(\, 1+ Z \,)^2$$

The functionality of the distance being expressed as a function of redshift involves the fact that in cosmological investigations distance can not be measured without sometimes making obtuse inferences with regard to a distance-redshift relation. Hubble's initial estimate was r ~ Z, with the constant of proportionality denominated *Hubble's constant* using the symbol H_0. We will discuss various more complex and more accurate formulations of this relation in later chapters.

To understand the previous formula, one must consider all that is involved in observing electromagnetic radiation from a distant object. The rationale for each is described in the following layout of the factors:

factors involved in *luminous flux, f(Z)*

total number of photons emitted by the object per second	×	energy of each emitted photon	×	proportion of emitted photons subtended by aperture	×	reduction in energy after redshifting each photon	×	proportion of total emitted photons that make it to the observer per second

$$\underbrace{\text{Rate} \times h\,c\,/\,\lambda_e}_{L_o} \times \Omega\,/\,4\,\pi\,r^2 \times \frac{1}{(1+Z)} \times \frac{1}{(1+Z)}$$

There are separate rationales for the final two factors of $1 / (1+Z)$, as indicated. The first of these involves the definition of redshift and the quantum theory of light. By definition redshift alters the wavelength of light such that the observed light will be redder (for optical radiation), i. e., have a longer observed wavelength, λ_o than the light that was emitted, λ_e from an object at redshift Z in accordance with the formula:

$Z \equiv (\lambda_o - \lambda_e) / \lambda_e$, such that,

$\lambda_o = (Z + 1)\,\lambda_e$.

In quantum theory luminous flux is partitioned into numbers of individual packets of 'quantized' radiant energy called 'photons' that are emitted and absorbed as the basic units of radiation in its interactions with matter. For each such photon, its energy is:

$E = h\,c\,/\,\lambda$

where Planck's constant, $h = 6.626 \times 10^{-27}$ erg seconds. So clearly the luminous flux will exhibit this same inverse proportionality with wavelength as does each photon. So converting this into the observed wavelength relative to the emitted wavelength, observed flux will be reduced by $1 / (1+Z)$ to reflect that each photon will carry that much less energy when it is observed than when it was originally emitted. This fact is independent of *why* or *how* the photon was redshifted. Rationale for the trailing factor is more complicated inasmuch as the same quantitative value has two explanations depending on *why* and *how* photons are redshifted. It doesn't involve the resultant energy of each photon, but rather how many there are. First let us understand what this factor involves in the standard model. In that model it stands quite simply for a commensurable reduction in the number of photons received per second by the observer relative to the number emitted by the source; it is associated with *time dilation*. This situation arises because that model embraces an expansion of the universe that *causes* the redshift. Ramifications of this explanation involve distant objects receding from the observer at a rate proportional to their distance such that the total number of photons emitted per second will be reduced to a slower rate at observation, where again this factor will be $1 / (1+z)$.

In the scattering model elaborated in this volume, there is a broadband absorption that in effect absorbs photons independently of their wavelengths with an absorption coefficient that reduces Lambert's exponential law of absorption to the factor involving $1 / (1+z)$. This effect will be derived and discussed in the next chapter.

Surface brightness is a measure of the total luminous flux averaged over the total observed solid angle subtended by an object.

We will use these metrics in testing predictions of the various cosmological models.

d. forces imposed on individual material charges

The effect of electromagnetic radiation on material substances is to force an acceleration of individual constituent charges of which the substance is comprised. The same basic phenomena occur whether the constituent charges are more or less loosely bound within atoms, molecules, crystalline structures, or are 'free' as in a diffuse ionic plasma. The acceleration of charges in a scattering medium results in either absorption of the incident radiation or emanations of 'secondary' radiation originating at the individual constituent charges associated with what is known as 'forward scattering'. Both of these phenomena are responsible for effects that have been perceived as 'cosmological'.

We will derive the primary dispersion formulas applicable to diffuse plasmas and the isolated neutral hydrogen clouds encountered in the intergalactic medium as a convenient reference for subsequent chapters for which the results will be essential. In the next couple of chapters we will discuss absorption. Then we will move on to apply the dispersion formulas to explore the effects of forward scattering in subsequent chapters. Finally, the applicability of these formulas will be re-examined to determine the impact of high temperature electrons for which relativistic effects are encountered in intergalactic plasma

Incident electromagnetic radiation will cause the individual electronic charges to experience a Lorentz force that will produce displacements of the charges. This force is given by:

$$\mathbf{F}_e = e \left(\mathbf{E}_{effective} + (\mathbf{v}_e / c) \times \mathbf{H}_{effective} \right)$$

where $e = 4.8 \times 10^{-10}$ statcoulombs (in cgs units that we will use throughout) is the individual electronic charge, $\mathbf{v}_e$ is the electron's velocity with respect to an observer of the scattered radiation and the vector operator symbol '$\times$' represents the cross product.[*] The effective electric field is related to the incident electric field $\mathbf{E}_i$ as follows:

$$\mathbf{E}_{effective} = \mathbf{E}_i + (4\pi/3) \, \mathbf{P}$$

The presence of the polarization, $\mathbf{P}$, is required because the force realized on a charge at any point in the medium is a combination of that due to an incident radiation field and *induced* polarization fields that oppose it. Refer to figure 36 where this induction process is illustrated.

The induced field is produced in surrounding regions of the medium because of an induced polarization of charges in the medium – positive charges migrate in one direction, negative ones in the other. $\mathbf{P}$ is known as the polarization of the medium. The contribution of the polarization from any *normally* appreciable region of space filled only by diffuse intergalactic plasma would be small of course, but in the vastness of intergalactic space it will nonetheless become extremely significant in an assessment of the associated optical properties.

We treat only the electric field here both with respect to induced polarization and in total; the effects of the magnetic fields could be treated analogously, but they can be shown to have no unique significance for the intergalactic medium in regions where there is no appreciable static magnetic field. We also consider only displacement of the negatively charged electrons and not that of the positive ions. The much greater mass of the positive nuclear ions present in the plasma (even though they are primarily single protons) will preclude

---

[*] This and other details of electromagnetic theory are discussed in Appendix A.

their accelerations becoming appreciable relative to those of much less massive electrons at frequencies with which we will be concerned..

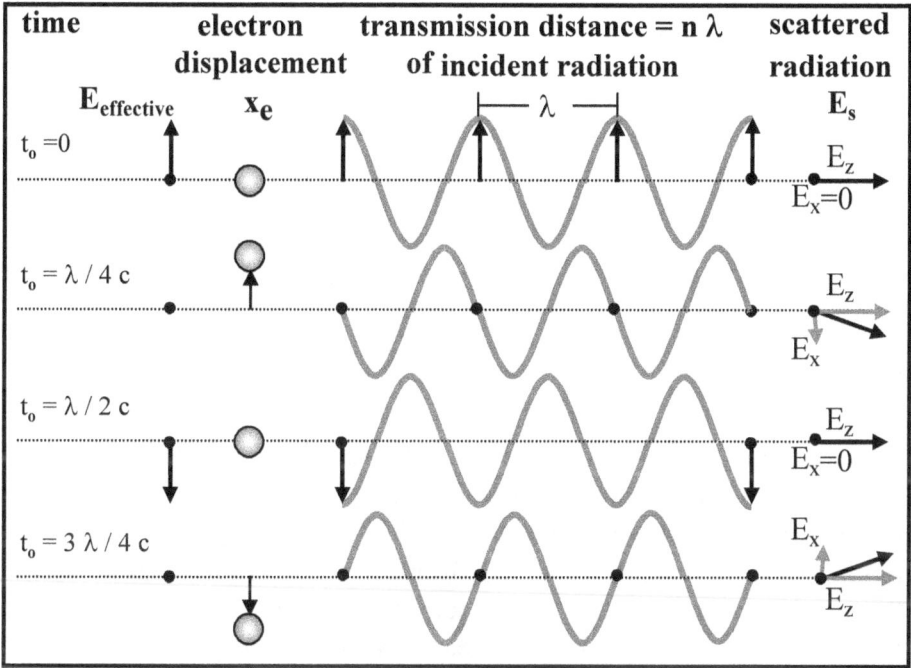

Figure 36: The creation of a scattered field by displacement of electrons

Each electron will experience, in addition to the Lorentz force, restoring and damping forces representing the medium's ability to restore the locally altered equilibrium and its tendency to dissipate energy. The resulting displacements will be along the x-axis under the assumptions of isotropy and homogeneity of the medium for the example incident monochromatic radiation shown in figure 35. The formulation of these forces on a single electron is represented by the following differential equation of classical electrodynamics:

$$m_e \, d^2x_e/dt^2 = e \, E_{effective} - k \, x - g \, dx_e/dt,$$

where the second and third terms on the right correspond to a proportional restoring force and a velocity-dependent damping force associated, respectively, with the harmonic restoring force on charges about a stable neutral position and an absorptive effect typically associated with collision probabilities while the incident wave is interacting with the electron, effecting joule heating. These coefficients will determine the magnitude of an inherent resonant frequency of the medium, and the absorption properties to be expected of the medium. The situation is considerably simplified by addressing merely a single resonant frequency; this simplification is appropriate for a fully ionized homogeneous medium in Maxwellian equilibrium.

Solution of the previous differential equation results in the displacement formula:

$$\mathbf{x_e} = (e/m_e)\, \mathbf{E_{effective}} / (\omega_o^2 - \omega^2 + i\,\gamma\,\omega\,), \text{ where}$$

$$\mathbf{E_{effective}} = E_o\, e^{\,(i\,\omega_o t - \frac{1}{2}\gamma t)}$$

$\mathbf{x_e}$ is a displacement that is in direct response to the incident radiation.

Figure 36 illustrates the mechanism whereby the *effective* field is augmented by these displacements caused by the incident electric field. At some distance removed from the electron that is affected by the incident radiation field, the electronic charge produces an electric field directed away from the charge. There will be a component of this field E_s that is along the direction of the incident radiation and a component at right angles to it as shown. The component that is along the direction of propagation will diminish with the distance from the electron according to the usual coulomb inverse square law. However, the transverse component of the field will diminish only as the inverse first power of distance. Thus, after any very appreciable distance E_s will itself comprise transverse radiation that is one quarter wavelength behind that of the incident electromagnetic wave.

As we will see in a later chapter, this scattering radiation from individual electrons in the medium will accumulate by coherent constructive reinforcement until eventually their combined effect is to nullify and replace the incident radiation altogether.

Both ω_o and γ will play key roles in interpreting observations through the intergalactic medium. From the above equations, we obtain the following:

$$\omega_o^2 = k/m_e - g^2/4\,m_e^2, \text{ and } \gamma = g/m_e\,, \text{ such that,}$$

$$\gamma^2 / 4 - (k / m_e) \gamma + \omega_0^2 = 0$$

This relation between k, γ, and ω_0 is shown in figure 37. It expresses relationships between the force constants and derived parameters that determine scattering behavior. Specifics of the scattering process determine these important parameters. With regard to the intergalactic medium, estimating γ based on knowledge of ω_0 will be of interest in a couple of instances in particular.

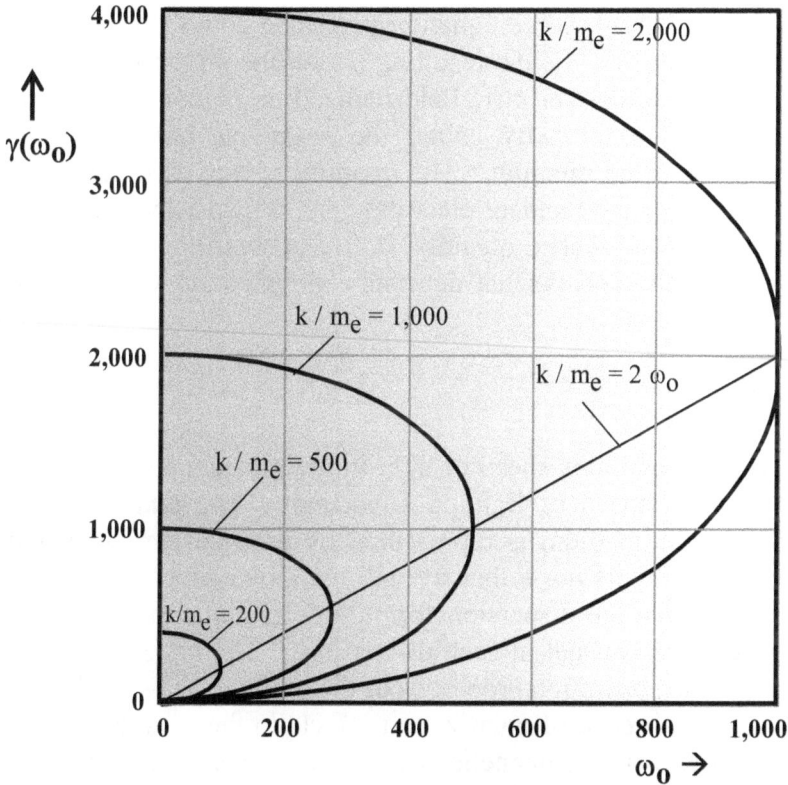

Figure 37: Relationship between resonant frequency and absorption constants

e. dispersion formulas

The induced *dipole moment* illustrated in figure 36 is a reaction to the incident force; it is also oriented along the x axis in an isotropic medium as appropriate to the intergalactic medium for wave functions like the one shown in figure 35. The reaction force on a single electron resulting from this displacement is given by:

$$\mathbf{P_e} = e\,\mathbf{x_e} = (e^2/m_e)\,\mathbf{E_{effective}} \;/\;(\omega_o{}^2 - \omega^2 + i\,\gamma\,\omega\,)$$

where P_e is the contribution of a single electron to overall polarization, **P**, of the entire medium at a given point in space.

The effect on the incident radiation of the interaction with the electronic charges in a material substance is that overall transmission characteristics may be significantly altered by introducing refraction, diffraction, and absorption. This dispersion is also associated with a wavelength dependent propagation velocity of the on-going radiation.

In the Lorentz force equation presented above, the introduction of scattered electric fields resulting from the induced polarization effects on the incident electric field realized throughout the medium is associated with a property called the *refractive index* or *index of refraction*, **n** of the medium. The quantity **n** was used earlier in the representation of the incident electromagnetic field wave functions. It is in general a complex quantity (i. e., possessing both 'real' and 'imaginary' components) that determine propagation velocities in the medium as follows:

$$\mathbf{n} = c\,/\,\mathbf{b},$$

where **b** is the complex speed of light in the medium and c is again the universal 'real' speed of light in a vacuum. The *phase velocity* of radiation in the medium is represented by the real part of **b** and the *group velocity* by its *norm* that we will mention again later. However, we will continue the dispersion formula derivations in reference to **n** rather than **b** as is usual in such discussions. The imaginary part of **n** determines the absorption properties of the medium as we will see.

In the absence of polarization effects in the medium, we would have had from electromagnetic theory that the induced electric field (the *electric induction*) in the medium, $\mathbf{D_r}$, would be determined as:

$$\mathbf{D_r} = \mathbf{E_i}$$

However, in a polarized or polarizable substance, this formula becomes:

$$\mathbf{D_r} = \epsilon\,\mathbf{E_i}$$

where ϵ is the *dielectric constant* of the medium, defined such that:

$$\epsilon \equiv n^2$$

For an isotropic medium ϵ is a scalar so that it can be assumed that the vector quantities $\mathbf{E}_{effective}$, $\mathbf{P}$ and $\mathbf{E}_i$, all employed earlier, as well as $\mathbf{D}_r$ are all aligned along the same direction. This will always be the case in an isotropic medium so that utilization of scalar ratios is meaningful. There is no reason to believe the intergalactic medium is not isotropic in this same sense over any appreciable interval. In this case:

$$\mathbf{D}_r = \mathbf{E}_i + 4\pi\, \mathbf{P},$$

as we have seen, and since vectors are aligned, we have a scalar result:

$$\epsilon = 1 + 4\pi\, P/E_i$$

Therefore, we have that:

$$\mathbf{E}_{effective} = (\, n^2 + 2\,)\, \mathbf{E}_i\, /\, 3$$

The presence of the factor of 1/3 in the effective electric field equation presented above is known as the *Lorentz-Lorenz correction*; the justification for this factor is somewhat involved and it will just be assumed in the present discussion. See for example Jackson (1962), Ditchburn (1963) or any other detailed text on electrodynamics. And the polarization field will be given by:

$$\mathbf{P} = (\, n^2 - 1\,)\, \mathbf{E}_i\, /\, 4\pi$$

The polarization field, $\mathbf{P}$, required in the determination of the effective electric field realized at points within the medium is defined as the dipole moment per unit volume as follows:

$$\mathbf{P} \equiv \rho_e\, \mathbf{P}_e$$

where ρ_e is electron density as discussed in chapter 3. By substitution for $\mathbf{P_e}$ from the equation derived above, we obtain the formula:

$$\mathbf{P} = \rho_e \, (e^2/m_e) \, \mathbf{E}_{\text{effective}} \, / \, (\omega_0{}^2 - \omega^2 + i\,\gamma\,\omega)$$

f. the Lorentz-Lorenz formula

By further substitution of the derived values we obtained above for $\mathbf{P}$ and $\mathbf{E}_{\text{effective}}$ into the preceding formula, we obtain a formula for the index of refraction as follows:

$$(\,n^2 - 1\,)\,/\,(\,n^2 + 2\,) = (4\pi/3)\,\rho_e\,(e^2/m_e)\,/\,(\omega_0{}^2 - \omega^2 + i\,\gamma\,\omega)$$

This is the *Lorentz-Lorenz formula* for the index of refraction applicable to media generally (including the intergalactic medium, an isolated neutral hydrogen cloud, or any of a broad class) exhibiting a single predominant resonant frequency. For cases where the index of refraction is very close to unity as is certainly the case for the intergalactic medium, the left-hand side of the previous equation becomes approximately $2\,(\,n - 1\,)\,/\,3$. This in turn results in the usual determination that is valid to a very high degree of accuracy, for which the *dielectric susceptibility*, $(\mathbf{n} -1)$, is assessed as follows:

$$\mathbf{n} - 1 \approx 2\pi \, (\,\rho_e\,e^2/m_e)\,/\,(\omega_0{}^2 - \omega^2 + i\,\gamma\,\omega)$$

The index of refraction, as we have said, is complex and can be expressed in the form:

$$\mathbf{n} \;=\; Re(\mathbf{n}) \,+\, i\;Im(\mathbf{n})$$

where $Re(\mathbf{x})$ and $Im(\mathbf{x})$ refer respectively to exclusively real and imaginary components of a complex argument, $\mathbf{x}$. From the formula for the dielectric susceptibility above, we obtain:

$$Re(\mathbf{n}) = 1 + 2\pi\,(\,\rho_e\,e^2/m_e)\,(\omega_0{}^2 - \omega^2)\,/\,((\omega_0{}^2 - \omega^2)^2 + (\gamma\,\omega)^2)$$

$$Im(\mathbf{n}) \;=\; -\,2\pi\,(\,\rho_e\,e^2/m_e)\,\gamma\,\omega\,/\,((\omega_0{}^2 - \omega^2)^2 + (\gamma\,\omega)^2)$$

78

These functions are plotted in figure 38 for nominal parameter values of $\omega_0 = \gamma = 87$, which are what one might naively expect for the intergalactic plasma. The forms of these two curves are quite general, but not without quite extreme domain peculiarities. For examples of the dependence on γ, for example, see figure 39.

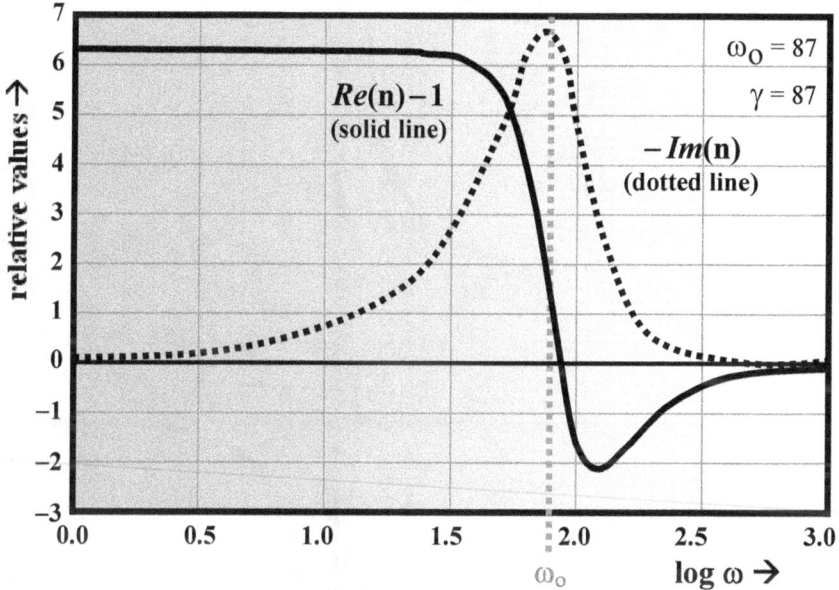

Figure 38: Real and imaginary components of complex index of refraction

In exploring the ramifications of these formulas, it is clear that behavior will be markedly different if the frequency ω of the radiation that is propagated through the medium is higher or lower than the resonant frequency, ω_0 that is a characteristic of the medium itself.

The magnitude and functional form of the index of refraction curves depend not only upon the relative values of ω and ω_0 as shown in figure 38, but intimately upon the relationships between and γ, ω, and ω_0 as well. These dependencies, particularly for the imaginary portion, have a profound effect on the luminosity of objects observed through the medium. As we will see in a couple of later chapters, it is the imaginary part of this complex index of refraction that affects the amount and nature of the absorption encountered by electromagnetic radiation propagated through such a medium. Counter-intuitively, the larger the value of γ becomes, the more nearly the medium will behave like a complete vacuum for which $Re(n) = 1.0$ and $Im(n) = 0.0$. This is readily apparent in figure 39.

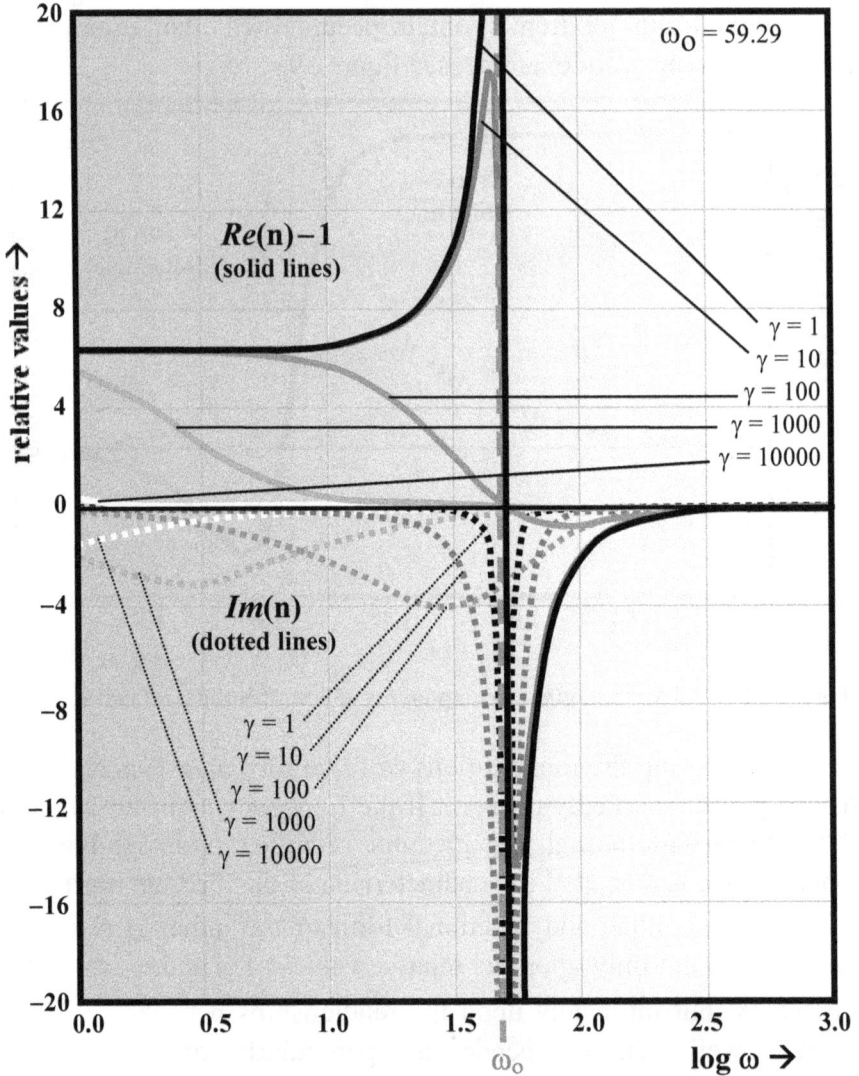

Figure 39: Range of variation in complex index of refraction with changing γ

As illustrated in figure 40, there are nine possible domains for which dispersion behavior could be basically different. Clearly, some of these alternatives are of much less consequence than are others. In particular in the vicinity of $\omega \cong \omega_o$, if γ is much greater than ω and ω_o, both functions shown in figure 39 are nearly zero. Furthermore, this quiescent neighborhood can be quite extensive in cases where $\gamma \gg \omega_0$ for an extremely large range of values of the incident radiation frequency, ω as shown in figure 41. If, for example, γ is two orders of magnitude greater than ω_0, then there will be a neighborhood four orders of magnitude on either side of the resonance frequency in which,

$$Im(\mathbf{n}) \propto \omega \propto 1 / \lambda$$

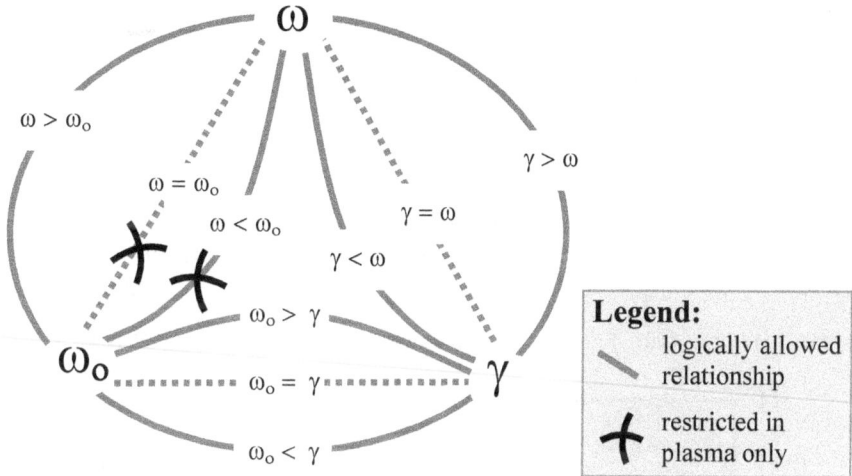

Figure 40: The allowed dispersion domains involving ω, ω_0, and γ

This non-dispersive feature of absorption applicable to plasma is one of the keys to an understanding of how the Hubble relation of the luminosity of objects at cosmological distances is accounted for in the scattering model described in this volume.

However, except for this region in the vicinity of ω_0 (whether vast or small depending on the value of γ) as described above and shown in figures 41a and 41b, the imaginary component of the index of refraction, and therefore absorption, will be extremely wavelength dependent. This factor will either be proportional to λ^2 (i. e., $1/\omega^2$) for $\omega > \omega_0$ or proportional to the inverse of λ (i. e., to ω) if $\omega < \omega_0$.

g. composite Lorentz-Lorenz formulas

Of course there is more than a single physical interaction taking place between electromagnetic radiation and the material entities in the

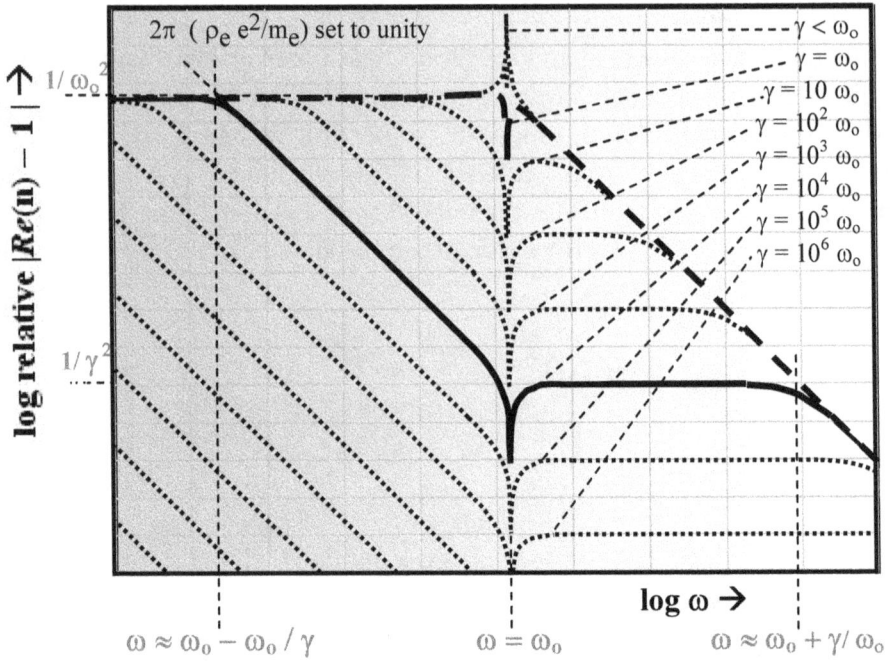

Figure 41a: The various parameter domains affecting $|Re(n) - 1|$

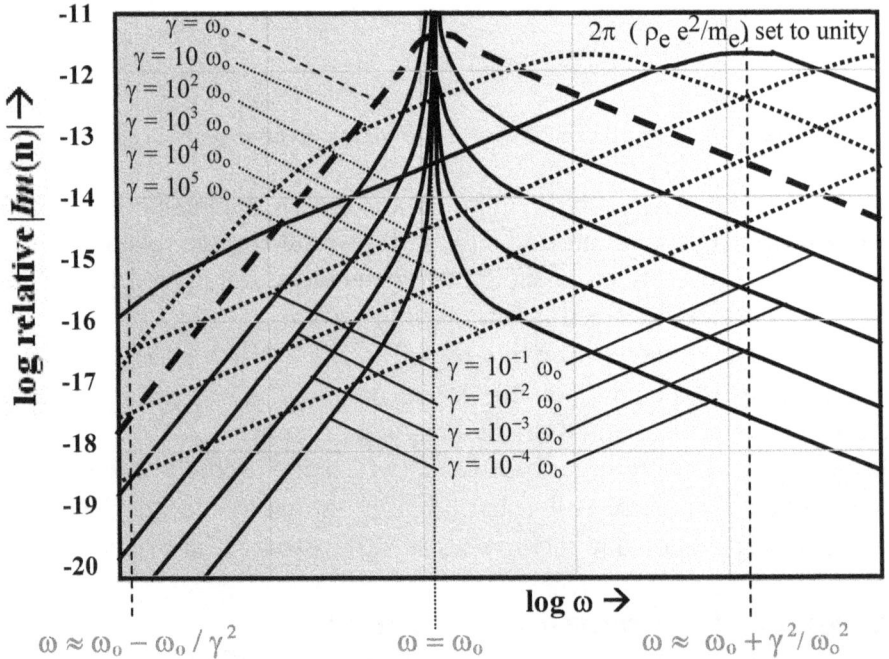

Figure 41b: The various parameter domains affecting $Im(n)$

82

intergalactic or most any other medium. For instance, each unique form of matter will interact uniquely, but only as differently as their functional interactions differ. Ionized plasma electrons that are relatively free of encumbrances exhibit one form of interaction in the vicinity of the plasma resonance; electrons bound in neutral atomic units exhibit quite another, with the uniqueness of the binding also having a major impact. These two noteworthy mechanisms operative in the hydrogenous plasma and lowest energy level bonding of the hydrogen atom will be involved in the description of this aspect of the intergalactic medium.

There are multiple differential equations like the one we solved above from which the electron displacement was determined. There will be one to account for each of the various forms of interaction forces. Each, otherwise identical, equation will be characterized by unique coefficients that give rise to their own resonant frequencies and/or absorption profiles. The displacement effects from the radiation passing through the medium will be additive and, therefore, the resulting scattered radiation will constitute a superpositioning of scattered fields. Naturally radiation of a frequency close to one of the resonances will be most affected by that interaction.

To the same valid approximation, the dielectric susceptibility formula can be replaced by the following summation:

$$\mathbf{n} - 1 \approx 2\pi \ (\rho_e \ e^2/m_e) \sum_s f_s \ / \ (\omega_s^2 - \omega^2 + i \, \gamma_s \, \omega)$$

The real and imaginary parts $Re_s(\mathbf{n})$ and $Im_s(\mathbf{n})$ of this result of the force equations are also, therefore, directly additive. So absorption in a frequency domain is the result of the sum of the absorption coefficients in the exponent. Therefore the luminous flux diminution caused by all of the various interactions will be the product of what would be caused by each. Similarly the real part of the index of refraction affecting the phase velocity of propagation will be a composite effect.

In a homogeneous medium, typically one would have:

$$\sum_s f_s - 1$$

Where there is one predominant process, the situation would be characterized by,

$$f_1 = 1 - \sum_{s\neq1} f_s \cong 1$$

But, in addition to differences in magnitude of f_s, differences in the individual parameters ω_s and γ_s of the various mechanisms may also substantially contribute to the observed behavior. These parameters may combine such that the real part of the net dielectric susceptibility may pertain almost exclusively to one mechanism taking place in the medium and the imaginary part result from quite another. Propagation characteristics would then be as appropriate to one mechanism and absorption characteristics appropriate to another. This is pertinent to the intergalactic medium in particular.

h. application of formulas to the intergalactic plasma

In the case of an hydrogenous plasma medium, the domain of applicability of the formula to the propagation of electromagnetic radiation is extremely to the high frequency side of ω_0, placing this useful domain far to the right of the right-most value ($\omega = 10^3$ Hertz) shown in figures 38 and 39. And since there is virtually complete absorption of radiation below this frequency, we can assume that $\gamma = 0$.

At the frequency ω_0 there is a Jell-O-like resonance of the medium as a whole in which the transverse wave motion is converted to a longitudinal (sound-like) action. Thus, electromagnetic radiation as such can only propagate through plasma at frequencies greater than this *radial plasma frequency*, ω_0. However, even 100 Megahertz radio signals correspond to values of ω that are on the order of 10^9 Hertz – many orders of magnitude greater than ω_0. The values of $\omega_0 = 87$ and 59.29 employed respectively in figures 38 and 39 are more or less directly applicable to intergalactic plasma although the value of γ is no doubt over-estimated.

At what is known as the 'plasma frequency', $\omega_p = \omega_0$, all of the transverse electromagnetic radiation is converted into longitudinal oscillations of the charges within the medium. This 'radial plasma frequency', as it is sometimes denominated in this context, is determined by the electron density as follows:

$$\omega_p = (\rho_e e^2/m_e)^{\frac{1}{2}} \approx 1.59 \times 10^4 \, \rho_e^{\frac{1}{2}}$$

84

This arrangement of basic parameters appeared independently as a leading factor in the dielectric susceptibility. Substitution from chapter 3 above, applicable to the intergalactic medium yields the following approximate restriction on the values of this resonant frequency:

$$2.0 < \omega_0 < 2.0 \times 10^3 \text{ Hz}$$

So that clearly for optical wavelengths (and for even the longest radio wavelengths currently employed in astronomical observations) there will be no observable spectral line attenuation attributable to this particular resonant frequency of intergalactic plasma. For optical wavelengths ($\lambda \approx 5000$ Angstroms $= 5 \times 10^{-5}$ cm) we have the relationship,

$$\omega^2 \approx 1.6 \times 10^{31} \gg 4.0 \times 10^4 > \omega_p^2$$

However, in later chapters we will discover the mechanism operative in hot plasmas whereby scattering by relativistic electrons produces a lengthening of wavelength of all electromagnetic radiation propagated through it. In the scattering model presented in this volume, a reduction of radiant energy via redshifting results from an angular deflection that, although compensated at each scattering event, is associated with an equivalent amount of energy being removed from the radiation. This 'removal' of energy from radiation is, in effect, an absorption by the plasma electrons involved in the scattering. At each interaction in the forward scattering process, we will find that the wavelength is increased by $\Delta\lambda_e \cong 6.12 \times 10^{-20}\ T_e$. This is accomplished by a Compton-like transfer mechanism that enforces the conservation of energy and momentum. This mechanism can be described similarly to explanations of other transfer mechanisms, i. e., using values g and therefore γ_C unique to this process.

Then by summing the contributions to the plasma dielectric susceptibility as discussed in the previous section, we obtain,

$$Re(\mathbf{n}) - 1 = 2\pi\ (\ \rho_e\ e^2/m_e)\ [\ f_0\ (\omega_p^2 - \omega^2\ )\ /\ ((\omega_p^2 - \omega^2\ )^2 + (\ \gamma_0\ \omega\ )^2\ )$$

$$+ f_C\ (\omega_C^2 - \omega^2\ )\ /\ ((\omega_C^2 - \omega^2\ )^2 + (\ \gamma_C\ \omega\ )^2\ )\ ]$$

Keeping in mind that $\gamma_C \gg \gamma_o$, we investigate the case where $\gamma_C \gg \omega_C$ is also the case. The final term in the bracketed factor will always be negligible in comparison with the first, so we can certainly ignore it throughout the extensive range for which $\omega_C \ll \omega$.

On the other hand, with regard to the imaginary component, a very different situation arises, where we obtain:

$$Im(\mathbf{n}) \; = \; -2\pi \, (\, \rho_e \, e^2/m_e) \, [f_o \, \gamma_o \, \omega \, / \, ((\omega_p{}^2 - \omega^2 \,)^2 + (\, \gamma_o \, \omega \,)^2 \,)$$

$$+ \, f_C \, \gamma_C \, \omega \, / \, ((\omega_C{}^2 - \omega^2 \,)^2 + (\, \gamma_C \, \omega \,)^2 \,) \,]$$

In this case, where we again assume that $\gamma_C \gg \gamma_o$, the second term will take on primary significance. First of all, the assumption is warranted because γ_o must be very nearly zero in order to effect the virtually complete absorption by the plasma resonance mechanism. Sso let us look at the range of possibilities for the frequency dependence of this second term. In figure 42 we illustrate the range of possibilities for the relationships between and among the parameters pertinent to absorption, γ_C, ω_C, and ω:

Since the mechanism of wavelength lengthening to be described later in this volume is unilateral for all wavelengths, therefore, the energy absorbed as an integral aspect of this mechanism must be extremely broadband throughout an extremely extensive region of the electromagnetic spectrum. As shown in figure 42, this broadband feature is accommodated when $\gamma_C \gg \omega_C$. There is no rationale for ω_C having a value other than ω_p, which is extremely small as we have seen. This form of plasma-determined absorption will, therefore, remain exclusively dependent on the second term in the bracketed factor of $Im(\mathbf{n})$ above, as long as we also have that $\gamma_C \gg \omega$.

Therefore, throughout a broad range of the electromagnetic spectrum we can characterize the index of refraction of intergalactic plasma as follows:

$$|Re(\mathbf{n})| \; = 1 + 2\pi \, (\, \rho_e \, e^2/m_e) \, / \, \omega^2$$

$$|Im(\mathbf{n})| \; = 2\pi \, (\, \rho_e \, e^2/m_e) \, / \, \gamma_C \, \omega$$

86

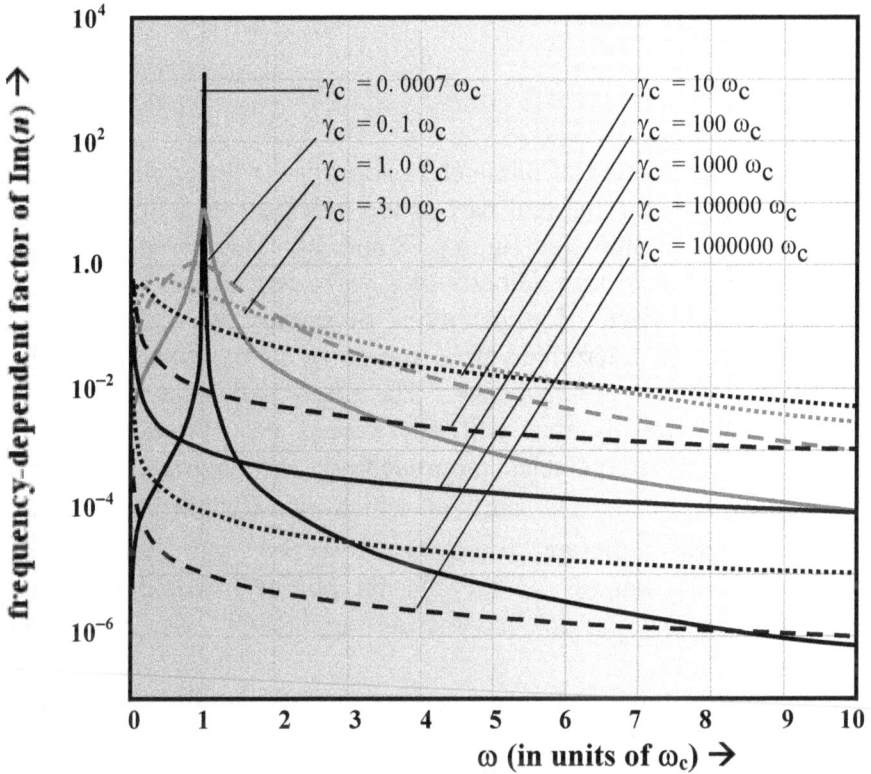

Figure 42: The emerging form of the plasma absorption factor *Im*(n)

Since neither f_C nor γ_C appear in any other context, we have defined γ_C to include whatever impact f_C might have had in its own right. In an attempt to obtain a value for γ_C, we note that based on later results the observed redshifting effect and energy balance equations will be maintained if (on average) the following holds:

$$\gamma_C \sim 2\pi \, c \, / \, \Delta\lambda_e \cong 3.08 \times 10^{30} / \, T_e$$

Therefore, for electromagnetic radiations with which we will be interested, the plasma parameters in the Lorentz-Lorenz dispersion formulas assure that the index of refraction of the intergalactic medium will be approximately $n = 1.0$, which almost identical to what would be appropriate to a vacuum. Only at what are considered 'cosmological' distances will its effects be otherwise observed. For visible light, component values are on the order of:

$$|Re(\mathbf{n}) - 1| \approx 10^{-38}$$

$$|Im(\mathbf{n})| \approx 10^{-37}$$

For the domain of parameters appropriate for application to the intergalactic plasma, the real part of the index of refraction is actually less than unity as shown in figures 38 and 39. This corresponds to the *phase* (as against *group*) velocity of a wave that exceeds the speed c. Of course the speed of transference of energy via electromagnetic radiation will be determined from the *norm* of the *complex* index of refraction, not just the *real* part. The norm will always produce an actual group velocity that will not exceed c. However, only for extremely low frequencies approaching the *plasma wavelength* ($\lambda_0 \equiv 2\pi c / \omega_0 \sim 2 \times 10^9$ cm), at which wavelength the transverse electromagnetic radiating energy is absorbed and converted into longitudinal oscillations of the plasma charges, would one expect *anomalous dispersion* to become an appreciably observable local phenomenon.

As illustrated in figure 41, there are two families of curves for the two distinct domains on opposite sides of $\omega = \omega_0$. Certainly $\gamma_C \gg \omega_0$ recommends itself as what must pertain to produce the observed broadband behavior. On the other hand, the fact that there must be complete absorption at ω_0 suggests that $\gamma_0 \ll \omega_0$ must pertain. Accepting this criterion with $\gamma_0 \sim 0$ accommodates the maximal absorption at $\omega \le \omega_0$ as well as accommodating the minimal absorption for all shorter wavelengths. Thus there are two concurrent operative processes in the plasma.

What is at issue in integrating these very different alternatives?

Values of γ are typically experimentally determined parameters that involve intricacies of the mechanisms. If the primary operative phenomenon was plasma resonance, clearly the steep functionality of $Im(\mathbf{n}) \propto 1 / \omega^3$ derivative to $\gamma < \omega < \omega_0$ would precipitate no absorption of radiation at shorter wavelengths.

However, in addition to the plasma being a *plasma* with its own unique resonance, there is a very different additional dispersion process that takes place in a 'hot' plasma, the exploration and ramifications of which are what many of the succeeding chapters of this volume will be about. There is a unique forward scattering process applicable to

88

extremely high temperature plasma that results in the redshifting of any radiation that propagates through it. The stealing of energy that is immeasurable in any single scattering interaction is nonetheless an ineluctable energy siphoning mechanism whose effects are magnified by coherent forward scattering.

But first we will discuss details of the alternative expected refraction and absorption profiles for a plasma.

i. application of formulas to neutral hydrogen 'clouds'

Throughout the intergalactic plasma there are, in addition to scattered field galaxies and clusters of galaxies of various shapes and sizes with their denser plasma, isolated islands of neutral hydrogen and halos of similar content surrounding foreground galaxies. These latter, familiarly called "Lyman alpha forests", do not obscure much of our view of emitted photons in optical and longer wavelength segments of the electromagnetic spectrum. However, at a minimum these clouds do absorb radiation at and below the spectral emission/ absorption lines of the neutral hydrogen in the clouds. Ultimately this obscures observations using these wavelengths from high redshift objects. Principal among the absorption lines that will be discussed in more detail is the Lyman-α line with wavelength λ_α=1216 Angstroms, corresponding to a radial frequency $\omega_\alpha \cong 1.6 \times 10^{16}$ Hertz.

This frequency is about fifteen orders of magnitude higher than the approximate value of the radial plasma frequency, ω_0 of the intergalactic plasma. Of course this ultraviolet radiation frequency is also somewhat higher than that of visible light. So for most observations, rather than photons having frequencies *above* the resonance frequency, they will in this case be *below* it. That is a very significant aspect of the observed absorption patterns. It is the domain to the left of the interesting kinks in figures 38, 39, and 41 but far to the right of any effect attributable to the intergalactic plasma resonance. It is of interest in quasar and "Lyman break galaxy" observations as we will see.

Chapter 5

Refraction of Radiation in Dispersive Media and Its Implications to Astronomy and Cosmology

Much of the study of optics involves an understanding of the concept of refraction. Certainly astronomical instrumentation depends intimately upon the refractive properties of lenses used in the various apparatuses, but these more usual applications of optical refraction will not be our primary concern with regard to cosmological effects that can be attributed to refraction in, and through, the intergalactic medium. We will compare this to observed gravitational lensing effects.

Refraction of electromagnetic radiation is a typical phenomenon associated with a dispersive medium. Whenever light passes through regions of differing index of refraction a deflection of the observed angle to the source emitting the light results. This situation can arise also in a completely ionized plasma medium, especially in cases we have already mentioned for which there are regions within clusters of galaxies for which electron densities are orders of magnitude higher than the average for the intergalactic medium as a whole.

In intergalactic space one would not typically expect refraction to become very appreciable. Electron densities in diffuse plasma, particularly in any intergalactic region between clusters, are so low that measurable deflection of the light path would require rich plasma cores and extreme distances to detect. The fact that such circumstances exist does, however, make this a meaningful consideration.

a. the speed of light in a medium and its implications

Fermat's principle states that light will take the path of shortest propagation time. That is, of course, not always the shortest distance between the point of emission and that of observation. This *principle of least action* maximizes distances through regions of lower refractive index relative to distances traveled through regions with higher index of refraction because the velocity through any medium is inversely proportional to this index. Ray tracing by insertion of lenses or other optical devices between points of emission and observation verifies alteration of paths taken by light in accordance with this principle. So the velocity of light must be known at each point throughout a medium in order to determine the path that light will take through it.

The formula that was presented earlier for the applicable speed of light was, $\mathbf{b} = c / \mathbf{n}$, where $\mathbf{b}$ is the *complex* speed of light in the medium, c the velocity (universal speed) of light in a vacuum, and $\mathbf{n}$ the *complex* index of refraction of the medium. Since $\mathbf{b}$ is complex, there are issues to be resolved: For example, if $Re(\mathbf{n}) < 1$, as it will be for $\omega > \omega_o$, is $|\mathbf{b}| > c$ implied? This would contradict Einstein's special theory of relativity that explicitly dictates that no velocity can exceed c.

We explained earlier that the real part of $\mathbf{b}$ corresponds to the 'phase' as against 'group' velocity of light in the medium. By *group* we mean what is involved with a *wave packet* or *photon*. This *group velocity* is how fast electromagnetic energy can be transferred through space; it corresponds to the *norm* of $\mathbf{b}$, written as $|\mathbf{b}|$, given by:

$$|\mathbf{b}| \equiv | \; Re(\mathbf{b}) + i \; Im(\mathbf{b}) \; | = \sqrt{ Re^2(\mathbf{b}) + Im^2(\mathbf{b}) }$$

Another issue involves division by the complex quantity $\mathbf{n}$. This is typically handled by determining an inverse for $\mathbf{n}$ using the complex conjugate $\mathbf{n^*}$ as follows:

$\mathbf{n}^{-1} \equiv \mathbf{n^*} / (\, \mathbf{n} \cdot \mathbf{n^*})$, where by definition,

$\mathbf{n^*} \equiv Re(\mathbf{n}) - i \; Im(\mathbf{n})$

This yields:

$\mathbf{n}^{-1} = Re(\mathbf{n}) \, / (\, Re^2(\mathbf{n}) + Im^2(\mathbf{n}) \,) + i \; Im(\mathbf{n}) / (\, Re^2(\mathbf{n}) + Im^2(\mathbf{n}) \,)$

Therefore,

$$|\mathbf{b}| = c \mid \mathbf{n}^{-1} \mid = c \,/\, \sqrt{Re^2(\mathbf{n}) + Im^2(\mathbf{n})}$$

In the previous chapter we obtained the approximations,

$$|Re(\mathbf{n})| \cong 1 + 2\pi\,(\,\rho_e\,e^2/m_e)\,/\,\omega^2$$

$$|Im(\mathbf{n})| \cong 2\pi\,(\,\rho_e\,e^2/m_e)\,/\,\gamma_C\,\omega$$

So that,

$$|\mathbf{b}| = c\,/\,\sqrt{1 + 4\pi\,(\,\rho_e\,e^2/m_e)\,/\,\omega^2 + (2\pi\,\rho_e\,e^2/\,m_e\,\omega\,)^2\,(\,1/\,\gamma^2_C\,+ 1/\,\omega^2\,)}$$

And since we have assumed $\gamma_C \gg \omega$ in accordance with arguments presented in the previous chapter, and ω is exceptionally large for any wavelength band typically employed in astronomy in any case, we can approximate this as:

$$|\mathbf{b}| \cong c\,/\,\sqrt{1 + 4\pi\,\rho_e\,e^2/m_e\,\omega^2}$$

$$\cong c\,(\,1 - \rho_e \times 4.0 \times 10^{-20}\,)\ \text{cm/sec.}$$

So the velocity of light through the intergalactic plasma will in all cases be less than the velocity in a complete vacuum. However, it will differ from the universal velocity of light in vacuum by less than about one part in $10^{20\text{th}}$ over any appreciable distance. The approximation applies to visible light observed through rich clusters of galaxies where its effect will be most noticeable.

In situations that arise for which gravitational lensing becomes appropriate due to galaxy clusters intermediate to an observed galaxy, for example, the much denser plasma of the cluster core might produce commensurable angular distortion effects. Such distortions must be considered along with whatever distortions are due to gravitational lensing from which inferences are being made with regard to the amounts of 'dark matter' associated with clusters, etc..

b. Snell's law and some of its ramifications

In application of Fermat's principle mentioned above, Snell's law provides a simple example sometimes denominated the 'law of sines'. It specifies that whenever electromagnetic radiation propagates from a medium of index of refraction n_i into another characterized by a different index of refraction, the light path will be bent in such a way that the parameters n_i and θ_i shown in figure 43 satisfy the relation,

$n_i \sin \theta_i = n_{i+1} \sin \theta_{i+1}$

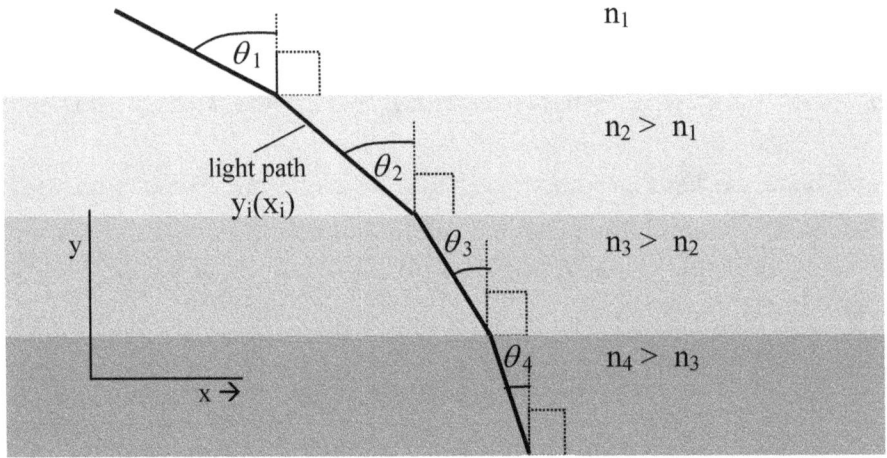

Figure 43: Snell's law of refraction with stratified indexes of refraction

There is a similar relation and specification for the angles of reflection at the interfaces that we will have less occasion to use. It is significant, however, that for light passing from a medium of higher index of refraction to one of lesser index, there is a 'critical angle', $\theta_{critical}$, below which there is no transmission. This critical angle is:

$\theta_{critical} = \sin^{-1} (n_2 / n_1)$,

where it is assumed that $n_1 > n_2$. At this (or any greater) angle for which the arc sine is undefined, all light will be reflected by the boundary. If the angle is less than $\theta_{critical}$ some light will be reflected by the boundary, with the rest refracted in passing through it. Other than this sometimes significant caveat, there is directional reversibility.

In any case, the unidirectional application of Snell's law can be extended throughout stratified media such that the relation holds even between non-contiguous layers. In effect intermediate index values, whatever they happen to be, are of little consequence, θ_N being fully determined by θ_1. And in particular even for continuous variations in the index due to the similar continuous variations in the electron density of the medium, the following generalization holds:

$$n(x_i,y_i) \sin \theta(x_i,y_i) = C_i, \text{ a constant}$$

The constant is associated with a curve $y_i(x_i)$ determined by the initial, or any other, angle along the path . Clearly, the slope of this curve is:

$$y_i' \equiv dy_i/dx_i = \cos \theta(x_i,y_i)$$

Thus we obtain a differential equation for the curve as follows:

$$y_i' = \sqrt{(C_i / n(x_i,y_i))^2 - 1}$$

This equation does not yield analytic functions in all cases, even for some fairly usual functions of $n(x_i,y_i)$. The equation is associated with what is called the 'brachistochrone' problem of the calculus of variation that involves finding the path of minimum transit time.

To formulate this, one must determine a time of transit for each segment along the path. This involves determining the length of each segment and a propagation speed for that segment by which to divide. As we have seen, the speed of light, b at each point in a medium is given as $b = c / |\mathbf{n}|$, where $|\mathbf{n}|$ is the modulus of the index of refraction identified in section a. This norm is approximately equal to the real value $n \cong |\mathbf{n}|$ in instances for which we will be interested in refraction.

Therefore, the amount of time required for light to propagate from the source to the observer through an intermediate medium of varying index of refraction is the integral of time taken along the path,

$$t_i = \int_{source}^{observer} (1 / b(x_i,y_i))\, ds_i = (1 / c) \int_{source}^{observer} n(x_i,y_i))\, ds_i$$

The integration parameter, ds_i is an infinitesimal distance along the light path for which in Cartesian coordinates we have:

$$ds_i = \sqrt{1 + y_i'^2} \; dx_i$$

Before attempting to solve for $y_i(x_i)$, we will look at situations for which the solution of such problems has particular meaning. Some of these pertain directly to cosmological topics as we will see.

c. common electromagnetic refraction phenomena

Refraction produces several interesting and readily observable effects. Because the earth's atmosphere produces anomalous refraction of electromagnetic radiation, we observe a flattened sun long after it has *in actuality* dropped well below the horizon at sunset and before it emerges above it at sunrise. In such situations, as light travels through the atmosphere its path is curved around the earth's surface by the variations in the index of refraction with altitude that it encounters.

Nor does such phenomena apply exclusively to earth's atmosphere. The sun also has an atmosphere (chromosphere) that must similarly produce effects like those in our atmosphere. Refraction of light from stars situated beyond the sun must certainly occur as well as the more celebrated gravitational effect. In addition, for the case of massive galactic clusters where gravitational lensing has been noted to collaborate inferences of 'dark matter', the plasma electron density at their cores is orders of magnitude greater than for the surrounding space, so that a similar atmospheric type effect may also be present.

We will investigate the relative magnitude of such refraction effects where gravitation has been shown to produce a significant deflection of light. Wherever there is a massive object, generally there is also a gaseous atmosphere surrounding it whose appreciable radial gradient produces refraction effects. It is essential to disambiguate the effect of gravitation on the one hand and that of atmospheric refraction on the other. Only by employing associated theoretical considerations of both disciplines can one accurately predict the magnitude of the measurable combined effect.

In the bending of solar radiation that allows the sun to be visible even when it is *actually* well below the horizon at sunrise and sunset there is, as with other refraction phenomena, the situation that light is refracted differentially by wavelength. This causes a setting or rising sun to display horizontal bands of slightly differing color. The most extreme example of this is when only the green light from the sun that is below the horizon reaches the viewer, resulting in a momentary green dot or flash. We will consider whether such dispersive properties of refraction might be observed as unique effects during solar eclipses.

Variation in the electron density of the atmosphere, with the much higher density at sea level, is what produces deflection of light paths associated with the sunrise and sunset phenomena. Just before sunset the solar disk, which at higher elevation angles is approximately a circle subtending half a degree of arc, considerably flattens. This is because light from the lower limb, which is nearly a full degree below the horizon at that time must travel a greater distance through the densest atmosphere. Therefore, it is refracted upward by a larger amount than is the case for the upper limb. So when the lower limb appears right at the horizon the upper limb will not have been affected quite so much. This is shown in figures 44 and 45.

d. refractive properties of earth's atmosphere

As suggested by the shading in figures 44 and 45, the index of refraction $n(x,y)$ of an atmospheric medium depends intimately upon its electron density, $\rho_e(x,y)$, which will in general be proportional to mass density of the atmospheric gases. As previously noted for radiation that is not in the immediate vicinity of a resonance frequency ω_0 where significant absorption takes place, the index of refraction can be characterized as a simple function of electron density as follows:

$$n[x,y] = 1 + 2 \pi \rho_e[x,y] (e^2/m_e) / (\omega_0^2 - \omega^2)$$

$$= 1 + \rho_e[x,y] (\lambda_0^2 e^2 / 2 \pi m_e c^2)(1 + \lambda_0^2 / \lambda^2 + \dots)$$

In application to earth's atmosphere the electron density will depend predominantly upon a single parameter, $\rho_e(h)$, where h represents altitude. The wavelength λ_0 is the primary neutral atomic resonance of the atmosphere in the vicinity of the wavelengths being observed.

The index of refraction of dry air at a temperature of 15 degrees Celsius and standard pressure at sea level is given as,

$$n(0) = 1.000272643 + 122.88 / \lambda^2$$

in the *Fourth Edition of the Handbook of Chemistry and Physics* where wavelength is measured in Angstroms. Resulting indices of refraction for a couple of specific wavelengths are the following:

$n(0) = 1.0002863$, at $\lambda = 3{,}000$ Angstroms

$n(0) = 1.0002752$, at $\lambda = 7{,}000$ Angstroms.

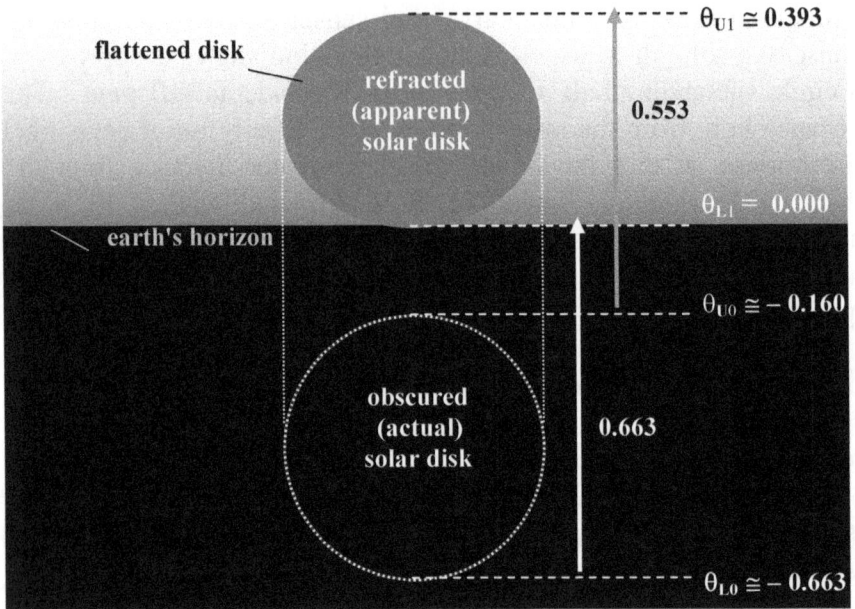

Figure 44: Apparent compression of solar disk at sunset

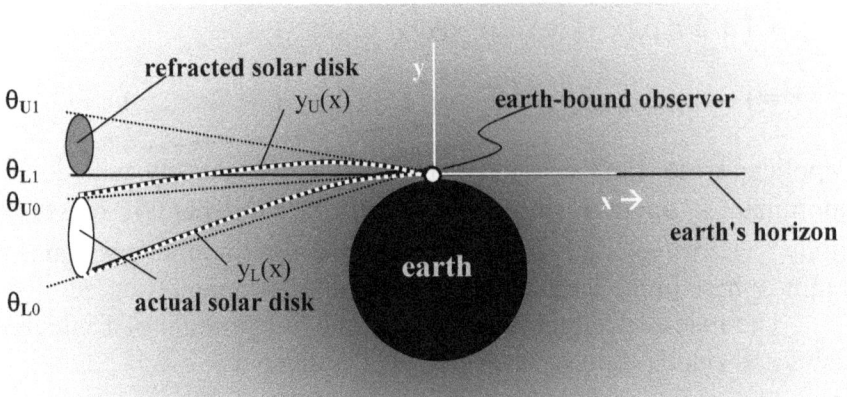

Figure 45: Angles to actual and observed solar disk

Naturally electron densities, ρ_e of gaseous material occurring in strata of differing mass density differ proportionately. At sea level under conditions similar to those given above the mass density is:

$\rho_m(0) = 0.001225$ gm cm^{-3}

The density decreases very nearly exponentially with altitude above sea level throughout the troposphere as indicated by NRLMSISE model data shown in figure 46. The dashed line that has been added to this plot is a best fit to the exponential equation approximation. The equation is seen to be extremely accurate out to an altitude of 90 km at which point the atmosphere is nearly a million times less dense:

$\rho_m(h) = 0.001225 \times 10^{-0.06\ h}$

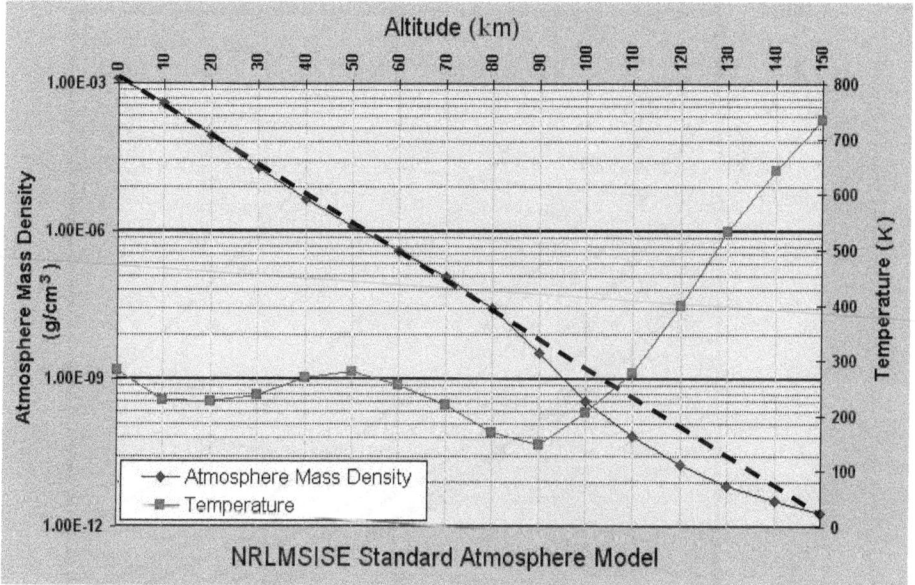

Figure 46: Atmospheric density and temperature with altitude data

If we measure altitude from the center of the earth, which is the center of the approximate symmetry of the atmosphere, we have that:

h $\rightarrow$ $r - r_o$, where $r_o = 6{,}365$ km.

To obtain the index of refraction as a function of this revised altitude parameter, we note that the elemental composition of the atmosphere is essentially the same throughout the first one hundred kilometers above sea level. The quantity $n(0) - 1$ must be proportional to mass density at each altitude implying that at $\lambda = 5{,}000$ A:

$$\rho_e(0) = 3.688 \times 10^{-3} \, \pi \, m_e \, c^2 / e^2 = 4.123 \times 10^{11} \, cm^{-3}$$

From which we have at this wavelength that:

$$n(r) = 1.0 + 0.0002776 \times 10^{-0.06(r-r_o)}$$

The differences in the functionality of this atmospheric index of refraction for various wavelengths are plotted as $n(r,\lambda)$ in figure 47. This is the dependence shown in equations above.

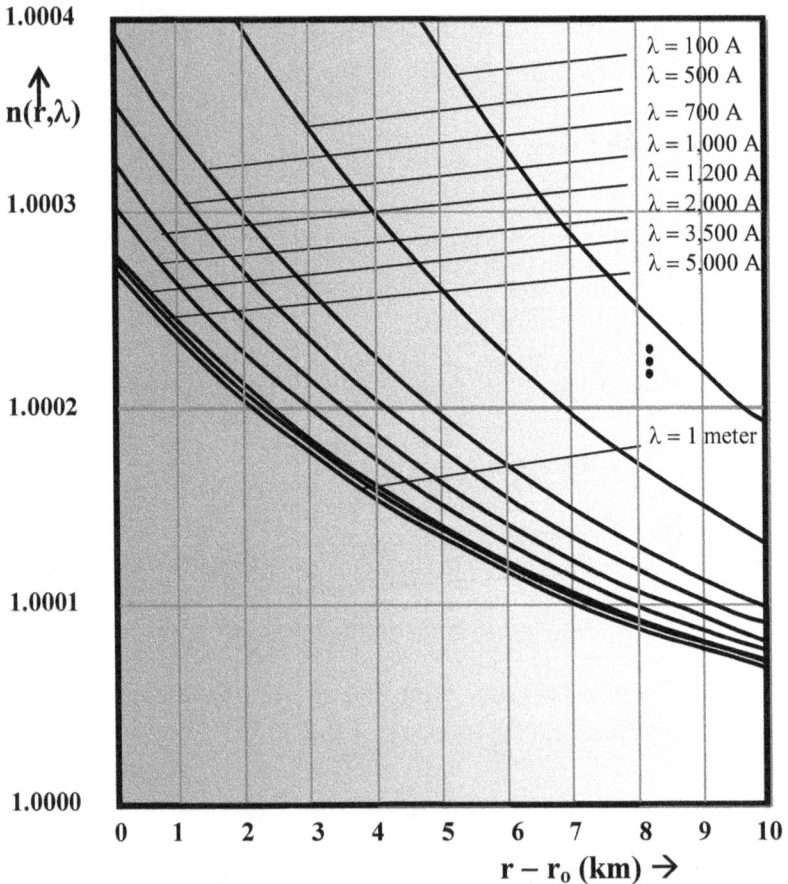

Figure 47: Atmospheric index of refraction for various wavelengths

We now have sufficient data from which to obtain a solution to the penultimate problem we set out to solve, namely obtaining values for the angles of refraction $\Delta\theta = \theta_{L1} - \theta_{L0}$ and $\Delta\theta = \theta_{U1} - \theta_{U0}$ that are shown in figures 44 and 45. Since the brachistochrone differential equation for an exponential density does not lend itself to immediate solution for an analytic function, we have used a recursive algorithm described in detail in Appendix B to perform straight-forward numerical integration to solve this and related refraction problems. A full range of refraction deflections at various viewing angles from the earth's surface calculated with this algorithm are provided in figure 48.

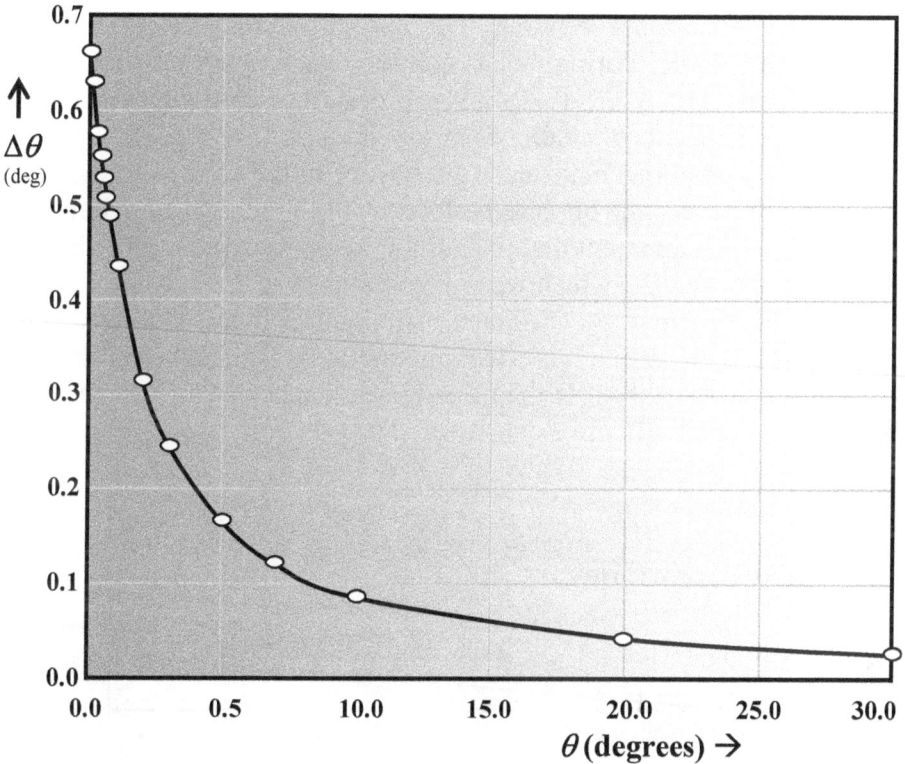

Figure 48: Atmospheric deflection due to refraction at various viewing angles

It should be clear that for an observer at a sufficient distance from the earth, this determined angle of refraction would be enough to allow observation of a star directly behind the earth. If it were aligned with the center of the earth for the observer, it would produce a ring of light around the earth as shown in panel a of figure 49. If the observer were further removed, without even being able to discern the earth, one could observe double images of stars with the refracted image displaced by up to 0.6623 degrees. These effects are illustrated in panel b of figure 49. They would, in several respects, be similar to the effects of gravitational lenses in deep space – effects we will consider later.

e. effects of refraction in the solar chromosphere

Our next task is to assess the extent to which refraction might, in fact, affect the results of the well known confirmations of Einstein's general relativity in the bending of starlight around the sun. Clearly, predicted gravitational bending that has largely been validated by observations made during solar eclipses such as those made by Campbell and Trumpler (1922) and are described in Shapley's 'source book' (1960) are much smaller than the amount of refraction we have found for observations made near the surface in the earth's atmosphere. Figure 50 illustrates the observed effects from Campbell and Trumpler's effort where it is apparent that in addition to an observed effect, there is a considerable scatter, which we will investigate to see whether, and to what extent, this may be attributed to refraction in the chromosphere. But whatever the refractive effects of the solar atmosphere, they are evidently considerably less than for our atmosphere, and that must be because of a lower effective density of electrons.

panel a, 'Einstein ring'

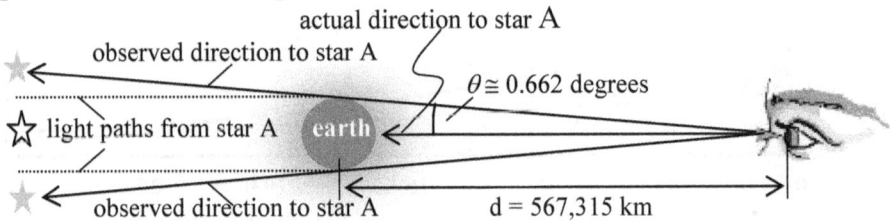

panel b, double images

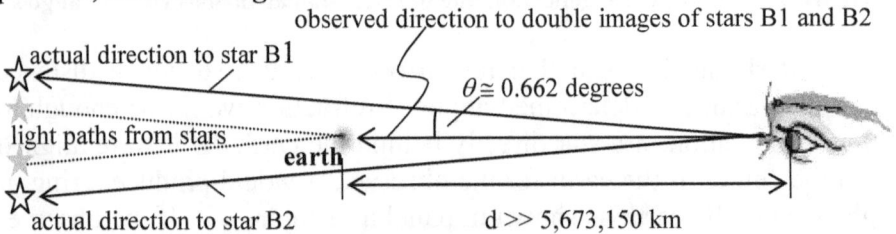

Figure 49: Lensing effects earth's atmosphere would have from deep space

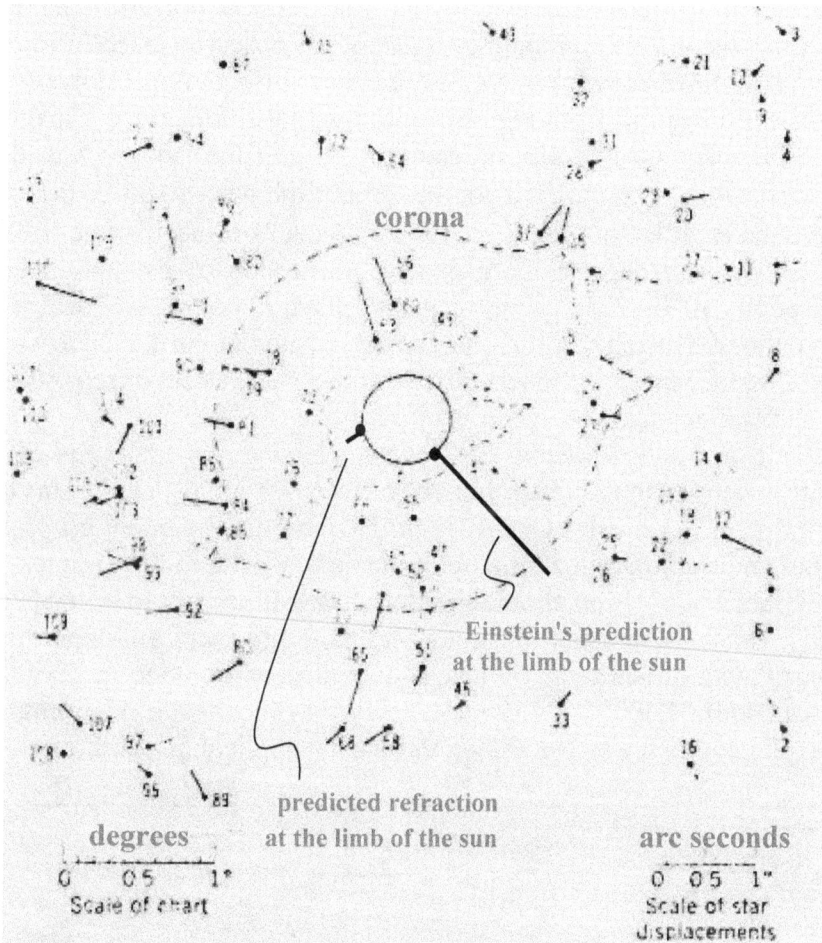

Figure 50: Deflection of starlight during eclipse of 1922

The reader might think that vagaries in the quality of data present in figure 50 must have improved considerably over the next decades. That improved accuracy was not obtained quickly, although ultimately predicted gravitational effects have been demonstrated very accurately, of course, but primarily due to fairly recent developments in very large array radio telescopes (Iess, 1999) with imaging of spacecraft in the vicinity of heavy planets whose atmospheres at the distances of the measurements are minimal. Nonetheless, in Dicke's summary of the results through 1964 quoted in Misner et al (1973), he stated, "It appears that one must consider this observation uncertain to at least 10 percent, and perhaps as much as 20 percent," the error generally resulting in an underestimate of the observed effect. Nor was a rationale for the magnitude of these errors forthcoming. However, in 1973, extrapolating from regions with minimal refraction, a University of Texas team successfully proclaimed: "The final value obtained for the deflection, extrapolated to the solar limb was (0.95 ± 0.11) L_E where the error is 1σ and $L_E = 1''.75$ is Einstein's value" (Brune, 1976).

We will defer a more general discussion of the gravitational deflection of light to a subsequent chapter where we deal with gravitational lensing. In this chapter we pursue our current endeavor of assessing the similar, if much smaller in this case, effect of refraction in the sun's 'atmosphere'.

Esser and Sasselov (1999) obtained various measures of the electron density in the lower regions of the solar atmosphere. This data is shown in figure 51. Clearly there is some disagreement on precise values and even the functionality of the density with radial distance, but assessments are within about an order of magnitude out to several solar radii. Most assessments illustrated in this figure represent upper limits rather than estimates of the actual overall density. See in particular Guhathakurta (1998). Esser and Sasselov acknowledge this limitation as applicable to the larger among these data, applicable as follows:

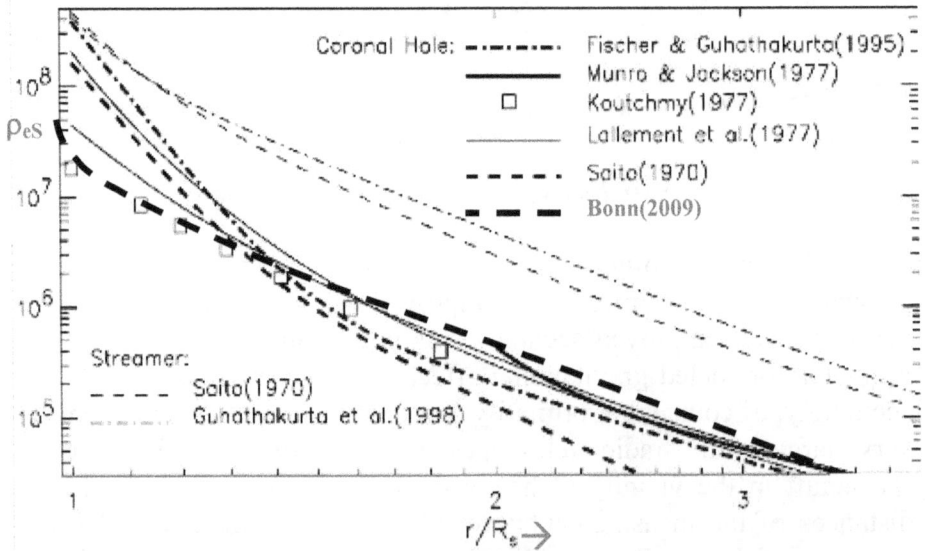

Figure 51: Electron density measures for the solar atmosphere

"...in particular to coronal holes rather than solar atmosphere generally. Comparing these atmospheric densities to coronal electron densities derived from polarization brightness measurements in the region from about 1.1 to several solar radii, it is shown that there is a discrepancy between the two sets of densities. The atmospheric electron densities are in agreement with a density of maximum 10^7 cm^{-3} at 1.1 R_S. The polarized brightness densities given in the literature are typically 5 x 10^7 cm^{-3} or higher. It is shown that this discrepancy might be due to an overestimation of the coronal electron densities below 1.5–2 R_S."

At about 2,000 km above the photosphere the electron density decreases roughly exponentially and then flattens out somewhat. The dark dashed line has been superimposed on figure 50 by the current author to illustrate the distribution used in our analyses. It is merely a very rough empirical fit to the data of Esser and Sasselov taking their caveats into account.

Assuming a similar functionality for wavelength dependence of the index of refraction, with $\rho_{eS}(r=R_S) \cong 3 \times 10^7$, we would have at the wavelength $\lambda = 5,000$ A:

$$\rho_{eS}(R_S) \cong 7.225 \times 10^{-5} \, \rho_{eE}(r_o) \; cm^{-3},$$

From which we obtain for the index of refraction at this wavelength:

$$n(r) \cong 1.0 + 0.0000000202 \times 10^{-2\sqrt{r/R_S} - 1},$$

where the form of the equation is applicable to the electron density curve, drawn in as the heavy dashed line in figure 51. Here the radius of the sun, R_S is taken as 695,500 kilometers.

Again the brachistochrone differential equation does not lend itself to immediate solution for an analytic function in the case of this electron distribution. We have therefore used the same recursive algorithm approach to a straight-forward numerical integration we used to obtained a solution in the case of earth's atmosphere as detailed in Appendix B for these analyses. In this case we obtain a maximum deflection of 0.13885 arc seconds at the solar limb, shown in figure 50 for comparison with the gravitational deflection prediction. It is about 8 percent of Einstein's estimate for gravitational bending, but enough perhaps – particularly with the variations due to chromospheric and corona phenomena – to clutter observations near the limb nonetheless.

This is particularly obvious when one considers the impact of relatively small regions of much more intense electron density in the dynamics of the solar atmosphere. It is also interesting that the predicted refraction is within Brune's budgeted error.

Of particular significance is the fact that the refraction effect is diminished much more rapidly than is the gravitational effect, which drops off as the inverse first power of the 'impact' distance of the light path from the center of mass. The index of refraction of the atmospheres of massive objects tend to diminish exponentially so that the effect at the limb drops off exponentially for observations made through higher altitudes. See for example, figure 52.

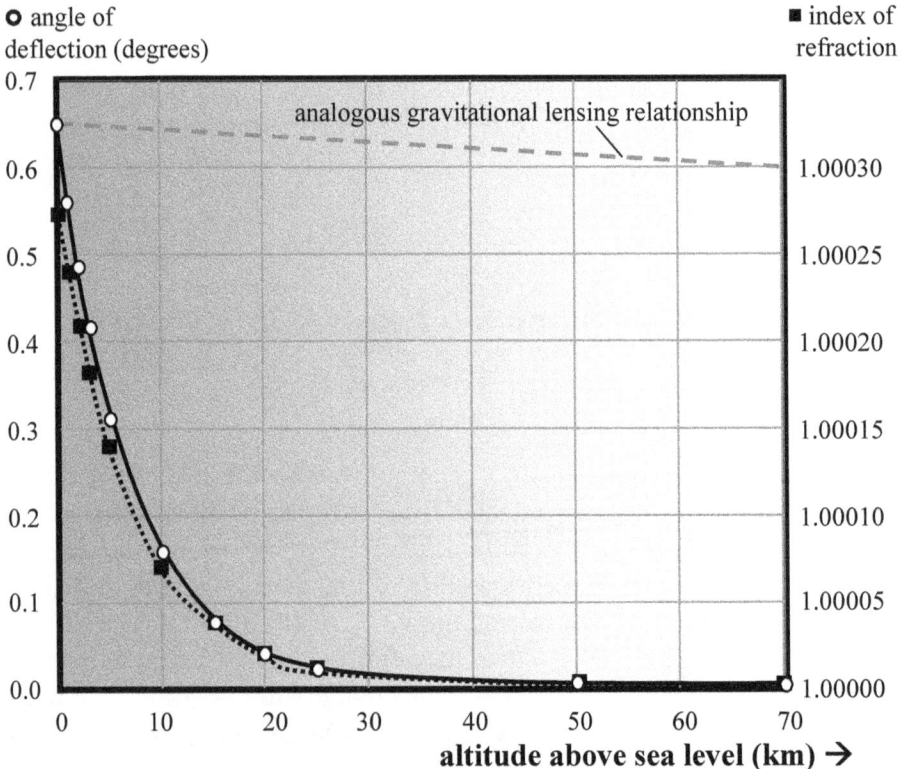

Figure 52: Index of refraction and associated deflection angles of observations made through various altitudes in earth's atmosphere

This illustrates the very major difference between refraction effects and those which are attributed to gravitational deflection of light. If the earth were sufficiently massive (which obviously it is nowhere near) to produced a gravitational deflection equivalent to the atmospheric deflection at the limb (surface), the gravitational effect would be only ten percent less when viewing past the earth at the altitude of 70 km as shown by the dashed line in figure 52. On the other hand, at 200 km above the solar surface, which is less than 3 ten thousandths of the solar radius, the refraction effect produces only a deflection of 0.03394 seconds of arc.

It should also be obvious from the modest magnitude of the effect of the solar atmosphere that wavelength dispersion will not be observable.

f. expected refraction effects in the intergalactic medium

But our primary interest is cosmology. As we have discussed, the intergalactic medium does not constitute a uniformly dense plasma with a uniform electron density. The central core regions of galactic clusters are associated with electron densities that are orders of magnitude greater than those regions between clusters for example. There are also regions of ionized hydrogen gas in various states of ionization in Lyman alpha forests, and as halos of quasars and galaxies. But none of these seem sufficiently dense in themselves to afford appreciable refraction effects.

In figure 52 we illustrated maximum deflection angles for observations made through the earth's atmosphere at various altitudes. These deflections diminish extremely rapidly with increasing altitude, but this decrease is clearly because of the associated decrease in index of refraction that diminishes as a direct function of the density of the atmosphere. The density of the atmosphere of a distant planet orbiting another star could conceivably be determined using analyses suggested in figure 50 of double images of a more distant star, the unique features of whose spectra could be identified as being duplicated in two images. But because atmospheric refraction layers are typically so very thin, objects larger than point sources such as stars or quasi-stellar objects that appear as mere points of light would not be effected.

In Issue 2.36, November 16, 2004 of *The Astrophysics Spectator*, we find the statement:

"The point gravitational lenses of stars are also observed, but these lenses are rare, and they are only found by systematically observing changes in the brightness of a million very distant stars every night. The image produced by a stellar lens is too small to resolve with a telescope, and the time delay associated with a stellar lens is too short to measure. The only observable effect of a stellar lens is the increase in apparent magnitude of stars seen through the lens."

It is worth noting that it might be extremely difficult to determine whether such double images of these point sources in our own galaxy were caused by refraction or gravitational deflection.

With regard to applications to cosmology itself, Cowie and Perrenod (1978), and many others more recently, have provided data for electron densities in rich galaxy cluster cores, which peak quite dramatically at the center as we will see in a later chapter. However, the peak densities of $\rho_{eCl}(0) \approx 0.013000288$ cm^{-3}, even in the 'richest' cluster core plasmas is seven orders of magnitude less than for the solar atmosphere and more than 13 orders smaller than earth's atmosphere. So it is fair to say that the intracluster plasma will not in itself produce any observable deflections that could be attributed to refraction.

However, this does not mean that within such rich cores, or elsewhere in the universe, there are not objects which accrete unto themselves more dense atmospheres from this plethora of material. If that were to occur, then much more distant objects beyond foreground objects in the cluster cores – that one assumes would be unobservable because of their limited size – might indeed produce even major deflections like those hypothesized in figure 49 and perhaps supporting more extensive double imaging. Much more typically, however, since these atmospheres would also most likely be constituted of a very thin layer exhibiting exponentially decreasing effects with angular distance from the object, primarily 'point' sources such as quasi-stellar objects might be seen in double images.

Chapter 6

Absorption of Radiation in Dispersive Media and What Is to Be Expected in a Redshifting Intergalactic Plasma

The absorption of energy from electromagnetic radiation that occurs in a dispersive medium is associated with the transfer of motions established within material constituents of the medium that persist after the immediate influence of the incident electromagnetic fields has ceased. This phenomenon has previously been thought to be the only nonconservative forces at work in the scattering process. In the very strictest sense, of course, that is indeed the case, although we have identified another previously unsuspected aspect of such induced and enduring motions that applies within a redshifting medium.

Where there are bound electrons, the individual atoms and molecules possess indigenous resonant frequencies corresponding to higher energy levels to which electrons may transition by absorbing energy from the radiation. Incident radiation with frequencies near this resonance value is particularly susceptible to being absorbed. In chapter 7 we will discuss that topic in detail.

A similar situation arises in a completely ionized plasma medium except that the only truly resonant frequency is now associated with a macroscopic resonance of the plasma itself. At this radial plasma frequency the medium becomes completely opaque, converting transverse electromagnetic waves containing energy into energetic longitudinal oscillations of the plasma itself.

a. Lambert's law

Absorption can be measured as a reduction in luminous flux of radiation passing through the absorptive medium. In general observed luminous flux will be proportional to the integral of the energy intensity over all wavelengths included in the incident radiation. As we have seen, at a particular wavelength, the luminous flux is defined in terms of the energy in the field strength vectors as follows:

$$S(d,\lambda) \sim \tfrac{1}{2} (\mathbf{E}(d,\lambda)^2 + \mathbf{H}(d,\lambda)^2), \text{ where}$$

$$\mathbf{E}(d,\lambda,t) = \mathbf{E}_o(d,\lambda)\, e^{\,i\,(\omega t - 2\pi\,\mathbf{n}\,d/\lambda)},$$

with a nearly identical expression for the magnetic field strength vector. Their squares are defined as the products of their complex conjugates so that, $\mathbf{E}(d,\lambda)^2 \equiv \mathbf{E}(d,\lambda,t)\,\mathbf{E}(d,\lambda,t)^*$, etc.. Since we have,

$$i\,(-\,2\pi\,\mathbf{n}\,d/\lambda\,) = -\,2\pi\,i\,(d/\lambda)\,(Re(\mathbf{n}) +\ i\,Im(\mathbf{n}))$$

The factors involving $Re(\mathbf{n})$ cancel in the product of conjugates. So,

$$S(d,\lambda) \sim \tfrac{1}{2}\,[\,\mathbf{E}_o(d,\lambda)^2 + \mathbf{H}_o(d,\lambda)^2\,]\, e^{\,-4\pi\,Im(\mathbf{n})\,d/\lambda}$$

For convenience, let us define,

$$S_o(d,\lambda) = \tfrac{1}{2}\,[\,\mathbf{E}_o(d,\lambda)^2 + \mathbf{H}_o(d,\lambda)^2\,]$$

to simplify the resulting expression as,

$$S(d,\lambda) = S_o(d,\lambda)\, e^{\,-4\pi\,Im(\mathbf{n})\,d/\lambda}$$

This equation represents the intensity of a monochromatic incident plane wave into which any field can be decomposed as was described in chapter 4. The fact that in general the index of refraction is complex results in luminous flux having a decreasing exponential dependence on distance. This can be seen by direct substitution of the complex expression for the imaginary part of the index of refraction, $Im(\mathbf{n})$ into the above expression appropriate to the flux of an incident monochromatic wave functions that was presented in chapter 4.

According to *Lambert's law*, the applicable luminosity of plane waves incident upon a dispersive medium is experimentally observed to decrease as an exponential function of the distance d propagated through the medium. This is expressed by the following formula:

$$S(d) = S_0 \; e^{-2a\,d}$$

where *a* is sometimes referred to as the *absorption coefficient*.

By comparison of luminosity formulas presented above and by substituting the imaginary component of the index of refraction derived earlier into that formula, we see that the absorption coefficient is in general a function of wavelength. However, since there are separate domains applicable to ionized plasma and neutral atoms in a gas such as the HI and HII clouds encountered intermittently throughout intergalactic regions, we will discuss these two cases separately.

b. optical depth and related concepts

The 'optical depth', involving the degree of 'transparency' or antithetical 'opacity' of a medium, is a measure of 'attenuation' or absorption that takes place in a medium. If one were to shine a light through any medium, after propagating to the optical depth of that medium, the intensity of the emitted light would drop to 1/e of its original intensity in accordance with Lambert's law independent of any other loss due to the inverse square law or loss of photon energy due to redshifting. This too is due to the scattering that takes place in the medium. There are various types of scattering: Rayleigh, Mie, Thompson, coherent forward scattering, Compton, Bragg, etc..

Rayleigh scattering involves scattering of electromagnetic radiation by particles much smaller than the wavelength of the light. It occurs when light travels through any transparent media, but usually pertains to gases. For example, Rayleigh scattering of sunlight in clear atmosphere is the main reason why the sky is blue. It results from wavelength-dependent deflection of light that contributes diffuse, as against direct, sunlight. Mie scattering involves scattering from spherical objects whose radii are larger than the wavelength, but to somewhat similar effect.

Of course Thompson scattering is much more directly applicable to scattering that occurs in plasma and to the general concept of coherent forward scattering that we will discover to effect a redshift when it occurs in a 'hot' (relativistic) plasma. So we will rely on some of his contributions when we get to that discussion.

To varying degrees attenuation will be introduced by any and all scattering phenomena. This attenuation associated with absorption will typically result in subsequent re-emission in a random direction a short period of time after absorption. Thus, there will be diffuse light penetrating considerably beyond the optical depth of the medium, but generally it will not be 'coherent'. This means the re-emitted photons cannot interfere constructively or destructively with photons emitted by the original source or with other re-emitted photons. Imaging will be obscured beyond that depth. See figure 53 for an illustration of associated concepts. We will discuss those concepts specifically involved with coherent scattering in a later chapter.

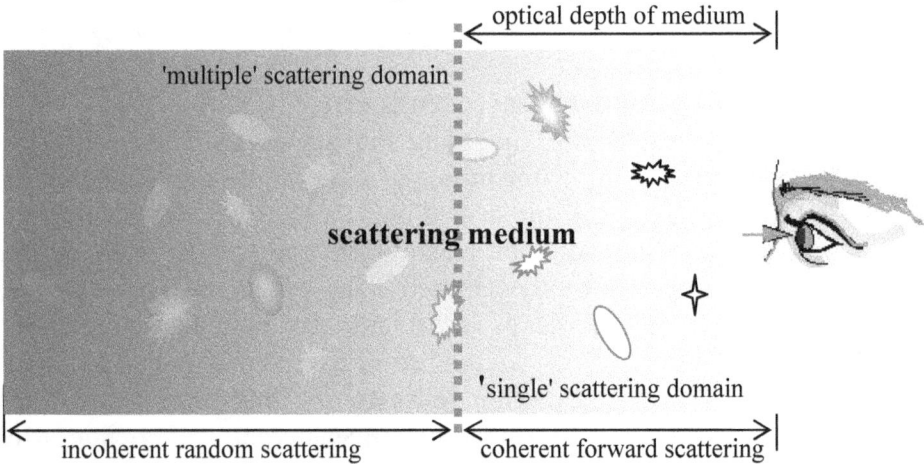

Figure 53: Illustration of a few scattering medium concepts

Clearly, 'single' scattering that occurs within one optical depth in a medium, supports coherent superpositioning of scattered light to effect the forward scattering by which we are enabled to observe images of objects even through very dense media that ultimately would absorb all light penetrating much beyond its optical depth. Absorption does not entirely obliterate light passing through such a medium, but beyond the optical depth light will be diffuse such that the medium is no longer completely transparent. As shown in the figure, a certain amount of obscured imaging takes place a little beyond the optical depth, but increasingly the light emitted will be randomly scattered.

112

One must add that the designations "multiple" and "single" scattering are not very precise. The single-scattered category entails stretches of forward-scattered radiations that may involve millions of coherently scattered but constructively reinforced unidirectional replacement of photons in a straight line to the source of the radiation. The optical depth is the distance at which absorption takes its toll on this process by absorbing too many of the forward scattered photons to sustain the process.

Multiply-scattered radiation designates the situation for which absorption (as against Thompson scattering) becomes the significant phenomenon, with *re-emission* in random directions following the absorption taking place rather than scattering per se. Obviously multiply-scattered radiation does not support imaging, nor therefore 'observation' of an object, as against merely determining that there is a light source in the general direction of the object of interest.

c. broadband absorption in a plasma medium

Far from the immediate vicinity of the radial plasma frequency domain, there will still be minimal absorption in accordance with the electromagnetic luminosity formulas presented earlier. In particular we must explore the alternative possibilities associated with the various domains of coefficient values suggested by figure 40. The imaginary component of the index of refraction and Lambert's absorption coefficient differ considerably for the various alternative cases.

Although there are theoretical models from which to estimate values of γ and the associated *optical depth* of a medium, for all intents and purposes it is an experimentally-determined parameter. For a homogeneous medium with a single resonant frequency, one could fit luminosity data measured at one wavelength and then verify the value as consistent with absorption measurements of the same object made at other wavelengths when the spectrum of the object type is known.

Let us explore the possibilities of absorption characteristics associated with various alternative determinations of the parameter γ. Probably the most natural possibility to come to mind would be that since we expect absorption to be minimal, one might (erroneously) infer that the value of γ must be relatively small. In this case:

$$a_{pl}(\lambda) = 2\pi \, Im(\mathbf{n}) \, / \, \lambda \approx (\rho_e \, e^2 \, / \, 2 \, \pi \, m_e \, c^3) \, \gamma \, \lambda, \text{ for } \gamma << \omega \text{ with } \omega_0 \approx 0.$$

The relationship $S(d,\lambda)/S_o$ shown in figures 54.a and 54.b is what would be expected in that case.

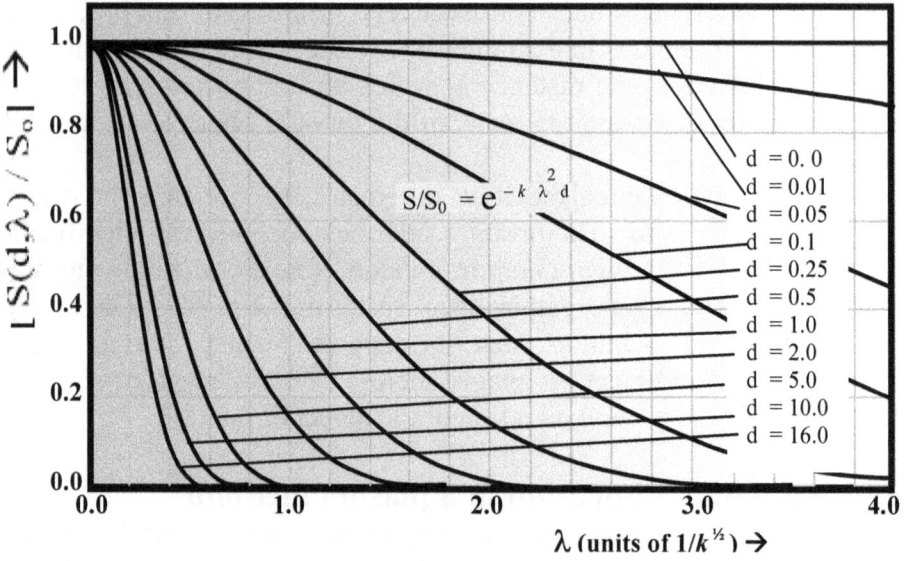

$$S/S_0 = e^{-k\lambda^2 d}$$

d = 0.0
d = 0.01
d = 0.05
d = 0.1
d = 0.25
d = 0.5
d = 1.0
d = 2.0
d = 5.0
d = 10.0
d = 16.0

λ (units of $1/k^{1/2}$) →

Figure 54.a: Dependence of total absorption on wavelength for various values of propagation distance

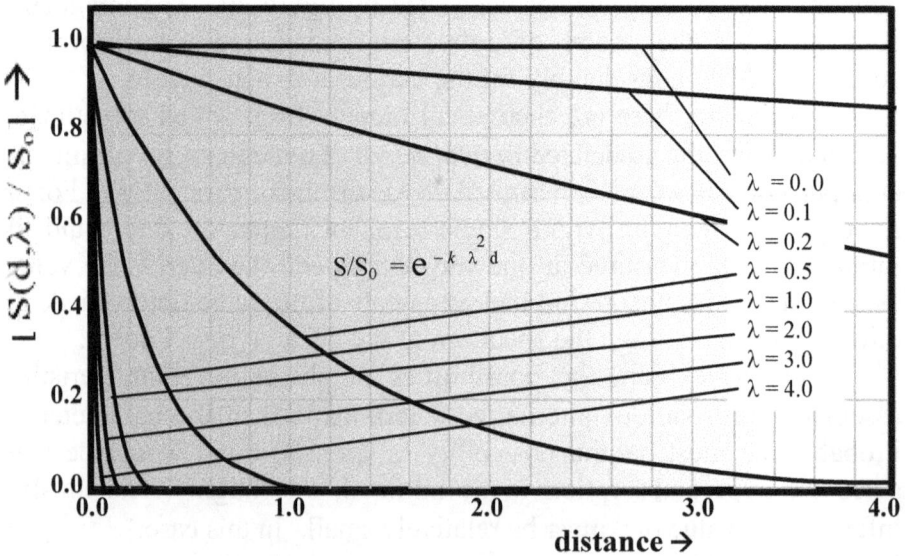

$$S/S_0 = e^{-k\lambda^2 d}$$

$\lambda = 0.0$
$\lambda = 0.1$
$\lambda = 0.2$
$\lambda = 0.5$
$\lambda = 1.0$
$\lambda = 2.0$
$\lambda = 3.0$
$\lambda = 4.0$

distance →

Figure 54.b: Dependence of total absorption on propagation distance for various values of wavelength

However, as we discussed earlier, γ_C is much greater than the resonant frequency ω_0. An extremely different value of Lambert's absorption coefficient obtains when γ_C is greater than ω. Absorption takes on a totally different functionality in this case for which we have:

$$a_{p2}(\lambda) = 2\pi \, Im(\mathbf{n}) / \lambda \approx 2\pi \, (\rho_e \, e^2 / m_e \, c) / \gamma_C \cong 0.053 \, \rho_e / \gamma_C$$

The unique aspect of this absorption coefficient is that it does not involve the wavelength at all and therefore would not result in the traditionally expected wavelength dependent dispersive *absorption* effects. A fuller explanation of the extent to which γ_C is associated with conservation of energy in diffuse intergalactic plasma must await conclusions to be drawn in later chapters. But, to summarize here, it is associated with mechanisms of the forward scattering process in high temperature plasma. In particular there is an associated transfer of momentum and energy (the wavelength dependence of photons) from the radiation into altered mechanical motions of electrons in the plasma at intervals whose lengths are inversely related to wavelength. For this process conservation of energy and momentum requires that,

$$\gamma_C \cong 1.48 \times 10^{27} \, \rho_e \, sec^{-1}$$

Later we will identify constraints on ρ_e and electron temperature that effect a situation for which Lambert's formula becomes,

$$S(d) \cong S_o \, e^{-H_o d}$$

H_o is Hubble's constant; it is approximately $7.14 \times 10^{-29} \, cm^{-1}$. Thus, in the scattering model to be presented in subsequent chapters of this volume, we have that,

$$e^{-2a d} = e^{-x H_o d} = 1/(\not{Z}+1)^x$$

where $\not{Z}$ is the amount of redshift incumbent on light having traveled a distance d through the intergalactic medium. The parameter x is used in the preceding equation merely to indicate the effect of slight variations in the value of γ_C on absorption.

Let it not be supposed that this is presented here as some sort of minimalist explanation of how cosmological redshift can be effected by a 'tired light' model. This volume will address that problem thoroughly in later chapters. After dispensing with the expectations of absorption we will get on with that discussion. The preceding formula, which is plotted in figure 55, is merely one of the significant side effects of the associated redshifting mechanism we will derive and describe.

However, when we come finally to comprehend and compare predictions of the scattering model presented in this volume with predictions of various versions of the standard cosmological model, it will be the data for observed luminous flux of objects observed at cosmological distances that is perhaps most pertinent. The flux diminution formula embraced by all models is the following,

$$S(d) \sim S_o / (Z+1)^2 \, d(Z)^2$$

where the assessment of distance as a function of redshift, $d(Z)$ is model-dependent. Of course the fact of the inverse square of distance aspect (however assessed in terms of redshift) in reducing inherent luminosity S_o is the traditional classical physics implication.

The two redshift dependent factors involving the inverse of $Z+1$ that appear in any explanation of observed luminosities at cosmological distances require separate explanations. One factor is a direct consequence of quantum phenomena whereby redshifted radiation will exhibit lower energy because of its lengthened wavelength. That is an explanation common to any and all current cosmological models. Rationale for the other identical factor in the denominator of the above equation is intimately tied to the mechanism responsible for redshifting. In established standard cosmological models there is a required time dilation factor associated with the conjectured recessional velocity of the observed object (or universal expansion). In the scattering model, this same factor arises because of unique absorption characteristics we have just described. Refer to pages 65 and 66 for further explanation.

d. effects of redshift on plasma absorption

The issue of the effect of redshift on absorption in intergalactic plasma must be considered in somewhat more detail. The phenomenon is not encountered with other media. There is, according to the

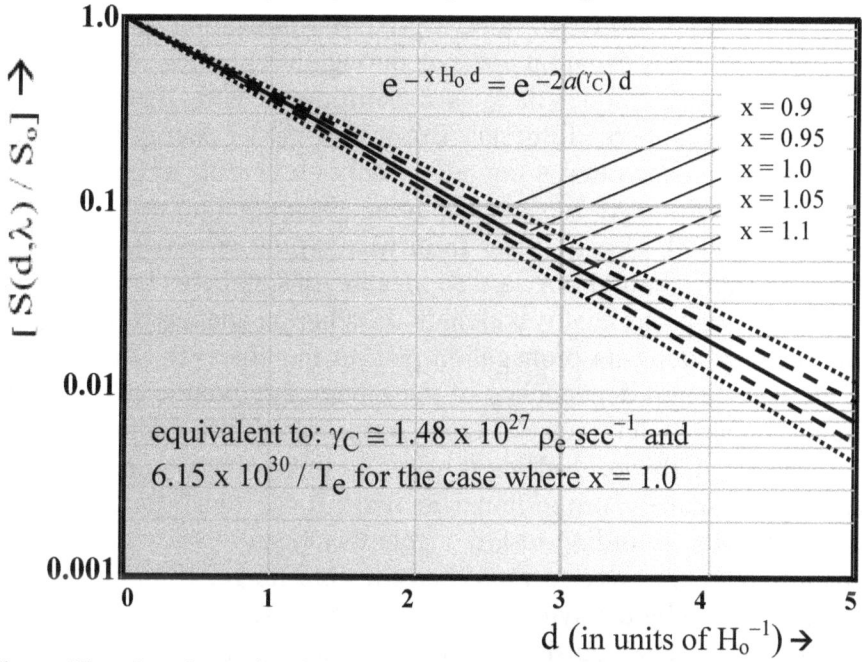

Figure 55.a: **Luminous flux loss in propagating through intergalactic plasma**

Within the figure:

$$e^{-x H_o d} = e^{-2a(\gamma_C) d}$$

x = 0.9
x = 0.95
x = 1.0
x = 1.05
x = 1.1

equivalent to: $\gamma_C \cong 1.48 \times 10^{27} \rho_e \sec^{-1}$ and $6.15 \times 10^{30} / T_e$ for the case where x = 1.0

Axis labels: $[S(d,\lambda) / S_o] \rightarrow$; d (in units of H_o^{-1}) $\rightarrow$

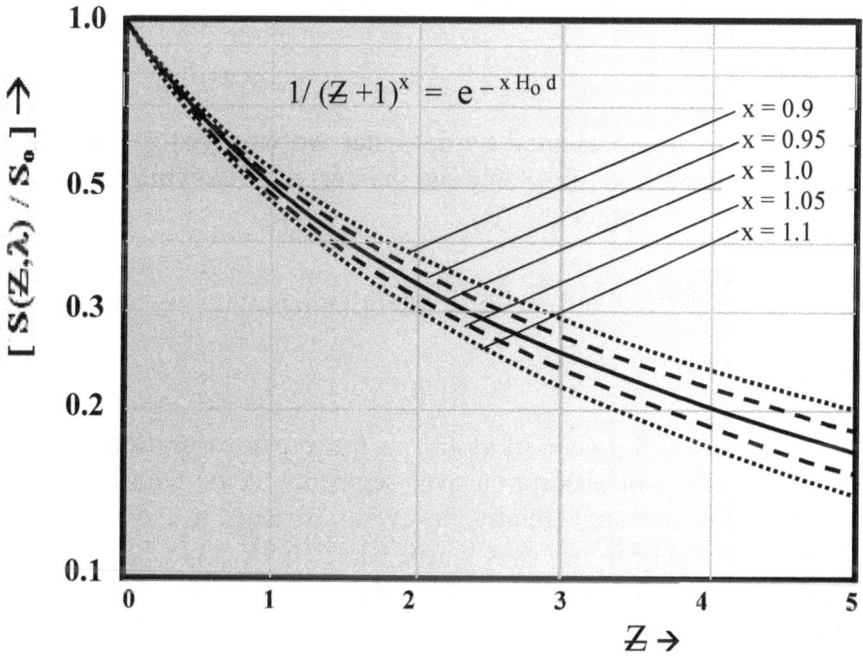

Figure 55.b: **Luminous flux loss in intergalactic plasma as a function of redshift**

Within the figure:

$$1/ (Z +1)^x = e^{-x H_o d}$$

x = 0.9
x = 0.95
x = 1.0
x = 1.05
x = 1.1

Axis labels: $[S(Z,\lambda) / S_o] \rightarrow$; $Z \rightarrow$

scattering model addressed here, a gradual lengthening of wavelength as a function of propagation distance through hot plasma. According to any cosmological model there is redshifting; it will, therefore in all cases, affect Lambert's formula. But in this model energy is absorbed by the medium such that a tiny effect only observable at cosmological distances results. At sufficiently great distances luminosity-distance relationships will clearly differ from what would otherwise have been expected from a straightforward application of Lambert's law.

The facts are that wavelength of radiation changes (very nearly) continuously along its propagation path to the observer. Nonlinearity of the wavelength dependence of the index of refraction results in the strange apparent luminosity versus distance/redshift relationship that we introduced above. Suffice it to say at this point that distance is at least approximately proportional to redshift for redshifts appreciably less than unity according to any viable theory and varies as the log of redshift (plus one) over a more considerable range. Since redshift is also referred to by the symbol 'z,' we have chosen the more distinctive symbol, Z to distinguish it from a third coordinate value at this juncture. As Hubble found, the following approximate relationship holds between distance, d, and redshift Z:

$$Z \cong H_0\, d,$$

Distance here, is represented by d. Later we will find that for larger values of redshift (i. e., for $Z \sim 1$ and greater) the following relationship will be much more precise:

$$d(Z) = H_0^{-1} \ln(Z + 1), \text{ with the differential,}$$

$$\Delta d = H_0^{-1} (Z + 1)^{-1} \Delta Z$$

Thus, for the case of $a_{p1}(\lambda)$ in the previous section, the total cumulative effect of absorption over segments of the path that can be considered infinitesimal (relative to overall distance to the source) will differ considerably from cases where wavelength could be considered essentially constant over the entire interval as assumed in previous sections.

This notion can be formulated as the products of the effects along each segment of the path as follows:

$$S(d,\lambda) = S_o \prod_{i=1}^{D/\Delta d} e^{-2\,a_{p1}(\,\lambda(d\,))\,\Delta d}$$

By substitution for the expression $d(Z)$, this becomes,

$$S(Z,\lambda) = S_o \prod_{i=1}^{Z/\Delta Z} e^{-(k/2)\,\lambda(Z\,)^2(Z+1\,)^{-1}\Delta Z}$$

Here we have defined the parameter k for compatibility with 'distance' measured as redshift, so that:

$$k \equiv H_o^{-1}\,\gamma_c\,\rho_e\,e^2 / 2\,\pi\,m_e\,c^3 \cong 0.2\,\gamma_c$$

For simplicity the value of $k = 1.0$ is used in the figures below.

The wavelength at a point along the propagation path segment that is a distance denoted by the redshift Z away from the observer of the radiation with wavelength λ can be expressed as follows:
$$\lambda(Z) = \lambda / (Z+1)$$

where $\lambda(Z)$ is the wavelength at the distance Z back along the path. Z is the redshift occurring between that point on the light path and the observer. Thus, we arrive at the formula:

$$S(Z,\lambda) = S_o\,e^{-k\,\lambda^2\int_0^Z (Z+1)^{-3}\,dZ} = S_o\,e^{-k\,\lambda^2[\,1-(Z+1)^{-2}\,]}$$

The product of the leading constant, the observed wavelength, and redshift-dependent factors in the exponent must produce a value on the order of unity in order for the ratio $S(Z,\lambda)/S_o$ to have an *interesting* value intermediate between unity and zero. Figures 56.a and 56.b illustrate the two dependencies of $S(Z,\lambda)$. Significantly, for large redshifts there is very little additional impact, contrary to what one might intuitively have thought would be the case. In fact, most all of the absorption has taken place within a redshift of 2 of its observation. This is readily apparent in figures 57.a and b. where the flux ratio is plotted against redshift.

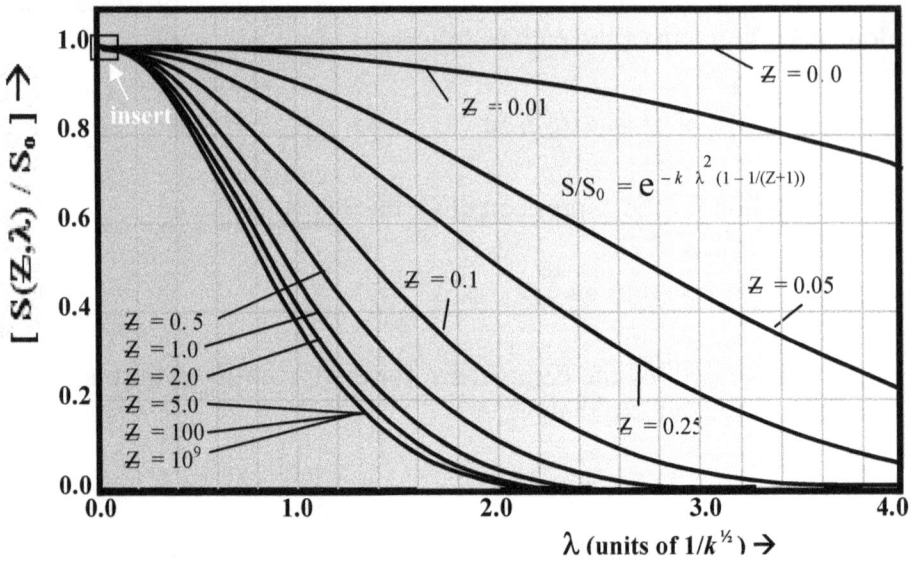

Figure 56.a: Dependence of total absorption on *observed* wavelength for various values of redshift

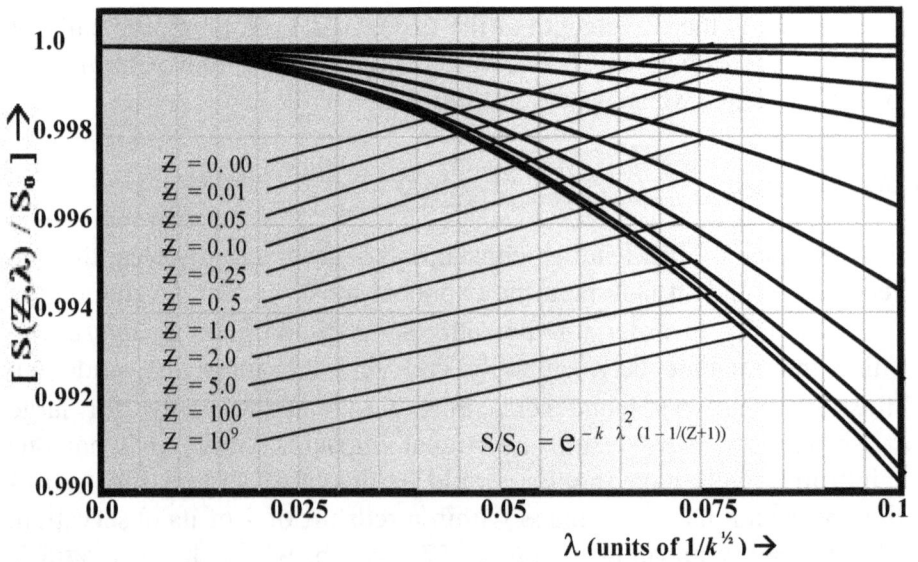

Figure 56.b: Inset for small values of *observed* wavelength for various values of redshift

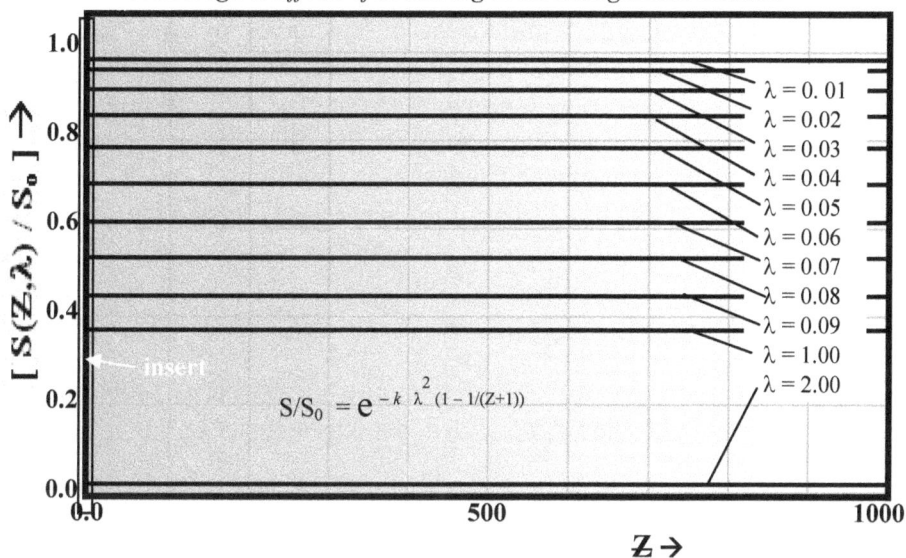

Figure 57.a: Dependence of total absorption on redshift for various values of *observed* wavelengths

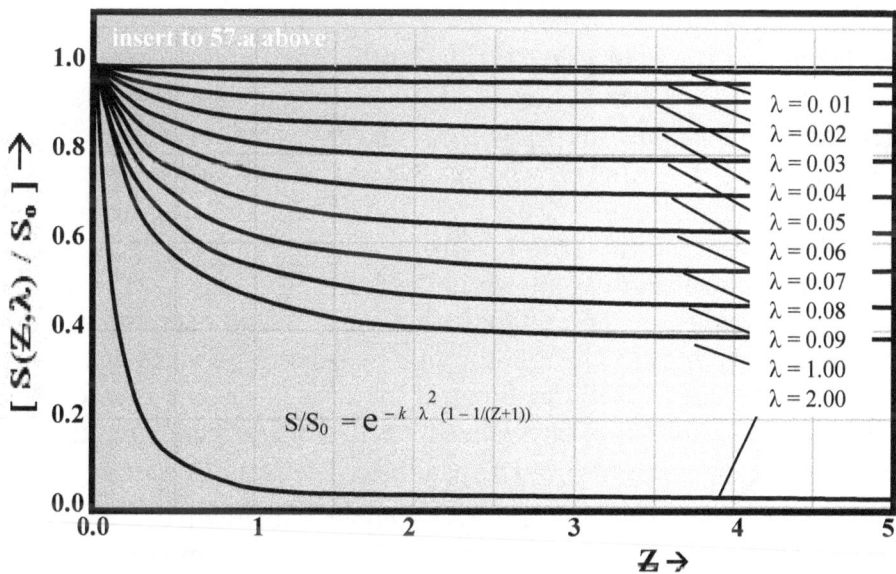

Figure 57.b: Insert for small values of redshift for various values of *observed* wavelengths

The wavelength emitted at the source of the radiation (sometimes referred to as the 'rest frame') and the wavelength that is detected at observation play somewhat symmetric roles in considering the amount of absorption that occurs in the intervening intergalactic medium. Clearly the impact of redshift on wavelength can be viewed from that other perspective – that of the impact on *emitted* radiation. In that case, the wavelength at a point along the path would be:

$$\lambda(\mathcal{Z}) = \lambda_e \cdot (\mathcal{Z}+1)$$

From this perspective, we obtain a somewhat different functionality using the same expression for $a_{p1}(\lambda)$ in the exponent of Lambert's la:

$$S(\mathcal{Z},\lambda) = S_o\, e^{-k\lambda^2 \int_0^{\mathcal{Z}} (\mathcal{Z}+1)\, d\mathcal{Z}}$$

$$= S_o\, e^{-k\lambda^2 [(\mathcal{Z}+1)^2 - 1]}$$

Here λ_e is the *emitted* (rather than the *observed*) wavelength. But since we are primarily interested in effects associated with $a_{p2}(\lambda)$ rather than the traditional $a_{p1}(\lambda)$, we will proceed with that objective.

e. deriving plasma properties from absorption data

Clearly, redshift would have a significant wavelength-dependent impact on absorption no matter which of the pertinent wavelengths one considers, whether the wavelength that is emitted or that which is observed. The form $a_{p1}(\lambda)$ of Lambert's absorption coefficient resulting from the Lorentz-Lorenz formula is very strongly affected by wavelength as shown in the preceding figures. This is not observed. Although the methods employed in this exercise will be useful in analyzing effects of absorption by neutral hydrogen at various depths of the intergalactic medium in the next chapter, this is clearly not the form of absorption realized by the intergalactic plasma, no matter what the value of ρ_e. The parameter γ is precluded from values that would render the coefficient $a_{p1}(\lambda)$ viable as alternatives.

This dependence is dramatically different than that indicated by straightforward application of Lambert's law for a dispersive medium. This difference is in large part due to the fact that redshift (as derived in up-coming chapters of this volume) is proportional to the exponential of the distance in accordance with the following relationship:

$$Z+1 = e^{+H_o\,d}$$

Therefore, in analyzing the effects of plasma absorption it is Lambert's absorption coefficient $a_{p2}(\lambda)$ identified in section c of this chapter and shown in figure 55 above that is the viable alternative. In this case the total cumulative effect of absorption over segments of the path when they are considered infinitesimal is much simpler. Here we have just:

$$S(d,\lambda) = S_o \prod_{i=1}^{D/\Delta d} e^{-2\,(a_{p2}(\lambda)/\lambda)\,\Delta d}$$

And, of course, as we saw, $a_{p2}(\lambda)/\lambda$ is independent of the wavelength λ. So we define:

$$a_{po} \equiv 2\,a_{p2}(\lambda)/\lambda = 4\,\pi\,(\rho_e\,e^2/m_e\,c)/\gamma_c,\ \text{for}\ \gamma_c \gg \omega.$$

By similar substitutions to the exercises concerning the impact of redshift on absorption we have just completed, we obtain:

$$S(Z,\lambda)/S_o = e^{-a_{po}\,H_o^{-1}\int_0^Z (Z+1)^{-1}\,dZ} = e^{-a_{po}\,H_o^{-1}\,ln(Z+1)}$$

If the leading constant factors in the exponent were to reduce to -1, then we would have

$$S(Z,\lambda)/S_o = 1/(Z+1)$$

This is what we assumed earlier where we determined $\gamma_c \cong 1.48 \times 10^{27}$ $\rho_e\ sec^{-1}$. But let us now look at what is involved in the constraint:

$$a_{po}/H_o = 1.$$

In a chapter 11 (see page 215 in particular) we will derive Hubble's constant from properties of the intergalactic plasma medium. There we will find that:

$$H_o = (3/2) \, (\, k \, h \, e^2 / \, m_e{}^3 \, c^5 \,) < T_e \, \rho_e > \cong 7.14 \times 10^{-29}$$

Using this derivation and concurrence with absorption data, we obtain:

$$a_{po} / H_o = [4 \, \pi \, (\, \rho_e \, e^2 / \, m_e \, c \,) / \, \gamma_c] / [(3/2) \, k \, h \, e^2 / \, m_e{}^3 \, c^5 \,) < T \, \rho_e >]$$

From this constraint we obtain for the plasma absorption parameter:

$$\gamma_c = 4 \, \pi \, m_e{}^2 \, c^4 / \, k \, h \, T_e \cong 6.15 \times 10^{30} / \, T_e$$

We will use this value of γ_c that we will find to be completely compatible with observed luminosity diminution data at cosmological distances. (Refer to chapter 12, figures 112 and 113 on pages 248 and 249 in particular.) The average dynamic pressure of the intergalactic medium involves the product of electron temperature and density of the plasma. We will find in later chapters that in order to effect Hubble's constant, this product would have to be:

$$\rho_e \, T_e \cong 4.13 \times 10^3 \, K \, cm^{-3}$$

Since we have obtained two expressions for γ_c, one in terms of the density along a light path, and the other interms of the electron temperature along that same light path, we have a consistency check on the determination of γ_c as follows:

$$1.48 \times 10^{27} \, \rho_e = 6.15 \times 10^{30} / \, T_e$$

From this, we do indeed obtain:

$$\rho_e \, T_e \cong 4.13 \times 10^3 \, K \, cm^{-3}$$

These values of ρ_e and T_e are not averages throughout space, but synchronously applicable to the scattering events that occur along each particular light path to a cosmological distance.

It is gratifying that the same product required to match Hubble's distance-redshift relation also accommodates the observed diminution of luminous flux from distant objects viewed through plasma. this provides a level of assurance that plasma redshift is indeed the operative phenomenon.

Given the parametric relation for Hubble's constant that was provided above based on a derivation that is forthcoming in later chapters, we have determined a consistency relationship of intergalactic plasma parameters that matches two separate types of observation.

f. absorption in neutral hydrogen

Let us consider once more the Lorentz-Lorenz formulas, now for the case of neutral hydrogen clouds and then apply the results to Lambert's law more or less as we did for the intergalactic plasma medium. We'll substitute the Lyman-α resonance frequency, ω_α for the previously employed plasma frequency, ω_0. The frequency $\omega_\alpha = 2\pi c / \lambda_\alpha$ of the Lyman-α (1216 $\overset{o}{A}$) line is what will be substituted to obtain Lambert's absorption coefficient, $a_{L\alpha=}(\lambda) = 2\pi\, Im(\mathbf{n}) / \lambda$, where:

$$Im(\mathbf{n}) = -4\pi^2 \left(\rho_e\, e^2/m_e \right) \gamma\, \omega / \left((\omega_\alpha^2 - \omega^2)^2 + (\gamma\, \omega)^2 \right)$$

Right at the line the coefficient will become:

$$a_{L\alpha=}(\lambda) \cong -2\pi\, (\rho\, e^2 / m_e c^3) / \gamma, \text{ for } \lambda = \lambda_\alpha$$

This will typically exhibit a small negative value corresponding to minimal absorption except in the immediate vicinity of the resonant spectral frequency. Absorption will become a maximum when $\lambda = \lambda_\alpha$; but, if γ is small, then absorption will become extremely large at this wavelength as indicated in the panels of figure 41 on page 82.

The absorption coefficient exhibits a symmetric functionality about the resonance line. However, this takes on a totally different form depending on whether γ is greater than or less than ω_α.

Below the line (in wavelength), i. e., for $\omega_\alpha < \omega$, it becomes:

$$a_{L\alpha-}(\lambda) \cong -(\rho\, e^2 / 2\pi\, m_e c^3)\, \gamma\, \lambda^2, \text{ for } \lambda < \lambda_\alpha \text{ and } \gamma < 2\pi c / \lambda.$$

For wavelengths above the Lyman-α line, i. e., for $\omega_\alpha < \omega$, it becomes:

$a_{L\alpha+}(\lambda) \cong -(\rho\, e^2 / 4\pi^2 m_e c^3)\, \gamma\, \lambda_\alpha{}^4 / \lambda^2$, for $\lambda > \lambda_\alpha$ and $\gamma < 2\pi c / \lambda_\alpha$.

Of course Lambert's law involves the product, $-[2\pi\, a_{L\alpha}(\lambda)/ \lambda]$ d as an exponent. So let us explore in a little more depth what this product looks like in the case of the Lyman-α absorption line. Figure 58.a shows the range of possible values of this exponent for various values of γ. This is similar to figure 41.b except that the symmetry is now complete. The figure is specific to Lyman-α absorption using a density of 10^{-5} atoms cm^{-3}. The plotted curves pertain to the entire exponent applicable to Lambert's law of absorption.

Once again we face the issue of assigning an appropriate value to the parameter γ – this time as applied to the neutral hydrogen clouds. Again, one must use observations to guide that estimation process. Figure 58.b sheds some light on the problem in as much as it indicates the amount of absorption to be expected at the peak wavelength λ_α as a function of γ. Knowing the extent of absorption in light having passed through a neutral hydrogen cloud of a known density ρ_h and extent d_h, we can estimate the value of γ_h. Since the extent of these clouds, although many light years across, do not approach 'cosmological distances' there will be no redshifting to take into account in transit.

So for example, if an absorption line occurs in the spectrum of an observed object at the Lyman-α wavelength that reduces expected luminosity by some factor in that region, then the exponent in Lambert's equation must have been increased accordingly. Thus we can make educated guesses once we have an estimate of the extent of the cloud and its atom density. The diagram of figure 58.b provides the means for doing that. Indeed, it seems reasonable to conjecture that in the neutral hydrogen clouds occurring in intergalactic space, the absorption parameter takes on values:

$$\gamma_h \ll 10^{18}$$

This range of values distinguishes the parameter applicable to neutral hydrogen from that which applies to the intergalactic plasma regions for which, of course, the absorption mechanism differs considerably and absorption is many orders of magnitude less.

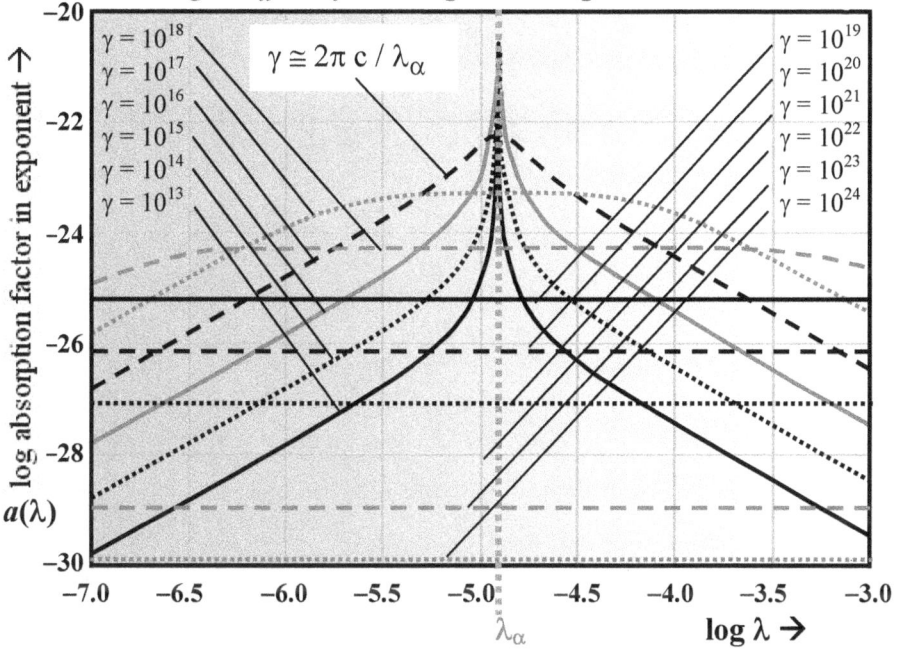

Figure 58.a: Range of values of Lambert's absorption factor

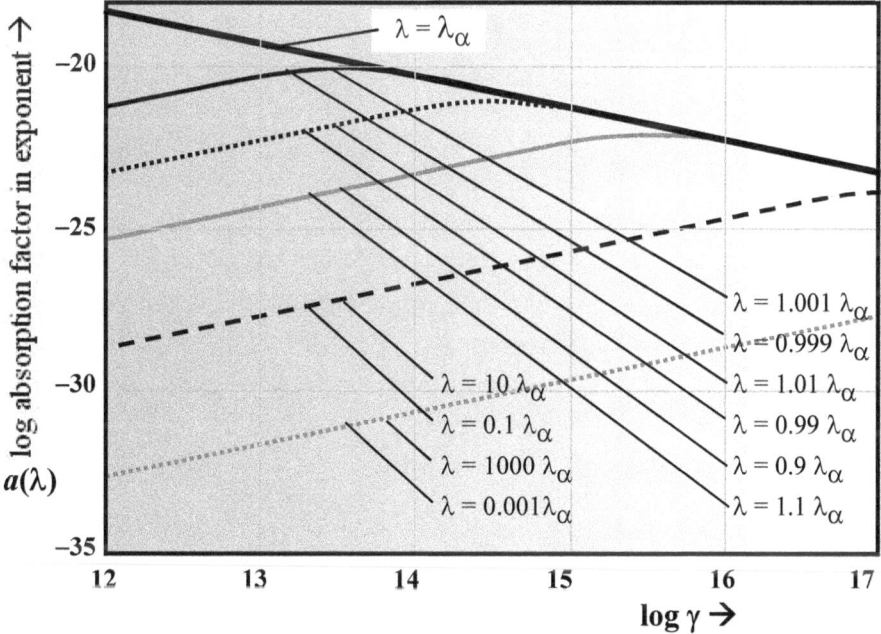

Figure 58.b: Factors in the exponent of Lambert's law applied for Lyman-α line

Chapter 7

Absorption by Neutral Hydrogen
in the Lyman-α Forests

Absorption in the vast intergalactic regions is most notable in consideration of the strong hydrogen Lyman alpha emission/absorption line in the foreground of distant quasars and Lyman break galaxies. This chapter will describe and explain that phenomena in some detail.

a. distinctive spectra of quasi-stellar objects (QSOs)

In figure 14 on page 31 an extended spectrum of a quasar was illustrated. A couple of further examples of specific quasars at very different distances were also shown over a lesser extent of their spectra in the panels of figure 15. It is the extremely jagged absorption lines on the short wavelength side of the Lyman-α peak that are readily apparent in the more distant QSOs that will interest us here. This distinctive characteristic of distant quasars, has been denominated "the Lyman-α forest"; it becomes increasingly apparent at higher redshifts. Figure 59 shows the details of this pattern from a base of 1,103 to 1,140 Angstroms for the quasar HE 2217-2818 at a redshift of 2.4. These figures support the argument that very short wavelength emissions from the quasar source are absorbed in passage to our observation point here in the Milky Way.

The primary explanation has been that in passing through intergalactic space these short wavelength radiations become redshifted to become the same wavelength as the Lyman-α resonance frequency of neutral hydrogen atoms intermediate between the quasar and the observer. The radiation is absorbed in the process of transitioning hydrogen atoms to their next higher energy levels.

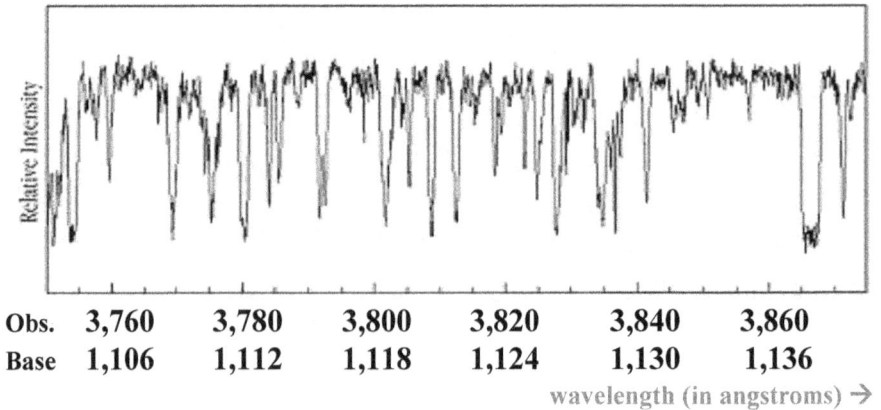

| Obs. | 3,760 | 3,780 | 3,800 | 3,820 | 3,840 | 3,860 |
| Base | 1,106 | 1,112 | 1,118 | 1,124 | 1,130 | 1,136 |

wavelength (in angstroms) →

Figure 59:   Lyman-alpha forest associated with QSO HE 2217-2818 at z=2.4

The terms 'forests' and 'clouds' that are used to refer to this phenomenon are, of course, mixed metaphors for what occurs in the foreground of these remote quasars and other galactic structures. The density of these neutral hydrogen clouds has seemed to most cosmologists to distinguish more remote regions from what we regard as our 'local' universe, and to support the notion of some sort of 'evolution' in the intergalactic regions of the universe.

b. quantum phenomena involved in absorption by neutral hydrogen clouds

In the previous chapter we addressed absorption phenomena as it pertains to the intergalactic medium. That included discussions of the more extensive plasma regions and intermittent hydrogen clouds. Although the intergalactic plasma is no doubt comprised of the same elemental constituents as the neutral hydrogen clouds that occur throughout this same domain, the dispersion effects differ considerably.

The radial plasma frequency at which transverse waves are converted into longitudinal oscillations of the plasma itself is not the sole electronic resonance in intergalactic regions. In the intervening hydrogen clouds photons with frequencies in the vicinity of atomic

130

spectral lines are swallowed up in transitioning the atoms to their higher energy levels. This phenomenon, perhaps best described by 'the Bohr atom', involves quantum energy transitions of electrons captured in the coulomb potential energy wells of the proton nucleus of the hydrogen atom. In this conceptualization shown in figure 60, bound electron energies are limited to specific quantum energy levels:

$$E_n = - m_h \, e^4 \, / \, 4\pi \, h^2 \, n^2 = - 2.18 \, (1/n^2) \times 10^{-11} \text{ ergs} = - 13.6 \, / \, n^2 \text{ eV}$$

where n is an integer n = 1, 2, 3, ...,∞, $m_h = 1.6726 \times 10^{-24}$ gm is the mass of the hydrogen atom, and h = 6.626×10^{-27} erg seconds is Planck's constant. The minus sign has to do with these energies being associated with bound states

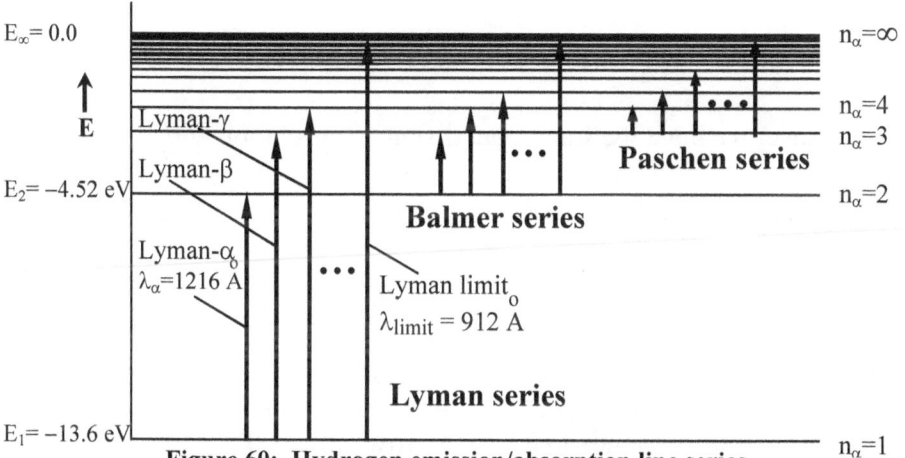

Figure 60: Hydrogen emission/absorption line series

The Lyman energy transition series is labeled in particular along with the Balmer and Paschen series that play much lesser roles, with culpability for smaller bumps and squiggles in the spectra of stars.

If the hydrogen atom is in its lowest energy state E_1 with a single electron in its lowest 'orbit', the atom can transition to the next higher energy state E_2 by absorbing a photon whose energy is given by,

$$\Delta E_\alpha \ = E_2 - E_1 = h \, c \, / \, \lambda_\alpha \cong 1.65 \times 10^{-11} \text{ ergs}$$

However, absorption implied at $\lambda \cong \lambda_\alpha$ will not be a single line of infinitesimal width, but will be spread somewhat for two reasons.

131

c. the line width of atomic spectra

One reason for there being more than mere infinitesimal gaps in the spectra caused by absorption by neutral hydrogen is the inherent line width of atomic emission and absorption spectra that is explained in terms of the quantum uncertainty principle. Since a photon of electromagnetic energy involves on the order of 10^7 wavelengths, a photon of visible light will be about 30 centimeters in length and will pass a given point in about 10^{-9} seconds. We can address the minimum action principle of quantum theory from the product of the uncertainty in time *times* the uncertainty in energy always being less than Planck's constant, h, or from the perspective of the product of uncertainties in complementary variables of position and momentum of the photon. The same result will obtain in either case, so using the former we have:

$$\Delta E \cong h / \Delta t \cong 6.625 \times 10^{-18}$$

In any case, we have the functional differential relationship,

$$\Delta E(\lambda) = h \, c \, \Delta\lambda / \lambda^2 , \text{ so that}$$

$$\Delta\lambda_\alpha \cong 4.93 \times 10^{-8} \text{ cm}$$

Therefore, emission/absorption will be spread somewhat with a half-width of about 10 angstroms, i. e., from about 1211 to 1221 $\overset{\circ}{A}$. This, in any case, still corresponds to a very narrow emission/absorption line.

This quantum-determined absorption per centimeter can be represented in a classical Lorentz-Lorenz-Lambert framework by assigning a value to γ of $\gamma_\alpha \cong 1.4 \times 10^{14}$. Refer to figure 61 for a result applicable to a nominal value of $\rho_h = 2 \times 10^{-6}$. The density is probably higher in these clouds requiring less distance to obtain total absorption..

However, in addition to this extremely narrow natural line width, there is an effect caused by thermal motions of the atoms in any such HI cloud. This is the well-known Doppler line broadening of emission lines in the chemical spectra of thermal sources. This phenomenon is caused by the Doppler effect on the radiation due to the various motions of the objects involved in the emission/absorption process. The Maxwell-Boltzmann distribution of energies in a thermal medium that was illustrated in figure 33 for free electrons pertains to all

132

substances that are in a stable equilibrium, although the specifics of the distribution of velocities will depend intimately upon the mass of the entities involved. In the case of a hydrogen gas, each atom will have its emitted radiation uniquely Doppler-shifted, when viewed by observers stationary with respect to the mean velocity of the emitting substance, which velocity we will assume in our discussion to be zero. Similarly, the light that is absorbed will be that which is at the Doppler-shifted frequency. So, of the multifariously emitted photons that will be observed, all will have slightly differing frequencies and associated wavelengths. At high temperatures this will produce a considerable 'broadening' of the natural width of the spectra reflecting the range of the velocities in the distribution as shown in figure 62 below.

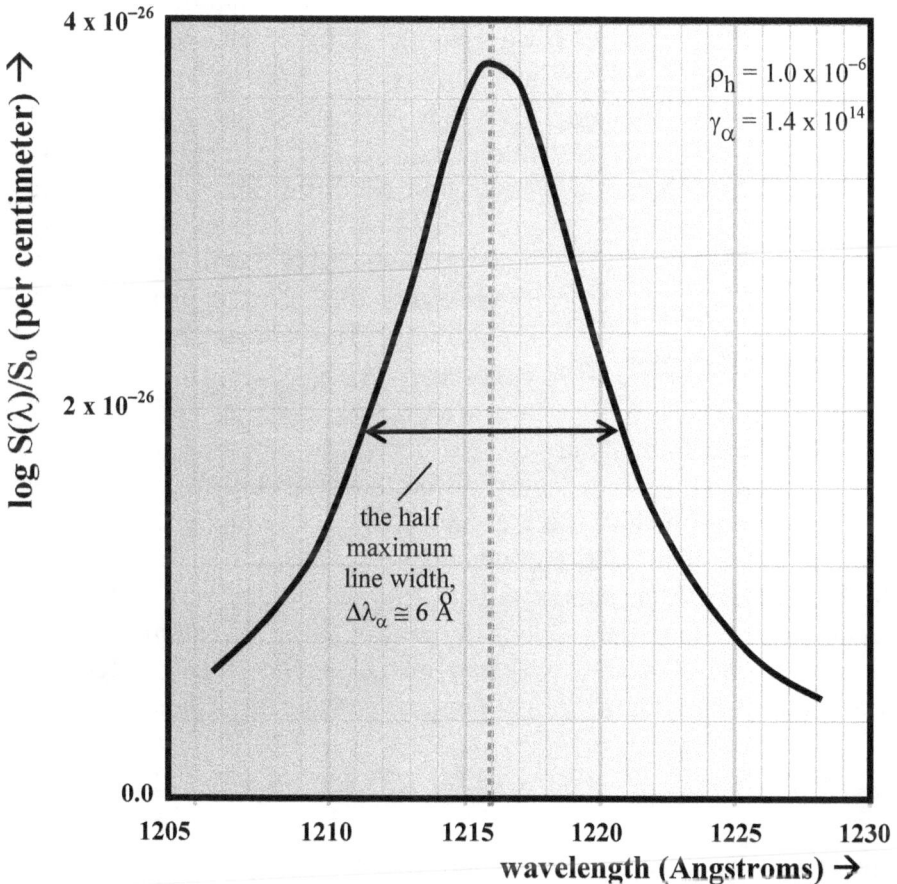

Figure 61: Absorption factor showing the natural line width of the hydrogen Lyman-α line

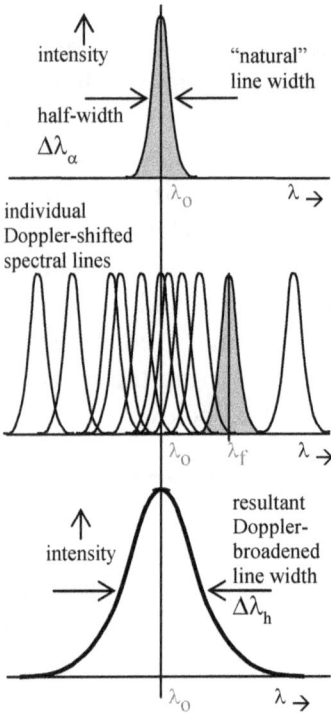

Figure 62: Spectral line broadening effects

Each atom will be traveling at a unique velocity loosely determined by the kinetic temperature of the cloud. We should expect this temperature to be just below the ionization temperature of hydrogen – somewhere in the vicinity of 10^4 K corresponding to a Lyman-α energy transition. In this range of temperatures, the Maxwell-Boltzmann distribution will place the majority of constituent atom velocities between about $\beta = 10^{-5}$ and 10^{-3} relative to the speed of light. The distribution of the individual Doppler wavelength shifts will be distributed accordingly. So emission and absorption lines will be spread more considerably by this effect. Either Lyman-α emission or absorption occurring for individual atoms in a cloud that is in thermal equilibrium at a temperature of 10^6 K will be observed at wavelengths that are Doppler broadened accordingly. The average shift will be approximately $\Delta\lambda \cong \pm 2.2 \times 10^{-7}$ cm or 22 Å.

When this is added to the natural width, it gives a total half-width of 54 angstroms, i. e., from about 1162 to 1270 Å.

Doppler line broadening differs considerably from the natural line width in as much as it is a composite effect involving ensembles of photons rather than each individual photon identically. In this effect, the spectrum of each photon is uniquely shifted one way or the other due to vagaries in radial motions of emitting/absorbing atoms toward or away from the observer, as shown in the second panel of figure 62, with broadening a summation of the diverse velocity-dependent effects.

$$\Delta\lambda_h \cong \Delta\lambda_\alpha + \beta_{average}\,\lambda_\alpha \cong 6.0 \times 10^{-8} + 1.83 \times 1.216 \times 10^{-10}\, T^{\frac{1}{2}} \text{ cm}$$

The composite absorption coefficient will be $\gamma_h = \gamma_\alpha + \gamma_T$.

'Transverse Doppler' will always be negligible with regard to the total increase in the width of the spectral line. The very slight tendency of the broadened line to be shifted to the right (i. e., toward longer wave-lengths) on this account will be insignificant. Only half the square of a usually small fraction of the transverse velocity divided by the speed of light is involved in transverse Doppler and the atomic velocities are much smaller than for free electrons at the same temperature. There is only a single instance involved in each particular energy state here, as against repeated instances involving 'secondary' emissions in longer distance transmission via many extinction intervals in a very hot thermal plasma in the intergalactic regions between hydrogen clouds that we will discuss in later chapters.

d. comparing emission-absorption profile data

The operative phenomenon thought to account for 'Lyman–α forests' largely as associated with quasars observed at high redshift involves the absorption of radiation from the background quasar at the frequency of the foreground HI Lyman-α line. The observed relationship between the absorption and emission of the Lyman-α line in neutral hydrogen (HI) clouds is shown in figure 63. Clearly, the Doppler half-width is about 30 km/sec, corresponding to a velocity of $\beta_f = 10^{-4}$ similar to our calculations above, but with a temperature of about 4.0×10^6 K rather than the 1.0×10^6 K that we used in that case. Bold dotted lines have been added to the figure to represent what we would have predicted. The observed data shown in the figure is from Kellermann (1993). Both absorption and emission occur at the same frequency in such a mix of neutralized ions and hydrogenous plasma. So obviously there are indeed clouds at temperatures in excess of 10^6 K in intergalactic regions which yet posses some neutral HI. Absorption profiles are more jagged than profiles for emission at this temperature – evidently this is caused by rapid re-emission from within the cloud since the temperature is well above the minimal ionization level.

There are also hydrogen HI protogalaxies such as those for which a spectrum is provided in figure 64 (below). Here the line width is much narrower – more like 12 angstroms corresponding to a cooler emission environment of about 10^4 K in temperature that corresponds more closely to the finer absorption structures shown in figure 59 and cited by Smith (1990) and Boroson et al. (1991). So there seems to be a more or less continuous range of hydrogen possibilities from the high temperature intergalactic plasma to cooler protogalaxy formations.

But let us look at what we should expect in the way of absorption profiles that would be caused by HI clouds between an observed source and its observation. Suppose there is a remote object for which the emitted spectrum provides a uniform intensity over some region short of, but including, the Lyman-α line of the electromagnetic spectrum as shown at the top right of figure 65. When viewing this object through an HI cloud, one would expect absorption as shown in a second instance of the spectrum. If another cloud exists in the same path but significantly closer to the observer such that an appreciable redshift occurs, observations would be as shown for the third spectrum.

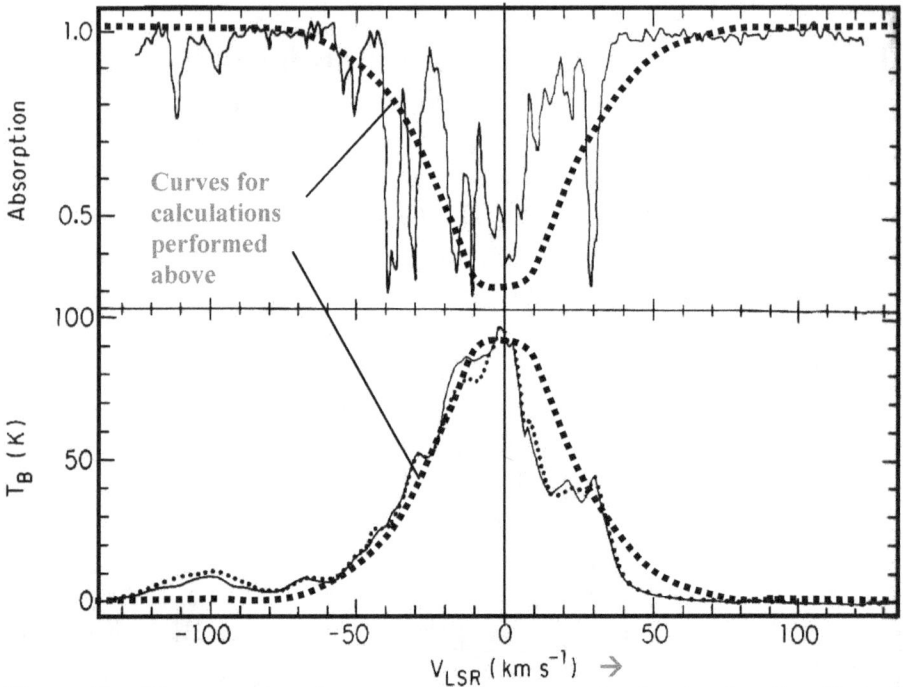

Figure 63: The operative phenomena of Lyman–α clouds, showing the effect of HI clouds on the observed spectra. (Kellermann, 1993, p. 98.)

Clearly the intensity, line width, and separation of absorption lines in the region below the Lyman-α emission wavelength tells us a great deal about the nature and distribution of the associated HI clouds. However, it must be kept in mind that the apparent frequency of occurrence of these lines is determined by the amount of redshift occurring between clouds, *not* the distance between them. Since redshift increases exponentially with distance, the frequency of these lines will also increase exponentially even if the clouds are regularly spaced.

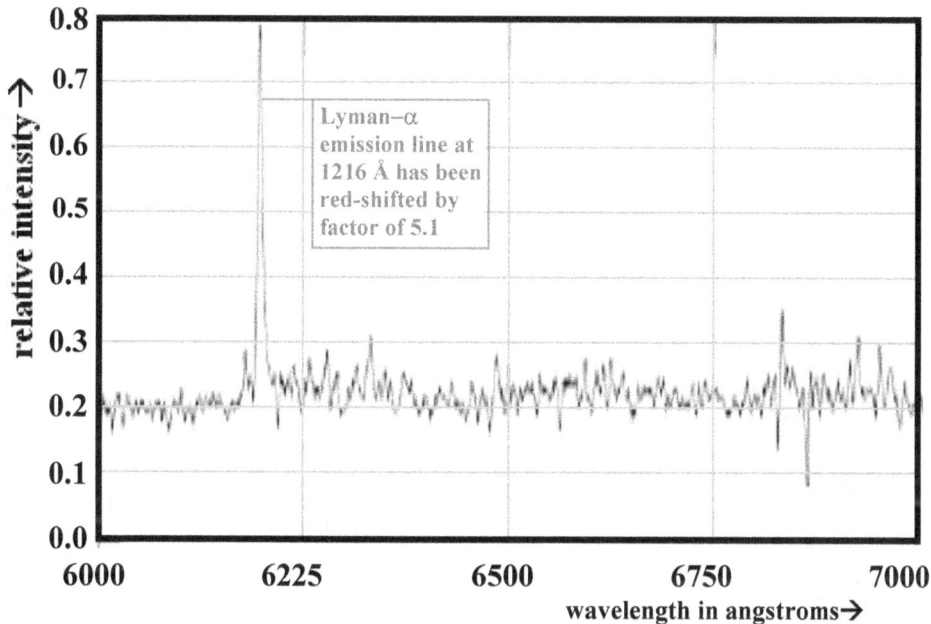

Figure 64: **The measured redshift of a young galaxy in a cluster near TN J1338-1942 which is at z = 4.1. (ESO – phot-11b-02, April 02)**

Figures 15 and 59 illustrated this trend in which the number and intensity of Lyman-α absorption lines increases from z = 0.158, to 2.4, and 3.62. In panel a of figure 15 we saw that at an emission wavelength of 1050 angstroms there was a broad absorption 'line' with a half-width of about 21 angstroms with a spike of emission occurring at the center of the trough. Since the absorption is nearly complete, we can infer a large cloud at a temperature of about 10^6 K. There are a few other anomalous depressions in the spectrum suggesting that more dispersed smaller clouds may have been encountered as well. Figure 59 provides evidence of many very brief encounters and a couple of more major encounters in this short segment of the spectrum, but nothing that completely obliterates the emission spectrum. In panel b of figure 15 we saw many instances of complete obliteration over narrow intervals such that there is zero transmission at these specific wavelengths, but very few cases involving very appreciable breadth.

137

As redshift approaches z = 6 (shown in figure 66 below) we see that there is virtually complete absorption of all emissions shorter than λ_α.

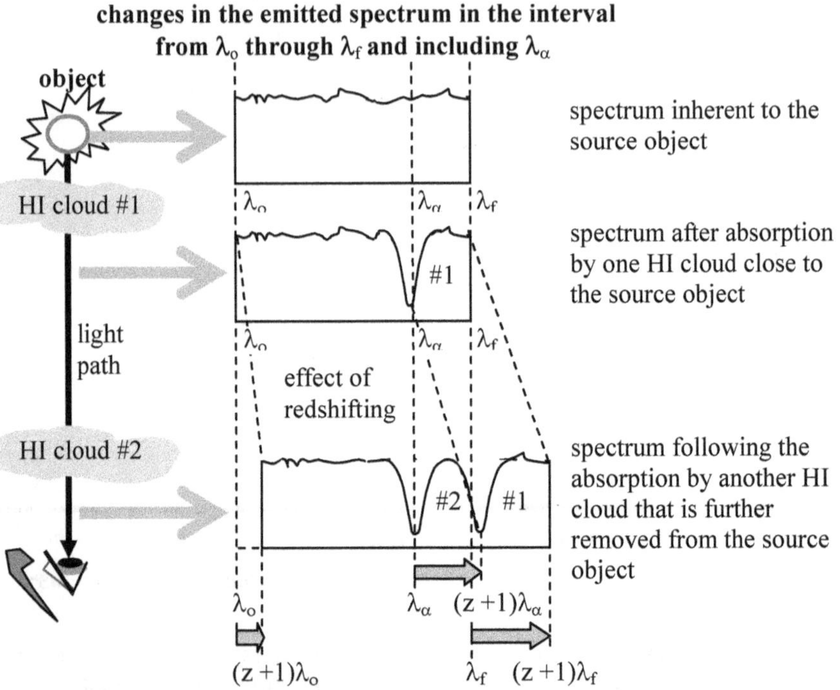

changes in the emitted spectrum in the interval from λ_o through λ_f and including λ_α

object

HI cloud #1

light path

HI cloud #2

spectrum inherent to the source object

λ_o λ_α λ_f

#1

spectrum after absorption by one HI cloud close to the source object

λ_o λ_α λ_f

effect of redshifting

#2   #1

spectrum following the absorption by another HI cloud that is further removed from the source object

λ_o λ_α $(z+1)\lambda_\alpha$

$(z+1)\lambda_o$ λ_f $(z+1)\lambda_f$

Figure 65: **Depiction of operative phenomena of observations over distances for which redshifting is appreciable with intervening HI clouds**

In chapter 3 we discussed the 'Gunn-Petersen trough' and what it implies relative to the characteristics of the intergalactic medium. Here we see it as the proof of neutral hydrogen in intergalactic regions. At extreme distances (redshifts) the electromagnetic emission spectra beneath the Lyman alpha wavelength decreases exponentially. Virtually all radiation below 912 angstroms is completely absorbed by neutral hydrogen. Refer back to figure 60 to see why neutral hydrogen is not selective in absorbing all radiation below this wavelength.

Similar phenomena occur in what are called 'Lyman-break' galaxies as shown in figure 67 where it is apparent that for galaxies observed at high redshift, there is an ever deepening and widening 'trough' where transmitted radiation from the galaxy beneath the Lyman-α line is increasingly obliterated.

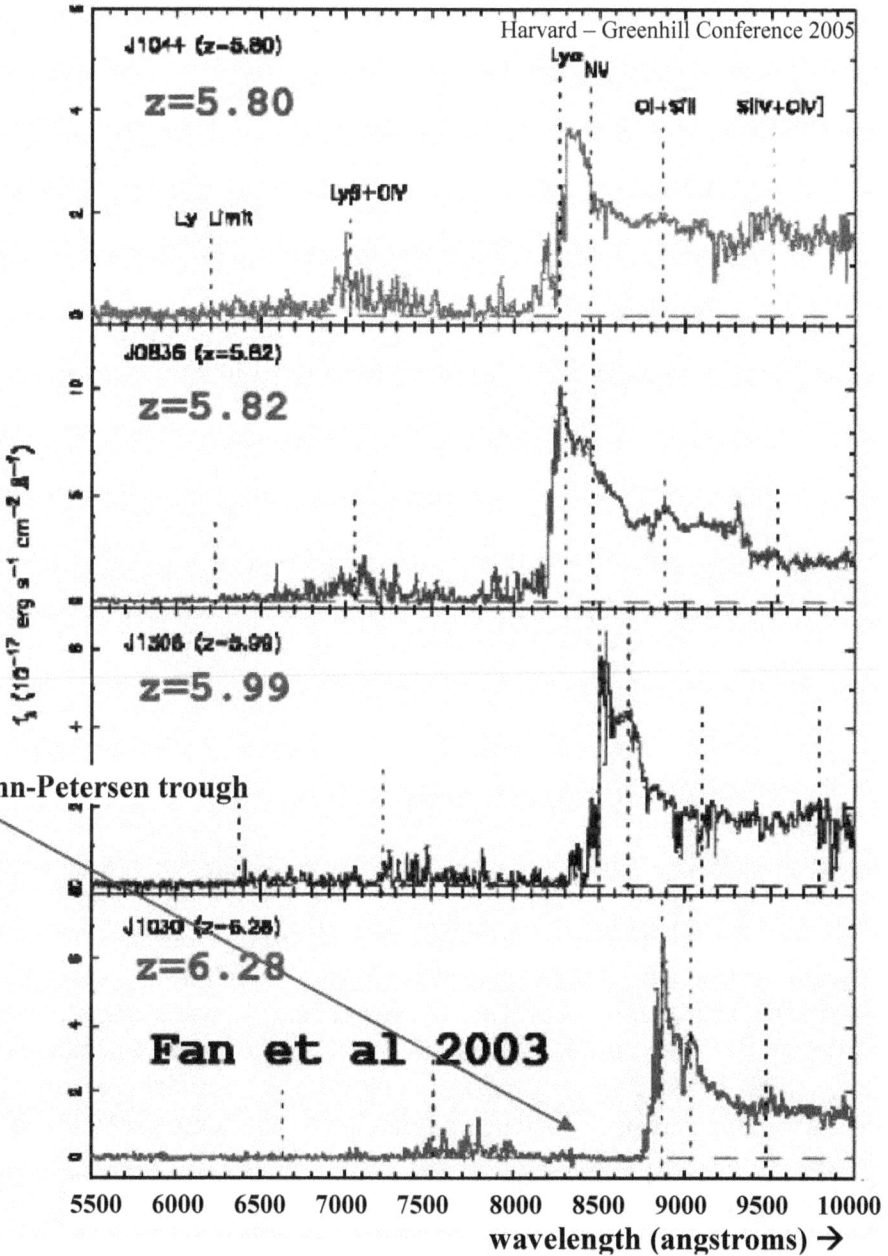

Figure 66: The Gunn-Petersen trough as it affects QSO spectra

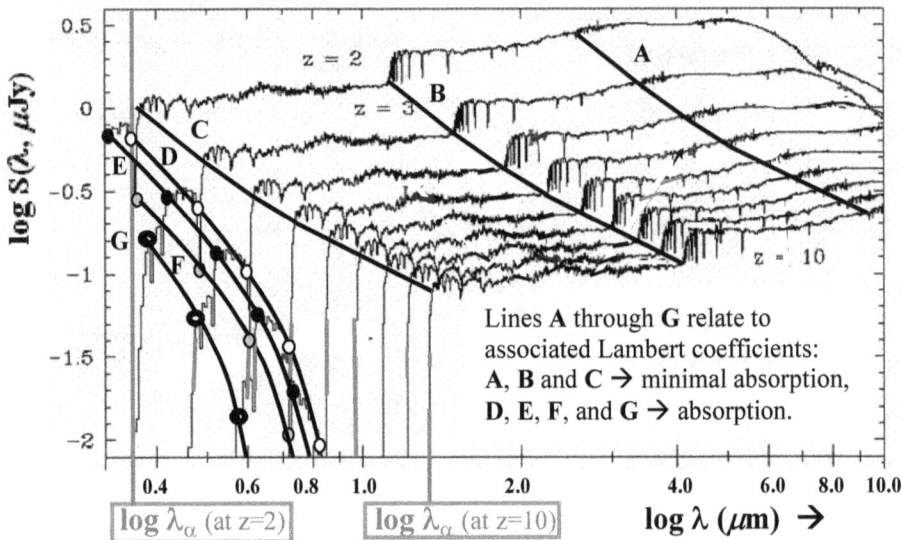

Figure 67: Typical Lyman-break galaxy spectra vs. redshift, with IRAC GOODS Legacy team survey limits indicated. Figure is derived from M. Dickenson, GOODS Legacy team (and STScI) Original at: *http://ssc.spitzer.caltech.edu/documents/compendium/xgalsci/*

e. anomalous effects of redshift on Lyman-α absorption

Naturally no two quasar or galaxy spectra are exactly alike so differences between the various spectra cannot be attributed exclusively to Lyman-α absorption. Of course also, each encounters a unique field of hydrogen clouds. The straighter lines A, B, and C added to the plot at the top right of figure 67 illustrate that absorption *per se* does not appreciably alter the spectrum in this region other than as discussed relative to broadband (wavelength-independent) plasma absorption. However, the lines D, E, F, and G at the lower left indicate an exponential increase in absorption versus redshift for features in the spectra at short wavelengths when plotted on the log scale for λ, where

$$\lambda = (z+1)\, \lambda_{base\_feature}$$

The parameter $\lambda_{base\_feature}$ is the wavelength of a particular feature in the spectrum of an object in the base (or emission, z = 0) frame of the object. The wavelength and observed flux from individual features at

140

various increasing redshifts are indicated by the lines A through G that have been added to figure 67. These lines make it clear that very different absorption phenomena are involved in regions above and below the base value of λ_α.

Expected ratios of observed-to-initial flux density, $S(d,\lambda) / S_o$ due to absorption by the intergalactic plasma differ considerably from those for absorption by neutral atoms as we discussed in the previous chapter. But both expressions are based on the relation,

$$S(d(z),\lambda) = S_o(z,\lambda) \, e^{-4\pi \, Im(\mathbf{n}) \, d(z) / \lambda}$$

where $Im(\mathbf{n})$ is the imaginary component of the complex index of refraction, the impact of which is described earlier and illustrated in figures 33 through 41, 53, 57 and 58. Now we have:

$$Im(\mathbf{n}) = -2\pi \, (\, \rho_h \, e^2/m_e) \, \gamma_h \, \omega \, / \, ((\omega_\alpha^2 - \omega^2\,)^2 + (\, \gamma_h \, \omega\,)^2\,)$$

$$= -(\, \rho_h \, e^2/m_e) \, \gamma_h \, c \, / \, \lambda \, \{ \, [\, 4\pi^2((c/\lambda_\alpha)^2 - (c/\lambda)^2\,)^2\,] + (\, \gamma_h \, c/\lambda\,)^2 \,\}$$

Here ρ_h, γ_h, and λ_α differ considerably from the plasma values and γ_h differs from γ_α (the natural line width coefficient) as we have seen also.

We mentioned in reference to plasma absorption, and will derive in considerable detail in immediately succeeding chapters, that the scattering model described in this volume involves a relationship between the distance that light propagates through a hot plasma, $\delta(z)$ and the redshift incurred in transit along that path, as follows:

$$\delta(z) = H_o^{-1} \, \ln(z+1)$$

Notice that luminous flux, $S(\delta(z),\lambda(z))$ is altered by distance (and therefore redshift) even without the occurrence of any absorption at all because of the inverse square relationship of luminosity with distance, $\delta(z)$ in classical physics. It is altered also by quantum energy diminution caused by the very fact of increases in wavelength, a parameter that is inversely proportional to radiational energy. This latter effect introduces a factor of $z + 1$ into the denominator in addition to the factor involving the inverse square of distance.

But as was shown in regard to figure 55.b on page 117, there is yet another factor of z + 1 that appears in the denominator because of the broadband (wavelength-independent) absorption by the plasma. This form of absorption is directly associated with forward-scattering processes to be explained in the next few chapters. The functionality of that total dependence on redshift is exhibited in the less dramatic lines A through C with slight upward curvature of on the log scale of figure 67. In this domain for which $\lambda_{base} \gg \lambda_\alpha$ and $\lambda_{base} \ll \lambda_\alpha$ there is virtually no absorption by neutral hydrogen or other elements. This functionality of the diminution of luminosity is given by:

$$S(d(z),\lambda(z)) / S_o = 1 / (z+1)^2 \, \delta(z)^2 = H_o^2 / (z+1)^2 \ln^2(z+1)$$

However, in dealing with quasar spectra it is absorption by neutral atoms with which we are concerned. Only one of the factors of $1/(z+1)$ involves absorption. The inverse square relationship with distance has to do with fewer of the emitted photons being subtended by an aperture as a function of distance – no photons are lost to this mechanism. The quantum effect merely acknowledges that there is less energy present in a redshifted photon, but the photon itself has not been obliterated. The remaining plasma absorption factor that balances the energy equations is the only aspect resembling traditional absorption *per se* in the above equation. So in what follows, we will include only that factor in the absorption analyses.

In addition to the plasma absorption effect, in the immediate vicinity of the Lyman-α (or other) neutral hydrogen line, there is a much more comprehensive absorption process that substantially alters the functionality. The spectral features below the Lyman-α line slide continuously up into, and past, the Doppler broadened line to which it is particularly vulnerable to absorption, causing additional lines to crop up in the spectrum as was shown in figure 65.

Although the absorption process that involves the $1/(z+1)$ factor discussed above that is appropriate to the domain where $\lambda_{base} > \lambda_\alpha$ is operative also for $\lambda_{base} \le \lambda_\alpha$, we assume that it is independent of this other neutral hydrogen process. The rationale for this assumption is that redshifting as well as associated broadband absorption phenomena are unique to the hydrogenous plasma that exists *between* (and perhaps in the midst of) the hydrogen clouds. The phenomenon to be described

142

now takes place in the neutral atoms themselves and involves any neutral hydrogen encountered along the light path. The thickness of the clouds is much less than the separations between them, or equivalently, the percentage of neutral hydrogen atoms that are mixed throughout the plasma is quite small. Thus, the radiation will have been shifted by some, however tiny, amount when it encounters the next cloud (or atom) along its path. In the equivalent mixed state scenario, the results will be the same even if accomplished more contemporaneously.

So we integrate the cumulative effect of absorption over segments of the path using the same analyses that we used to determine the effects of redshift on alternative possibilities of plasma absorption in the previous chapter. We begin by again accepting a result to be obtained in a later chapter, namely an infinitesimal distance increment,

$$dr(z) = H_o^{-1} (z+1)^{-1} d(z+1)$$

Then using the Lorentz-Lorenz formula in Lambert's law above and the $1/(z+1)$ plasma absorption factor, we obtain the total absorption as a product of the neutral hydrogen absorption factors along the path:

$$S(d(z),\lambda) / S_0(z,\lambda) = [\ 1 \ /(\ z + 1 \ )] \prod_{i=1}^{d(z) \, / \, \Delta d} e^{ - f(\, \lambda_{base}, \, z+1 \,) \, \Delta d(z+1)}, \text{ where}$$

$$f(\lambda_{base}, z+1) \, \Delta z = \{ \ 8\pi^2 \ (\ \rho_h \ e^2/m_e \ H_o^{-1}) \ \gamma_h \ / \ c \ (\ \lambda_{base} \ (z+1) \ )^2 \ \} \ \Delta(z+1)$$

$$/\{[4\pi^2 \ c^2 \ ((1/\lambda_\alpha)^2 - (1/\lambda_{base} \ (z+1))^2 \ )^2 \ ] + [\ \gamma_h \ / \ \lambda_{base} \ (z+1)]^2\}$$

The product of exponential functions becomes a summation of their exponents. So that when the distances along individual path segments are considered infinitesimal (relative to the overall distance to the source), the summation can be replaced by integration. Thus, we arrive at the formula:

$$S(z,\lambda_{base}) / S'_o = e^{-\ln(z+1)} - \int_1^{z+1} f(\lambda_{base}, \zeta) \, \Delta\zeta$$

The natural logarithm of $S(d(z),\lambda) / S_0(z,\lambda)$ is plotted as a function of redshift in figure 68 below. We have used, $\rho_h = 10^{-7}$, and $\gamma_h = 8 \times 10^{14}$ in this figure. The density value reflects that there is a fraction of the intergalactic medium that is comprised of neutral hydrogen. This value is probably too high, but it is only the form of the solutions that we are interested in here. The absorption constant reflects a temperature that is also probably too high, but adjusting these two parameters will not destroy the obvious form of the solution.

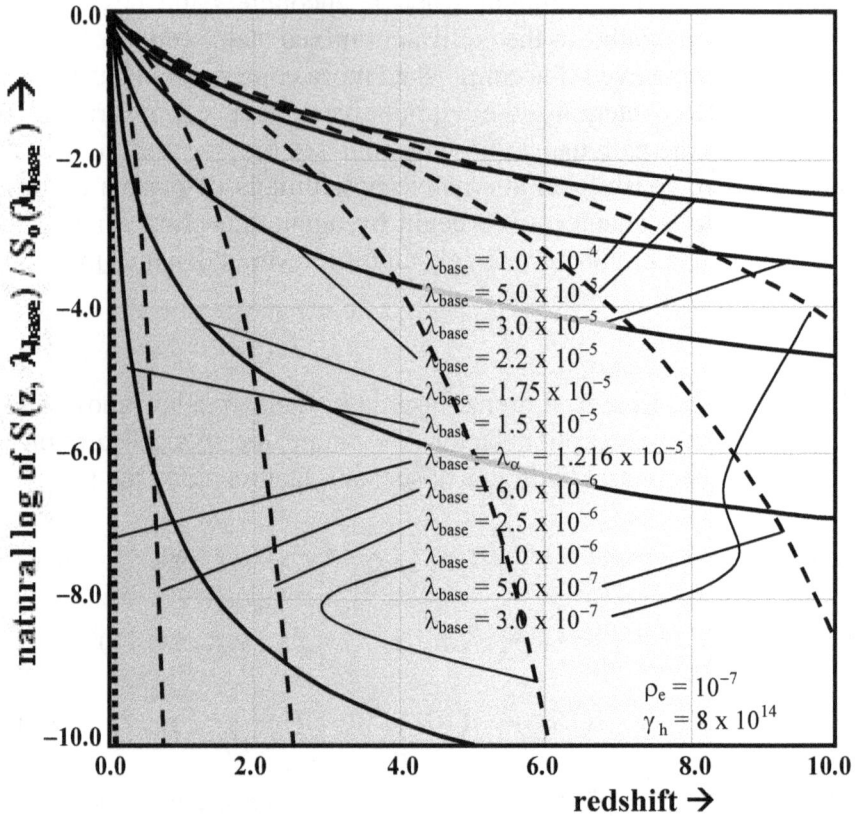

Figure 68: Form of predicted absorption characteristics by neutral hydrogen on both sides of the Lyman-alpha absorption line

There are two domains with very different absorption profiles in the equation above, that pertinent to base wavelengths that are below the Lyman-α line, and those that are above. This cleavage occurs right at the line. In this plot we have shown the form for $\lambda_{base} = 3 \times 10^{-7}$ cm, which is somewhat unrealistic because the Lyman break at 9×10^{-7} cm, but it illustrates the form. It is also of note that the Lorentz-Lorenz formula can be (and should maybe have been) extended to include more than the single resonance frequency. However, it would not have changed much in the domain for which we are interested, with virtually total absorption possible below the Lyman break.

What is illustrated in particular in this plot is the exponential increase in absorption below (but *not above*) the Lyman-α line. This predicts/collaborates the lines D through G in figure 67.

In figure 69 we have plotted the very same equation as a function of λ base for various redshifts. Certainly this plot shows values that are so tiny that they could never actually be measured, but a feel for the extent of the predicted total obliteration of the spectrum that is what is observed (as shown in figure 66) is provided. Of course this only accounts for obliteration down to the Lyman-α break (dotted vertical line); below that it is not constrained by quantum levels.

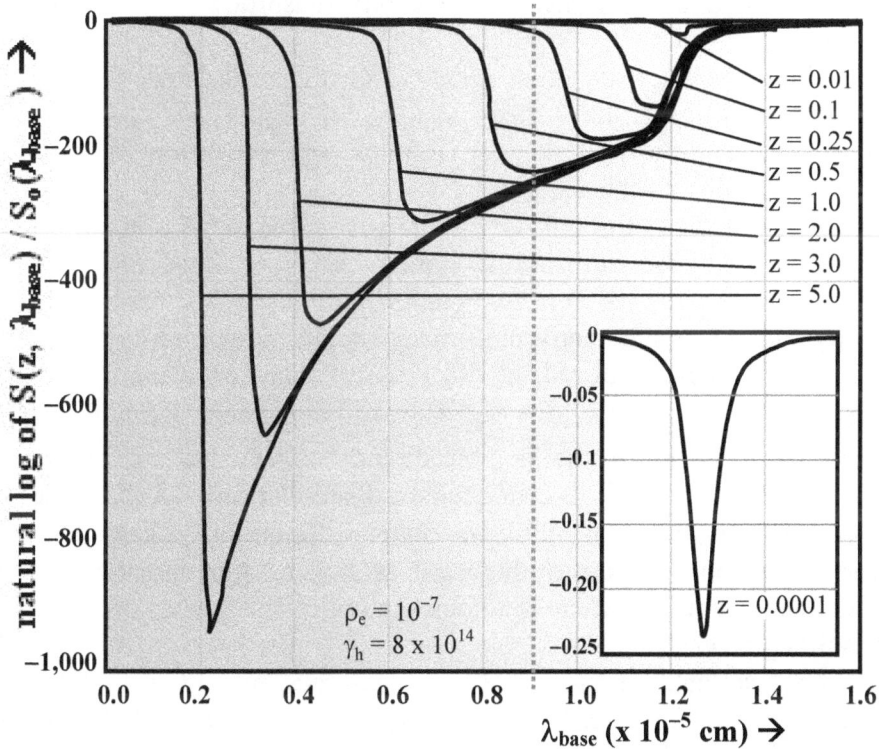

Figure 69: The Gunn-Petersen trough of neutral hydrogen absorption as implied by the Lorentz-Lorenz formula and Lambert's law

In these figures there is no attempt to precisely model individual absorption lines as they would show up at random in the spectra depending on the particular nature of each neutral hydrogen cloud that is encountered or to the particular separations between them. It is simply handled probabilistically based on a uniform column density of neutral hydrogen. But clearly, at these extremes it hardly matters whether the absorption occurs in bunches or continuously, and it is clear that the existence of increasing absorption at low redshifts and extreme absorption at larger redshifts does not depend on evolution of the intergalactic medium. A uniform medium accounts fully for the accelerated absorption. Nor should one think that the absorption takes place primarily at high redshifts. It does not. The most vertical aspects of the absorption profiles occur the closest to the observation point. The fact that there is so little absorption apparent in closer quasars is due to the fact that the spectrum between the Lyman-α line and the Lyman break wavelength have been quickly redshifted out of harms way.

This characteristic functionality clearly exhibited in the absorption profile of the galaxy spectra of figure 67 has been demonstrated as following directly from absorption processes in a uniform medium. At the various values of $\lambda_{base} \leq \lambda_{\alpha}$ that have been connected by lines D through G for increasing redshift in figure 67 clearly have the same form as the similar curves in figure 68. The illustrated prediction is for two unique domains of absorption characteristics to an inherently uniform spectrum by neutral hydrogen absorption over a range of redshifts of 0 to 10. The vast disparity that occurs at the transition $\lambda_{base} = \lambda_{\alpha}$ is readily apparent. Figure 70 converts this to a form to that is more directly comparable with figure 67. The very same break in characteristic features occurs at the same Lyman–α base wavelength and the predicted differences in phenomena appear on the opposite sides of this break as shown. Agreement is also obtained with figure 66 where total obliteration of spectra has occurred in redshifts greater than 5.

The explanation presented here assumes merely a uniform density, ρ_h of neutral hydrogen atoms anywhere along the path, whether within the clouds or mixed in whatever way with the plasma.

The explanation does *not* presuppose an obscurely theorized evolution of neutral hydrogen density becoming more diffuse or less ionized as time goes by even though the resultant absorption increases nonlinearly with 'lookback time'. Although agreement with observation is based on a uniform density of neutral hydrogen clouds – at least a

146

uniform column density at large scales – how these neutral atoms are distributed doesn't affect the absorption pattern very much.

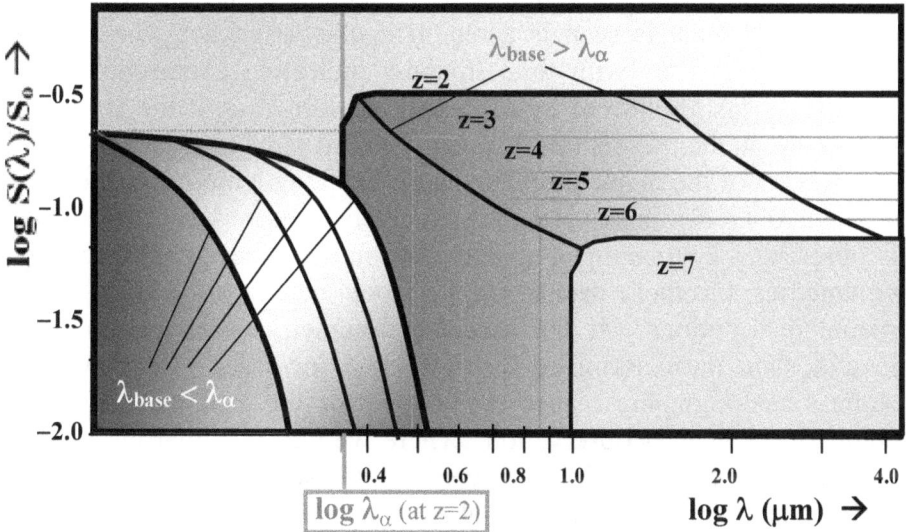

Figure 70: Symbolized effect of the predicted absorption characteristics of neutral hydrogen shown in figures 68 and 69 on uniform spectra on both sides of the Lyman-alpha absorption line showing a very basic agreement with the observations shown in figure 67.

This absorption effect also is independent of the behavior of electromagnetic transmission characteristics through more extensive distances of intergalactic plasma. These dissimilar states of similarly constituted substance produce independent effects. The hydrogenous plasma regions absorb very little radiation of astronomical interest other than to supplement the general diminution of luminosity. The neutral hydrogen regions on the other hand will produce virtually no redshifting by the mechanism to be described in the following chapters but will always be involved in absorbing radiation with wavelengths below the Lyman-α line.

f. redshift distribution of neutral hydrogen clouds

Observations of the redshift distribution of neutral hydrogen clouds are key to deciphering whether 'evolution' and/or expansion is occurring that affects this absorption phenomenon. We have treated it in this chapter as being fully accounted by the scattering model with a completely uniform distribution. However, see Monier et al. (1999) for actual observation data on sizes, shapes, and distributions of these clouds. Also, Ashmore (2008) cites several references to justify his claim that, "What is seen is a region in the past where the average separation increases with time (thereby indicating expansion?) *[The question is his.]* followed by a gradual slowdown until the Hydrogen clouds are, in the present epoch, evenly spaced."

One of the major problems with accepting a model, particularly a 'standard' model, is that it tends ineluctably to become a filter on what is observed. For example, spatial distributions depend intimately on volumetrics, but these depend on the model one employs in assessing spatial distributions. If the accepted volumetric assumptions prove invalid, then the determined distribution is incorrect. Later we will address cosmographic metric predictions for the various models for comparisons against observation. There we will find that the volume metric for the currently favored version of the standard model increases with redshift much more rapidly at intermediate redshifts than does the comparable metric for the scattering model although at small redshift ('present epoch') they are virtually identical. Consistently applying the scattering model Euclidean volumetrics rather than those of the currently favored version of the standard model reduces separations of the clouds at the higher redshifts so that the uniform distribution is consistent with the data when applied to the scattering model as we have done above.

Lehner et al. (2007) identify distribution characteristics of cold ($\leq 10^5$ K) Lyman-α regions (NLAs with narrow line absorption, $\Delta z \leq$ 40 km sec^{-1}) and warmer ($\sim 10^5$ to 10^6 K) regions (BLAs with broad line absorption, $\Delta z \geq$ 40 km sec^{-1}). They note median and mean Δz-values 15%-30% higher at low z than at high z, and suggest that the number density of BLAs to NLAs at redshifts less than about 0.4 is about a factor of 3 higher than for redshifts between 1.5 and 3.6. It is difficult to unravel this data from the quagmire of standard model metrics we will discuss, but it is apparent that effective broad line absorption is strongly affected by the volumetric characteristic of that assessment.

Chapter 8

The Unique Constraints
of Forward Scattering

The forward scattering effect of homogeneous isotropic media involving many constituent electrons is an instance where coherence of electromagnetic fields plays an all-important role. Individual effects of polarization symbolized in figure 36, page 73 accumulate because of constructive interference. The superposition principle guarantees that scattering effects can be handled independently of ongoing incident radiation. We will look into mechanisms whereby, rather than just gradually absorbing incident radiation, intermediate scattering electrons actually effect the replacement of the incident radiation by similar radiation with a different phase and slightly different velocity.

The operative replacement mechanism preserves the ability to resolve images of distant objects. It involves the summation of the individual effects of the various scattered electromagnetic fields from intermediate electrons to effect replacement of the incident radiation.

a. characterization of the scattered field

So we will consider the electric field emanations of the induced electron oscillations of the individual constituents of the scattering medium. The radial component of these resulting field emissions diminish as the inverse square and higher order powers of the distance from the constituent electrons as shown earlier. The radial component will, therefore, be negligible after a modest distance with respect to the wavelength of the incident radiation. This is in contradistinction to a transverse component, one of whose terms diminishes as only the inverse *first power* of distance. This component of scattered electric fields radiated as transverse radiation by an accelerated electron will be oriented in the opposite direction (i. e., *out of phase*) from that of the incident radiation field that induced it. Refer to pages 523 through 529 of Ditchburn (1963) or similar text for an in-depth discussion of the radiating dipole applicable to this analysis, where it is shown that the transverse component of the radiation from a dipole of moment $\mathbf{P_e}$ is:

$$E_e(\omega,r) = -(\omega^2/c^2)\,|\mathbf{P_e}|\,\cos\alpha\,/\,r$$

$$= -(\omega^2/2\pi\,\rho_e\,c^2)\,|\,\mathbf{n}-1\,|\,\cos\alpha\,/\,r$$

It is assumed that $r \gg \lambda$, where r is the distance from the electronic charge and $\lambda = 2\pi c / \omega$. The angle α is the scattering angle of the radiation with respect to the direction of the incident radiation.

The intensity of the radiation with substitution of parameters for the general expression of the index of refraction will be:

$$I_e(\omega,r) = E_e(\omega,r)^* \, E_e(\omega,r)\,/\,2$$

$$= E^2_0 \,\{(e^2\,\omega^2/m_e\,c^2)^2 / ([\,\omega_0^2 - \omega^2\,]^2 + [\,\gamma\,\omega]^2)\}\,[\cos\alpha\,/\,r\,]^2$$

The scattering angle will always be extremely close to $\alpha = 0$ for coherent forward scattering because of the constructive interference in this direction as we will show. These considerations ultimately dictate that there will be cancellation of all fields arriving from outside an angular domain that, for the intergalactic medium, will be extremely narrow so that we can assume that $\cos\alpha = 1$ and of no consequence to the realized field strength or intensity at distant locations.

However, although the field strength and intensity of the individual scattered radiation are relatively insensitive to slight angular and even to small linear distance changes in the location of assessment,

150

this is not the case with regard to the phase of the radiation. Very small differences in propagation distance and angle produce major changes in the phase that are essential to the following analyses.

The reader might well ask, "What does all this have to do with what will actually be observed when incident radiation passes through a scattering medium?" All too frequently that question has been glibly answered in the negative. To clarify why that is significantly incorrect with regard to intergalactic plasma, we will investigate the processes whereby images of objects are resolved even when they are observed through a scattering medium. This capability pertains, of course, to transmissions through distances that are less than the optical depth of that medium as defined previously. Clearly however, as we will show, even in this case the intermediate electrons will have some affect.

b. coherence of scattered radiation from separated events

In order for radiation from separated sources to interfere, wave functions from these sources must be similarly polarized and pass the same observation point within the *coherence length* of individual wave functions. The coherence length is the effective linear dimension of a photon; it is on the order of 10^7 times the wavelength of the radiation. (For clarification of this fact concerning the nature of photons, refer to the discussion on page 112.) For separations larger than this amount, an entire photon of radiation from one source will have passed a given location before the other arrives. In that case neither positive nor negative interference phenomena will occur. The intensity of radiation at the location will just be the simple result of accumulation of separate photons from the two sources over a designated period of time.

In figure 71, r_i and r_j are respective distances of two scattering electrons e_i and e_j from the observation location P_o. Here we illustrate the situation for which the scattered wave functions from two such electrons *do* interfere. In this case, the electron scattering phenomena are precipitated simultaneously by oscillations caused by the incident radiation. Let the angle between the two electrons from the perspective of P_o be α, with t and t+τ the respective times of arrival at P_o for the scattered radiation from the two electrons. The following implications result from the geometrical relations:

$r_j = c\,(t+\tau) = r_i \tan \alpha$, where $r_i = c\,t$, and:

$$\tau \; = t \, (\, 1 \, / \cos \alpha - 1 \,) = t \, (\, 1 - \cos \alpha \,) \, / \cos \alpha$$

$$= t \sin^2 \alpha \, / \, (\, 1 + \cos \alpha \,) \cos \alpha$$

$$\approx \tfrac{1}{2} t \sin^2 \alpha, \text{ for } \alpha \cong 0.0$$

With wave phenomena the phase of the waves originating at separate sources affects the net field strength realized at a given location. So, whenever the value of $c \, \tau$ is an integral multiple of λ, the two waves will constructively (additively) interfere. This constraint affects the required separation between two electrons if they are to constructively interfere. These constraints ultimately determine the collaborative affect of intermediate scattering electrons on the radiation observed at a given location. Fenn et al. (2000) describe use of this effect in phased array radar antenna systems driven off a common feed.

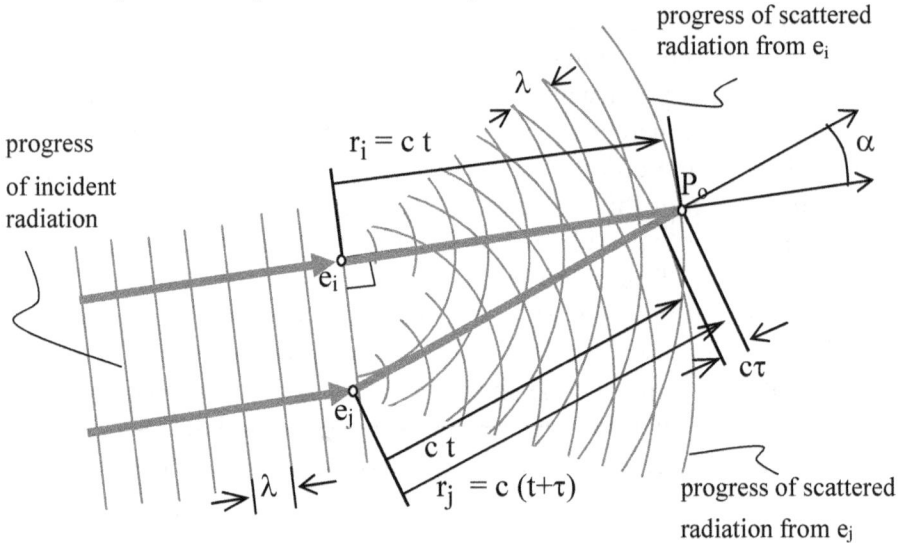

Figure 71: Illustrating conditions for coherent forward scattering

Consider, for example, the scattered electric fields from the two electrons situated such that they both begin to experience the effects of the incident plane wave radiation at the very same instant as shown in figure 71. If $\alpha \cong 0.0$, as we are positing initially, and will confirm as a necessary consequence, then at P_o a distance $c \, t$ from the electron e_i we will have for the total field strength of the scattered radiation at P_o:

$$E_{P_0}(\omega,ct) = E_{e_i}(\omega,ct) + E_{e_j}(\omega,c(t+\tau))$$

$$\cong K(\omega)\, e^{\,i\omega\,(t_0-\,t)}\,(1 + e^{\,-i\omega\,\tau}\,)\,/\,c\,t$$

where c t is the perpendicular distance to the plane of constant phase of the incident radiation and $K(\omega)$ includes all those factors in $E_e(\omega,r)$ above that exhibit approximately equal values as long as $\alpha \cong 0.0$:

$$K(\omega) = E_e\,(e^2\,\omega^2/m_e\,c^2\,)\,(\omega_0^2 - \omega^2\,)\,/\,[(\omega_0^2 - \omega^2\,)^2 + (\gamma\,\omega)^2\,]$$

Even at large distances with the constraint of a phase difference, $\tau \ll t$, appreciable effects result. The instantaneous illumination intensity of this net scattering radiation at P_0 becomes a sinusoidal function of whatever that location-dependent phase difference happens to be:

$$I_{P_0}(t,\tau,\lambda) = [\,E_{e_i}(ct,\lambda) + E_{e_j}(c[t+\tau],\lambda)\,]^* \,[\,E_{e_i}(ct,\lambda) + E_{e_j}(c[t+\tau],\lambda)\,]\,/\,2$$

$$= (\,K(\omega)\,/\,c\,t\,)^2\,[\,2\; + \;e^{\,-i2\pi\,c\,\tau\,/\,\lambda} + e^{\,+i2\pi\,c\,\tau\,/\,\lambda}\,]\,/\,2$$

$$= (\,K(\omega)\,/\,c\,t\,)^2\,(\,1 + \cos 2\,\pi\,c\,\tau\,/\,\lambda\,)$$

In terms of angular separation of the two electrons from the perspective at P_0 this becomes the following as shown in figure 72.

$$I_{P_0}(t,\alpha,\lambda) \cong (\,K(\omega)\,/\,c\,t\,)^2\,[\,1 + \cos(\,\pi\,c\,t\,[\sin^2\alpha\,]\,/\,\lambda\,)\,]$$

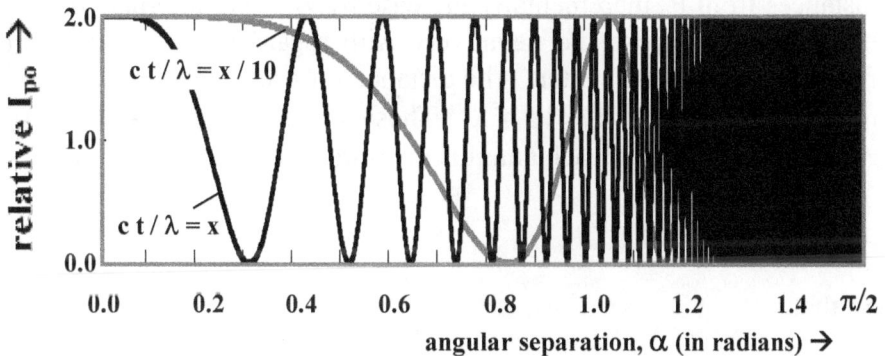

Figure 72: Illumination from two separated scattering events

153

In these equations, of course, the square of the complex factor $K(\omega)$ is:

$$K(\omega)^2 = \{(e^2 \, \omega^2/m_e \, c^2)^2 \, / \, ([\, \omega_0^2 - \omega^2 \,]^2 + [\, \gamma \, \omega]^2) \}$$

where the general form of the index of refraction is used.

The final term of the trailing factor of the resulting illumination, I_{P_0} is affected by the relative wave phase, τ of the two induced wave functions. Depending upon the angular separation of the two electrons, the resultant effective intensity of the scattering field at P_0 will vary between the extremes of zero and twice the intensity of a single scattering electron as shown. As the distance increases, in addition to the inverse square impact, the predominant central illumination will be narrowed. At appreciable distances it will be very narrow indeed.

Furthermore, rather than just two individual secondary sources interfering, we must consider electrons throughout the entire areas on the planes of incident radiation. Clearly scattered radiation from those electrons closest to the point P_0 will constructively interfere whereas radiation from electrons that are further away will increasingly interfere destructively and therefore not contribute much to overall intensity at P_0. In figure 73, an annulus of the medium is identified in the thin vertical planar sheet whose thickness is assumed to be much less than the wavelength of the radiation, λ. Throughout the indefinitely-extended sheet incident radiation is assumed to arrive simultaneously from the left instigating scattering by imbedded electrons.

Notice that the annulus corresponds to points near an x,y plane that are all approximately equidistant from P_0. Successive annuli of synchronized scattering emanations from embedded electrons whose distances from P_0 incrementally increase by $\frac{1}{2}\lambda$ will alternate between interfering positively and negatively with regard to radiation coming from electrons at the center. The distance for each such annulus can be represented as, $r_n = z + \frac{1}{2} n \lambda$. These zones are effectively *fresnel zones*. The radii ρ_i of each annulus measured from the center satisfies:

$$\rho^2_n + z^2 = (z + \frac{1}{2} n \lambda)^2$$

So that,

$$\rho^2_n = n z \lambda + (\frac{1}{2} n \pi \lambda)^2 \cong n z \lambda, \text{ if } z \gg \lambda \text{ as it will be.}$$

Then if we consider the area within each zone, we obtain:

$$A_n = \pi (\rho^2_n - \rho^2_{n-1}) = \pi\, z\, \lambda$$

So that the area of each zone is the same as for any other and, therefore, the intensity contributions from each zone will be the same even though they are increasingly narrower as α increases with wave phase.

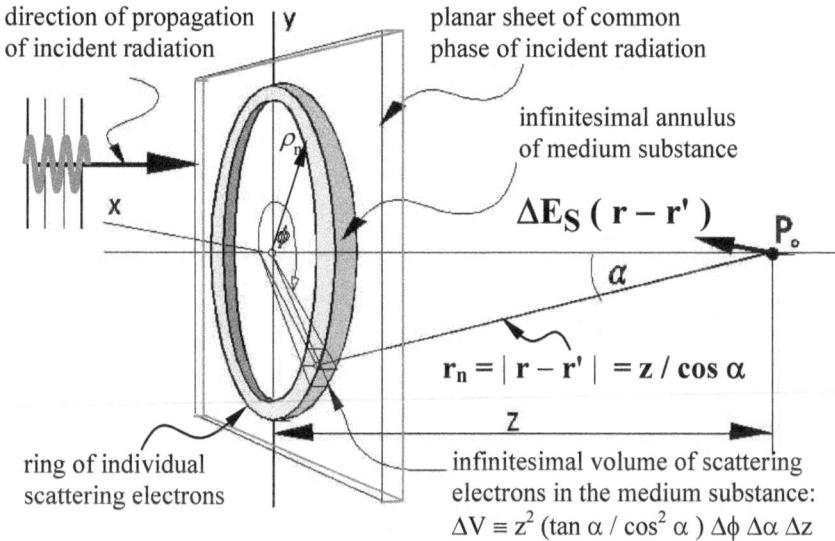

direction of propagation
of incident radiation

planar sheet of common
phase of incident radiation

infinitesimal annulus
of medium substance

$$\Delta E_S (r - r')$$

$$\alpha$$

$$r_n = | r - r' | = z / \cos \alpha$$

ring of individual
scattering electrons

infinitesimal volume of scattering
electrons in the medium substance:
$$\Delta V \equiv z^2 (\tan \alpha / \cos^2 \alpha)\, \Delta\phi\, \Delta\alpha\, \Delta z$$

Figure 73:   Geometrical considerations for integrating effects of forward scattering

Since the phase varies even within each 'half-period zone', the resultant field strength from each is $2/\pi$ what it would be if the phase were the same throughout the zone. The phase of the contributions of successive annular zones alternate with amplitude slowly decreasing as α increases, so that there is a series which, when added together, sums to $1/\pi$ times the effect that would result from only the first zone if the phase were to have been identical throughout. Thus, by using this factor, we need only include in our analyses the first zone for which,

$$\tan \alpha = (\lambda / z)^{\frac{1}{2}}$$

This procedure greatly simplifies integrating the effects of scattering.

We have yet to address a means to simplify linear integration all the way from the source to the point of observation. We do that now.

c. concept of 'extinction' in a scattering medium

Statements of what is called the *extinction theorem* (Born and Wolf, 1980; Wolf, 1971, etc.) all aver that in an extensive medium, the incident radiation will be totally replaced by forward scattered radiation after having propagated to a distance within the medium, δ_0, known as the *extinction interval*. This distance at sea level in our atmosphere, for example, is on the order of only a millimeter or so. In the intergalactic medium, in accordance with the sparse electron density of this medium as presented earlier, it can be many thousands of light years.

Throughout any scattering medium, in any case, the radiating fields are continually being replaced. The question is, how long and how far does that process require? Replacement fields, following a first replacement, will propagate at the velocity dictated by the index of refraction, $\mathbf{n}$, of the medium which thereafter completely characterizes propagation through the medium. Thus, comprehensive tests of the *Second Postulate* of special relativity, as for example Brecher (1968), have had to take this process into account in confirming that postulate to a high degree of accuracy with pulsar data obtained from within our own galaxy and more distant globular clusters. Brecher performed his analyses using the extinction distance formula whose derivation will be developed here. If this concept is already understood by the reader, there are yet aspects of the methodology that need to be understood.

We will use the extinction distance, δ_0, after which incident radiation is replaced by *forward-scattered* radiation, to further restrict a volume enclosed by a conceptual horn shown in figure 74 that we will refer to as the *'coherency domain'* of the point P_0. Scattered radiation emanating from charges within the surface of this horn can legitimately be considered as limiting the integration calculation that determines a total composite forward-scattered field $\mathbf{E}_{net}(\mathbf{r})$ at P_0, thus effecting the extinction process. The horn is defined as being elongated just enough so that there will be replacement of the incident field by the similar forward scattered field, which is out of phase with it at the point, P_0.

$$\mathbf{E}_{net}(\mathbf{r}) = \mathbf{E}_i(\mathbf{r}) + \mathbf{E}_S(\mathbf{r}) = -\mathbf{E}_i(\mathbf{r})$$

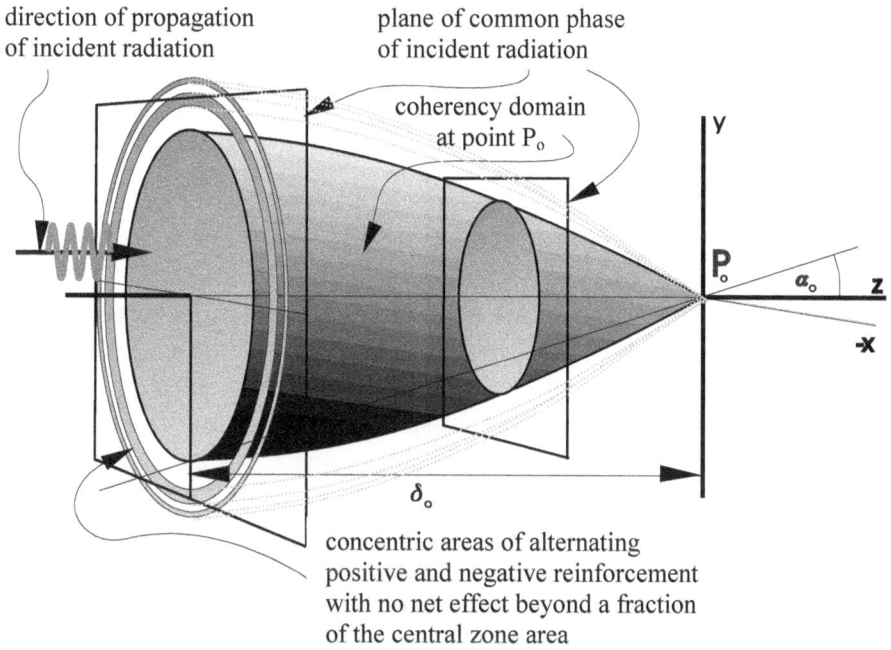

direction of propagation of incident radiation

plane of common phase of incident radiation

coherency domain at point P_0

concentric areas of alternating positive and negative reinforcement with no net effect beyond a fraction of the central zone area

Figure 74: Coherency domain in a scattering medium

To effect eventual replacement of the incident radiation, we must have that the scattering field eventually overpowers the incident field, $E_i(\mathbf{r})$:

$$E_S(\mathbf{r}) = -2\,E_i(\mathbf{r})$$

where $E_S(\mathbf{r})$ is the resultant of scattered fields derived earlier. It is:

$$E_S(\mathbf{r}) \equiv (1/\pi) \int_{V'(\text{horn})} \Delta'\,E_S(\mathbf{r}-\mathbf{r}')$$

where $\mathbf{r}$ is evaluated at P_0 and $\mathbf{r}'$ specifies the position of electrons within the infinitesimal volume of integration denoted in figure 73 as $\Delta V'$. By substitutions like those performed for individual electrons above, a scattered electric field strength, $\Delta'E_S(\mathbf{r}-\mathbf{r}')$ can be obtained for each infinitesimal segment of volume in the coherency horn as follows:

$$\Delta'E_S(\mathbf{r}-\mathbf{r}') = \Delta V'\,(\omega^2/c^2)\,(\mathbf{n}-1)\,E_i\,/\,2\pi\,|\,\mathbf{r}-\mathbf{r}'\,|$$

157

$$\Delta'E_S(\phi,\alpha,z) \cong \Delta V'(\phi,\alpha,z) \, (\, 2 \, / \, \lambda^2 \,) \, |Re(\mathbf{n}-1)| \, E_i \cos \alpha \, / \, z$$

The net scattered field at any point within the medium can then be obtained as the integral all these infinitesimal fields.

d. integrating coherent scattering effects

Clearly the net effect of incident radiation includes the contributions of due to polarization of the medium (scattered electromagnetic fields induced by the incident radiation) resulting from all intermediate electrons in the medium. This contribution is determined in large part by phase considerations of the various components received from each scattering electron throughout the medium as discussed above. Ultimately coherent reinforcement of all of the various forward-scattered fields over the total volume of integration results in sufficient scattered field intensity to effect the replacement (or what is called *extinction*) of the incident radiation by the virtually identical scattered fields that have an opposing phase.

Geometrical considerations are typical of integration schemes to be employed throughout this treatise. It is, of course, simplified by considerations just discussed whereby the angular bounds of integration can be reduced to include only the first half-period zone as elaborated in figure 74. Importantly, although the result of the integration we will obtain is well known, understanding the constraints of integration illustrated in these figures is key to understanding the alterations that will be required to accommodate for the similar, *although significantly different*, process in a hot plasma in the next chapter.

By restricting the angle as also described above, one can proceed using only the first fresnel zone as if all scattered radiation from within this angle shares a common phase. Thus the overall range of integration will be greatly reduced by the allowed angular restriction identified above as $\alpha = \tan^{-1} [\pi \lambda]^{\frac{1}{2}}$. Similarly, the length over which integration must take place need only extend from zero to the extinction distance of the medium, δ_0 rather than the all the way to the source of the incident radiation. So that an integration, which must take into account entire planes of synchronized scattering events induced by incident plane wave radiation and further extended to include planes from the point where evaluation takes place to the source of the radiation, is greatly simplified. The envisioned integration procedure obtains an accurate value for net electromagnetic scattering fields experienced at any point P_0.

Note the usual assumption in any scattering analysis, that the effects of scattered fields from charges closer to the point at which evaluation takes place than several wavelengths will cancel. This typical assumption is certainly realized statistically in the intergalactic medium for which the likelihood of there even being a scattering electron within this radius is zero to a very high degree approximation.

We will perform this integration here to determine an extinction distance for the intergalactic medium, which is essential to determining the cosmological effects of scattering.

e. determining the extinction distance and base angle

Having established simplifying limits of integration appropriate for the horn illustrated in figures 73 and 74, with δ_0 defining the extinction distance at which total extinction is accomplished and α_0 the angular limit at the distance δ_0, we will proceed by instantiating the parameters of integration defined in figure 73, we proceed as follows:

$$dV'(\phi,\alpha,z) = z^2 \, d\phi \, \tan\alpha \, \frac{\partial}{\partial\alpha} \tan\alpha \, d\alpha \, dz = z^2 \, \frac{\tan\alpha}{\cos^2\alpha} \, d\phi \, d\alpha \, dz$$

$$E_S = [4 \, |Re(\mathbf{n}-1)| \, E_i / \lambda^2] \int_0^{\delta_0} \int_0^{2\pi} \int_0^{\tan^{-1}(\lambda/z)^{\frac{1}{2}}} (\cos\alpha / z) \, z^2 \, \frac{\tan\alpha}{\cos^2\alpha} \, d\phi \, d\alpha \, dz$$

$$= [8\pi \, |Re(\mathbf{n}-1)| \, E_i / \lambda^2] \int_0^{\delta_0} \int_0^{\tan^{-1}(\lambda/z)^{\frac{1}{2}}} z \, \frac{\tan\alpha}{\cos\alpha} \, d\alpha \, dz$$

$$= [8\pi \, |Re(\mathbf{n}-1)| \, E_i / \lambda^2] \int_0^{\delta_0} z \left[\frac{\tan^2\alpha}{(1 + 1/\cos\alpha)} \right] \Big|^{\tan^{-1}(\lambda/z)^{\frac{1}{2}}} dz$$

$$= 4\pi \, |Re(\mathbf{n}-1)| \, \delta_0 \, E_i / \lambda, \text{ for } z \gg \lambda$$

Since we must have that $E_S = -2 \, E_i$ to effect replacement, we obtain:

$$\delta_0 = \lambda / 2\pi \, |Re(\mathbf{n}-1)|$$

In this way, the accepted average extinction distance formula is obtained appropriate for a plasma (or any other) medium. Clearly, a similar analysis applies to any point within any isotropic homogeneous medium. Brecher (1968) and others have assumed this formula in their analyses of extinction in intra- and near extra-galactic plasmas. It is certainly reasonable to apply it to the intergalactic plasma as well.

To assess the value of the derived extinction distance in terms of properties of the intergalactic plasma medium, we must substitute the parametric expression for $Re(\mathbf{n}-1)$ obtained earlier. It was:

$$Re(\mathbf{n}-1) = 2\pi \left(\rho_e \, e^2/m_e \right) (\omega_0{}^2 - \omega^2) / ((\omega_0{}^2 - \omega^2)^2 + (\gamma \omega)^2)$$

We obtain differing values under different assumptions:

$$\delta_{o1} \approx [m_e \, \lambda / (4\pi^2 \, \rho_e \, e^2)] (4\pi^2 \, c^2 / \lambda^2), \text{ if } \gamma << \omega$$

$$\delta_{o2} \approx [m_e \, \lambda / (4\pi^2 \, \rho_e \, e^2)] \gamma^2, \text{ if } \gamma >> \omega$$

The leading factor in either case is:

$$m_e \, \lambda / (4\pi^2 \, \rho_e \, e^2) \cong 1.00 \times 10^{-10} \, \lambda / \rho_e$$

So that we obtain the alternatives:

$$\delta_{o1} \cong 3.56 \times 10^{12} / \lambda \, \rho_e, \text{ if } \gamma << \omega$$

$$\delta_{o2} \cong 2.19 \times 10^{44} \, \lambda \, \rho_e, \text{ if } \gamma >> \omega$$

The former of these alternatives must certainly apply for all radiation of interest, in which case, we have that $\omega >> \gamma >> \omega_0$ as illustrated in figure 41.a on page 82 above. Then, applying the first of these formulas to optical wavelengths (5000 Angstroms will be used as representative of optical wavelengths as it was earlier in assessing a nominal value of $\mathbf{n}$ for this medium) results in:

$$\delta_o = 7.12 \times 10^{17} / \rho_e$$

This may be a distance of as much as thousands of light years for visible light with expected values of electron density. It is appreciable by any account, but still on the order of a thousand times less than the distance to our cosmological neighbor Andromeda and many thousands of times more diminutive than distances for which redshift is even considered a marginally reliable indicator of distance.

Notice also that the principal angle at the base of the coherency domain at this distance for optical wavelengths and a plasma density of 10^{-6} cm^{-3} certainly justifies the assumption $\alpha_o \cong 0.0$. It is:

$$\alpha_o = \tan^{-1}(\lambda/\delta_o)^{\frac{1}{2}} \cong 8.4 \times 10^{-13} \text{ radians}$$

So that for visible light the conceptual 'horn' depicted in figure 74 would be extremely narrow indeed – even use of the term 'needle' would grossly over estimates its breadth. However, because of its extreme length the base may still be on the order of 10^9 cm across, i. e., over five thousand miles. Even background microwave radiation has an extinction distance in excess of a light year or 10^{18} cm.

If we look at the total number of scattering electrons N_c involved in the extinction process within a single coherency domain (refer again to figure 74), we must first assess its volume V_c for which we obtain:

$$V_{horn} = \frac{1}{2} \pi \lambda \delta_o{}^2 \cong 5.1 \times 10^{35} \lambda / \rho_e{}^2$$

In the case of visible light, the number of electrons involved in effecting extinction in the intergalactic plasma will be:

$$N_c \approx 10^{36} \text{ electrons.}$$

Despite the low density, this is many orders of magnitude more than are involved in the replacement process in our atmosphere at sea level.

f. characterization and limitation on spectral invariance

However strange the forward scattering processes might seem to those who have not been familiar with them, the replacement of radiation by extinction does not in itself alter the characteristics of the incident radiation.

Investigation into the generally observed invariance of the spectra of electromagnetic radiation after having propagated to a considerable distance from its source through a scattering medium by Wolf and others (1971, 1972) led Wolf (1986) to his discovery of a general scaling law. That law specifies the conditions under which this traditional cornerstone of spectroscopy remains valid. Understanding this law allowed Wolf (1987 and 1989), James et al. (1990), and others to discover source correlations, which by virtue of violations to these specified conditions generate radiation whose spectrum differs from that which was originally emitted. Using the well-known analogy between the processes of emission and scattering of electromagnetic fields, Wolf et al. (1989) were able to show that predicted modes of spectral variation of electromagnetic fields in spatially distributed primary sources apply also for randomly distributed scattering media whose constituents act as secondary sources of radiation more or less as we showed in earlier sections of this chapter.

In other investigations they were able to construct distributed secondary sources of electromagnetic radiation that actually exhibited exceptions to this rule. However, single-scattered emanations used in the investigations do not support forward scattering through multiple extinctions that is essential to undistorted image viewing. The degree of alteration of the spectra was produced by the introduction of a commensurably appreciable scattering angle. Their results bear consideration with regard to conclusions we will obtain farther on when we see that there is an angular *convergence* rather than a *divergence* of *scattered* radiation in a relativistic thermal medium.

By a simple analogy between distributed primary sources of radiation emissions and secondary dipole sources associated with the constituents of a scattering medium, one can derive the analogous forward scattering 'scaling law' for the invariance of spectra propagated to what is referred to as the 'far zone' through a scattering medium. The isomorphism of the analogy to *primary* sources is so complete that other than the differences we will identify as associated with the relativistic motions of high temperature plasmas, the same conditions for invariance result. This might seem (prematurely as it turns out) to suggest that radiation should also penetrate a thermal medium without experiencing the slightest spectral variation.

This premise has been generally presumed with demonstrations of the validity of the presumption readily available. For example,

Fraunhofer diffraction patterns of lines in the solar spectra correlate extremely closely with emission spectra obtained in the laboratory for the associated elements other than for the minor redshifts apparent in the limb of the sun for which various explanations have been proposed (for which we will later add one more). This is true also of other stars within our galaxy and neighboring galaxies with any differences in the spectra typically being accounted for directly as Doppler effects due to peculiar motions of the sources.

Specifically, in demonstrating invariance, Wolf showed that if the degree of spectral coherence, μ, realized throughout a distributed source obeyed the condition:

$$\mu(\,\mathbf{r'}_{ij}\,,\lambda\,) = \mu(\,|\,\mathbf{r'}_{ij}\,|\,/\,\lambda\,)$$

then radiation from the source would remain spectrally invariant into the 'far zone'. *Far zone* in this case is defined as a distance of remove, d for which d $\gg \lambda$, where λ is the wavelength of the emitted radiation. In the preceding equation $\mathbf{r'}_{ij}$ characterizes the vector separation between constituent charges, p_i and p_j of the distributed source whose positions are defined by the coordinate vectors, $\mathbf{r'}_i$ and $\mathbf{r'}_j$ as we did above. The dependence of spectral coherence on $\mathbf{r'}_{ij} \equiv \mathbf{r'}_i - \mathbf{r'}_j$ is a statement of statistical homogeneity. The constituents, p_i and p_j are assumed to be members of an ensemble, $P^{(o)}$ associated with the distributed (secondary) source. See figure 75 for an illustration of the parameters involved. In our earlier discussion, a cross section of the coherence domain with a nominal width as shown in figure 74 would correspond to an ensemble, $P^{(o)}$, for a secondary radiation source, where p_i and p_j correspond to individual scattering electrons within a cross section. The equation above is the formal statement of Wolf's scaling law. Lambert's radiation law (Born, 1980) applicable to blackbody sources and other usually-studied thermal sources is a special case of this law for which:

$$\mu(\,\mathbf{r'}_{ij}\,,\lambda\,) = \sin(\,2\,\pi\,|\,\mathbf{r'}_{ij}\,|\,/\,\lambda\,)\,/\,(\,2\,\pi\,|\,\mathbf{r'}_{ij}\,|\,/\,\lambda\,)$$

The form of this relationship is shown in figure 76; clearly it relates to what resulted from constraints of forward scattering shown in figure 72.

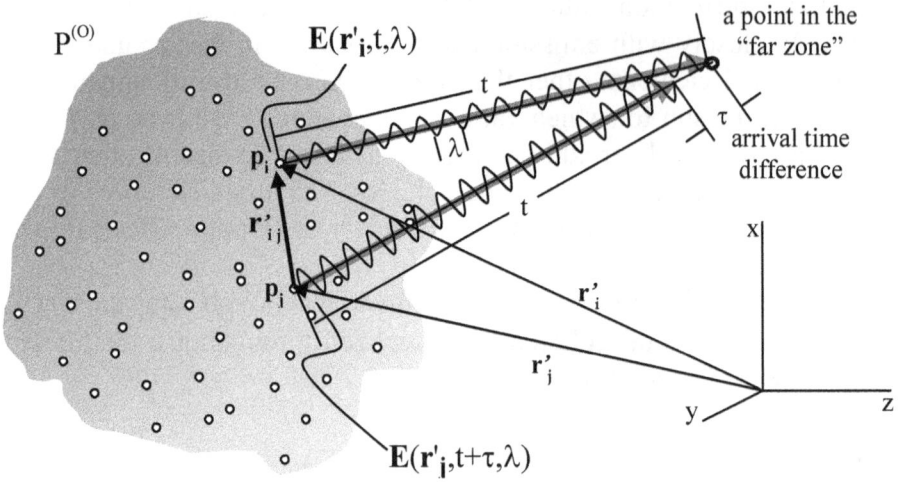

Figure 75: Illustrating conditions for distributed source coherence

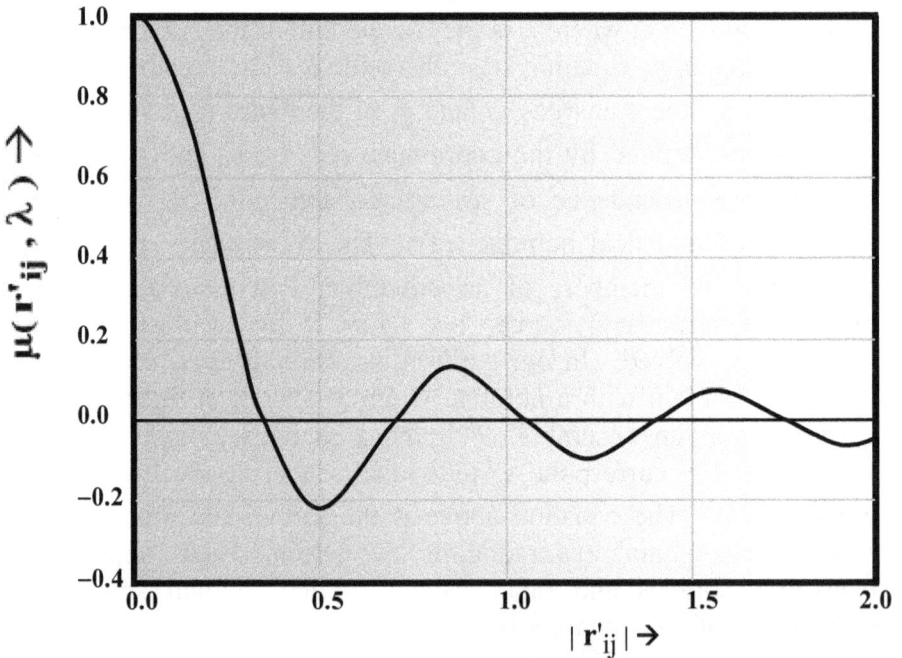

Figure 76: Spectral coherence applicable to thermal sources

The degree of spectral coherence is related to *cross spectral density*, $W(\mathbf{r'}_i, \mathbf{r'}_j, \lambda)$, as follows:

$$W(r'_i,r'_j,\lambda) \equiv \int_0^\infty E(r'_i,t,\lambda)^* \cdot E(r'_j,t+\tau,\lambda)\, d\tau$$

where $E(r'_i,t,\lambda)$ is the electric field emanating at time t from the point r'_i; in our case the formula is to be applied to $E_e(r'_i,t,\lambda)$, the scattered field emanating from the electron denominated by p_i. Throughout any extended source, $P^{(o)}$, the following formulas apply:

$$W(r'_i,r'_j,\lambda) = S(\lambda)\, \mu(\,r'_{ij},\,\lambda\,),\text{ where}$$

$$S(\lambda) \equiv S(r'_i,\lambda)\ \text{ or }\ S(\lambda) \equiv S(r'_j,\lambda)\ \text{ whenever}$$

$$S(r'_i,\lambda) = S(r'_j,\lambda),$$

and where in turn

$$S(r'_j,\lambda) = \int_0^\infty E(r'_i,t,\lambda)^* \cdot E(r'_i,t+\tau,\lambda)\, d\tau$$

The latter expression is the *spectral intensity* or *spectra* of the radiation. So that the degree of spectral coherence is obtained as follows:

$$\mu(\,r'_{ij},\,\lambda\,) = \int_0^\infty E(r'_i,t,\lambda)^* \cdot E(r'_j,t+\tau,\lambda)\, d\tau\ /\ \int_0^\infty E(r'_i,t,\lambda)^* \cdot E(r'_i,t+\tau,\lambda)\, d\tau$$

Thus, ultimately Wolf's law specifies the coherency conditions between electromagnetic fields emanating from the constituents, p_i and p_j of a distributed source.

In the analyses performed by Wolf et al. (1989) of scattering media by analogy with primary distributed sources, they considered only a single extinction/re-radiation process (as has been done here). They also very explicitly excluded consideration of the motions of the individual dipoles; that limitation was adequate to support their results.

Predictions of Doppler-like shifts in spectra in Wolf's later investigations (1987 and 1989) depended on observations of scattered light that was not aligned with the direction of propagation of the incident radiation. Therefore, these predictions pertain only to single-scattered radiation and do not support an accumulation of the effects of forward scattering involving repeated extinction processes.

It might seem, therefore, that their predicted null result in the 'forward' direction after a single extinction is significant for the forward scattering situation we are considering. In as much as spectra remained unchanged after a single extinction in this direction, there might seem to be no new phenomenon introduced by conjoining multiple extinctions that could effect a change after many.

They employed, however, the usual practice of excluding from the formal statement of their predictions all physical phenomena whose effects are assumed to be negligible and whose inclusion would only obscure an otherwise simple relation. Thus, an effect that remains negligible on one occurrence might still accumulate to a measurable effect after many, especially if the value of that *negligible* amount could be shown to always possess the same sign.

The condition indicated by Wolf's scaling law will not be precisely satisfied by a *primary* thermal source because of individual radial Doppler shifts associated with the distinct radial motions of the constituents, p_i and p_j, which has been repeatedly demonstrated to result in the well known spectral line broadening shown earlier in figure 62 on page 134. Transverse Doppler shifts will also occur, of course, but those will be of second order in β_i and therefore typically insignificant relative to the radial Doppler effects which will be first order in β_i, where $\beta_i = v_i / c$ and v_i is the velocity of the component p_i.

The effects of the radial motions in a *secondary* source with a zero mean velocity distribution will be reduced to fourth order effects in β_i after repeated extinction, whereas the transverse Doppler effects will remain second order in β_i at every extinction. Thus, these effects may accumulate without limit. We will pursue this avenue of investigation. It will prove to be extremely significant. Although these transverse Doppler effects are nearly always negligible in any one instance, they are additive and therefore may accumulate to significance. In the case of the extremely large number of extinctions and large free electron velocities that pertain in particular to the intergalactic medium, we will show that they accumulate *very* significantly, ultimately producing what is called "cosmological" redshift.

Chapter 9

Relativistic Effects in High Temperature Plasma

All implications of scattering by the intergalactic medium that we have examined thus far have derived primarily from its electron density. Implications of its high temperature – although less dramatic in the sense that we have largely been able to ignore them so far,

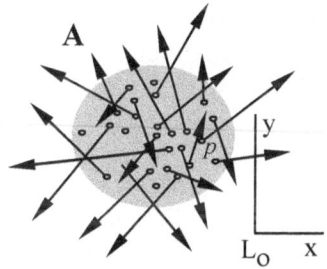

Figure 77: Distances traveled by electrons after scattering event

result in qualitative differences that are unprecedented among all the previously conceptualized scattering phenomena. These effects result directly from the thermal velocities of the plasma electrons that are not inconsiderable with respect to the speed of light, nor are distances traveled by these particles in the time interval of a single extinction as shown in figure 77. It is a well understood, but rarely acknowledged, fact that the impact of such high temperature effects on forward scattering would always exhibit the same sign. Because of this fact, rather than a miniscule effect that occurs at each extinction but cancels after many or is negligible in any case, these tiny effects may accumulate appreciably. In fact with increasing numbers of extinction events that occur when light is propagated from more and more remote regions of the universe, this becomes the most significant cosmological fact of all. It becomes increasingly dramatic as instruments improve.

a. significance of the transverse component of electron velocity

Special relativistic considerations implied by the extreme temperature of the plasma medium require modifications to the previous forward scattering analyses. These modifications are required to take into account primarily those effects introduced by the transverse components of the plasma electron velocities, i. e., components of their velocities perpendicular to the direction of propagation of the incident radiation, or parallel to the x,y plane in figures 35, 36, 73, 74, and 77 above. The reasons that only the transverse components need be considered are threefold.

In the first place the aberration of angles, which contribute so substantially to the required difference in treatment, involves primarily the transverse velocity components. In the second place, the effect of time (and therefore wave phase) differences introduced because of the relative radial velocities of electrons that become the secondary sources of the scattered radiation and an electron that detects it on successive extinctions will cancel to within *fourth order* effects in β_r. And, the effect will be non-null only when there is also a transverse component of relative velocity. (The parameter β_r is introduced here to represent the ratio of the electrons' radial velocity divided by the speed of light.)

Effects of relative transverse velocities on the other hand are *second order* in β_t at each extinction independent of whatever the radial component might also have been. We define β_t to represent the ratio of the electrons' relative transverse velocity divided by the speed of light.

Finally, the crux of the matter is that transverse velocities are associated with second order transverse Doppler effects. These involve *unilateral* redshifts in contradistinction to *signed* wavelength shifts that cancel after several extinction intervals. Any such radial redshift would be canceled by a precisely equivalent blue shift and vice versa in a statistically stationary medium.

For these reasons, the following analyses have been simplified to include only the transverse components of electron velocities characterized by what we now define more simply as $\beta \equiv \beta_t$ without thereby losing generality. Wherever we speak of an electron velocity without explicit denomination of components in the following analyses, the transverse component of that velocity is, thereby, implied.

b. concepts of 'simultaneous' wave planes and coincident observation

Considerations paramount to coherency of the electromagnetic fields scattered from electrons in a non-vacuous medium are based on the simultaneous realization of a common phase of the wave function throughout the plane at any cross section of the conceptualized coherency domain. This was illustrated in figures 73 and 74. It was implicit also that simultaneity of phase applied to each electron in the infinitesimal volume about P_0 (*observers* in this case). Simultaneity is, however, a concept that is suspect from the outset when relativistic velocities are involved. Relativistic treatment, therefore, requires taking into account various Lorentz transformations in determining coherency domains for each of the individual electrons involved in the small volume of the medium around the point, variously denominated as P_{β_0}, P_{β_1}, P_{β_2}, or etc., at the tip of the horn of each domain. Each designation P_{β_t} refers to the same physical location, but as denominated in the frame of reference of the electron moving at the relative velocity v_t with respect to an *infinitesimal* volume about the co-moving statistically stationary point P_0 within which the various electrons instantaneously 'coincide' in forwarding scattered radiation.

Coincident observers (electrons within the infinitesimal volume about P_0, that may actually be quite large but which otherwise satisfies conditions as an infinitesimal volume because of the tremendous distances between successive extinction phenomena) share an instantaneous '*now*' in special relativity. But each is inevitably in very basic disagreement concerning observed time intervals and distances to remote events in their respective Lorentz reference frames. We will see that these differences ultimately necessitate unique coherency domains for each of the various *coincident* electrons that happen instantaneously to co-occupy any small volume under consideration in a high temperature medium like the one realized in intergalactic space. This is shown in figure 78. In this case, the electron scattering event identified as p in the domain of P_0 is the same as that identified as p' in the domain of P_β. You will notice by comparison with figure 77 that the dotted arrow from p to p' is a velocity vector indicating how far the electron p travels in the time it takes the scattering event to be detected at P_0 in the frame of reference L_0. P_β is defined as the same point as P_0

$\equiv P_{\beta_o}$, but in a frame of reference stationary with respect to the electron whose scattering event is variously denominated p and p', and as such that electron is stationary within the domain **A'**. As shown, p and p' are the only 'event'(s) shared between the two domains, implying that the alternatively-denominated (and relatively situated) event is the *only* one affecting both the coincident detecting electrons at P_o and P_β. Clearly, if β were appreciably larger they might share *no* scattering events in common.

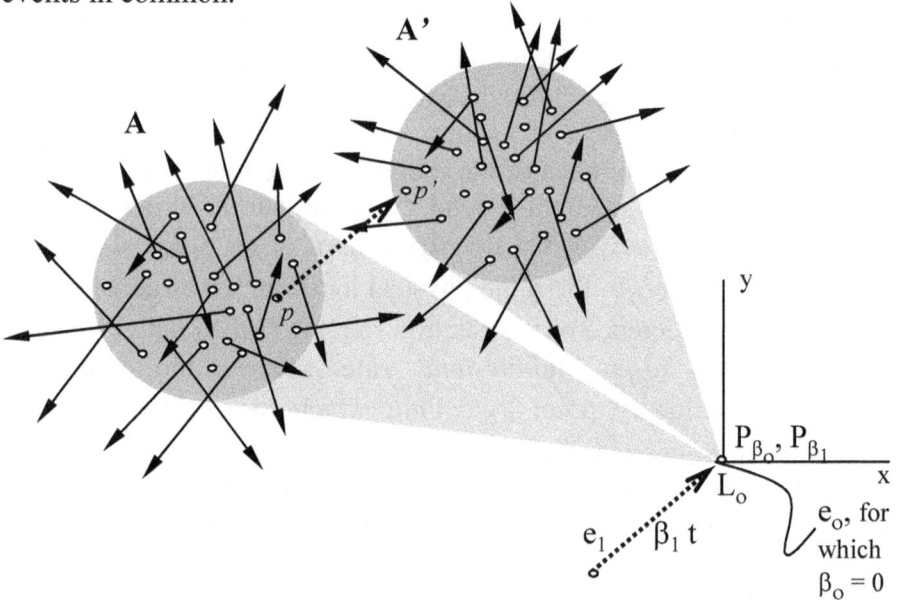

Figure 78: Scattering planes of two electrons from their perspectives at the tip of their respective coherency domains

     The secondary electromagnetic radiation that emanates from the various relatively moving coincident electrons in the small volume (at P_o, P_β, or etc.) will be coherent with the secondary emissions of any other at that point. This is similar to the fact that scattered radiation from respective coherency domains **A** or **A'** in figure 78 would exhibit coherent effects on respective electrons e_o and e_1 at P_o. The associated polarization field of the medium will, therefore, bear no direct witness to the diversity of the 'clocks' (or the required treatment of wave propagation time intervals) for the individual components of which the domain is constituted.

It is significant that in special relativity observers cannot *directly* discern the *relative* velocity of an *isolated* source of radiation, although, as with Hubble's hypothesis, inferences are frequently made with regard to such velocities – some valid, some invalid. Intelligent observers are restricted to such indirect inferences, such as their rationalizations from otherwise extraneous considerations, including the known stationary spectra of the elements, etc.. From such inferences we can deduce a comparative frequency for the radiation being emitted and thence, by modeling the type of operative Doppler effect, infer an associated velocity. *It must always be understood however that such inferences are not the observations.*

The inevitable limitation on our ability to directly observe certain phenomena has given rise in this case to the possibility of alternative explanations to the question of "*Why is there a redshift of distant galaxies?*" that do not necessarily invoke the radial Doppler effect.

c. locally stationary reference frame and local time

It is meaningful in the present context of forward scattering to speak of the *local time*, t, of the small volume at P_0 in which the scattered light is 'detected' by affected electrons. This will group a variety of events as though all the various detections were but a single event in the lengthy chain of the forward scattering process. This is true even if each such volume is comprised of many nearly coincident electrons in relative motion just as for those in **A** and **A'** of figure 78. This local time is analogous to the *co-moving local time* employed in general relativity. (Refer to any of the many good discussions of this topic as for example Singh, 1970.) This time is defined as the reading of a clock that is stationary with respect to an observer moving with the mean velocity of the constituent electrons of the volume. Local time is the assumed time value that would be given by a clock associated with a Lorentz reference frame, L_0, that is stationary with respect to the mean flow velocity of the constituents of the medium at that particular location. L_β is defined as the Lorentz reference frame of an observer (electron) moving with a velocity, $v \equiv c\beta$ (assumed as described above as at right angles to the direction of the incident radiation), all such relative velocities being measured with respect to L_0.

Coincidence serves as a synchronization mechanism for all the coordinate frames including the various time values; all their space-time measurements will be referenced to zero at the location/time of coincidence within the infinitesimal volume about the variously denominated point, P_o, P_{β_1}, P_{β_2}, or etc.. This is usual in special relativistic treatment. In our analyses the mean velocities of the sample volumes, themselves distributed throughout the intergalactic medium, are all assumed to be zero with respect to the original source and ultimate observer of the forward scattered radiation. Therefore, all coordinate values in any L_o are commensurable with those of any other.

It might seem natural to suppose that under Hubble's hypothesis one would more reasonably assume the relative co-moving radial velocity characteristic applicable to that hypothesis. However, we consider the stationary distribution instead, assured that subsequent developments will justify this assumption for the reader and if not, he or she can adapt the analyses quite easily in accordance with a composite model of choice accommodating Hubble's hypothesis.

A coincident but variously moving electron that detects forward-scattered radiation near the point, P_β in L_β, would be in very basic disagreement with his coincident 'peer' electronic observers. This disagreement concerns the phase of the incident radiation emanating from the various points on any mutual plane at the base of the horn shown in figures 73, 74, and implied by figures 77 and 78 which is being detected by the *other* electrons during coincidence.

Detection of scattered radiation from a mutually simultaneous phase of incident radiation emanating from various uniquely specified points on the planar surface is, however, compatible with the Lorentz transformation of the special theory that does support a mutual superpositioning of forward scattered radiation as we will see. This is an obvious but non-trivial fact. As will be demonstrated, the scattered radiation which is variously detected will have emanated in general from mutually exclusive areas on cross sectional surfaces of coherency domains. This implies that the various scattered fields detected by each electron would effectively constitute incoherent light with respect to any other electronic observer in a different frame of reference. Nonetheless the resulting accelerations of the *coincident* but relatively moving electrons will produce coherent in-phase radiations, which will combine in L_o to contribute mutually to extinguishing the original incident radiation at P_0 just as it does in a low temperature medium that was described earlier, i. e., images *do* get transmitted through a plasma.

d. applying relativity to coherent forward scattering

Let the value, $t_0 = -\delta_0/c$, be the local clock time at which the currently-detected in-phase forward scattering occurred throughout the base plane in the analyses discussed earlier in reference to figures 73 and 74. This is the clock time of the incident wave front reaching the base plane of the coherency domain for an electron that is stationary with respect to L_0. The time t_0 is that at which scattered light had to have left the center of the base plane at $(0,0,-\delta_0)$ in order to be detected *now* (i. e., at $t_0 = 0$) by a relatively stationary electron at $P_0 = (0,0,0,0)$ at the tip of the coherency domain for a *stationary* electron.[1] Slightly more negative emission times, i. e., longer required transmission times, will be associated with observed points on the plane further from the center as follows:

$$t = -(\delta_0^2 + x^2 + y^2)^{1/2} / c$$

$$\approx t_0(1 + (x^2 + y^2) / 2\delta_0^2)$$

The latter binomial approximation applies as long as $x^2 + y^2 \ll \delta_0^2$. Coherency considerations impose this restriction to an extreme degree in the intergalactic medium as was demonstrated earlier.

But throughout each cross sectional phase plane of the coherency domain, as everywhere else in a hot plasma medium, electrons will be experiencing rapid thermal motions. Most electrons residing in the plane at $t_0 = -\delta_0/c$ (for example) will not be permanent residents. (See for example, figures 77 and 78.) They will disperse to widely ranging locales by the time their radiating fields have been detected as shown in those figures. It is a premise of relativity that motion accommodates equally valid treatment from either perspective. The instantaneously coincident electrons in the region about P_0 have diverse origins as implied in figure 78 and illustrated now in figure 79.

---

[1] We use the convention for which (x, y, z) refers to a point in space and (x, y, z, t) to an event in spacetime.

e. Lorentz transformation equations

For an electron with relative velocity characterized by β in the positive x direction with respect to the locally stationary volume about $P_0 = (0,0,0,0)$ in L_0, identified as $P_\beta = (0,0,0,0)$ in L_β also, the Lorentz-associated clock time and location of the scattering events occurring at the base plane as shown earlier must be determined according to the associated Lorentz transformation equations as follows:

$$t' = (t - \beta x / c) / (1 - \beta^2)^{1/2}$$

$$x' = (x - \beta c t) / (1 - \beta^2)^{1/2}$$

$$y' = y, \text{ and } z' = z$$

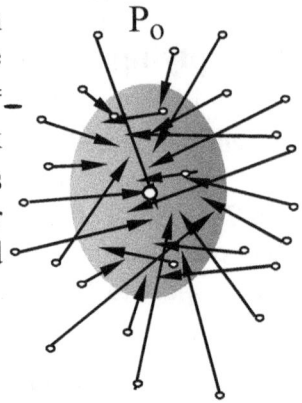

the immediate neighborhood of

P_0

Figure 79: Situation with the various electrons that 'coincide' in the vicinity of P_0 at time $t_0 = 0$

This set of Lorentz transformation equations will be denominated by the symbol $\boldsymbol{L}(E,\beta)$ indicating that the coordinates of an event, $E'=(x', y', z', t')$ observed in L_β but situated at $E=(x, y, z, t)$ in L_0 shall be mapped by the above set of equations as follows:

$$(x', y', z', t') \equiv \boldsymbol{L}(x, y, z, t, \beta)$$

At the point on the base plane for which $x = 0$ and $y = 0$, the following relation between the time values results:

$$t' = t_0 / (1 - \beta^2)^{1/2}$$

f. relativistic aberration equation

Refer to figure 80, where the relationships of values obtained from the Lorentz transformation $\boldsymbol{L}(E,\beta)$ of 'simultaneous' scattering events on a base plane in L_0 and its inverse, defined as $\boldsymbol{L}^{-1} \equiv \boldsymbol{L}(E,-\beta)$ obtained by reversing the perspective on the relative velocity, operating on events in L_β are illustrated. The constant phase ('simultaneity') planes required for coherency considerations are illustrated as having

174

been rotated with respect to the z = constant plane. The figure preserves the geometry in each frame of reference including relative distances (and time intervals). The directions to observed events are 'rotated' due to the relativistic aberration of light. The geometry of these angles, corresponding to lines of sight to the corresponding events is preserved in the figure. The corresponding angles, denominated θ and θ' respectively, to the same physical location of any point on the plane are related by the relativistic aberration formula:

$$\cos \theta' = (\cos \theta - \beta) / (1 - \beta \cos \theta)$$

This results from substitution of the implicit trigonometric relations:

$$\cos \theta = x / c\,t, \text{ and } \cos \theta' = x' / ct'$$

for the corresponding event coordinate values in the associated Lorentz transformation equations. For $x = 0$ with angle of incidence, $\theta = \pi/2$, we obtain for $\psi \equiv 3\pi/2 - \theta'$, $\sin \psi = -\beta$ as shown in figure 80.

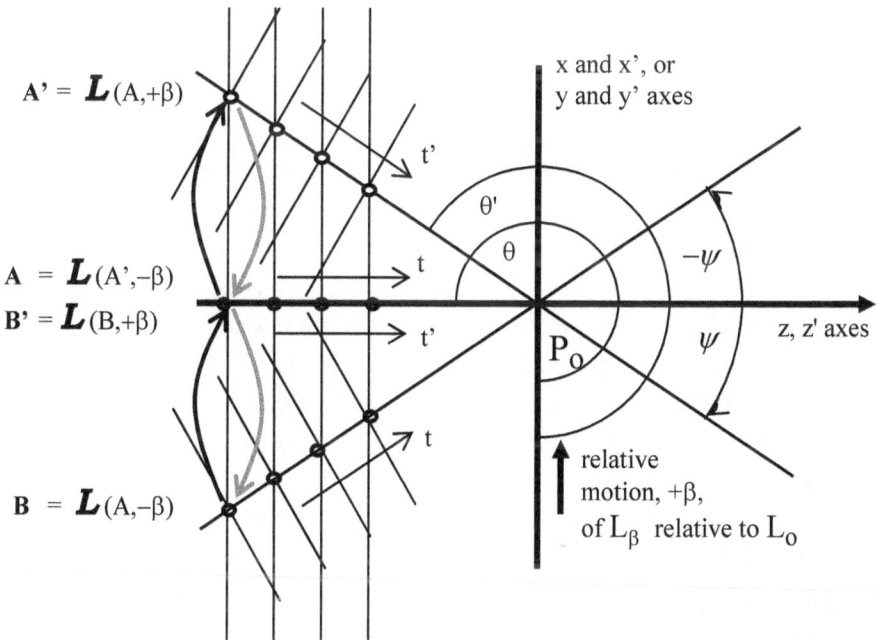

Figure 80: **Temporal relationships of events on wave plane surfaces of incident radiation resulting from the Lorentz transformation**

g. application of Lorentz equations to forward scattering

From the temporal Lorentz equation in section e above we see that for scattering along the centerline of a *stationary* electron's coherency domain where $x = y = 0$ and $z = -c\,|\,t\,| = -\delta_0$, we have:

$$t'_{x'>0} - t_{x=0} \approx -1/2\ \beta^2\ \delta_0\,/\,c$$

This difference is negative because z is negative. It indicates that light would have to have been emitted in the more distant past in order for it to be observed at P_β at the time $t' = 0$ for the *moving* electron if the constancy of the velocity of light is to be preserved as it must be. This is easily seen in figure 80 where the distance from point p' in L_β differs from that for p in L_0. For electrons with the root mean squared velocity of a high temperature thermal medium and the average extinction distance appropriate to a very diffuse medium, this would constitute an appreciable difference in transmission time relative to that for a 'stationary' electron. It would typically amount, in fact, to a difference of many hours for the average extinction distance in the intergalactic medium. And since the coherency time interval associated with the passage of a photon is on the order of 10^{-9} seconds, electrons at P_β will *not* be mutually involved in the forward scattering events of the same photon on the same phase plane as correctly implied by figure 80.

The value of x' obtained from the Lorentz transformation with β positive for this same centerline point is positive (rather than 0) owing algebraically to t_0 being negative and physically to the electron's having been at a more negative relative x axis position at the time that the earlier scattering event occurred. Consistent with the location and transmission time given by the Lorentz transformation equation, the relativistic aberration effect implies that an electron moving at right angles to the line of sight would detect the scattering from the angle $\psi = \sin^{-1} \beta$ to the normal of the plane in his frame of reference.

For forward scattering this is further complicated, however, by the fact that the area on the base plane of the coherency domain in L_0 that must be aligned with the phase plane of the incident radiation is not so-aligned in L_β. This fact precludes Lorentz-transformed counterparts of a coherency domain in L_0 constituting 'coherency' domains in the various frames, L_β. Refer to figures 80 and 81 where relativistic aberration, i. e., the angle ψ, results in a tipping of the coherency domain plains relative to the phase planes of the incident radiation.

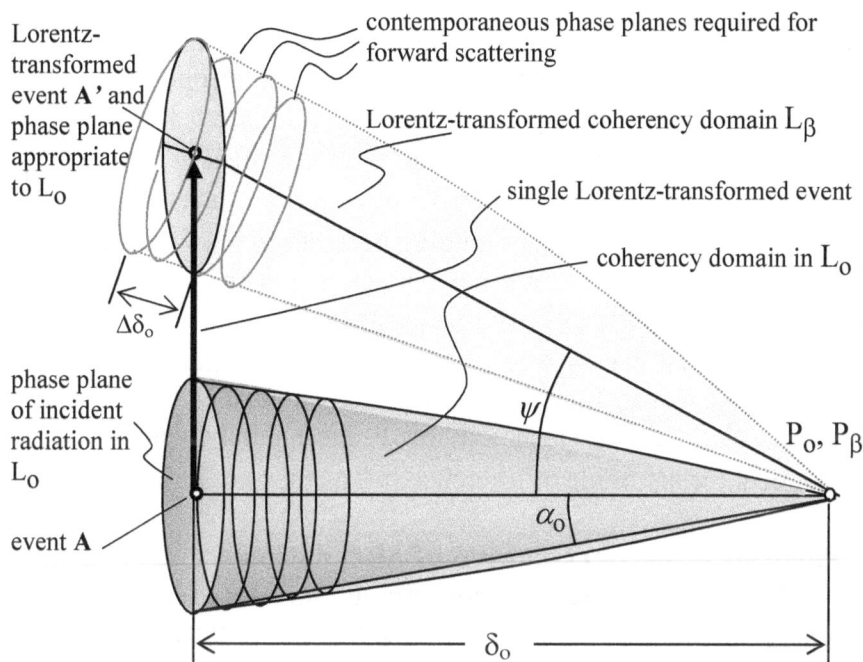

Figure 81: **Geometrical relationship between the coherency domain in L_0 and its Lorentz-transformed counterpart in L_β**

The Lorentz-transformed domain will not satisfy the coherency requirements for forward scattering, namely $\Delta\delta_0 < \lambda / 4$, where

$$\Delta\delta_0 = \delta_0 \sin \alpha_0 \sin \psi = \beta \, \delta_0 \sin \alpha_0 \,,$$

where α_0 is the half-angle of the base plane of the coherency domain. In actual fact, this criterion will be exceeded by a factor of well over 10^{10} for optical radiation.

The second postulate of the Special Theory (universality of the speed of light in vacuum, which applies *between* scattering events) also implies, however, that the 'moving' electron could simultaneously have detected scattered light from the same incident radiation. There will be a coherency domain in his own frame of reference involving the same planar surfaces, including base plane events that occurred at x'=0, y'=0, z'= −δ_0. This similar set of events will be observable many hours earlier (in fact, at the same time as the other events in L_0) by merely 'looking' in the sensible direction perpendicular to the surface in *his own* Lorentz frame of reference. Clearly, it is in this direction that the geometrical analyses of a central Fresnel zone and a coherency domain similar to that shown in figures 73 and 74 will apply. In L_β the light scattered from the center of a similar (but unique) disk in the plane at x' = 0, y' = 0 and z' = − δ_0 will arrive after a time interval, |t'$_0$| = δ_0/c following scattering. We are changing perspectives now relative to the illustration in figure 81. We are replacing β with −β in looking at coherency from the perspective of L_β rather than L_0. Now rather than transforming from events A to A' as shown, we are transforming from B' to B, as shown in figure 80. This accommodates similar analyses considered above for L_0 since both the distance and the velocity of light are identical for his frame of reference just as they are for any other.

h. reversing the perspective of aberration

As discussed earlier concerning forward scattering in a *stationary* medium, scattered photons from angles much greater than the principle angle α_0 of the coherency domain will destructively interfere, canceling effects that would otherwise occur when they finally arrive. Using the Lorentz transformation of this location (and direction) from which the relatively moving electron would detect the *same* incident radiation from the same plane (although different area on it as shown in figure 82) at the moment of coincidence, we obtain:

$$x' = 0; x = - \beta \delta_0 / (1 - \beta^2)^{1/2},$$

$$y' = y = 0; \text{ and } z' = z_0 = -\delta_0$$

Thus, in L_0 a 'counter time dilation' would be implied if it were assumed that both observed the same scattering events while in coincidence − which they won't. Because, of course, the stationary electron could not detect the light from these scattering events until hours later than that from his own centerline. Again, there would be an aberration of comparative angles, this time in the opposite direction, so that the electrons in L_β would not detect those photons while in coincidence with L_0. Refer to figures 80 and 82 for illustrations of the symmetry of the observations of 'simultaneous' scattering planes. Scattering from the circular neighborhood of the point B in the area A_β on the planar surface in L_β of figure 82 will exhibit all the features realized around the centerline of the coherency domain whose base area is A_0 for the 'stationary' electron in L_0.

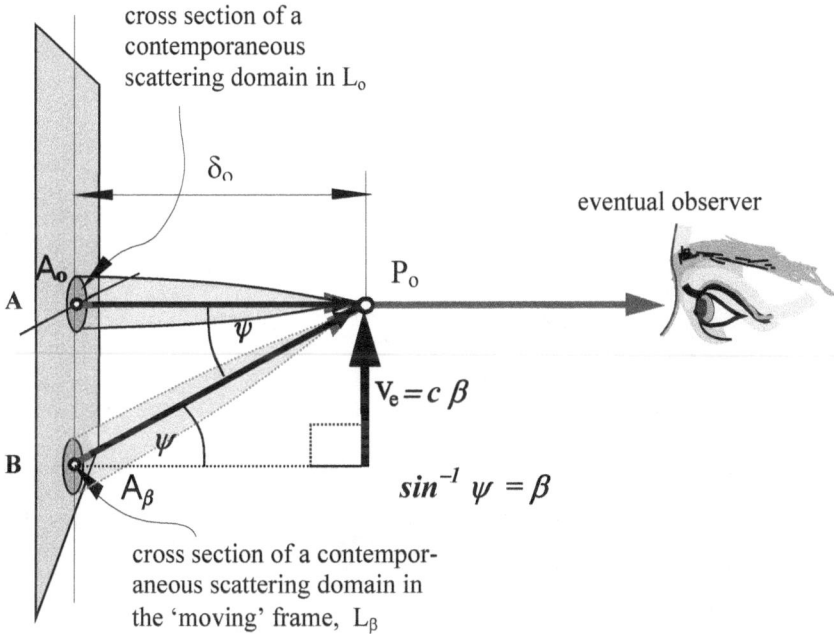

Figure 82: **Geometrical and temporal relation between scattering events for relatively moving electrons**

Figure 83 illustrates this more specifically for the coherency domain as it must be realized in L_β. As the electron characterized by the velocity β progresses to its coincident position at P_0, various planar surfaces will effectively constitute the scattering surfaces whose effects combine at the base of the horn at t'=0 in his frame of reference. These cross sections will occupy positions in L_0 as shown, which will in combination constitute an inverse Lorentz mapping of the coherency domain in that frame of reference as suggested in figure 83.

Notice that there is no pretense that the scattering events which are detected are the very same in both reference frames; clearly they typically will not be, as was shown in figure 78. But – and this is an important point – although there are Lorentz time differences to the respective points on the plane, these differences are entirely accounted for by differences in propagation times in the two frames if the same radiation were to be assumed as being detected while in coincidence. In other words the time difference is *not* in the emission clock times from areas A_o and A_β in figure 82 or points A and B in figure 80, nor, therefore, in the phase of the incident radiation at the time of those scattering events that will so variously be detected! *

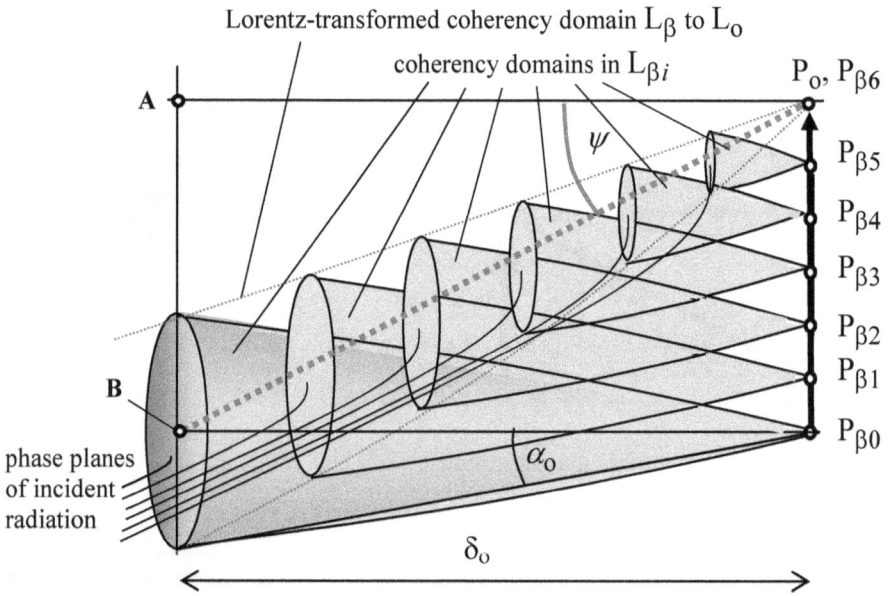

Figure 83: **Geometrical relationship between the coherency domain in L_o and its Lorentz-transformed counterpart in L_β**

---

* We will not digress at any length into subtleties of which electron's clock runs slow in this case or equally absurd questions concerning which observer would observe forward scattered light arriving from directions which are not normal to the surfaces of constant phase of incident radiation in his own frame of reference. Attempting resolution of which 'observer's' clock time values are dilated while the other's are assumed to be correct is pointless here. A temporal Escher *strange loop* diagram results from the relationships in such attempts. Suffice it to say that there are antinomies beyond the scope of the present volume and that in those cases where such apparent (or real) conflicts arise, realization of a contemporaneous common phase essential to forward scattering is precluded by many orders of magnitude. There is, however, a very straight-forward interpretation of the Lorentz transformed values in

i. combining effects for individual electrons

Thus there will be similarly shaped domains of *simultaneously* observable scattering events from among all those that occur throughout the $z = -\delta_0$ plane of the incident radiation in L_0 at $t = t_0$. All of these events will not, as we have seen, be simultaneously observable by any *one* electron 'observer' at P_0. However, in-phase scattering events occurring on the same base plane in L_0 will be independently detected by the variously moving electrons while in coincidence at the origin P_0 of L_0 at t=0 as was shown in figure 78. This scattering phenomena will originate in domains that are at various angles, ψ, to the centerline in L_0 as shown in figures 82 and 83.

It is worth considering whether the relatively large velocities and associated angles ψ, whose sines are equal to β, will be sufficiently large that assumptions of an incident plane wave rather than spherical wave front should be reconsidered. However, it can be shown, as we will further on, that this fact does not invalidate the analyses. When the distance to the source is much greater than the extinction distance, the adjustments due to modifying the geometry by employing a spherical wave front emanating from the original source of the radiation is of negligible consequence.

It is also important to note that the electrons included at the variously denominated point, P_0, as shown in figure 84, will exhibit the complete range of thermal velocities. Some will be predominantly radially-oriented along the line of sight of forward scattering unlike the transverse velocity vector implied in figures 80 through 83. We have attempted to depict this in figure 84 below.

---

such cases which resolves apparent paradoxes: Rather than implying contradictory time dilations and spatial contractions, they must be interpreted as transforming unique but commensurable spatial coordinate values and associated light propagation times for respective scattering events occurring at the 'same time' in each frame but from unique points on a coplanar surface as shown in figures 71, 73, and 74. In other words, the *apparent* relativistic aberrations are *real*! The two observers observe a unique set of scattering events while in coincidence which are spatially separated as required by the aberrant angles. The more usually accepted interpretation of the distortion of space/time between observers as a geometrical rather than physical electromagnetic interaction-related fact over-constrains the simultaneity and co-locality of two events. The scattering of a photon from electrons within one small volume of space and detection of the same photon by electrons within another small volume at a considerable remove in multiple unique Lorentz frames would expressly be denied by a rigorous application of usual interpretations. (See Bonn, 2008.)

The problem in any attempt to depict this phenomena as we have is that the planar surfaces will appear variously tipped depending on the specifics of the motion of the particular electron relative to P_0 as was shown in figures 80 through 83 and further elaborated in figure 84. Each individual coherency domain will have a base plane at a different distance as well as different angle from P_0 in L_0. These base planes will all have originated with the same phase of the advancing incident wave, however, and all of these domains will involve the same shape and size in their respective frames of reference as shown.

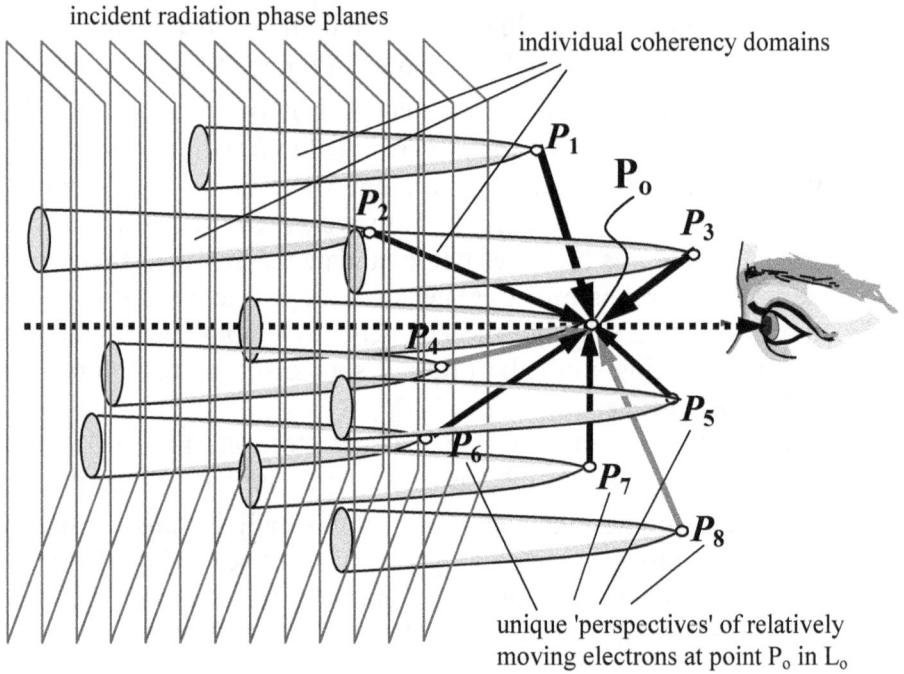

incident radiation phase planes

individual coherency domains

P_1

P_0

P_2

P_3

P_4

P_5

P_6

P_7

P_8

unique 'perspectives' of relatively moving electrons at point P_0 in L_0

Figure 84:   Suggestion of a composite view of the various scattering domains for the variously moving electrons

As stated above, the origination of the incident radiation detected while the electrons at P_0 are in coincidence will be from both various distances and angles relative to an observer in L_0. However, since these detected scattered radiations pertain to different stages of the progress of the same phase plane of the incident radiation, they will combine coherently to effect a resulting in-phase oscillation of the various electrons to effect on-going forward scattering. All coincident electrons at P_0 whose values of $v^2_{transverse} = v_x^2 + v_y^2$ are equal (as for example, P_0 and P_7, and P_5 and P_8 in figure 84) will have coherency domains originating on the same plane in L_0. This is integrated in figure 85 from the perspective of L_0. The scattering effects from all of these coherency domains on each progressing scattering plane must be included in the determination of extinction in this case. In figure 85 the domains shown in figures 84 as all being perpendicular to the incident wave phase planes are shown from the 'locally stationary' perspective.

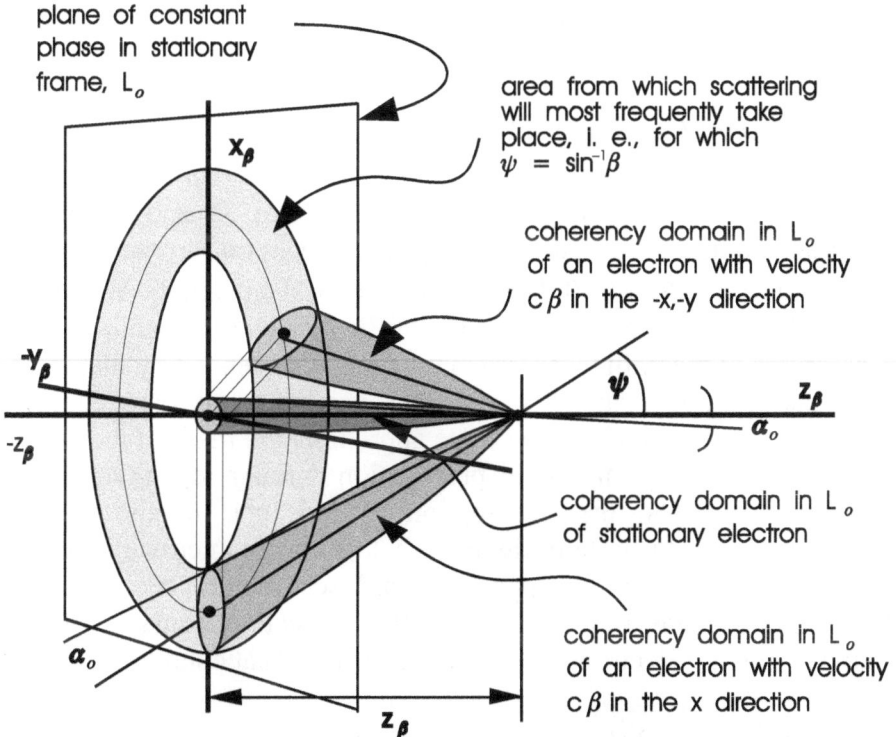

Figure 85: Depiction of unique coherency domains applicable to relativistic electrons

It should be noted that the depictions in both figures 84 and 85 take considerable license with what each of the electrons experiences since from their perspectives, each domain is aligned perpendicularly to the incident phase planes. In that sense, figure 85 depicts more of an inverse Cherenkov radiation effect of a high-speed electron through the intergalactic plasma medium. It is an effect of the medium that is being detected by the electron, not an effect of the motion of a particular area of the planar surface. The areas from which secondary radiation originates on a constant phase surface is no more in one frame of reference than any other since the radiation derives from an ensemble of electrons, no one of which will typically be stationary with respect to any other detecting electron. It is a premise of relativity that the detection of radiation preclude knowledge of the velocity of the source(s) in any case. So the direct Lorentz transformation as illustrated in figure 81 does not apply. Each electron is 'looking at' a scattering event as legitimately at right angles and 'stationary' in his own frame of reference as in any other.

j. geometrical considerations

It was determined that the principal angle α_0 of coherency domains applicable to analyses shown in figures 73 and 74 above would be extremely small in the intergalactic medium. In fact, it is on average on the order of 10^{-12} radians for optical radiation. However, even that tiny angle to the base plane of the coherency domain will be many orders of magnitude larger than the angle subtended by that same base plane from the original source. This is what reduces any concern about the assumption of a plane wave approximation made at the outset in chapter 4. Certainly after propagation through even very few extinction intervals the assumption would not invalidate our conclusions. We illustrate in figure 86 below, the geometrical relations that must be realized to support forward scattering. As discussed above, this remains true in each of the respective domains of the relatively moving electrons in the vicinity of P_0. That condition is that,

$\delta_0 \left(1/ \cos \alpha_0 - 1 \right) < \lambda/4$ and $r_0 \left(1/ \cos \chi - 1 \right) < \lambda/4$.

By approximating for small angles, we have that:

$$\chi^2 / \alpha_0{}^2 \leq \delta_0 / r_0$$

To the extent that this condition is easily met as long as $r_0 \gg \delta_0$, the surfaces of constant phase will, in fact, be so nearly planar as to not require further explanation.

184

Constraints on the principal angle α_o of each coherency domain remain as they were defined earlier to assure that the distances from the two scattering events shown in the figure below would differ by no more than approximately $\lambda/4$, where again, λ is the wavelength of the radiation. Essentially the same constraints must apply to the angle χ from the original source.

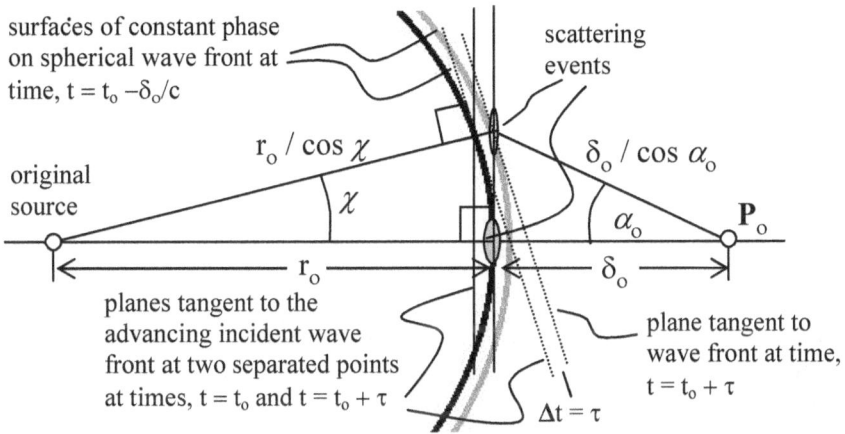

Figure 86: Geometrical considerations that allow a spherical wave front to be treated as approximately a planar wave front

k. relativistic time constraints

For the coherent forward scattering analyzed earlier, differences in path length (travel time) for two photons had to be less than about 10^{-16} seconds (implied by the quarter of a single wavelength constraint in the optical region) in order to contribute substantially to forward scattering. This is what restricted the angular coherency domain to such an extremely small base angle of α_o for each electron within the infinitesimal region at P_o. Certainly, with this constraint, depicted in figures 83 through 85, one might think that the radiation detected by one electron at P_o in the intergalactic medium could virtually never be considered coherent and in phase with that detected by any other electron. The radiation detected within the region about coincident points, P_o, $P_{\beta 1}$, $P_{\beta 2}$ at $t_o = t_{\beta 1} = t_{\beta 2} = 0$ in each Lorentz reference frame of figures 83 and 85 involve unique scattering domains. This condition results because of the transmission time interval difference for photons scattered once within an extinction distance from a given area on a phase plane that are detected by the two separate electrons. For cases where one is stationary with respect to L_0 and the other has a relative motion characterized by β, the difference may exceed 10^8 seconds, i. e. many years, as was shown above. This is due to the extreme extinction distance (i. e., $z = -\delta_0$), the narrowness of the coherency domain, α_0, and extreme thermal velocities, β. It is easily verified by substitution of approximate values for α_0 and δ_0 that this time difference equation presented in sections d, e and g above is approximately:

$|\Delta t| \sim 10^{11} \beta^2 \approx 10^9$ seconds.

For two electrons in such a high temperature medium to be affected by the very same scattering events from their central location at the tips of respective coherency domains at P_0 is virtually impossible. The requirement for forward scattering of $|\Delta t| < 10^{-8}$ seconds, would demand that the absolute magnitude of the tangential components of their velocities differ by less than 10^{-9} centimeters per second. This is an *astronomically* unlikely situation in any reasonably localized region of such a low density (or any other) medium. But as suggested in the previous chapter, this does not preclude forward scattering; it merely assures that the fields detected while in coincidence will originate from unique (rather than shared) scattering events at various angles to the general direction of the original source. But in each case from the same phase plane and perpendicular distance from the advancing incident plane wave in respective Lorentz frames.

l. angular convergence of the detected radiation

In figure 87, panel a, we illustrate the extent of the distribution of angles from which scattering might be realized assuming a Maxwellian distribution of velocities in a thermally stable medium. The two panels in this figure also illustrate that in addition to the dearth of electrons whose velocities are similar enough to allow them to share coherency domains, there is an aberrant off-center focus of origination to the scattered fields detected by the electrons. The curve in panel a of the figure portrays the Maxwellian distribution of the arcsines of the angle ψ shown in figures 80 through 83 and 85 that correspond to transverse velocities of the electrons. These values pertain to the centerline angles of the coherency domains of electrons at P_β from the perspective of L_o: The plots are applicable to $T = 4 \times 10^7$ K and involve the following expression without specific normalization:

$$N(\sin^{-1} \beta) \Delta v = \{\rho_e\, m_e\, v_e\, (\sin^{-1} \beta)^2 /(2\pi\, k\, T)\}e^{-E(\beta)/kT}\Delta v$$

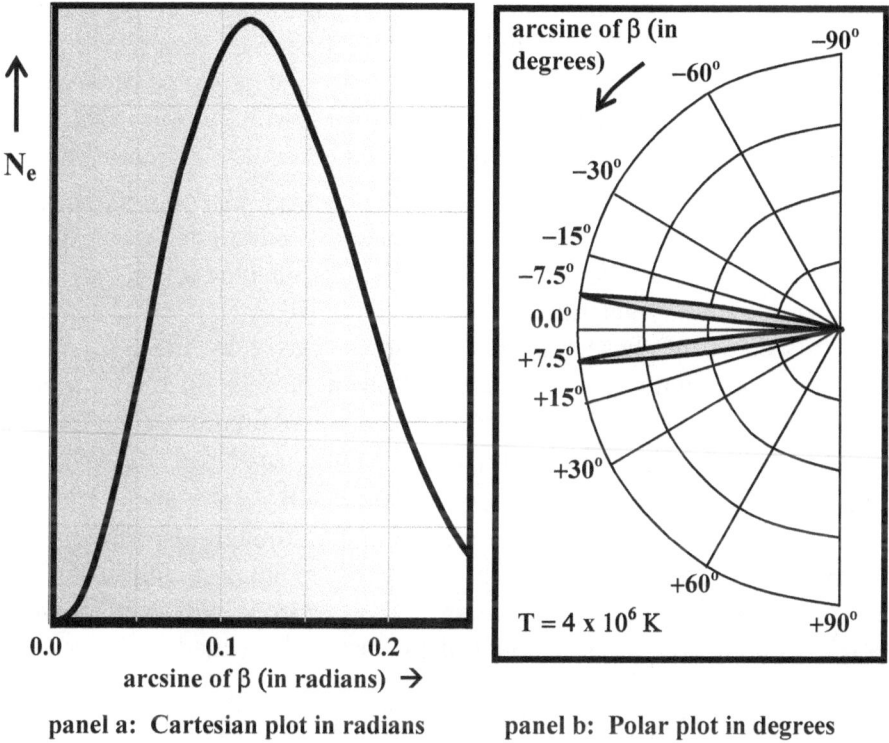

panel a: Cartesian plot in radians　　　panel b: Polar plot in degrees

Figure 87:　Frequency of electrons obtaining forward scattering intensity from various angles for $T = 4 \times 10^7$ K

where $E(\beta)$ is the relativistic expression for electron energy discussed in chapter 3 with regard to properties of the medium. Because it is the transverse velocities $(v_x^2 + v_y^2)$ we are considering here, there are two degrees of freedom and the expectation energy in this plane is $<E_{trans}>$ $= k\,T$ rather than the $3/2\, k\, T$ as was discussed in that similar context earlier for total electron energy. All electrons at P_0 whose values of v_x^2 $+ v_y^2$ contribute equal amounts of intensity to forward scattering, the composite of the areas on each scattering plane must be included in the determination.

Clearly, the number of electrons per cubic centimeter ($N \sim 10^{-24}$) that would have sufficiently similar velocities to share coherency domains justifies the assumption that *all* these coherency domains realized by the electrons at P_0 will be mutually exclusive. In contrast, the molecules in air at sea level on earth with a temperature of about 300 K will detect scattered light from the same cross section of a single coherency domain. This is true for such a wide range of velocity differences at this temperature and density that virtually *every* molecule in any infinitesimal volume about P_0 would detect (be affected by) the very same scattering events. This then, is a very essential difference between forward scattering in the intergalactic medium and other media which have been studied and whose dispersion characteristics have been experimentally verified.

Because of these relativistic effects caused by the high-speed electrons, the situation encountered in the intergalactic medium will essentially result in totally unique coherency domains for each electron in the vicinity of P_0. Therefore, in the vicinity of P_0 at t=0 in every frame of reference, virtually every electron will detect scattered light that has derived from vastly separated unique (although respectively the closest) events on a common, constant-phase, incident radiation surface as was shown in figures 84 and 85. The electrons will each possess their own unique coherency domains as further illustrated in figures 78, 80 and 82.

However, regardless of how much their velocities differ they will *simultaneously* contribute in their own way to the in-phase oscillations at P_0 since they are synchronously coordinated *ipso facto* by the immediate coincidence of all of the detecting electrons and the constraints of forward scattering. But those electrons whose absolute transverse velocities (speeds) are most similar to the mean will, because of their much larger numbers, generate the preponderance of secondary induced radiation which will be further filtered by the wavelength coherence considerations due to redshifting effects yet to be described.

Other than the detected frequency of the radiation to be described, the net scattered field experienced by each, however-moving, electron at P_0 does not change substantially from what it was determined to be earlier in obtaining the extinction distance in a low temperature medium. To understand this one must merely consider the electrons identified in figures 77 and 78 whose relative motion takes them outside of the coherency domain prior to their scattered radiation being detected. The radiation will nonetheless have derived from scattering events with relative times and positions similar to that for an electron with no motion relative to L_0. It is the *principle of relativity* that ensures this.

The velocity of light is the same no matter what the motions of the source. One can easily verify by transforming back from the event p' as detected by a coincident electron e_β at P_β that this would be the case for e_0 at P_0 in L_0 as depicted in figure 78 and subsequent figures. Although e_β and e_0 simultaneously detect different scattered radiation at P_0 in L_0, these respective scattered radiations will be in phase with each other since they both originated at a similar phase plane of the same incident radiation and distance from the variously denominated point P_0. Thus they satisfy the same forward scattering constraints as we encountered earlier for cooler media.

Obviously the incident radiation that will be cancelled by extinction at the point of the respective coherency domains will be the same since it is coincident at the point P_0 without having experienced interference along the way. So both (all) scattered radiations from the various coherency domains contribute to the ongoing radiation at the point P_0, P_β, etc.. Each electron's contribution to the net scattered field at P_0, therefore, is not substantially altered from what it was determined to be as the result of the integration performed earlier to obtain the extinction distance in a low temperature medium.

This is because the sum of scattered fields signified by figures 73 and 74 above affects *each* electronic charge at P_0 *individually*. In the cases implied in the figures above, each electron is again individually affected although in these cases by a completely distinct ensemble of scattering events. So the sum of all these integrals affected by the electron density in furthering the next stage in the propagation of the radiation by forward scattering will be substantially unchanged and the extinction distance will remain essentially unchanged accordingly.

The number of electrons involved in *each* domain is similar to what it would be in a cooler medium, although more electrons are included in the total of the composite regions that contribute to forward scattering in this case. Extinction of visible light in the hot plasma of the intergalactic medium can only occur after propagation through hundreds or thousands of light years and ultimately upwards of 10^{23} electrons have gotten involved in the scattering in each extinction interval similar to what would happen in our own atmosphere.

m. wavelength characteristics of the detected radiation

So far we have concentrated on the effects of relativistic velocities on the geometry of the coherency domains of the various contemporaneous *detections* of scattered radiation from the perspective of the various electronic constituents of a small volume of the medium about the variously denominated point, P_0. The induced radiation of the affected electrons that ultimately effect forward scattering from that point forward must now be considered in more detail. There are accordingly two aspects to be considered in this case: Characteristics of the radiation that is detected by each electron, and the characteristics of the radiation emanating from the region about P_0.

In figures 80 through 85 we have shown the necessity of accepting that there are unique detection situations for each of the various electrons at the variously denominated point, P_0. In figure 85, the fact of each coherency domain analogous to figure 74 originating at varying distances and angles from P_0 was illustrated. This was done more or less from the perspective of the individual electrons (to the extent that that is possible in a single diagram). Figures 82 and 83 illustrated the base plane geometry for a single electron moving at right angles to the line of sight. In figures 84 we attempted to combine perspectives of all the variously moving electrons. Figure 85 combined the viewpoints of those electrons whose detected scattering radiation derives from the same perpendicular distance from P_0.

These diagrams each depict aspects of Lorentz transformation equations in their own unique ways to give some physical insight into what is happening with regard to the unique time and distance measures coming out of those equations. Electrons moving toward the original distant source will detect Doppler 'blue shifted' radiation, i. e., radiation with a shorter wavelength. Those moving away will detect a Doppler redshift, i. e., a longer wavelength to the same radiation.

However, electrons with transverse velocities will also experience redshifting, but this will be the second order *transverse Doppler effect*.

The pertinent relativistic wavelength formula from which all these effects can be calculated is the following:

$$\lambda_s = \lambda_i \, [\, 1 + (\, \mathbf{v \cdot u}\,)\, /c \,]\, [\, 1 - v^2/c^2 \,]^{-1/2}$$

Here $\mathbf{v \cdot u}$ is the 'dot product' of the vector velocity of the electron $\mathbf{v}$ and the vector direction to the source, $\mathbf{u}$. The velocity squared is, of course:

$$v^2 = v_x^{\,2} + v_y^{\,2} + v_z^{\,2}$$

Like the relativistic aberration formula for angular distortions, this equation also derives directly from the Lorentz transformation equations provided earlier in this chapter.

It is a unique consequence of Einstein's relativity that motion at right angles to the source (whether primary or secondary) results in a lengthening of the detected radiation wavelength. The effect is called 'transverse Doppler', and its formula is:

$$\lambda_s = \lambda_i \, [\, 1 - v^2/c^2 \,]^{-1/2}$$

which results directly from the above formula with $\mathbf{v}_T \cdot \mathbf{u} = 0$. Here $\mathbf{v}_T$ is the velocity in the plane perpendicular to the source, whose square is, $v_x^{\,2} + v_y^{\,2}$.

The 'radial Doppler' wavelength effect in Einstein's relativity pertains to the effect of radial motion toward or away from the source. It's formula is:

$$\lambda_s = \lambda_i \, [\, 1 \pm v_z /c \,]\, [\, 1 - v^2/c^2 \,]^{-1/2}$$

This formula does, in fact, incorporate both the radial and transverse effects since, if $v_z = o$ this gives the transverse effect. So that if the incident radiation is characterized by the wavelength λ_i, then the once-scattered radiation detected by an electron having velocity $\mathbf{v}$ relative to L_o will be characterized by the wavelength, λ_s.

n. wavelength characteristics of the emanating radiation

Next we must consider the effects, if any, of relativistic velocities of the detecting electrons on the *emanations* from that small volume about P_O. The difference in perspective is extremely significant here since in the special theory of relativity it is the relative state of two observers and not a presumption of absolute motion that is important.

As each electron at P_O is induced into oscillatory motions that precipitate its emission of sympathetic radiation as depicted in figure 36, it becomes the source of subsequent 'detections' in the path of continuing extinctions. By scattering, the electron 'detecting' radiation subsequently contributes to the incident radiation of the next extinction interval. It will exhibit an oppositely directed motion relative to a subsequent L_O from what it had in the current L_O. It is the perspective difference illustrated in figures 77 and 79. So now we have:

$$\lambda_S = \lambda_i (1 \pm v_z /c) (1 - v^2/c^2)^{-1/2}$$

$$\approx \lambda_i (1 \pm v_z /c) [1 + \tfrac{1}{2} (v_x^2 + v_y^2 + v_z^2)/c^2]$$

$$= \lambda_i [1 + \tfrac{1}{2} (v_x^2 + v_y^2 + v_z^2)/c^2 \pm v_z/c \pm \tfrac{1}{2} v_z(v_x^2 + v_y^2 + v_z^2)/c^3]$$

$$\cong \lambda_i [1 + \tfrac{1}{2} \beta^2]$$

Notice that in taking the product implied by the expression for λ_S above, terms involving the radial velocity, v_z will cancel statistically over repeated extinction intervals because of the statistical nullification due to sign changes in a 'stationary' medium. Therefore, this formula applies as long as we can ignore fourth order terms in electron velocity divided by the speed of light.

So in a relativistic plasma, extinction of incident radiation will result in a tiny alteration of the distribution of wavelengths from that of the original. This is an exception to Wolf's scaling law that was discussed in the previous chapter as applicable to cooler media. Radiation emanating from P_o will be very slightly reddened. This reddening will be by the slightest percentage of a wavelength, but it is always an increase. Thus, it will *not* 'average out' as the saying goes.

o. convergence of the emanating radiation

Again the constraints imposed by forward scattering determine the envelope of outward emanations. To constitute forward-scattered radiation in the locally stationary Lorentz frame of reference, the emanations from about P_0 must obviously be coordinated to within that same angle of α_0 if these emanations are to contribute to coherent interference phenomena at a subsequent extinction. This is precisely what was determined earlier when we ignored the effects of the velocities of the electrons. The fact that individual relativistic electrons at the next juncture may detect radiation from domains that do not include this particular P_0 (just as in figure 78 and subsequent figures where different coincident detections come from completely separate areas on a plane) does *not* imply a divergence of the emanations. Radiation from co-planar surfaces of constant phase will proceed in the generally 'forward' direction even in the *locally stationary* frame of reference, still contributing to forward scattering. The extinction process assures that they can not diverge. In the locally stationary frame of reference of the medium emanations must *all* proceed in essentially the same direction (to within plus or minus the angle α_0). Otherwise they will all cancel just as do radiations from outside a coherency domain as we saw earlier. Only those radiations directed straight ahead will contribute to the coherent amplification inherent in replacing incident fields by forward-scattered fields at the next juncture in a continuing extinction process.

As discussed above, a tenet of the special theory of relativity dictates that observers can not distinguish the velocities of the various constituents of the medium about any point other than by mere inferences from wavelength. Therefore subsequent electron detections will not be affected by the individual scattering electron velocities. The velocity of light (as indicated by its arrival time) measured at the next extinction point, P'_0 will be totally independent of the velocity of any and all of the various constituents at P_0.

Furthermore, radiation scattered from each individual electron will emanate in *all* directions, with only a *slight* intensity preference in the direction of propagation of the incident radiation as variously determined in the frames of the individual scattering electrons. This intensity preference includes a factor of the cos σ, where σ is the angle between the direction of propagation of the *incident* and individually *scattered* radiation fields as shown below in figure 88. Each electron's 'primary' direction of scattering will be unique for each electron at P_o because of the unique aberration associated with each. However, this will not appreciably reduce the collective intensity of the scattered fields propagated along the z axis in L_o since the directional intensity factor that will be realized in that direction is nearly unity for *all* contributing electrons as shown for small angles in figure 88. Quantum treatment does not substantially alter this picture except at extremely short wavelengths as we will show in the next chapter.

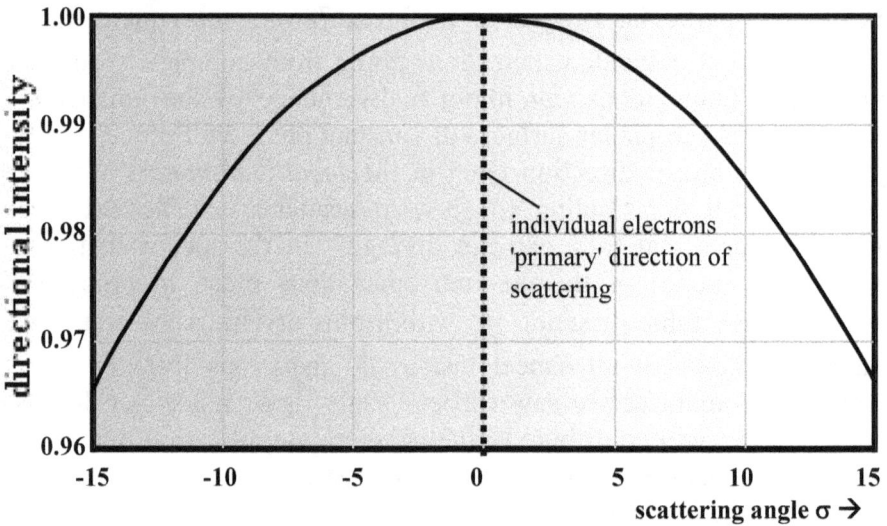

Figure 88: Slight angular preference of individual electron scattering

Thus only light proceeding within a cone angle of α_o with respect to the direction of the positive z axis in the locally stationary Lorentz reference frame will arrive at the next juncture with minimum transit time and sufficiently in phase to contribute to extinction. That is a constraint for interfering coherently at the next juncture as required for forward scattering that was derived earlier for a 'cold' medium as shown in figure 89. All other angles in the locally stationary frame of reference of the observer and medium will be excluded.

Also, for those who are one step ahead of us, the slight variation of wavelengths that result at each extinction will not 'grow', since too large of excursions will defy coherency constraints. The majority wins.

Excursions outside of the constraints will simply not contribute to the extinction process just as multiple scattering does not.

Probability densities replace intensities but to the same basic effect. The resulting reinforced directionality and uniformity of wavelength of the secondary radiation observed to emanate from the volume about P_0 is addressed specifically in figure 90. In figures 82 and 84 the perspectives of the stationary observer and of specific relatively moving electrons within the volume were both illustrated. The differences in their perspectives are to be understood in terms of the aberration of light associated with their relative motion.

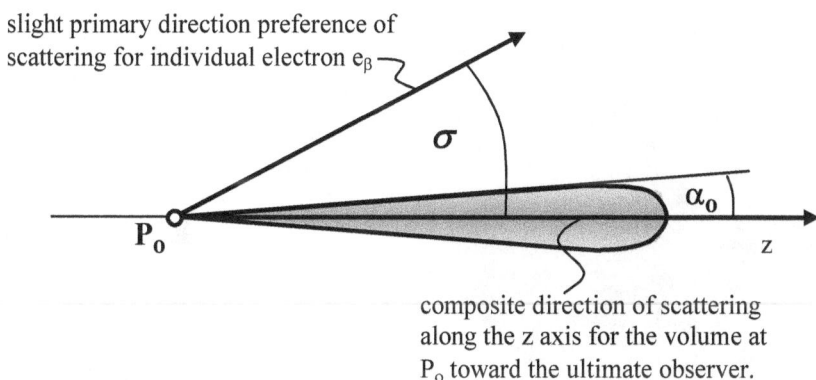

slight primary direction preference of scattering for individual electron e_β

σ

α_0

P_0

z

composite direction of scattering along the z axis for the volume at P_0 toward the ultimate observer.

Figure 89: Angular dependence of the continuing forward-scattered radiation

From the perspective of any observer of this scattered radiation, assuming for simplicity now that only one *moving* electron were involved, the observed light will have undergone an angular 'deflection' at the scattering event at P_0. This angle is a function of the individual electron's velocity relative to the ensemble that determines the forward scattering direction (alternatively relative to the medium or subsequent observer) and can be accounted as a transverse relativistic aberration effect. In essence the light scattered by every electron within the volume (except for the low probability case of electrons with no relative local velocity) will be responsible for some such velocity-dependent deflection as a criterion for forward scattering.

But in all cases the deflection (at P_0 in L_0) is with respect to the *general* direction of the incident radiation, *not* with respect to any particular forward scattering direction as shown the earlier figures. It is a special case that there is no such deflection involved when the scattering occurs in a low temperature medium, and even there, it is merely a matter of the deflection angle being limited to within the constraint of the angle α_0 rather than ψ. In other words it is merely the magnitude of the deflection angle, α_0 that effects a negligible result in those cases.

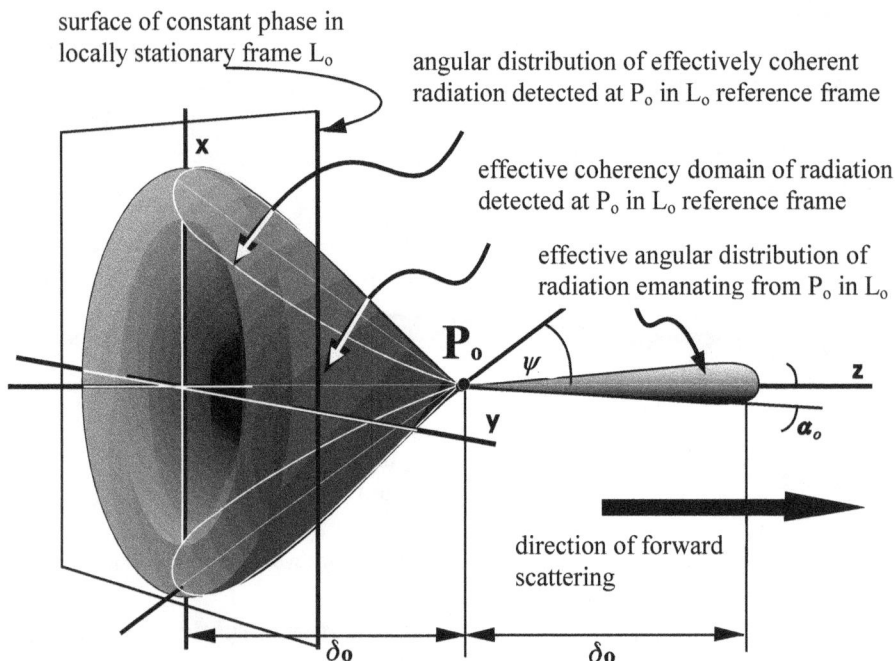

Figure 90: Effective coherency domain about (and scattering from) P_0 in L_0

This results, in the case of the intergalactic medium, in a *convergence* of radiation into P_0 from what is nearly a conical shell whose principal angle is determined by temperature and whose thickness is determined by the Maxwellian distribution of velocities. It is merely the same forward scattering considerations that established α_0 for more usual low temperature media. This is shown heuristically in figures 92 and 93 of the next chapter. The Maxwellian distribution of velocities will result in a conical shell of principal angle, $\psi = \sin^{-1} \beta$ from which the vast majority of forward scattered radiation will originate as determined by the distribution that was shown in figure 87.

Chapter 10

Implications of Conservation Laws to Scattering by Plasma Electrons

Scattering of radiation by high-speed electrons in intergalactic plasma certainly brings to mind what is called *Compton scattering*, although that has typically involved single photon/particle interactions.

a. Compton scattering formulas

Scattering of X-radiation by electrons was shown by Compton to produce a lengthening of the wavelength of the scattered radiation in accordance with the following formula as illustrated in figure 91:

$$\lambda_{final} - \lambda_{initial} = (h / m_e c) (1 - \cos \psi),$$

Here ψ is the angle of deflection of the photon; ϕ is dependent on the value of ψ; m_e is the rest mass of the electron; and h is Planck's constant. The origin of the preceding equation is a set of conservation constraints on energy and momentum laws as illustrated in figure 91.

This phenomenon does not depend on the electron being bound, of course. The resulting *change* in wavelength is *not* initial wavelength dependent and has an absolute maximum value called the *Compton wavelength* defined as:

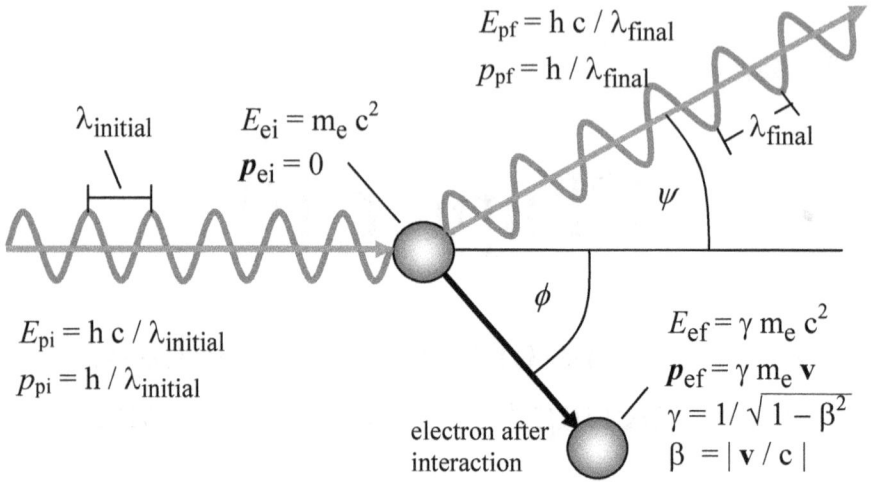

$E_{pf} = h\,c / \lambda_{final}$

$p_{pf} = h / \lambda_{final}$

$\lambda_{initial}$

$E_{ei} = m_e\, c^2$

$p_{ei} = 0$

λ_{final}

ψ

$E_{pi} = h\,c / \lambda_{initial}$

$p_{pi} = h / \lambda_{initial}$

ϕ

$E_{ef} = \gamma\, m_e\, c^2$

$p_{ef} = \gamma\, m_e\, \mathbf{v}$

$\gamma = 1/ \sqrt{1 - \beta^2}$

$\beta = | \mathbf{v} / c |$

electron after
interaction

Figure 91: Depiction of Compton scattering relations

$\lambda_C \equiv h / m_e c$

λ_C is approximately 0.02425 Angstroms. Expectation values of the wavelength change are very considerably less than this amount, of course; only for X-radiation and gamma rays would such quantities become significant after a single scattering. Although 'Compton scattering' typically pertains to X-ray deflection from electrons, the effect applies throughout the electromagnetic spectrum. Primarily only the relative magnitude of associated effects vary in accordance with $\lambda_{initial}$ as one would expect from one portion of the spectrum to the next with Thompson scattering applicable to the forward scattering we have discussed that will transition continuously to Compton scattering .

Clearly, the lengthening of the wavelength of incident radiation associated with this (or any similar) effect must be obtained as a direct result of an energy and momentum transfer from the electromagnetic photons to affected material electrons as Compton demonstrated. In addition, such interactions necessarily alter the subsequent direction of travel of the photon in addition to that of the electron. It is inevitably associated with a 'bending' of the light path. The electron will pick up the associated loss of energy in the photon in the amount of ΔE, where:

$$\Delta E = h\,c / \lambda_{initial} \left[\frac{\lambda_C\, (1 - \cos \psi)}{1 + \lambda_C\, (1 - \cos \psi)} \right]$$

198

There will also be a proportional increase in momentum since photon energy and momentum always remain proportional.

It is evident that whenever there is substantive interaction of electromagnetic radiation has with an electron an associated deflection of both the photon and the electron will result. The angular changes in these directions formulated by the Compton are the observable effects of a momentum transfer taking place during any such interaction.

In relativistic quantum mechanics, the scattering cross section of free electrons is given by the Klein-Nishina formula. For lower energy photons, the Thompson cross section that is directly applicable to forward scattering as approximated for the expected small angles as shown in figure 88 suffices for all but the shortest wavelength X-ray and gamma rays. Later we will show a discriminating refutation test of the cosmological theories based on observed redshifts in this domain.

b. applying conservation formulas to forward scattering

Before calling such a process into account as in some way responsible for effecting a cosmological redshift *per se*, let us depict where and how (if at all) such deflections are envisioned as occurring in the forward scattering process. Following the diagrams of 65 through 70, it is fairly obvious that there *is* a bending in the transmission path that occurs at each electron scattering event in the frame of L_0, i. e., once per extinction distance in the progress of radiation through a medium. This is illustrated again in figure 92.

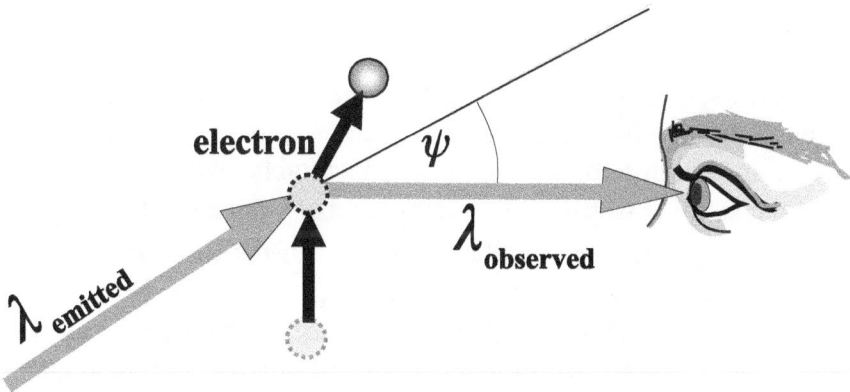

Figure 92: **Geometry of the Compton-like scattering effect on an individual plasma electron involved in forward scattering**

Clearly wherever there is such a deflection, the wavelength must be increased to satisfy the conservation laws for momentum and energy. The fact that any change in wavelength toward the red end of the spectrum must involve a proportionate deflection has seemed to most, however, to preclude effects such as Compton scattering from being immediately applicable to mechanisms of forward scattering. In fact reddening would become appreciable only after many extinction intervals by which we observe the distant regions of the cosmos. Because these changes in direction at each juncture would seem to preclude the straight-ahead progress required to effect the coherent imaging by which distant regions are observed, it has been debunked as a means of explaining cosmological redshift.

This negative thesis was developed by Zel'dovich to discredit "any conceivable version" of the *tired light* theory once propounded as an escape from Hubble's expansion hypothesis, Misner et al. (1973). He presumed to have demonstrated that there could be no physical phenomenon whatsoever that would effect an otherwise-unobserved momentum transfer from electromagnetic radiation to the material substance of an intergalactic medium to produce cosmological redshift as the result of such interactions. He emphasized that associated with any such phenomenon would be an inevitable angular dispersion. His comment in this regard was the following (1963):

> "If the energy loss is caused by an interaction with the intergalactic matter, it is accompanied by a transfer of momentum; that is, there is a change in direction of the photon. There would be a smearing out of the image; a distant star would appear as a disc, not a point, and that is not what is observed."

However, it was evident in discussions associated with figures 80 through 85 and 90 in the previous chapter, and 93 that the forward scattering in a high temperature plasma inevitably involves deflection through the angle ψ whose sine is approximately equal to $<\beta_e>$. It does this *without* imposing any angular *divergence* of the resultant radiation path that would produce image "smearing". This is because the necessary deflection angle associated with the momentum transfer is obtained from *convergence* of detected radiation back into the ongoing forward direction in the reference frame L_o, for example. So there is *convergence* rather than *di*vergence of the scattered radiation. Once detected in the various coherency domains it is again re-focused to the common direction as dictated by forward scattering constraints shown in the figures cited above. The deflection angle results as sort of an *inverse* relativistic aberration effect as was shown in figures 82 and 83 that refocuses the remaining energy rather than blurring it outward as Zel'dovich supposed. Figure 93 provides a fairly accurate depiction of this repeated process whereby an extinguished photon's energy is continuously recycled right back along a direct line of sight with energy and momentum losses at each stage as there must be. Of course rather than regimented cycles as shown for didactic reasons in the diagram, each electron along the path would contribute in its own scattering domain. A fairly broad swath of space is involved in forward scattering from distant regions as determined primarily by the dynamic pressure of the medium. Its width of this swath will be approximately,

$$w \cong \delta_o <\beta_e>.$$

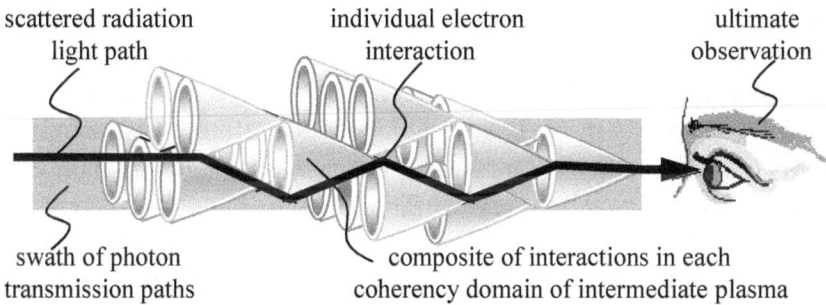

scattered radiation light path   individual electron interaction   ultimate observation

swath of photon transmission paths   composite of interactions in each coherency domain of intermediate plasma

Figure 93: Continual refocusing of electromagnetic energy through extinction via high energy plasma electrons

If the prospect of imaging photons that involves such a wide swath through space is difficult to comprehend, consider the staples of radio astronomy that uses widely dispersed interferometry. Coherent reinforcement of photons in that application uses antennas that are separated by many thousands of miles on earth as well as employing the Japanese HALCA satellite that was put in a 13,000 mile orbit above earth in 1997. Together with the ground based antennas very sharp images are thus obtained using HALCA. The images produced exhibit more than 100 times the resolution achieved by the Hubble Space Telescope. Thus, effects of wide swaths of coherent radiation can be brought to bear on an imaging process just as applies here.

So that is how Zel'dovich's concern is addressed by the scattering model presented here. *Yes*, the transfer of momentum from radiation to the constituents of the intergalactic medium *does* involve deflection. But, *no*, that *does not* necessitate any blurring at all.

To have conceived that billions of physical interactions of electromagnetic photons and material entities could have occurred in the process whereby we see the distant cosmos – as must happen in order for us to see everything we do see – with absolutely no loss of energy or associated decrease in momentum seems extremely naïve to this author. It is time to address the inevitable reality of the no-free-lunch aspects of the energy exchanges known to occur in dispersive media.

In the ordinary extinction process in low temperature media that was detailed earlier and depicted most specifically in figure 74, there were angles of deflection up to and including α that all contributed to the extinction process. How could we not suppose that there were momentum losses and, therefore, wavelength increases even associated with this scattering process? What was it that gave license to assume these interactions could be excluded from all know principles of physics? Primarily it was that any possible effect was completely immeasurable, since $\alpha \approx 10^{-18}$ radians for visible light in such cases. And so in everyday laboratory work we would have had a maximum wavelength increase of on the order of 10^{-28} cm – a truly immeasurable amount even after an astronomical number of extinction intervals.

The impact of large thermal velocities and extreme distances to 'cosmological' objects is, therefore, to produce both more appreciable deflection angles and numbers of extinction intervals whereby cumulatively measurable amounts of momentum and energy *can* be transferred from the radiation to the intergalactic medium. This still supports coherent forward scattering and associated imaging. We will show that this relativistic effect produces profound implications to our view of the cosmos via these Compton-like transfers of energy and momentum to the medium producing unilateral wavelength changes.

Notice that the effect with which we are concerned is different than the confirmed *Sunyaev-Zel'dovich microwave diminution* effect proposed as resulting from Compton scattering of background microwave radiation by high temperature ($\sim 10^8$ degree K) and higher density ($\sim 10^{-3}$ cm^{-3}) *intracluster* electrons we will discuss later.

(See Carlstrom, et. al., 2002, and also Birkinshaw, 1999.) This latter effect is concerned with the indiscriminate noncoherent scattering of background radiation, which would have a slight affect on the apparent brightness of the background in the direction of galaxy clusters as discussed by Zel'dovich and Sunyaev, (1969) and Sunyaev (1969). This effect is related to the plasma absorption phenomenon described in the earlier chapter. However, although obviously somewhat related to scattering being described here, this phenomenon does not address *forward* scattering *per se*. It is definitely not the operative mechanism from which cosmological redshift derives.

The deflection angle, $\sin^{-1} \beta_e$, illustrated in figures 83, 85, 91, and 92 above is a simple function of the relative velocity of the electron with respect to the assumed mean zero velocity of the electrons in the plasma medium. These velocities are distributed according to the Maxwellian distribution of plasma electron velocities as shown in the curves of figure 34, page 61 in chapter 3 and more specifically in figure 87, page 187 in the previous chapter. The average such velocity can fairly accurately be approximated by the root-mean-square given by:

$$< \beta > \ = (3 k T / m_e c^2)^{1/2}$$

Substitution of this value to obtain a meaningful estimate of the average deflection angle, i. e., $<\psi> \ = <\sin^{-1}\beta>$, to insert into the preceding energy conservation formulas produces the estimated average change in wavelength for the forward-scattered fields. The braces indicate the estimated mean of a parameter value. So that,

$$\sin^2<\psi> = <\beta>^2$$

And, since,

$$\cos^2 <\psi> = 1 - \sin^2<\psi> = 1 - <\beta>^2$$

$$(1 - \cos <\psi>) = \ <\beta>^2 / (1 + \cos <\psi>)$$

As long as $<\beta>^2 \ll 1$, which still applies fairly well even for the high temperatures in cluster regions of the intergalactic medium, the following approximation will hold:

$$(1 - \cos <\psi>) \approx 1/2 < \beta >^2$$

c. assessing the magnitude of wavelength change at each extinction

Thus, for the mean deflection angle $<\psi>$ of photons scattered by electrons in the intergalactic medium which will be statistically typical of each forward scattering event, we can obtain from the conservation formulas the mean of the wavelength increment at each scattering event. If we begin by defining the mean change in wavelength associated with each such event as:

$$< \Delta\lambda_e > \equiv \lambda_{scattered} - \lambda_{incident} \,,$$

then from Compton's formula above we have:

$$< \Delta\lambda_e > = (h / m_e c) (1 - \cos <\psi>) \approx (h / 2 m_e c) < \beta >^2$$

$$= 3/2 \, k \, T_e \, h / m_e{}^2 \, c^3 \cong 6.14 \times 10^{-20} \, T_e \text{ centimeters}$$

This individually immeasurable mean wavelength shift occurring at each extinction interval is independent of the propagation wavelength excepting for extremely short wavelengths that we will discuss, and is, therefore, not a redshift *per se*. However, it occurs at intervals, which are not equidistant since, as we saw above, these intervals depend on wavelength so that the cumulative effect will, in fact, exhibit the same wavelength dependence as Doppler changes, as we will show.

Whereas statistically extinction events will occur at regular intervals in low temperature media such as earth's atmosphere because wavelength changes remain insignificant, with high temperature media this is not the case. With wavelength changing at each scattering event, extinction distance shrinks throughout the *life cycle* of electromagnetic transmission from an original source to an ultimate observation. Knowing the extent of this wavelength change per extinction interval, we are now in a position to determine the statistical change in extinction distance after each scattering event, and thus establish the effective number of extinction intervals occurring in a given distance that light propagates through the intergalactic medium.

The total propagation distance can be assessed as a summation of all the individual extinction distances that occur in getting light from one point to another as follows:

$$r(n) = \sum_{i=0}^{n} \delta(\lambda_i)$$

where

$$\delta(\lambda_i) = m_e c^2 / (\rho_e e^2 \lambda_i) \cong 3.55 \times 10^{12} / (\rho_e \lambda_i)$$

So that $\delta(\lambda_i) = 3.55 \times 10^{17} / \lambda_i$ when electron density $\rho_e = 10^{-5}$ cm^{-3}. But we also have that:

$$\lambda_i = \lambda_{i-1} + <\Delta\lambda_e> = \lambda_{i-2} + 2 <\Delta\lambda_e> = ...$$

$$= \lambda_e (1 + i <\Delta\lambda_e> / \lambda_e) = \lambda_e (1 + i k T_e h / (m_e^2 c^3 \lambda_e))$$

where λ_e is the emission wavelength. So we obtain the expression:

$$r(n) = [3.55 \times 10^{12} / (\rho_e \lambda_e)] \sum_{i=0}^{n} 1 / [1 + i (6.14 \times 10^{-20} T_e / \lambda_e)]$$

The form of this equation is plotted in figures 94 and 95 for a dynamic pressure determined by values of approximately $\rho_e = 2 \times 10^{-5}$ cm^{-3} and $T_e = 2.5 \times 10^8$ K. These values are a compromise appropriate to the hot dense intracluster plasma gases and sparse distances between clusters.

It is necessary to present a family of curves, one for each value of λ_e. Clearly for shorter wavelengths a pronounced curvature appears. Notice the major difference in the distance scale in the two figures. To show that this curvature is of a logarithmic form of distance versus number of extinction intervals, refer to figure 96.

Clearly the logarithmic form will 'kick in' after differing numbers of extinction intervals for various types of astronomical sources of radiation. Naturally, it will become a useful relation only after many extinction intervals, sooner for X-rays and gamma rays, but requiring greater numbers with n >> 10^5 for ultraviolet wavelengths such as the Lyman-alpha emissions. Optical and longer wavelengths require appreciably more than that. These figures and discussion

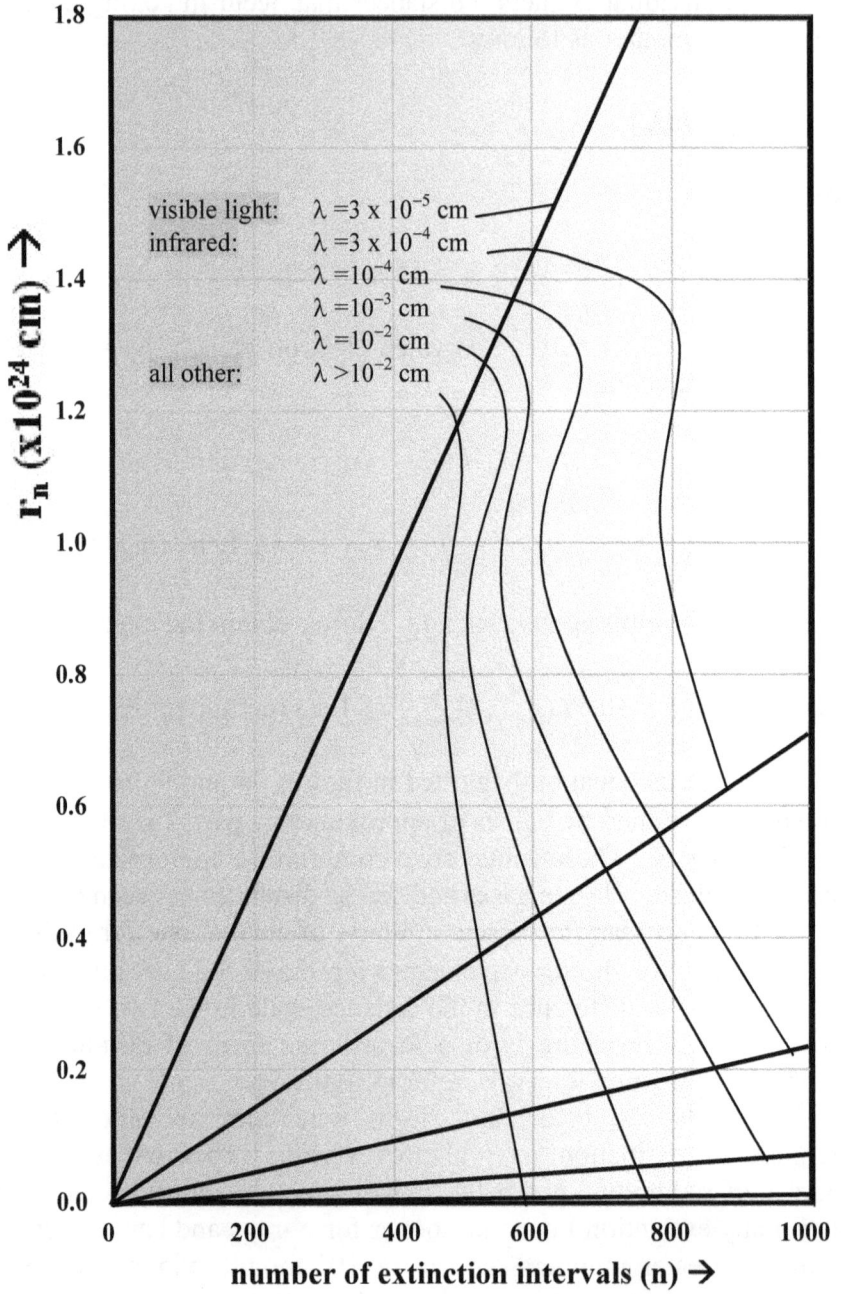

Figure 94: Distance versus number of extinction intervals (longer wavelengths)

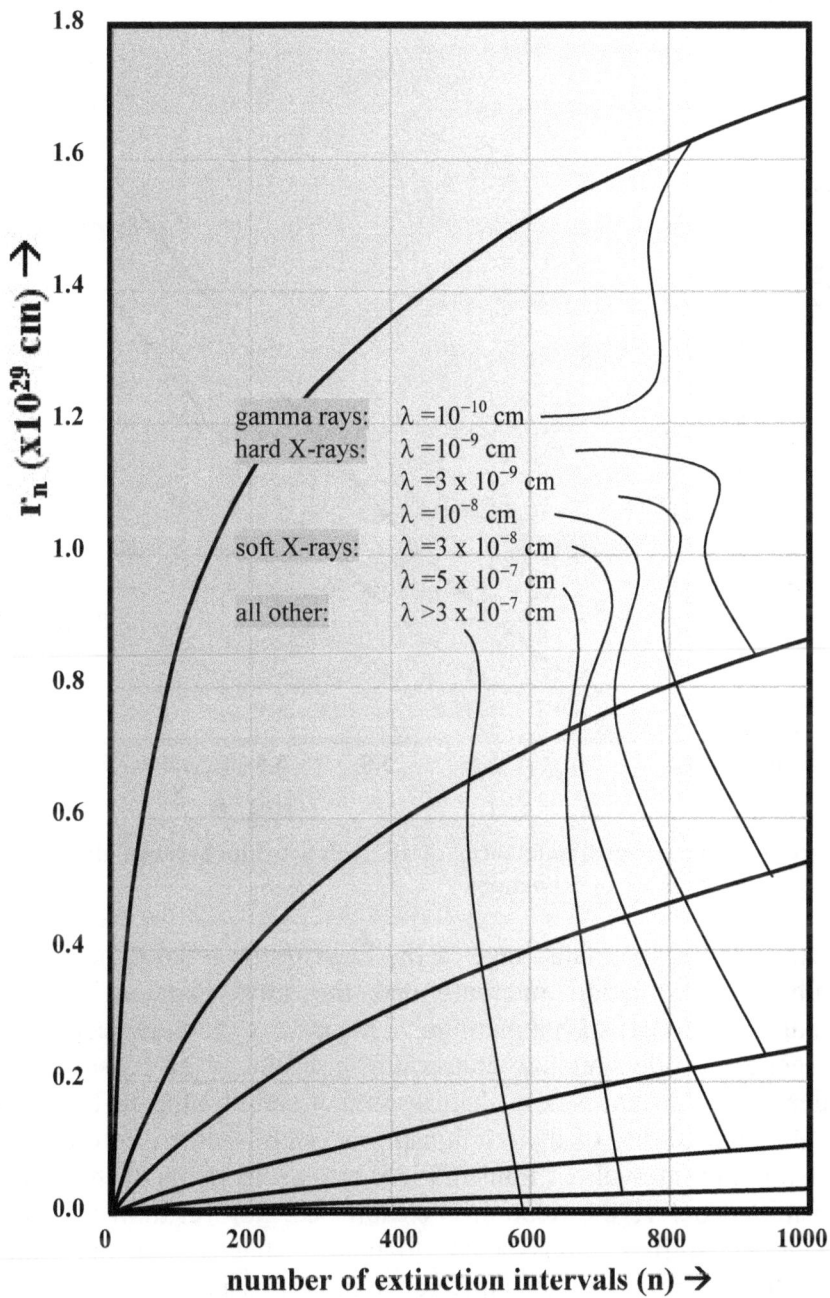

Figure 95: Distance versus number of extinction intervals (short wavelengths)

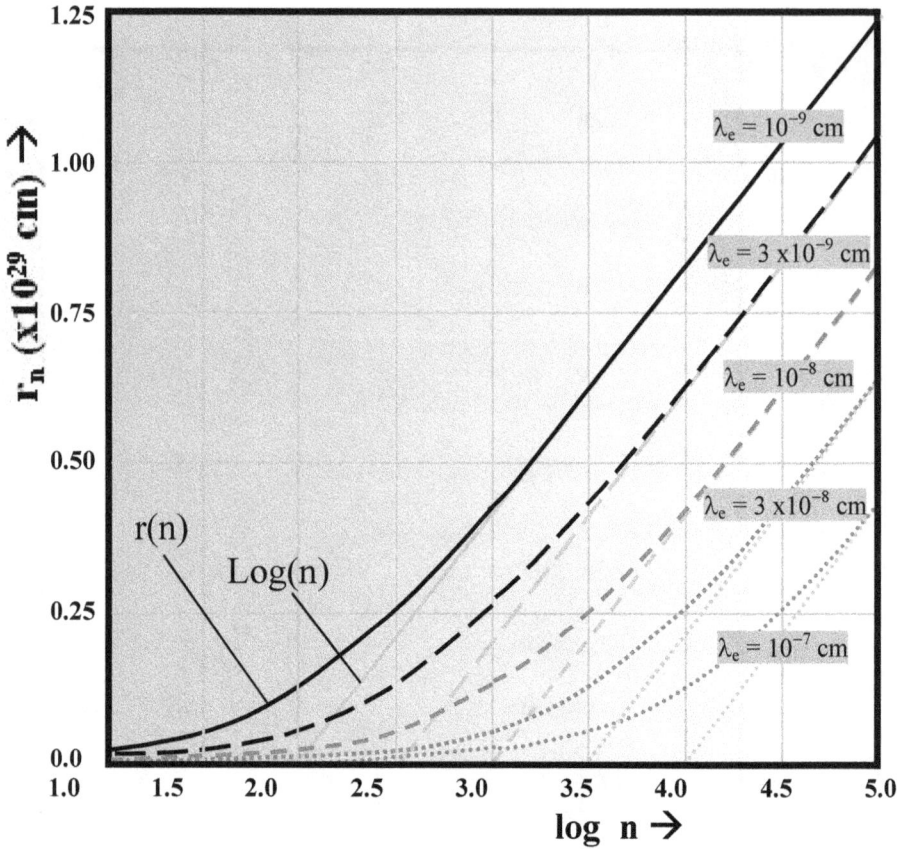

Figure 96: The logarithmic form of the relationship between distance and number of extinctions

illustrate the appropriate range of the logarithmic relation between the number of extinction intervals and the total distance that light propagates through such a medium. This is very much in keeping with the general understanding concerning usefulness of redshift as an indicator for distances less than several hundred Mpc in the optical range. The validity of the relationship is established based on number of extinction intervals of transmission, but *not* in terms of the distance transmitted before the relation is established, nor yet until the similar distance-redshift relations begins to hold. That caution may not be immediately apparent. This difference in emphasis is very germane to later discussions of the nature of the distance-redshift relationship.

Chapter 11

Derivation of a
Distance-Redshift Relation

'Redshift', for which the symbol 'z' is usually employed, is a unit-less parameter. It is defined as the ratio of the induced change in wavelength toward the red end of the spectrum (i. e., toward longer wavelengths with regard to visible light) divided by the wavelength of the emitted radiation. These definitions are the following:

$z \equiv \Delta\lambda / \lambda_{emitted}$, where

$\Delta\lambda \equiv \lambda_{observed} - \lambda_{emitted}$

A redshift is to be distinguished from a *blueshift* which is basically the same unit-less parameter except that instead of the observed wavelength being longer than the emitted wavelength, the reverse is true so that, as defined above, $\Delta\lambda$ would be negative.

a. extending energy conservation to a redshift relation

We've seen that scattering in a plasma produces a small mean change in wavelength, $< \Delta\lambda_e >$ once per extinction interval. After n such forward scattering instances, the wavelength will be increased proportional to n. We introduce the following incrementally increasing redshift-like parameter based on what we have found to occur at each extinction interval via forward scattering in the intergalactic medium:

$$\xi(n) \quad \equiv (\lambda_n - \lambda_e) / \lambda_e = 3/2 \ n \ k \ T_e \ h \ / \ (m_e^2 \ c^3 \lambda_e)$$

$$\cong n \ (6.14 \ x \ 10^{-20} \ T_e \ / \ \lambda_e)$$

This redshift-like parameter exhibits a linear relationship to the number of extinctions that have occurred. However, the wavelength dependence illustrated in the curves plotted in figure 97 clearly distinguishes this from a Doppler redshift. Each wavelength demands its own plot. But one could not expect a functionality that was independent of wavelength for redshift versus number of extinction intervals – as against distance – through intergalactic regions. This is because propagation distance is also wavelength dependent as we saw.

To understand observed redshifts of objects deep in space – what for *most* intents and purposes would appear indistinguishable from a *Doppler* redshift – let us consider the plots of distance versus number of extinctions n presented in figures 94 and 95 above. In figure 97 we present the associated functionality of redshift.

Merging the functionalities of $r(n)$ and $\xi(n)$ by plotting one versus the other, the dependence on number of extinction intervals is entirely eliminated. Since distance is a linear function of the log of n as was shown in figure 96, but inversely dependent on emission wavelength, a distance versus redshift plot produces an *extremely* similar plot for all values of wavelength. A virtually identical relationship holds from hard X-rays ($\approx 10^{-9}$ cm) through extremely long wavelength radio signals ($>10^7$ cm). In this way, we obtain a Doppler-like distance-redshift relation directly applicable to this broad range of wavelengths. The combined curve is shown in figure 98 below, where it is easily seen that the wavelength dependence is virtually eliminated for wavelengths greater than 10^{-10} cm. It is also clear that for wavelengths less than 10^{-9} cm, unique relationships pertain.

The logarithmic form applies to these curves also, but the distances at which they are finally established are greater than for longer wavelengths. There is clearly an observable distinction to be made at low redshifts, where for extremely short wavelength radiation as shown in figures 98 and 99, a fixed relationship between Z_n and r_n will not yet be established such that the ratio of Z_n over r_n is not a reliable indicator of distance. This results in objects characterized by such emissions experiencing less redshift than objects at the same distance characterized by longer wavelengths. Since there is no similar distinction according to any proposed variant of standard cosmological models, these exceptions to usual distance–redshift curves at short wavelengths provide a possible means of falsifying cosmological models. It would indeed be interesting to determine whether an effect could be observed with gamma ray bursts associated with galaxies at redshifts much less than unity. Is the distance at which a redshift relation becomes reliable greater, in fact, at short wavelengths or not?

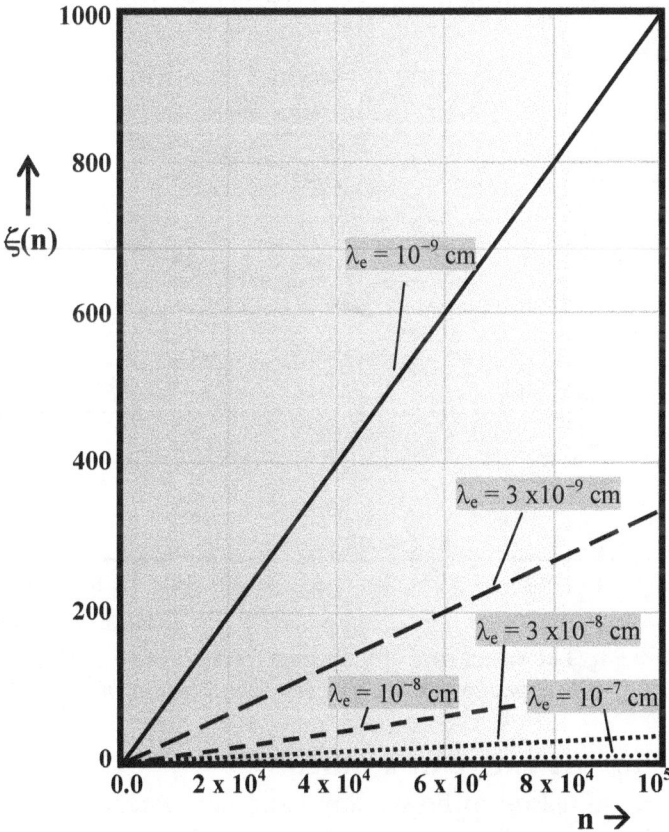

Figure 97: Form of relationship for "redshift" versus number of extinctions

In order to explore these interesting threshold phenomena we need to look in more detail at how distance accumulates with increasing numbers of extinction intervals for the various wavelength ranges. Figure 99 shows how uniquely this process takes place in the various short wavelength domains as compared with longer wavelengths. Figure 100 illustrates this in a little more detail.

It seems obvious that the wavelength at which this cutoff of unilateral applicability of a single function representing the distance-redshift relation is very dependent upon the temperature and density of the intergalactic medium. In these curves we have used $\rho = 2 \times 10^{-5}$ cm^{-3} and T= 2.065×10^{8} K. In a later section we will discuss the values of these parameters that closely match the Hubble constant for a more definitive representation of observations of this type.

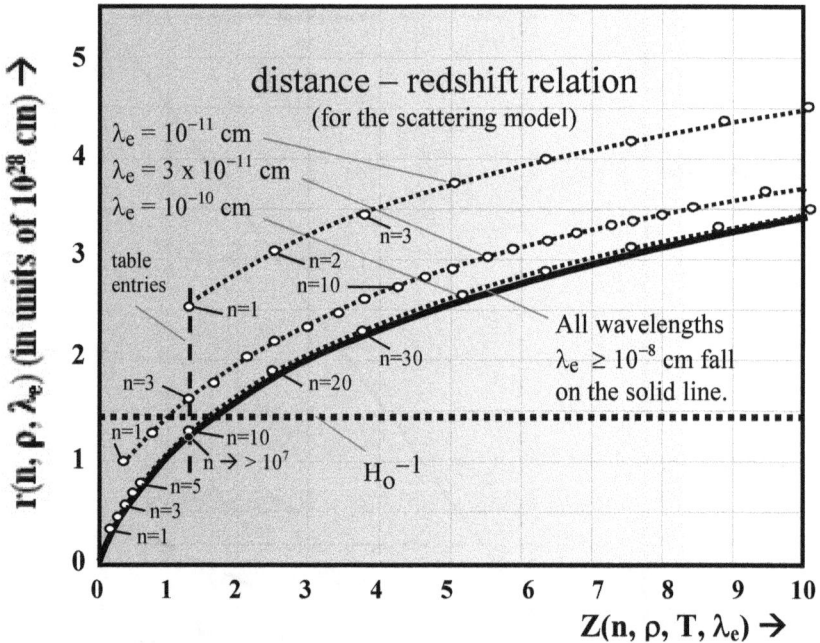

Figure 98: The emergence of a distance versus redshift relation
independent of wavelength over a broad range

b. exploring the range of applicability

A determination of how many extinction intervals are required to effect a given redshift and before Hubble's relation would apply with any validity are summarized in the table below. Together with the previously referenced figures these results tell us where the distance-redshift relation begins to have meaning for the various wavelengths.

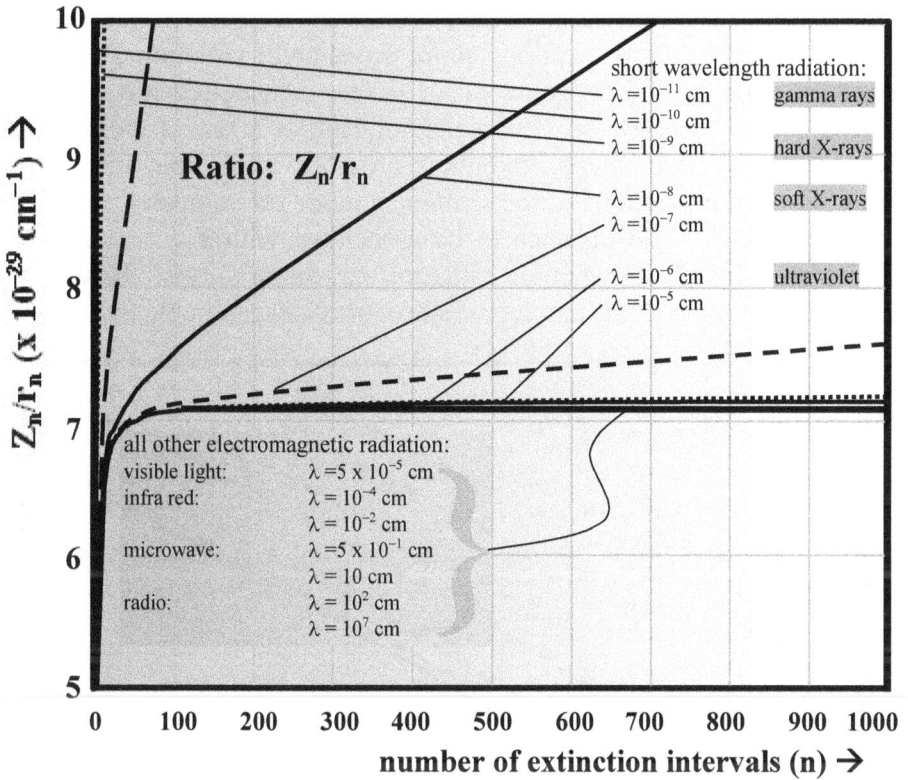

Figure 99: The broad range of applicability of a single curve for the distance-redshift relation and the scope of the dispartity at extremely short wavelengths

number of extinc-tion intervals, n	initial wave-length, λ (cm)	redshift, Z_n	distance, r_n (x 10^{28} cm)	ratio, Z_n / r_n (x 10^{-29} cm)
1	10^{-11}	1.268	2.560	4.592
3	3×10^{-11}	1.268	1.591	7.970
10	10^{-10}	1.268	1.277	9.927
100	10^{-9}	1.268	1.160	10.926
1,000	10^{-8}	1.268	1.149	11.035
10,000	10^{-7}	1.268	1.148	11.047
100,000	10^{-6}	1.268	1.147	11.048
1,000,000	10^{-5}	1.268	1.147	11.048
10,000,000	10^{-4}	1.268	1.147	11.048

Figure 100 illustrates where the distance-redshift relation begins to have meaning for the various short wavelength radiations, and the magnitude of errors that would occur if the distance-redshift relation were assumed to obey the same sense of Hubble's law at these short wavelengths. High energy radiation provides a possible means of distinguishing predictions of the scattering model as compared with the standard model. The distance to be associated with a given redshift differs significantly for gamma radiation and hard X-rays. However, redshifting of emitted ultraviolet radiation provides a reliable indicator.

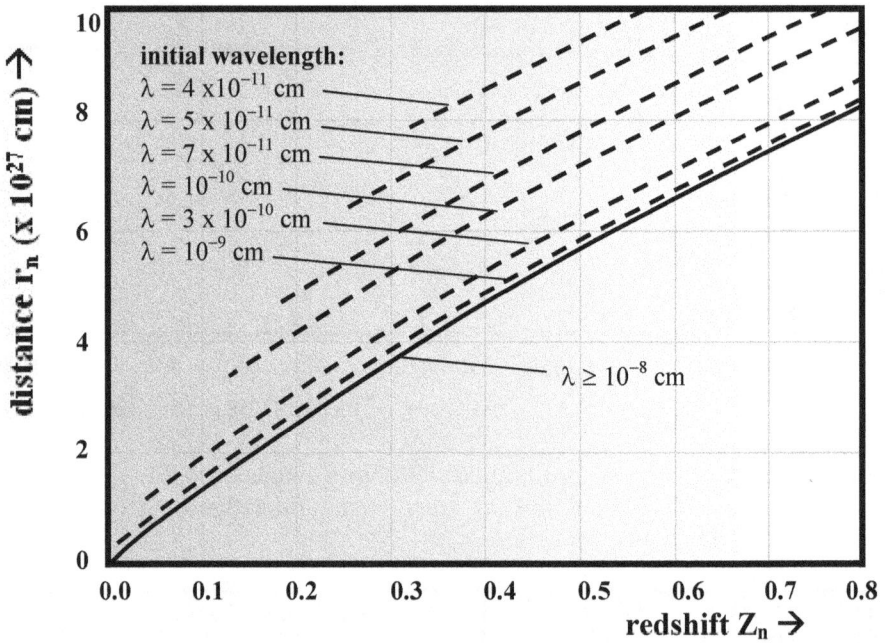

Figure 100: Distances and wavelengths at which meaningful distance-redshift relations begin

There is a wavelength dependent offset to all distance-redshift relations that was shown also in figure 98. This only manifests itself at these extremely short wavelengths. Objects of differing wavelength that happen to be at the same redshift may be at different distances, with the difference a function of initial wavelength. But other than this offset that is not extremely large when compared to truly cosmological distances even at wavelengths as small as 10^{-10} cm, we have succeeded in obtaining a unilateral distance-redshift relation.

c. matching Hubble's constant

With exceptions for extremely short wavelengths, we can define the effective amount of wavelength shift per centimeter produced by extinction. It is essentially the same for all radiation throughout a tremendously broad range of the electromagnetic spectrum as follows:

$$\Delta\lambda_{cm}(n) \equiv \; <\Delta\lambda_e> / \; \delta(\lambda_n)$$

where $\delta(\lambda_n)$ is, of course, the length of the nth extinction interval as discussed previously. By substitution for the two parameters on the right, we obtain the following assessment of the approximate amount of wavelength shift per unit distance:

$$\Delta\lambda_{cm}(n) \;\approx\; (\, h\, e^2\, \beta^2\, \rho_e\, / \, 2\, m_e^2\, c^3\,)\, \lambda_n$$

$$= 3/2\, (k\, h\, e^2/\, m_e^3\, c^5\,)\, T_e\, \rho_e\, \lambda_n \;\equiv H_{igm}\, \lambda_n$$

where $H_{igm} \cong 1.728 \times 10^{-32}\, T\, \rho_e$ cm^{-1} is defined to be analogous to Hubble's constant, H_0. This parameter involves exclusively universal constants except for the temperature and density of the intergalactic medium. Notice that only the product $T_e\, \rho_e$ affects the value of H_{igm}.

$\Delta\lambda_{cm}(n)$ is dependent upon λ_n that increases with increasing n. But by defining the amount of redshift per centimeter we eliminate the dependence on the explicit number of extinctions. In the limit this leads to a parameter that is indistinguishable from a Doppler redshift. It will indeed be wavelength-independent and therefore independent of the particular extinction interval, n to which it pertains. So we will drop the reference to n except to note with regard to the just previous discussions that for extremely short wavelength radiation this redshift indicator will be unreliable.

We thus define redshift per unit distance as follows:

$$Z_{cm}(n) \equiv \Delta\lambda_{cm}(n) / \lambda_n \; = H_{igm}$$

Notice that although $\Delta\lambda_{cm}(n)$ is merely a prorated amount of wavelength shift incurred over the entire n^{th} extinction interval divided by the interval length, the wavelength shift itself is actually accrued as wavelength-independent discrete incremental changes of statistically equal size, $< \Delta\lambda_e >$. Thus, only for distances large with respect to δ_o does it even make sense to talk of the change per unit distance as a continuously changing parameter as shown in figure 98. This is in somewhat the same sense that quantum mechanical transitions may be considered continuous as though they were the infinitesimals of the calculus when sufficient numbers of transitions are involved.

To obtain the associated 'continuous' wavelength shift (with the restriction of its appropriateness exclusively for differences in distance, Δd that are large with respect to δ_o but still infinitesimal with respect to other criteria as, for example, differences for which the wavelength is only infinitesimally changed) we notice that the change in wavelength with respect to a difference in distance is given throughout every appreciable interval by:

$$\Delta\lambda = H_{igm} \, \lambda \, \Delta r$$

So that in the limit for which wavelength changes can be considered infinitesimal,

$$\frac{\partial\lambda}{\partial r} = H_{igm} \, \lambda$$

Integrating this partial derivative such that radiation emitted at the wavelength λ_e at a distance r from the observer will subsequently be observed at wavelength λ_o, we obtain:

$$r(\lambda_e) \quad = H_{igm}^{-1} \int_{\lambda_e}^{\lambda_o} (1/\lambda) \; d\lambda = H_{igm}^{-1} \{ \ln(\lambda_o) - \ln(\lambda_e) \}$$

$$= H_{igm}^{-1} \{ \ln(\lambda_o / \lambda_e) \} = H_{igm}^{-1} \ln(1 + Z)$$

This formula provides the distance-redshift relation (independent of λ_e) that was shown for longer wavelengths in figure 98. It will be useful for comparisons to be made with other current cosmological models. Furthermore, for small redshifts, $Z \ll 1$ we have Hubble's useful approximation:

216

$$\mathcal{Z}(r) \approx H_{igm}\, r$$

This equation approximates the actual situation for cases when the distance is much greater than the average extinction distance but much smaller than the inverse of the intergalactic medium constant, $1/H_{igm}$. Limitations on the domain of applicability can be seen in figure 101.

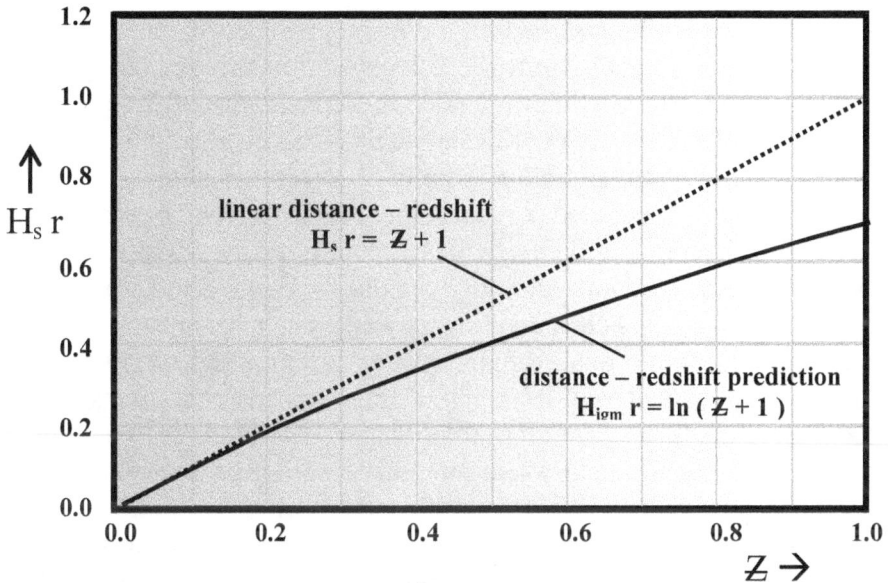

Figure 101: Predicted logarithmic relationship between distance and redshift

As we saw for optical wavelengths, the average extinction distance for light transmitted through the intergalactic medium δ_o is on the order of hundreds of light years. Even this distance, however, is more than a thousand times smaller than distances for which the hypothesized expansion velocities of distant galaxies are thought to predominate over usual Doppler effects of orbital galactic motions according to Hubble's hypothesis. In other words, observed redshifts of galaxies may involve a considerable contribution due to Doppler shifts caused by their local motions. Redshift has not, for this reason, been considered a legitimate distance metric for galactic distances less than several hundred million light years. That restriction remains. Thus, to the accuracy of the observed data, the wavelength increments at each extinction interval may indeed be considered 'infinitesimals'. On the other hand, distances for which redshift data has been obtained without relying on model-dependent redshift-luminosity data *per se* are orders of magnitude smaller than $1/H_0$.[*] We must also emphasize what was just discussed with regard to short wavelength objects such as hard X-ray sources and gamma ray bursts. For these objects redshift is probably not a reliable distance indicator until even larger distances are realized.

We have obtained, at this point, a predicted distance-redshift relation derived directly from the effects of forward scattering in an intergalactic plasma medium. Under the constraints of homogeneity and isotropy, the formula provides a nearly linear relationship for small redshift that was what Hubble originally observed and noted.

We have included caveats for the distances at which this relationship becomes reliable, which are in agreement with Hubble's acknowledged restrictions, although we did find some short wavelength dependence that bears consideration as well. The mechanism predicts a precisely logarithmic relationship between distance and redshift for all *cosmological* distances. It is interesting perhaps that the standard model predicts a very similar relationship but requires what many consider to be contradictory and unrealistic parameter values to obtain the similar predictions. We must review the data of observations of the redshifts at calibrated distances to various galaxies to determine how well all these predictions work. We will address that presently.

d. some background on the distance-redshift relation

It was an apparent linear relationship that precipitated Hubble's bold hypothesis of the possibility of recessional velocities determining a relativistic Doppler redshift of wavelengths. No other cause of such a redshift that adequately accounted for such phenomena was available although some believe Hubble's writings suggest that he was optimistic that ultimately a viable 'tired light' model would be found. The

---

[*] It will be demonstrated further on that there are distinctions to be made with regard to how distances were determined, whether by data on 'luminosity' or 'angular separations", etc.. Assumptions made in gathering and presenting such data are sometimes not compatible between the current scattering model and standard Big Bang models. This is because the data is typically gathered under the assumptions of high recessional velocities for which relativistic differences would become significant so that any valid comparisons between predictions involving these distance parameters must take into account these differences as we will discuss in more detail later on.

recessional velocity relation he came up with was the following, with the factor of c sometimes integrated into constant of proportionality:

$$Z_{galaxy} \approx V_{galaxy} / c \equiv \beta_{galaxy} \approx H_o \, r_{galaxy}$$

Hubble perceived H_0 to be a universal constant. However, the determination of its value has proven to be problematical in retrospect with a major change in its estimated value occurring once, with more minor changes occurring ever since Hubble's first bold conjecture. It is a matter of record that until very recently Hubble's constant had not been determined to within a factor of about 2 to everyone's satisfaction (Fukugita, 1993). Its measurement has gradually come to employ various techniques, most notably the use of SN1A supernova data. However, Roberts (1991) for example, obtained a value from observations of a double quasar using gravitational lensing, etc.. There have been at least two camps of proponents who emphasized respectively one type of measure or another for determining the value of H_0 to fairly high accuracy. But they did not agree with regard to which measurements were the valid ones and measurements within camps did not always agree either. This dilemma was addressed by authors of technical papers requiring use of Hubble's constant generally hedging their bets by employing a parameter h, whose fractional value is intended to adjust H_0 to the ultimately accepted value. This parameter h is defined as follows:

$$h \equiv H_o / 100$$

Propounded values of H_0 have recently ranged of 55 ± 10 to 90 ± 10 km sec^{-1} Mpc^{-1} in the most vocal extremes. Although more recent efforts seem to have narrowed it down. Its value has more recently been thought to be 70.1 ± 1.3 km c^{-1} sec^{-1} Mpc^{-1} given in traditional cosmological units rather than per centimeter (cm^{-1}) as used throughout this volume. Because of this recent consensus, we will sometimes substitute $h = 0.7$ where that seems reasonable.

Before proceeding further with explorations of the similarity of what we have referred to previously as H_{igm} and what is traditionally referred to as H_0, it is important that the most accurate value of H_0 be obtained. It is this value that will constrain our attempts to match observational data. Figure 102 is taken from John Huchra's web site that describes the history of this parameter (2008).

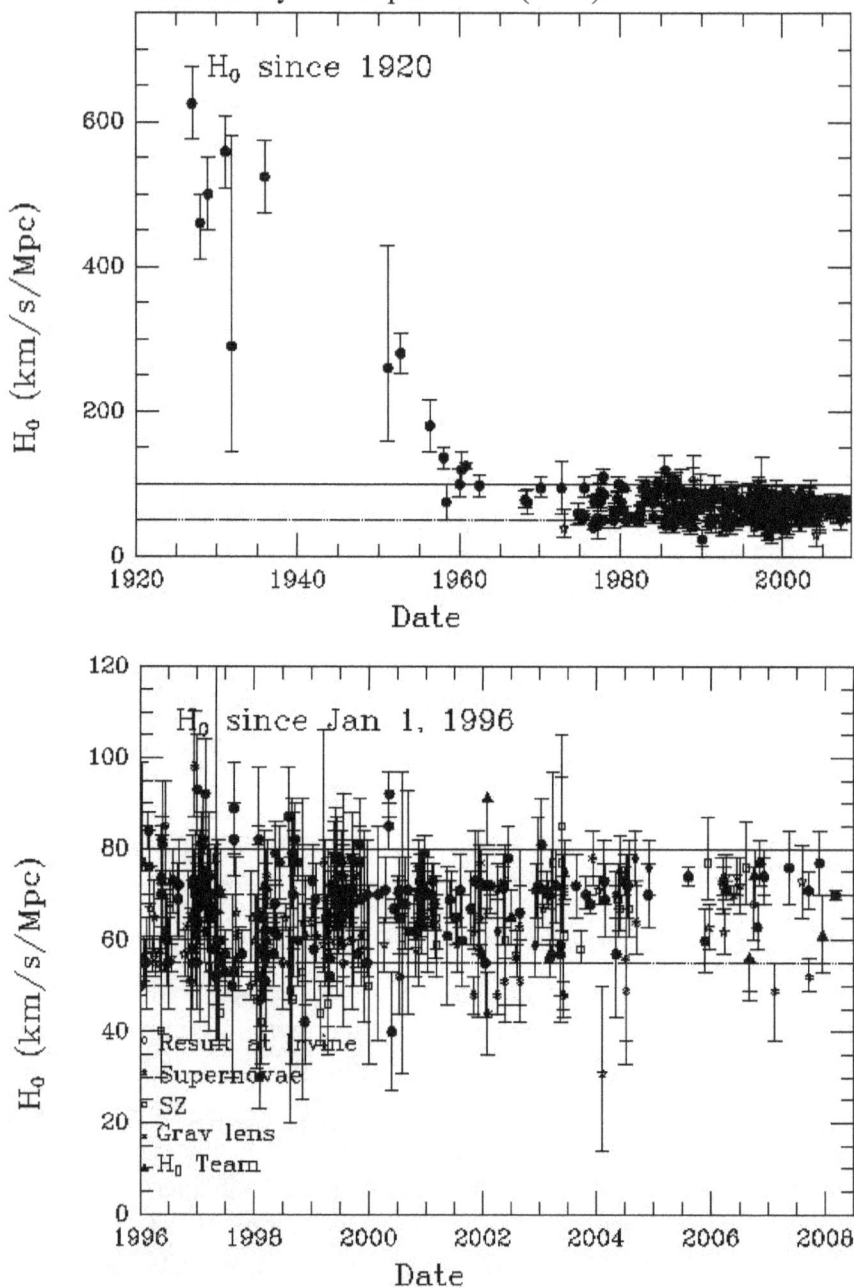

Figure 102: History of efforts to refine measurements of H_0

But even the constancy of the value of Hubble's *constant* is challenged, first as an evolving density-related parameter, but more recently by the perceived 'acceleration' in distant SNIA objects we will review presently. Similar confusions are encountered when attempting to determine the deceleration constant, which for theoretical reasons based in general relativity should (according to many) be precisely 0.5 cm sec.$^{-2}$, but has seemed in many observational contexts to be more nearly unity if employed at all. This parameter arises, of course, because whatever the origin of the extreme recessional velocities at great distances, the gravitation of the entire universe is perceived as capable of producing a deceleration if any sort of usual physical laws are to be considered applicable. The recently perceived contrary *acceleration* that has been claimed from more distant observations, although in a somewhat different context, will be discussed further on.

The implications of these recent changes to this scenario with the claimed *acceleration* as well as what is termed a 'jerk' to turn this around so as to then *decelerate* the expansion of the entire universe has been embraced by the standard cosmological model. This seems quite horrific no matter how one states the claims. The origin of our entire universe starting from some sort of conflagration, only to be followed by an unexplained inflationary expansion of incomprehensible proportions, to settle into a slower expansion, and then ultimately to accelerate only to decelerate again. Whoa! But we'll get into that later.

It should be clear that the linearity of Hubble's hypothesis with regard to recessional velocities of distant galaxies is not accepted for large velocities by any cosmological model because of the observations as well as theoretical relativistic considerations. Rather than a linear relationship, the relativistically legitimate formula becomes,

$$Z_{galaxy} + 1 = \sqrt{\frac{1 + \beta_{galaxy}}{1 - \beta_{galaxy}}}$$

From this, of course, one gets velocity-from-redshift as follows:

$$V_{galaxy} = c \, \frac{(Z_{galaxy}+1)^2 - 1}{(Z_{galaxy}+1)^2 + 1} \, .$$

Of course these formulas apply only within a particular Lorentz reference frame in the generalized theory of relativity. So a position has arisen among cosmologists that disavows the use of recessional velocity – as against *expansion* per se – as depicting the operative phenomena because of the uniqueness of the coordinate frame to which these formulas apply. But although this seems to be in vogue now, the same cosmologists do not seem averse to use of the notion of velocity 'boosting' (associated with Einstein's more restrictive *special* relativistic velocity addition formula) in discussing what has recently come to be considered an 'acceleration' of that expansion.

Let us ignore this for the present, however, to consider current estimates of Hubble's constant *as a constant*. It is this value that we will compare with the determination we have made for H_{igm} that was derived from the intergalactic medium for the scattering model. First we will convert the units usually employed for H_0 to the units of cm^{-1}:

$$km\ sec^{-1}\ Mpc^{-1}\ /\ c \equiv 1.06 \times 10^{-30}\ cm^{-1}$$

As was shown in figure 102, we can be fairly certain of the following value for H_0, which is the equivalent value in these units:

$$H_0 \cong 7.14 \times 10^{-29}\ cm^{-1}$$

All the caveats referred to in the preceding discussions apply to this now quite stable estimate.

Thus, if we are to assert that the scattering model accounts for observed cosmological redshift, we must maintain that:

$$H_{igm} \rightarrow H_o$$

$$H_{igm} \cong 1.728 \times 10^{-32}\ T\ \rho_e \rightarrow 7.14 \times 10^{-29} \cong H_o$$

In order for this association to apply, we would have to accept an average value of the dynamic pressure of the intergalactic medium that produces this rate of redshifting with distance. This requires an average value (indicated by brackets) for the product of temperature and density of the intergalactic medium such that:

$$<T\ \rho_e> \cong 4.13 \times 10^3\ K\ cm^{-3}$$

This value must neccessarily involve averaging intergalactic and intracluster values over total lines-of-sight distance to cosmological objects. Aspects of variations in density and temperature values were discussed in chapter 3, the extreme values of the product of these parameters in the cores of rich galactic clusters accommodate the magnitude of this averaged value. This conjecture will be collaborated in later chapters. If we tentatively accept this average product value for now, then we obtain concurrence with Hubble's approximate relation:

$$Z_{igm} (r_{igm}) \approx 7.14 \times 10^{-29} \, r_{igm}$$

We use the subscript on the redshift and distance here to remind us that the determination in this case involves the assumption of scattering in an isotropic homogeneous intergalactic medium rather than assumed recessional velocities as tentatively hypothesized by Hubble or as due to the commensurable expansion of spacetime as current purists insist. Thus, we have been able to predict Hubble's observed relation in at least the sense of being able to determine a realistic dynamic pressure constant for the intergalactic plasma that would account for the related phenomena. This has been achieved without accepting an 'expanding universe' hypothesis that has been the hallmark of all 'standard' models over much of the last century.

There is the issue of these values (particularly temperature) seeming too large for the vast regions of intergalactic space. However, we know that the largest numbers of observed galaxies actually occur in galaxy clusters for which the intracluster plasma gases exhibit dynamic pressures that are orders of magnitude larger than specified above. Much of the universe is observed through such clusters. So that this averaged value of H_{igm} does not correspond to a uniformly smooth redshifting function. There is considerable data that indicates that there are major peaks and valleys in the observed redshift density function in galaxy redshift surveys. We will analyze this anomaly specifically in chapter 17 where we will complete this discussion.

However, before we compare predictions of the current scattering model with the various versions of the standard model, we must look at the observational basis for assessing distance, which, in the case of the standard model, involves a theory-based difference in how distance is to be assessed. Since standard models have obtained a consensus among leading cosmologists, the data itself has come to absorb parameters appropriate only to the hypothesis of expansion. Therefore, in order to perform an accurate comparison, these must be backed out of the form of the data in which it is usually presented.

e. other examples of redshift occurring in ionized plasma

In derivations applicable to the domain to which this new-found method of redshifting is assumed to apply, we considered the properties of intergalactic plasma . For this approach to account for the observed cosmological effects, the weighted average of the dynamic pressure parameters indicated above must pertain in some sense throughout the cosmos. We will find that increased redshifting occurs in the more dense plasma regions in clusters cores that significantly increases the average value of H_{igm}. Thus, the dynamic pressure calculated above must be attained primarily along lines of sight passing through rich clusters of galaxies. The vast regions between clusters accommodate a somewhat reduced temperature such that hydrogen clouds, known to bespeckle these regions, are allowed to form. A statistical average produces the same net redshift, although with less uniformity. We will find from redshift survey data that cosmological redshift is by no means a uniformly distributed phenomena.

A plasma redshift is much more general than the specific application to which we have applied it. Figure 103 illustrates a range of situations for which plasma redshifting applies. For each application unique extinction interval distances and aberration angles would pertain, but the mechanism and class of phenomena would basically be the same in all such cases. In this diagram, in addition to showing a line that corresponds to the pressure characterized by the product value of density multiplied by the temperature of just over 4,000, a plasma domain that applies to a solar chromosphere is also noted. There is observed redshift phenomena in the chromosphere/photosphere/corona regions of the solar atmosphere and has been observed in other stars that remains unaccounted (or rather, has only been questionably-accounted for). Redshift of Fraunhofer lines observed in the limb of the sun have been inconsistently attributed to gravitational effects, but if it were due to gravitational effects, it would be even larger in the central disc of the sun where it is nil. We will not discuss that further in this volume, but it is an area worthy of further investigation.

It may be worth noting in this regard the work of Brynjolfsson (2005). In the reference he cites work with regard to a "plasma-redshift cross-section". He states that: "This new plasma-redshift cross-section explains the redshift of the solar Fraunhofer lines..." In addition he has also made claims similar to those put forward in this volume although he has come to his conclusions via a quite different route..

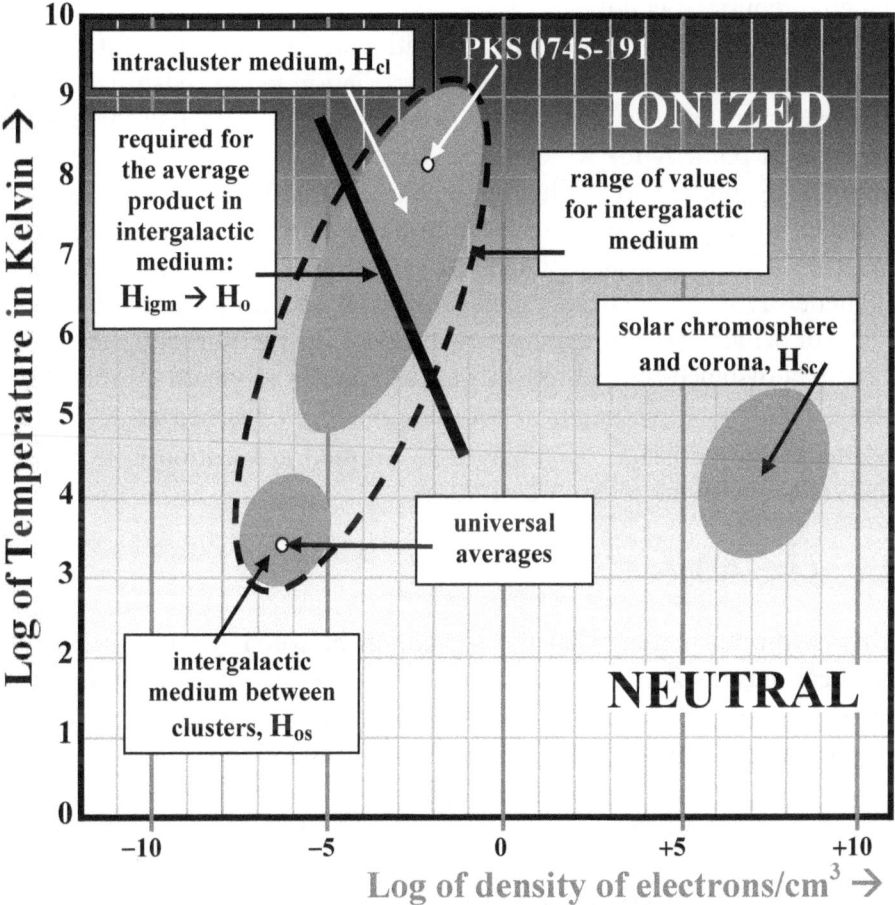

Figure 103: Several plasma redshift application domains

We will also have occasion to look into the excessive redshifts that occur in galaxy clusters. To account for this phenomenon 'dark matter' has been hypothesized to account for Doppler-inferred velocity dispersions much greater than are predicted (or would be reasonable) based on the luminous mass observed in these clusters using a straight forward application of the virial theorem. See chapters 16, 17, and 18.

f. arguments supporting a logarithmic relationship

There is something very compelling about the logarithmic functional form for the distance-redshift relation that we have found with regard to its role in observational cosmology. In fact, it is so compelling as to virtually be a logical necessity as the form of that relationship – whether that fact is generally acknowledged or not, which of course it is *not*.

To adequately understand the rationale for this claim, let us look at what is involved in electromagnetic radiation being redshifted along a propagation path from emission to observation. Suppose there is an observer at point A for which a telescope on earth would suffice as an instance. Suppose further that there is an ensemble of atoms in a star in a distant galaxy that we will refer to as point C that emits light of a specific wavelength associated with the spectra of the particular element involved. These atoms will emit electromagnetic photons, some of which will ultimately be observed by the telescope at A. If there is a distance-related redshift in the spacetime where all this takes place, then the wavelength of the radiation λ_A observed at A will be related to the emission wavelength λ_C emitted at location C that is in accordance with the redshift definition:

$$Z_{AC} = (\lambda_A - \lambda_C)/ \lambda_C$$

This is true no matter what the separation between A and C. But for physical reasons Z_{AC} must be a continuously increasing function of the separation $\overline{AC}$. So, let us define the redshift-related parameter $\zeta(r)$ as a continuous function of the separation $r = \overline{AC}$ as follows:

$$\zeta(r) \equiv Z_r + 1 = \lambda_A / \lambda_C$$

Since $\zeta(r)$ is continuous, we can choose A and C to have any separation and the relationship should still hold. Thus, we should be able also to place an observer at any point B along the light path from C to A, where $r_1 = \overline{AB}$ and $r_2 = \overline{BC}$. Refer to figure 104 where the illustrated radiation exhibits wavelengths and redshift as follows:

$$\zeta(r_1) = \lambda_A / \lambda_B \text{ and } \zeta(r_2) = \lambda_B / \lambda_C$$

Therefore, over the total distance for which $r = r_1 + r_2$ the following relation must apply:

$$\zeta(r_1 + r_2) = \lambda_A / \lambda_C = \zeta(r_1) \cdot \zeta(r_2),$$

And as a necessary consequence of this relation, it follows that:

$$\zeta(r) = e^{\alpha r} = e^{\alpha(r_1 + r_2)}.$$

And, therefore, of course:

$$r(\zeta) = \ln(\zeta).$$

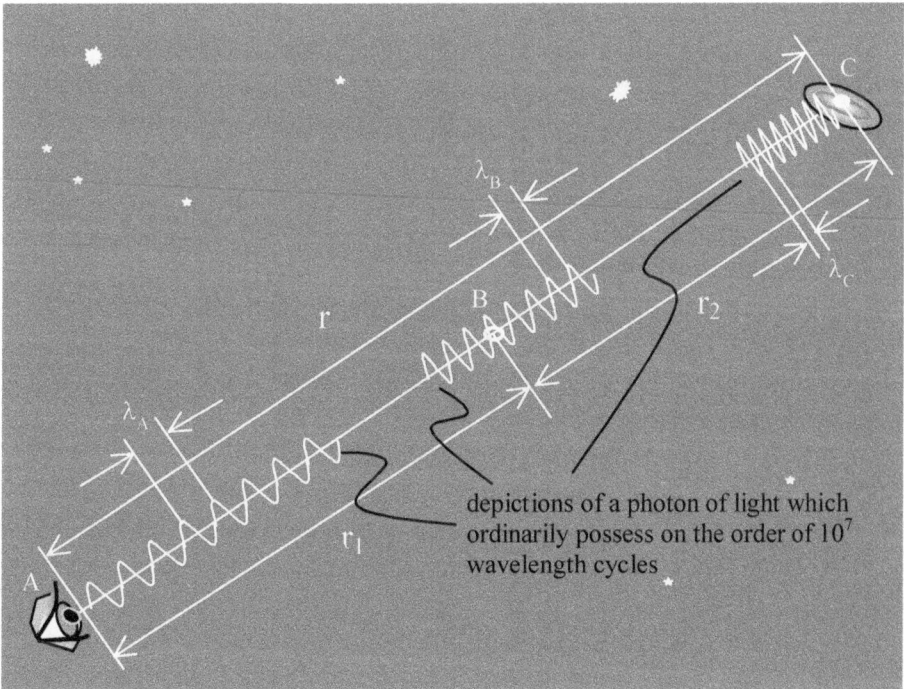

Figure 104: **Illustration of the significance of the logarithmic relationship between distance and redshift**

As we will see, the standard model embraces a broad class of disparate alternatives loosely united by adherence to Hubble's hypothesis and one form or another of Einstein's theory of general relativity. These alternative forms will be discussed in the next few chapters. We will see that there is a model of an expanding universe that predicts an exponential form.

It is worth noting that early in the previous century the Dutch astronomer Wilhelm de Sitter found the unique solution to Einstein's cosmological equation that resulted in a logarithmic form for the distance-redshift relation. The problem with his solution was that it implied and empty universe, but some of its predictions were very nearly equivalent to those of the scattering model. It is one of the simpler standard model alternatives, because of which it is frequently discussed for didactic reasons. Unlike the scattering model, however, its other predictions fail because it corresponds to an unrealistic empty universe in that paradigm. It is, therefore, generally disparaged as too naïve for serious consideration.

This short shrift given to the form as well as against its prediction failures seems somewhat ill-advised in light of the interesting fact that a key feature of this model (and only that particular version of the standard model) is that its distance-redshift relation satisfies the intuitively advantageous logarithmic form. This form is also, of course, what we have just found to result from our derivation of the distance-redshift relations appropriate to a fully functional universe with a substantive intergalactic medium in our scattering model. The empty universe version of the standard model is not considered viable for consideration as an ultimate cosmological solution for quite legitimate reasons, of course. We will consider these later. (Note, however, that none of these reasons apply to the scattering model.) But despite the failure of de Sitter's model, the implied logarithmic form of the distance-redshift relation has frequently been used by practitioners in analyzing data because it seems to fit a broad class of the observed redshifts of distant objects.

The seeming improbability, but nonetheless presumed, failure of the logic we have described contributes substantially to presumptions of supposed evolution in the development of our universe. But if one might just disconnect redshifting as an observed phenomenon (*whatever its cause*) from constraints imposed by *whatever causes it* according to one cosmological theory or other, then the logarithmic relationship to distance makes logical sense as we have shown above. We will be told, of course, that to presume that distances could be linearly added together if space itself were distorted in some nonlinear manner would itself be an improbability. But in what sense would it?

As an interesting exercise, it is worth considering what would be implied by a relationship *other than* one involving the logarithmic form. What is involved is whether or not the homogeneity of space could be made to apply to other relationships. In figure 105 we have drawn a situation similar to that of figure 104 except that space is such that line of sight distance is curved with the light path through space. In this case, in addition to observers similar to A and B in figure 104, we have observer B able to emit radiation from a separate source at the moment of his observation of the light from C which is set to resonate at precisely the same frequency (wavelength) as the radiation he observes. Let us analyze the possibilities here.

As before, we must now have that $\zeta(r_1) = \lambda_A / \lambda_B$ and $\zeta(r_2) = \lambda_B / \lambda_C) = \lambda_{B_1} / \lambda_C$. This would seem to apply by reason of the definition of redshift, if the source of the radiation of wavelength λ_{B_1} is indeed set up to equal that of λ_B. This could be verified by digital communication from observer B to observer A independent of the redshift impact on that link if A's antenna is properly tunable.

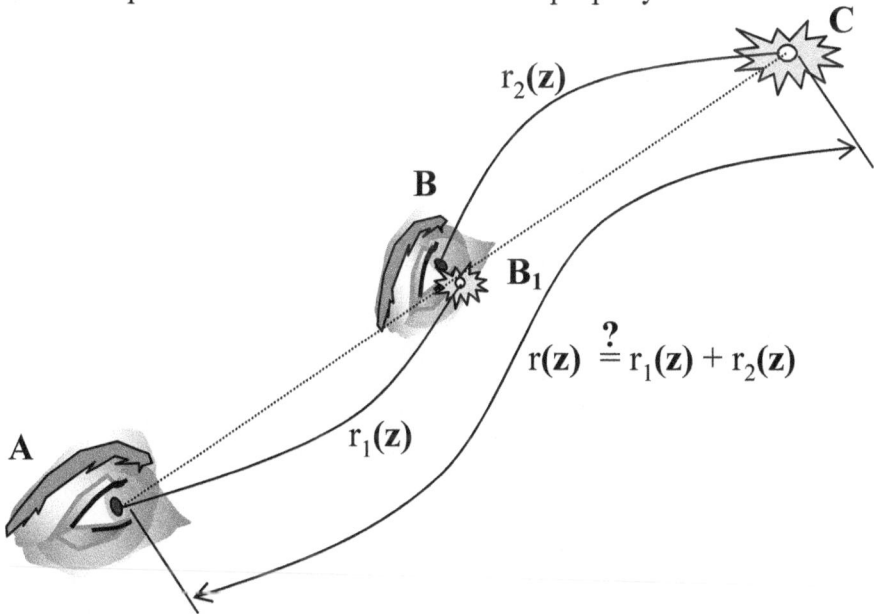

$$r(z) \overset{?}{=} r_1(z) + r_2(z)$$

Figure 105: Illustration of the need for testing the logarithmic relationship of distance and redshift in a nonlinear spacetime

Let us assume that there is some general formula applicable throughout space and time relating redshift and distance, as follows:

$$r \leftarrow f(Z+1) \quad \text{and} \quad Z+1 \leftarrow \zeta(r)$$

Then if the peculiar functionality of $f(x)$ and $\zeta(y)$ were independent of position in spacetime (i. e., if spacetime is indeed homogeneous, isotropic, etc.), the logarithmic/exponential relationships must apply.

Perhaps it is just intellectual snobbery that strives for this kind of reasonableness of the universe. But the author does not feel impelled to apologize for the modus operandi that characterizes the scientific method. The orderly sorting of facts into some sort of scheme that makes sense is what science is all about.

Chapter 12

Cosmographic (Metric) Predictions of the Standard and Scattering Models

The just preceding section notwithstanding, it is the author's contention that once one allows subjective beliefs of *how things actually 'are'* to distort observation, one begins to drift a bit from what can be considered *hard-nosed* (and thereby *'valid'*) scientific methodology. One can no longer *trust* the 'data' of observation in ways that are traditionally accepted as valid. And although it is generally recognized that the theory one chooses to *believe* will inevitably determine in large part the data by which one will justify that *scientific* theory – at a minimum constraining where one will look for that justification – one must strive to minimize the epistemological impact of that unfortunate situation. It should be noted that in the current treatise, although we have striven to define a redshift-distance relationship, there has been no attempt whatsoever to redefine distance or any other parameter to accommodate measurement as has had to be done with standard models – in particular with regard to the various metrics for distance that we will discuss in this chapter.

a. a variety of distance metrics, each with problems

Measuring astronomical distances becomes increasingly more difficult what with one tool after another succumbing to increasing inaccuracies as distances to which observations can be made increase.

Parallax that is a most useful tool to objectifying what we see in our daily lives with the *naked eye* based on the separation of our *two* eyes becomes useless even with the separation of instrumented observations at the extremes of earth's orbit about the sun. This gives a separation of some 186,000,000 miles, but it is not nearly enough for the difficult job that must be done in measuring the extreme cosmological distances to objects that have become visible with augmented instrumentation.

To compensate, 'standard candle' bases for viable inferences have been established concerning the nature of many of the objects that can be observed at these extreme distances as described in chapter 2. This has allowed reasonable estimates to be made concerning how far away individual objects of each type happen to be. These *types* are the standard candles that provide capabilities very like seeing a man at a great distance, where just knowing that the object *is* a man rather than some other object such as an excellent scale model toy soldier, allows one to determine its distance with a reasonable degree of accuracy. The standard candles that are used for astronomical purposes have applicability throughout different ranges of distance at which the key characteristics associated with the objects can be distinguished clearly.

Another problem that must be addressed as distances increase is that when redshifts become appreciable, the theoretical explanation of redshift increasingly intrudes into the formulas that apply to the various standard candles and quite significant variations arise for disparate cosmological models. Data that is gathered to distinguish predictions of the various models are typically presented with certain of the more general assumptions of the entire class of *standard cosmological models* included in the data itself. The primary example is that virtually all viable models accept the expansion of the universe in accordance with Hubble's hypothesis as a given. This is so uniformly accepted that redshift values are typically given in kilometers per second as though it were a direct measure of velocity. The implied extreme recessional velocities as redshift increases approach the speed of light certainly require relativistic treatment if the explanation proves correct. But that proof is in large part what is at issue. In the scattering model being presented here, of course, that velocity-related assumption does *not* apply and the implications of the associated relative velocities of the sources of radiation must be backed out of the data. To more fully appreciate what is involved, refer to figure 106.

a. Euclidean conception

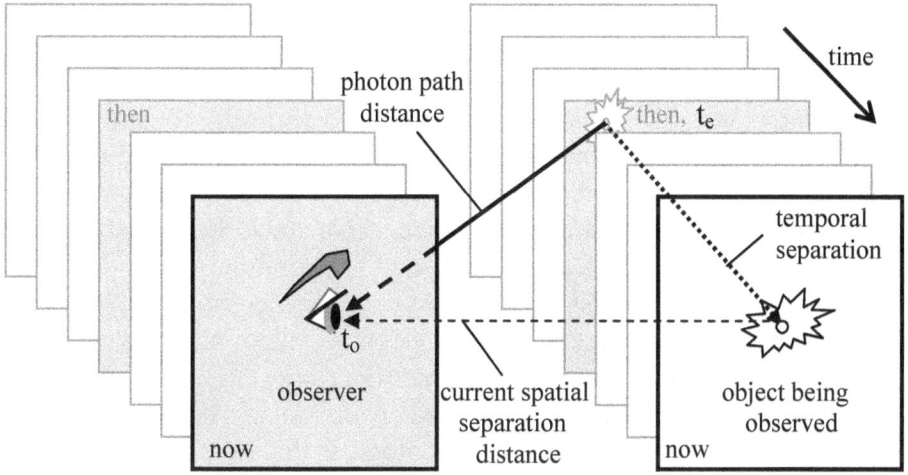

b. Einstein's conception

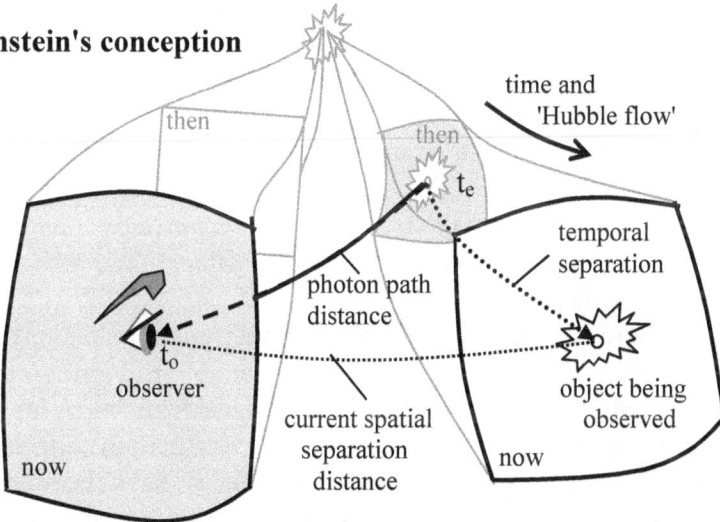

Figure 106: Alternative conceptions of spacetime metrics

Redshift is, of course, a direct measure of how much the wavelength values at the peaks and valleys in the intensity of the spectra of observed radiation differ from those which are expected of the kind of phenomena responsible for emitting that radiation, divided by the assumed wavelength of the radiation when it was emitted. Other than the inference that two such related spectra do indeed have a common cause, which is generally obvious, the distance value depends exclusively on the measurement of wavelength. Yet, of course, those who operate in this arena must deal with the prevailing presumption that redshift measures a recessional velocity that assumes expansion. To understand the degree to which it is unnecessary and illogical to have integrated this presumption into the data itself, consider related facts associated with thermodynamics where assessment of temperature has in most all cases until quite recently been achieved using measures of pressure of the associated substance. That situation did not give rise to a similar conflation in the definition of temperature.

Obtaining assessments of distance on a cosmological scale in any case does ineluctably depend upon a rather complex chain of theory-based explanations. It turns out that there are alternative definitions of 'distance' employed even in a single theoretical framework in various tests of cosmological theories as we will soon see. To accommodate associated complications – particularly if the phrase *recessional velocity* is disavowed as Harrison (1993) and others have done – an 'expansion factor', a(t) must be taken into account. In this process Hubble's relationship evolves considerably and there is no longer a single parameter to associate with *distance* using theoretical relationships between redshift and 'distance' in the standard model as suggested in figure 106. These alternatives include *lookback time* (multiplied by the speed of light), *angle-distance* employed when angular measurements are made, surface brightness, and *bolometric* (i. e., 'luminosity') distance.

But there is certainly no observational basis for standard candles of known *lookback time* or *angle distance* per se. The spectra of a star (or any thermal object), on the other hand is determined by its surface temperature which is in turn determined by association with its mass. Taxonomies of stars along lifecycle sequences have been obtained which provide estimates of their sizes and therefore of the total inherent luminosity as a basis for use as *standard candles*. This was shown in the Hertzsprung-Russell diagram in figure 1 of chapter 2. Without such approaches there would be no effective scientific discipline dealing with inherent luminosity. This absolute luminosity that is a measure of the intensity of the radiation emitted by a source, to the extent that we can know it, is typically involved in determining cosmological distances. So measuring luminosity is the master key to unlocking the secrets of cosmology. This will all be treated in detail for each of these metrics in the next sections of this chapter.

b. cosmographic underpinnings of the standard model

Complexities of the standard model are illustrated somewhat simplistically in figure 107 where curvilinear three-dimensional space is illustrated as a two-dimensional surface by analogy. Discussion of these parametric 'distances' will be based on treatment by Hogg (2000) and Peebles (1993). Later sections will compare predictions of various versions of the standard model with those of the scattering model.

Cosmography involves cosmological metrics which for standard models employs an evolving *expansion parameter*. Cosmological redshift is directly related to this scale factor a(t) that represents an evolving 'size' of the universe. At redshift z, the following pertains:

$$1 + z = a(t_o) / a(t_e)$$

where $a(t_o)$ is the size factor of the universe at the time the light from the object is observed, and $a(t_e)$ represents that same factor when the light was emitted from the object being observed.

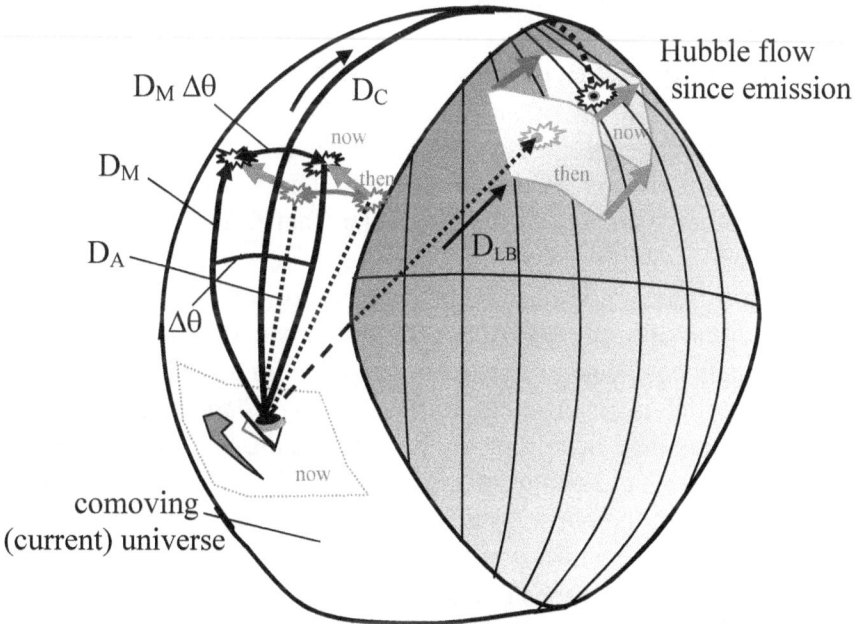

Figure 107:Cutaway view of distance metrics in standard models using analogy of curvature in three-space: Line-of-sight comoving distance D_C, comoving transverse distance D_M, angular diameter distance D_A, for assessing angular size, and lookback distance D_{LB}

The mass density ρ_o of the universe that we discussed in chapter 3 and the once-maligned cosmological constant Λ, whose introduction Einstein considered to have been his 'greatest mistake', are now both accepted by standard model theorists as universal properties affecting the evolution of metrics over time. Whether considered legitimate or not, a value must nowadays be assigned respectfully to Λ nonetheless.

As determined by the dearth of luminous baryonic matter that has been observed, there seems to be insufficient matter in the universe to 'close' Einstein's equations without Λ. That is, unless one introduces 'dark', i. e., unseen gravitational matter and even more mysterious 'vacuum energy' to propel perceived accelerated expansion. Currently an alternative of only thirty percent baryonic matter, the rest made up of mysterious dark matter, is favored. Various solutions with less mass are being actively debated. We will include curves for several of these alternatives in the plots below. Each represents a host of alternatives.

In deference to Einstein's having considered his greatest blunder to have been the introduction of Λ into his general theory, that we will discuss in a later chapter, Kochanek et al. (1996) have provided conclusive evidence that Λ cannot exceed a very small percentage of the total density of the universe. Thus, it would seem prudent in a scientific context to favor an alternative that eliminates Λ altogether. However, generations of cosmologists have recalled it unapologetically with the onset of any evidence that the standard model might be in jeopardy otherwise. Note in particular that Riess (2004) suggests splining an evolutionary trend for Λ to account for a perceived acceleration, jerk, and subsequent deceleration of expansion indicated by observation of high redshift SN1A supernovae over the last decades.

In any case cosmologists are coping with an evident reality that baryonic matter is insufficient to effect the *critical mass* required to 'close' the universe according to Einstein's time-honored formulation. Cosmologists don't attribute much to the small percentage of baryonic matter 'hidden' in the plasma state in intergalactic regions.

The basic quantities involving baryonic and 'dark' matter are both converted into dimensionless density parameters denominated Ω_M and Ω_Λ. This is accomplished by scaling the quantities in units of what Einstein defined as the 'critical density' of the universe ρ_o as follows:

$\rho_o \equiv 3\, c^2\, H_o^2\, /\, 8\, \pi\, G$ *Einstein's critical density*

The value of this parameter is somewhere near 5×10^{-30} gm cm^{-3}, but depends directly on the ultimately accepted value of H_o. Thus,

$$\Omega_M \equiv 8\pi \, G \, \rho_o \, / \, 3 \, c^2 \, H_o^2$$

$$\Omega_\Lambda \equiv \Lambda \, / \, 3 \, H_o^2$$

The subscripted "o" typically indicates that the quantity pertains to the value it possesses 'now' but in general these parameters are conceived in the standard models as having 'evolved' over time.

There is also a third density parameter subscripted here with an "R" that assesses the 'curvature of space'. In this context it is typically defined by the constraint,

$$\Omega_M + \Omega_\Lambda + \Omega_R = 1$$

According to theory, these parameters completely determine the geometry of our universe, supposing that at large enough scales it is homogeneous and isotropic. But there is an open issue concerning whether the universe is indeed 'matter-dominated', or if, as more recent discussion precipitated by the acceleration perceived in supernova data suggests, dominated instead by some mysterious vacuum energy.

For many years most general relativity theorists believed that the density triad Ω_M, Ω_Λ, Ω_R must have values 1, 0, and 0, respectively. The re-instatement of Λ as a legitimate parameter in Einstein's equations was thought by some to resolve the problem of there being a shortfall of baryonic mass to 'close' the universe as $\Omega_M = 1$ implies in Einstein's theory. Proponents of other possible allocations among the density parameters have become more vocal. Most notably arguments for (Ω_M, $1 - \Omega_M$, 0) and (Ω_M, 0, $1 - \Omega_M$) have recently predominated.

If Ω_Λ were to be zero, then the deceleration parameter q_o would have to be $\frac{1}{2} \Omega_M$. This deceleration parameter q_o can be derived as:

$$q_o = \ddot{a}(t_o) \, a(t_o) \, / \, \dot{a}(t_o)^2 = \tfrac{1}{2} \, \Omega_M \overset{?}{=} \Omega_\Lambda \qquad \textit{deceleration constant}$$

The final questionable equality would apply only for a matter-dominated universe with no substantial curvature. (Note that the double and single dots over a(t_o) refer respectively to double and single differentiation by time as typically employed in the differential calculus.) However, if $\Omega_\Lambda \neq 0$, q_o plays a considerably diminished role.

Our discussion of the cosmographic calculations appropriate to the standard model will feature seven density combinations, some of which are considered pathological, others somewhat more realistic:

model descriptions	Ω_M	Ω_Λ	Ω_R
Einstein–de-Sitter	1.0	0.0	0.0
all lambda	0.0	1.0	0.0
de-Sitter (empty)	0.0	0.0	1.0
low-density	0.05	0.0	0.95
high-lambda	0.20	0.80	0.0
intermediate model	0.27	0.73	0.0
concordance model	0.31	0.69	0.0

These seven models effect very different predicted phenomena as we will see. Notice that the first four of these models are already ruled out by observations according to most cosmologists, but a complete consensus of a suitable array of values has not yet been achieved.

The problem with the Einstein–de-Sitter model is that there is a dearth of observed baryonic matter to justify its closure criterion. Low-density models are more or less precluded by an obvious presence of matter in the universe. The final three models have each been discredited and virtually eliminated by work of Kochanek (1996) and others (even if one ignores Einstein's rejection) by establishing tight limits on the possible values of Λ, using gravitational lensing statistics.

Most cosmologists seem to be of the opinion that it is likely that the truth lies in a combination of values for these three parameters that is yet to be determined. However, if the current investigation is correct, the truth will be that the parameterization itself is totally incorrect.

c. comoving line-of-sight distance

This distance measure is an attempt to normalize a distance parameter between objects, taking into account differences in the expansion factor current at the two locations. Peebles (1993, pp. 310–321) refers to this metric as 'angular size distance', which is not the same as the parameter that he refers to as 'angular diameter distance' that we will discuss farther on. Hogg (2000) sometimes refers to this as 'line-of-sight comoving distance'. He defines a small increment to this 'comoving distance' δD_c between two nearby objects as a separation that remains constant with *epoch* if the two objects happen to be moving with the 'Hubble flow'. Put another way, "it is the distance between them which would be measured with rulers at the time they are being observed (the proper distance) divided by the ratio of the scale factor of the universe then to now."* It is 'proper distance' multiplied by redshift, since in expansion explanations, $1 + z = a(t_0) / a(t_e)$. Here t_e is the time at which the light was emitted.

This, perhaps simplest and most fundamental of distances in general relativity, is obtained by integrating line segments δD_c between nearby events along a light path from the observed object to the observer who is assumed to be at $z = 0$. However, in order to obtain a formula for determining its value we must first define the function:

$$E(z) \equiv \sqrt{\Omega_M (1 + z)^3 + \Omega_R (1 + z)^2 + \Omega_\Lambda}$$

This function is proportional to the time derivative of the logarithm of the scale factor, i. e., $\dot{a}(t)/a(t)$, taken with respect to redshift z.

From this definition, it is clear that $H(z) = H_0 E(z)$ is the Hubble constant as measured at a redshift of z. Since $dz = da$, it turns out that $dz / E(z)$ is proportional to the time-of-flight of a photon moving through the redshift interval dz, divided by the evolving scale factor along each segment of that path. Under the constraint of a constant speed of light, this integral results in the *proper distance* divided by the scale factor. This then defines the line-of-sight distance equivalent for the standard model. It is given by,

$$D_{C_{sm}} = H_0^{-1} \int_0^z dz' / E(z')$$

where H_0^{-1} is often defined as the 'Hubble distance'.

---

* Hogg (2000) notes appropriately that the word "proper" has a specific use in relativity. The *proper time* between two events is the time delay between the events in a frame of reference in which they take place at the same location. The *proper distance* in this same context is the spatial separation between two events in the frame of reference in which they happen simultaneously. It is the distance that would be measured by a ruler at the time of observation. The distance defined here is not, therefore, a *proper distance* in that sense. It is more properly, the *proper distance* divided by a ratio of the applicable scale factors.

As noted above, this line-of-sight distance is what, according to general relativity theorists, would be measured locally between events locked in the Hubble flow. It seems to be the accepted metric for measuring aspects of large-scale structure in our universe according to all variations of the standard model of cosmology. The functionality of this parameter with redshift is illustrated in figure 108 for several models.

What is the analogy for the scattering model?

Of course here, as with other measures, it is the path that light would take in traversing the distance via many extinction intervals divided by the speed of light. This, we have found, is quite simply,

$$D_{C_{sc}} = H_0^{-1} \ln(z + 1)$$

for distances large with respect to the extinction interval.

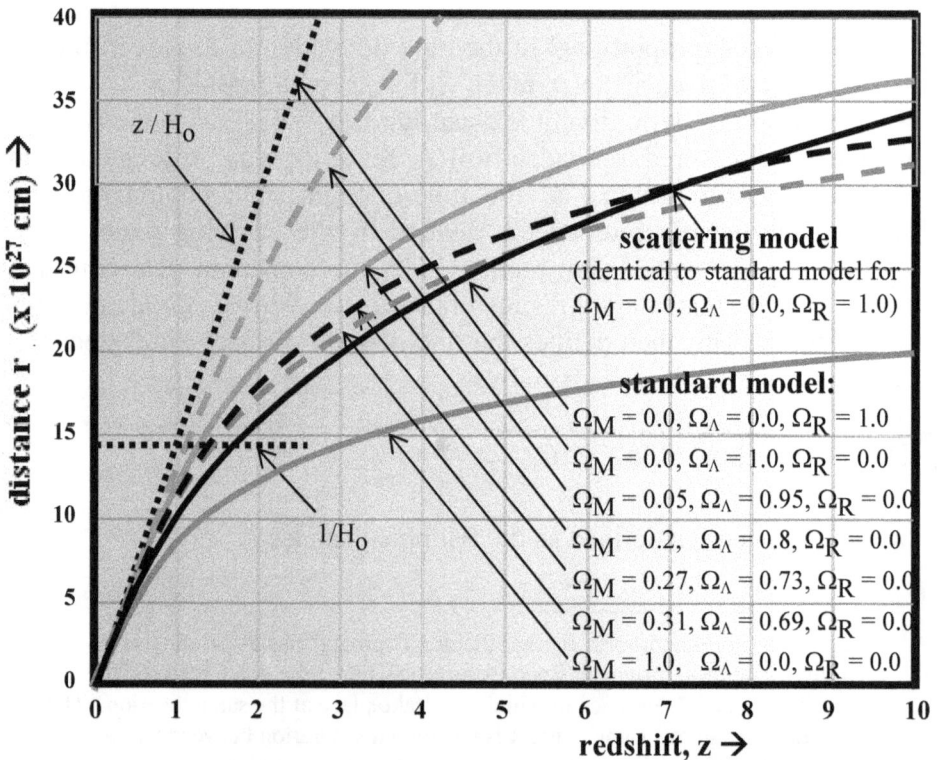

Figure 108: Line-of-sight comoving distance D_C versus redshift predictions for the *standard model* with various density parameter values (but with $\Omega_R = 0.0$ and $\Omega_M + \Omega_\Lambda = 1.0$) as well as for the *scattering model*

We have already established the physical basis of this logarithmic distance-redshift relationship predicted by the scattering model. The form of the relationship was illustrated earlier; it is included as the heavy solid curve in figure 108 below along with those of various subset 'world models' associated with the standard model. It is significant that the scattering model and the de Sitter model exhibit an identical functionality for this parameter.

d. comoving transverse (proper motion) distance

Let us now explore the impact of the Ω_R parameter to a little more depth to understand what density values might be – and why among the plethora of possibilities, only certain values *are* – anticipated as appropriate for the standard model in this regard. The Ω_R parameter, relates to the supposed curvature of spacetime, becoming more significant where observed angles are involved. The distance between two events at the same redshift (line-of-sight 'distance') that are separated on the sky by some angle $\delta\theta$ would traditionally be given by $D_c \, \delta\theta$ in Euclidean geometry. In the standard model, however, such distances involves the 'transverse' comoving distance D_M as follows:

$$D_{M_{sm}} = \begin{cases} H_0^{-1} \, (1 / \sqrt{\Omega_R}) \, sinh \, [\, \sqrt{\Omega_R} \, H_0 \, D_{C_{sm}}] & \text{for } \Omega_R > 0 \\ D_{C_{sm}} & \text{for } \Omega_R = 0 \\ H_0^{-1} \, (1 / \sqrt{|\Omega_R|} \,) \, sin \, [\, \sqrt{|\Omega_R|} \, H_0 \, D_{C_{sm}}] & \text{for } \Omega_R < 0 \end{cases}$$

Here the trigonometric functions *sin* and *sinh* address Einstein's alternatives for 'curvature of space'. The density parameter Ω_R is related to local mass-energy density or Einstein's stress–energy tensor.

When $\Omega_\Lambda = 0$, there is an analytic solution to these equations:

$$D_{M_{sm}} = 2 \, H_0^{-1} \, \frac{[2 - \Omega_M \, (1 - z) - (2 - \Omega_M)\sqrt{1 + \Omega_M \, z}]}{\Omega_M^{\,2}(1 + z)}, \text{ for } \Omega_\Lambda = 0$$

This distance is what Peebles (1993, pp. 320–321) refers to as "proper distance," which is in common usage, but inappropriate nonetheless as noted above. Also, note that some theorists, including Misner, Thorne & Wheeler (1973, pp. 782–785), prefer a derivation of this metric using a qualitatively different method that employs a 'development angle' χ that increases as the universe evolves. But the resulting expression is the same. Refer to figure 109.

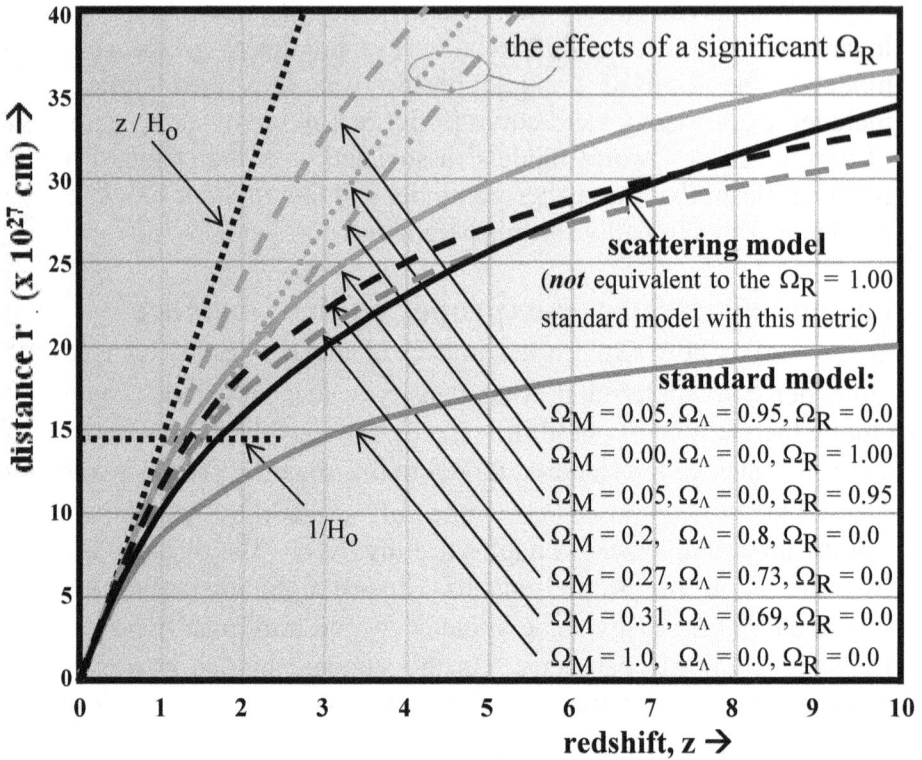

Figure 109: Comoving transverse (proper motion) distance D_{CM} versus redshift predictions for the *standard model* with various density parameter values as well as the *scattering model*

e. angular diameter distance

In mundane Euclidean metrics – assumed by the scattering model – angular separations of distant objects reliably decreases as the inverse of the distance to the objects. However, in the standard model angles play havoc as we noted above. The angular diameter distance D_A is a measure of the distance to astronomical objects when the light that is seen now was emitted. Under the aegis of the standard model it is the ratio of the comoving transverse distance to an angular dimension (measured in radians). This measure is used to convert angular separation in telescopic images to a spatial separation between objects. And since in the standard model this angular diameter distance diminishes substantially at large redshifts as illustrated in figure 110, its value beginning to diminish at about z ~ 1, it predicts that more distant objects would actually appear larger. In standard models this parameter is related to the previously defined transverse comoving (proper motion) distance by,

$$D_{A_{sm}} = D_{M_{sm}} / (1 + z)$$

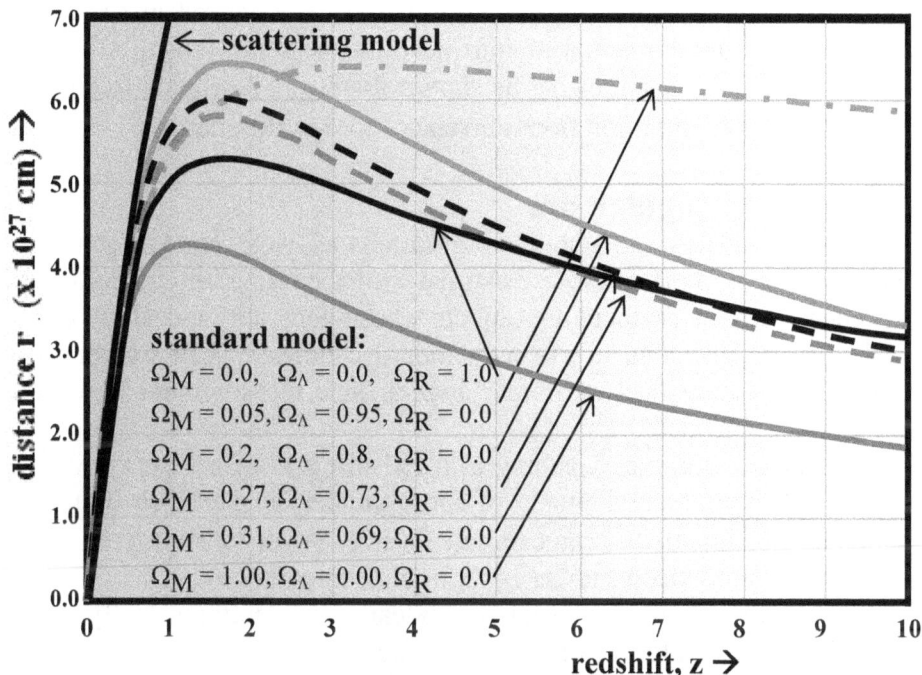

Figure 110: Angular diameter distance D_A versus redshift predictions for the *standard model* with various density parameter values. The analogy for the *scattering model* is also shown.

In the standard model there is also an angular diameter distance parameter D_{Aij} that applies between two objects at different redshifts of z_i and z_j. This measure has frequently been used in evaluating the effects of gravitational lensing. One cannot just subtract the two individual angular diameter distances D_{Ai} from D_{Aj}. The correct formula, for $R \geq 0$, is the following:

$$D_{Aij} = [1 / (1 + z_j)] \left[D_{Mj}\sqrt{1 + \Omega_R(D_{CMi} H_o)^2} - D_{Mi}\sqrt{1 + \Omega_R(D_{CMj} H_o)^2} \right]$$

Here D_{CMi} and D_{CMj} are comoving transverse distances to the objects at z_i and z_j. According to Hogg (2000) there is reason to believe that this equation may not apply for $R < 0$.

If it were possible to measure parallax for high redshift objects, the distance so measured would, of course, be a most useful check on this and other measures. It may one day be possible to measure parallaxes to distant galaxies using gravitational lensing. However, in such a situation, a modified parallax distance versus redshift relation would be required that takes into account both the redshifts of the source and the lens somewhat as shown for $D_{A_{12}}$. Adequate treatment would divert our attention from currently more useful metrics.

f. apparent angular size

It is difficult to obtain observations of large enough standard candle objects to effect precise estimates of their sizes at cosmological distances. However, there are objects whose generality as galaxy types makes them amenable to use as standard candles. Their tremendous diameters allow a fairly accurate angular size measurement based on the angles subtended.

If, for example, a galaxy at a redshift z is of a type known to have a half luminosity diameter of approximately d, then if its diameter subtends an angle θ, one can estimate its size. As distances increase the angle unilaterally decreases as 1/r in Euclidean geometry.

However, although such Euclidean considerations apply quite directly to the scattering model, theoretical considerations in standard model cosmologies dictate redshift dependence as specified for angular distance, D_A that was described and illustrated in figure 110 for various density value combinations. The parameter D_A that was illustrated there becomes an artifact in determining expected angular dimensions of an object as follows:

$$\theta_{sm} = d / D_{A_{sm}} = (d / D_{A_{sm}})(z+1)$$

Needless to say, the standard model prediction of angular size of objects of common length d viewed at great distances is one of the more counter-intuitive of many such notions deriving from that model. In figure 111 we illustrate the weird prediction that the angular size of objects should actually increase at large distances. Based on this prediction, at their current size and abundance, the sky should be completely tiled with galaxies at a redshift of $z \approx 20\,\Omega^{1/3}$ were there to be no evolution of these objects to preempt such a situation. To account for the disparity with this prediction, it has been hypothesized that the dimensions of 'standard candles' (the d in the above equation) evolve in a way that nullifies the prediction. Of course that jeopardizes the very concept of a standard candle.

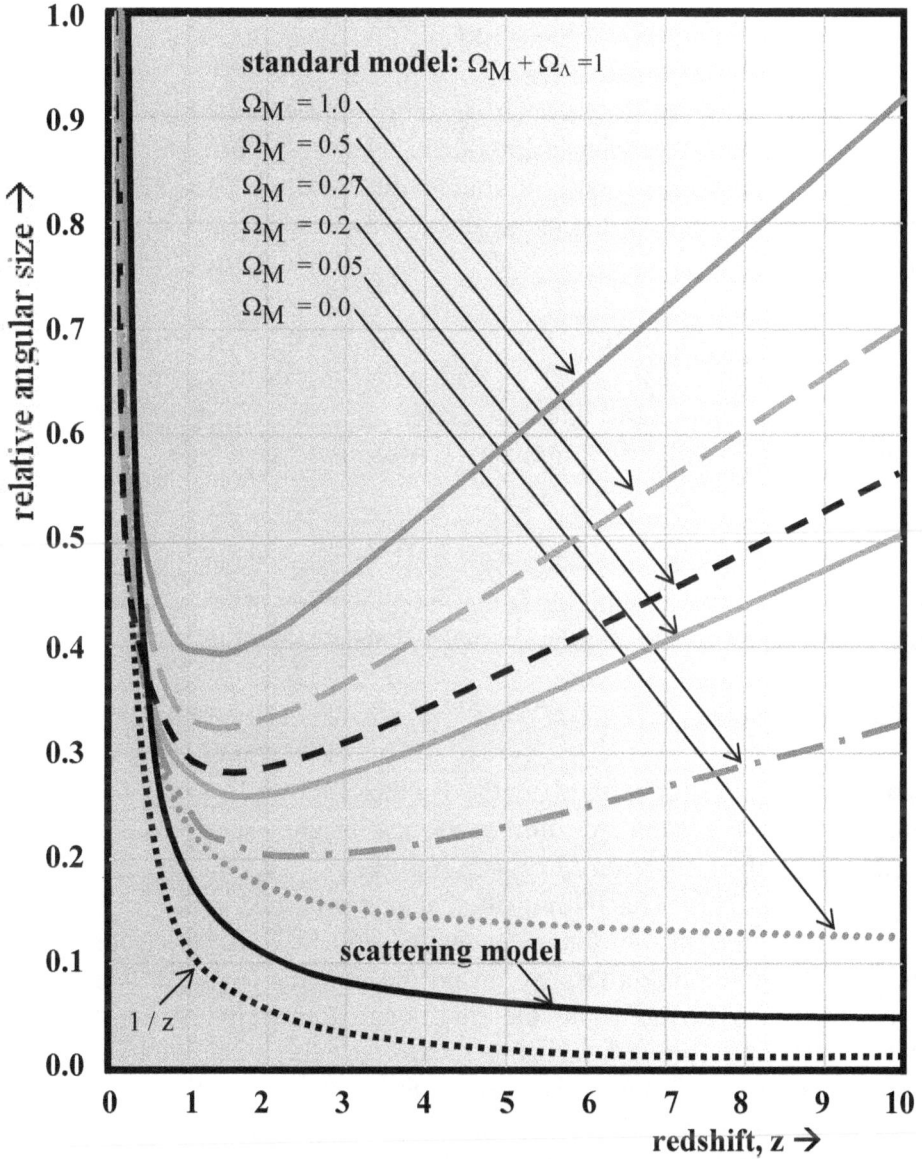

Figure 111: **Apparent angular size predictions of the *Standard Model* with various percentages of baryonic mass, along with the curve for the scattering model**

In the scattering model the straight-forward prediction is:

$$\theta = d \, / \, D_{C_{sc}} = H_o \, d \, / \ln(z+1)$$

No strange phenomena are predicted by the scattering model in this regard and 'distance' remains always Euclidean *distance*, no matter what the test to be made. The heavy line in figure 111 is what one would anticipate from Euclidean considerations discussed above for the scattering model. It predicts a Euclidean relationship in distance out to the extremes of observation. Of course, the bottom curve in the figure that corresponds to distance being linear with respect to redshift is not *spatially* Euclidean.

g. luminosity distance

Luminosity distance D_L is defined by the relationship between the observable bolometric (i. e., integrated over all frequencies, v)* flux S and inherent bolometric luminosity L of the object observed:

$$D_L \equiv \sqrt{L \, / \, 4\pi \, S}$$

In the standard models this distance parameter is related to comoving transverse distance and angular diameter distance as follows:

$$D_{L_{sm}} = (1 + z) \, D_{M_{sm}} = (1 + z)^2 \, D_{A_{sm}}$$

The latter relationship with diameter distance follows from the fact that according to the standard model surface brightness of a receding (redshifted) object is reduced by a factor $(1+z)^{-4}$, and the angular area goes down as D_A^{-2}. This luminosity distance is plotted in Figure 112.

Of course in the scattering model, and in Euclidean geometry generally, surface brightness would not be affected by redshift in this way, since angular area and the traditional luminosity of the object whose surface brightness is being measured exhibit inverted squared dependencies on distance. However, quantum energy effects reduce flux commensurate with each photon's increased wavelength as it is redshifted. Also, as was illustrated in figure 55.b on page 117,

---

* See the discussion of bolometric flux in chapter 2.

broadband absorption occurs in plasma that effects an additional $1/(z+1)$ flux diminution factor, so the scattering model incorporates that functionality.

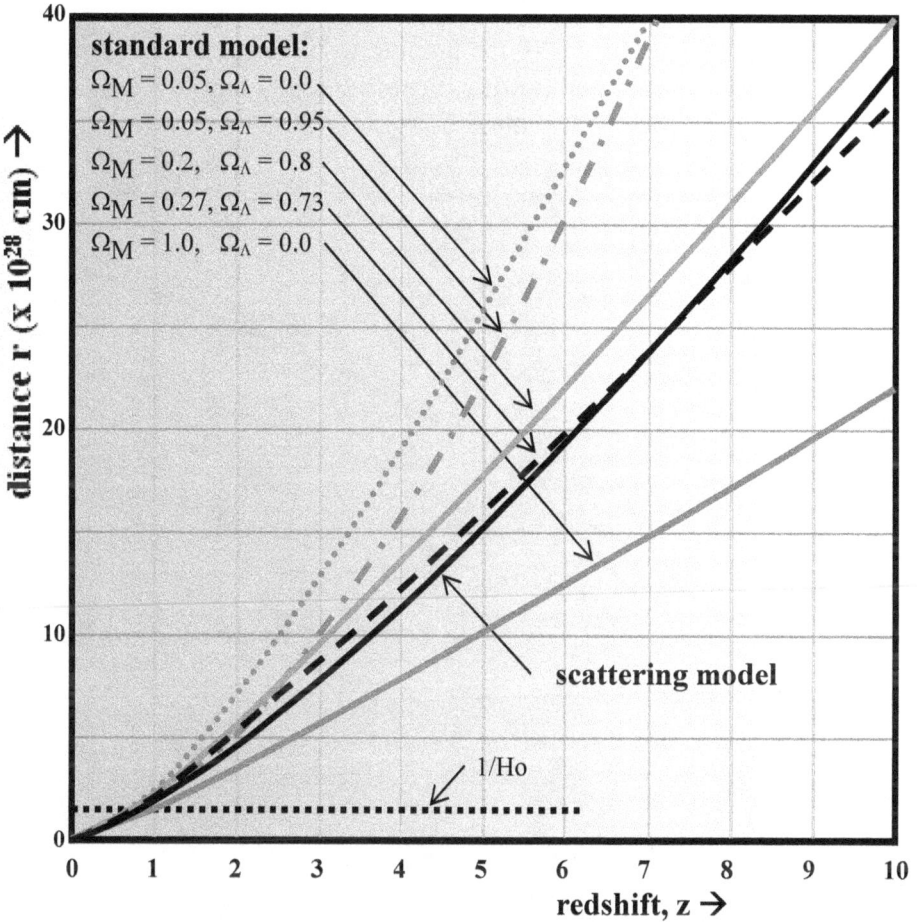

Figure 112: Luminosity distance (D_L in centimeters) *standard model* with various density parameter values as well as for the *scattering model*

$$D_{L_{sc}} = (1 + z) D_{C_{sc}} = (1 + z) \ln(1 + z) / H_0$$

The apparent magnitude of astronomical sources in photometric bandpass filters is defined to be a ratio of the apparent flux S of the observed source divided by the apparent flux of the bright star Vega when viewed through the same bandpass filter. The distance modulus D_M becomes a logarithmic relation defined as:

$$D_M \equiv 5 \log (D_L / 10 \text{ parsec})$$

It is the magnitude difference between an object's observed bolometric flux and what it would be if the luminous flux were from the same object located at 10 parsecs (once thought to have been the distance to the star Vega). This is perhaps the most frequently employed metric in cosmological research; it is plotted in figure 113.

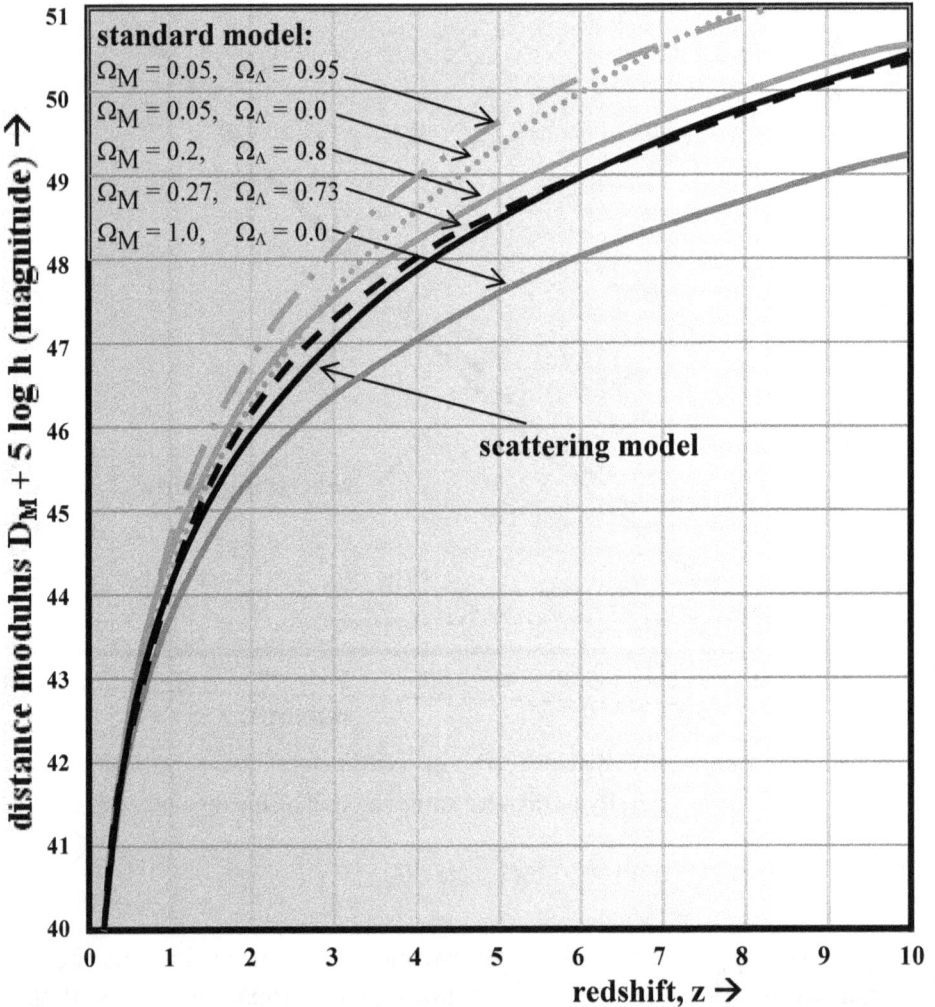

Figure 113: Luminosity distance modulus (D_M in relative magnitude) predictions for the *standard model* with various density parameter values as well as the *scattering model*

h. comoving volume

In standard models, the comoving volume V_C applicable to the comoving number densities of galaxies is the volume metric for non-evolving objects locked in the Hubble flow with positions constant with respect to redshift. It is *proper volume* with three factors of the ratio of the scale factors for *now* and *then*, or $(1+z)^3$, since $1+z$ is equivalent to that ratio as we have seen. Since the derivative of comoving distance with redshift is $1/E(z)$ discussed above, the angular diameter distance converts a solid angle $d\Omega$ into a *proper area*, and two factors of $(1+z)$ convert a proper area into a comoving area, the comoving volume element in solid angle $d\Omega$ and redshift interval dz is, therefore,

$$dV_{C_{sm}} = H_0^{-1} (1 + z)^2 D_A^2 [1 / E(z)] \, d\Omega \, dz$$

where D_A is the angular diameter distance at redshift z and $E(z)$ is as defined above.

For the standard model, the integral of the comoving volume element from the present to the redshift z gives a total comoving volume, all-sky, out to the redshift z as follows:

$$V_{C_{sm}} = \begin{cases} [4\pi H_0^{-3} / 2\Omega_R] \, [\, (D_{CM} H_0) \sqrt{1 + \Omega_R (D_{CM} H_0)^2} \\ \qquad - (1/\sqrt{|\Omega_R|}) \, \text{arcsinh} \, [\, \sqrt{|\Omega_R|} \, D_{CM} H_0], \text{for } \Omega_R > 0 \\[2mm] 4\pi D_{CM}^3 / 3 \, , \qquad\qquad\qquad\qquad\qquad\quad \text{for } \Omega_R = 0 \\[2mm] [4\pi H_0^{-3} / 2\Omega_R] \, [\, (D_{CM} H_0) \sqrt{1 + \Omega_R (D_{CM} H_0)^2} \\ \qquad - (1/\sqrt{|\Omega_R|}) \, \text{arcsin} \, [\, \sqrt{|\Omega_R|} \, D_{CM} H_0], \quad \text{for } \Omega_R < 0 \end{cases}$$

H_0^{-3} is sometimes referred to as the "Hubble volume".

In general, of course, we have the geometric relations:

$$dV_C(z) \equiv \frac{\partial}{\partial z} V_C(z) \, dz \, ,$$

and in spherical coordinates, a differential solid angle is the traditional, $d\Omega \equiv 2\pi \sin \theta \, d\theta \, d\phi$, where Ω here is not to be confused with the densities Ω_M, Ω_Λ, or Ω_R. This form of differential solid angle is used when integrating over a spherical (constant radial distance) surface in Euclidean geometry.

For the scattering model dV(z) retains the form of the Euclidean volumetric relationship, $dV(z) = r(z)^2 \, d\Omega \, dr(z)$, which becomes:

$$dV_{C_{sc}} = H_o^{-3} \, (\ln(1 + z))^2 \, [1 / (1 + z)] \, d\Omega \, dz$$

So 'comoving' volume in the scattering model becomes merely,

$$V_{C_{sc}} = [4\pi \, H_o^{-3} / 3] \, (\ln(1 + z))^3$$

The comoving volume element and its integral are both used frequently in predicting observed galaxy survey data we will review presently. Again there is a wide disparity in expectations, from which eventually observations should enable refutation of erroneous models. The comoving volume element curves are plotted in figure 114.

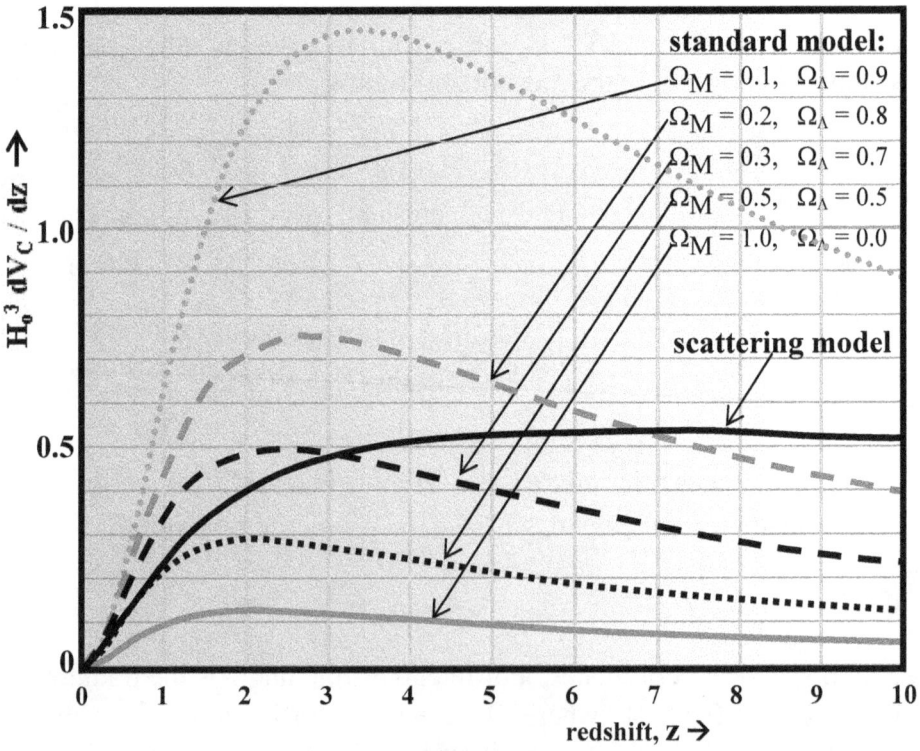

Figure 114: The dimensionless comoving volume element with several density parameter variations for the standard model

250

In figure 115 we illustrate the integral of $dV_C(z)$ with curves for the total volume out to a given redshift. In this figure, panel b provides a clearer view of the behavior for redshifts below $z = 2$.

i. 'lookback' time and distance

In the standard model the 'lookback time' t_{LB} to an object is the difference between the age of the universe now (at observation) and the age t_e of the universe when photons being observed *now* were emitted. It has been a primary tool for predicting the properties of high-redshift objects with evolutionary models, such as passive stellar evolution for galaxies, etc.. Besides which, it is the closest thing to what might be considered the 'actual' distance in cosmology.

We pointed out earlier that $E(z)$ is the time derivative of the logarithm of the scale factor $a(t)$ and that the ratio of scale factors is proportional to $(1 + z)$. So the product $(1+z) E(z)$ is proportional to the derivative of z with respect to the lookback time so that:

$$\frac{dz}{dt_L(z)} = (1+z)\, E(z)$$

Thus, we have that:

$$t_{LBsm}(z) = -\int_{t_o}^{t_e} dt'_{LB} = \int_0^z t_H \, dz' / (1 + z')\, E(z')$$

where t_H is defined as $1/(H_o c)$, the "Hubble time". Converting this to a distance, we obtain for lookback distance,

$$D_{LBsm}(z) = c\, t_{LBsm}(z)$$

The concern here is with the age of the universe $t(z) = t_e$ at the time the light was emitted, so the integration is from the current time for which $z = 0$ backward in time toward the big bang for which $z = \infty$. There are analytic solutions to this equation in some cases. The lookback time/distance is plotted in figure 116. It includes the more direct scattering model redshift-distance relationship, which is simply:

$$D_{LBsc}(z) = (1/H_o)\, \ln(1 + z)$$

Panel a:

standard model:
$\Omega_M = 0.1, \quad \Omega_\Lambda = 0.9$
$\Omega_M = 0.2, \quad \Omega_\Lambda = 0.8$
$\Omega_M = 0.3, \quad \Omega_\Lambda = 0.7$
$\Omega_M = 0.5, \quad \Omega_\Lambda = 0.5$
$\Omega_M = 1.0, \quad \Omega_\Lambda = 0.0$

scattering model

panel a

panel b insert

relative values →

redshift →

Panel b:

standard model:
$\Omega_M = 0.1, \quad \Omega_\Lambda = 0.9$
$\Omega_M = 0.2, \quad \Omega_\Lambda = 0.8$
$\Omega_M = 0.3, \quad \Omega_\Lambda = 0.7$
$\Omega_M = 0.5, \quad \Omega_\Lambda = 0.5$
$\Omega_M = 1.0, \quad \Omega_\Lambda = 0.0$

scattering model

insert from panel a

panel b

relative values →

redshift →

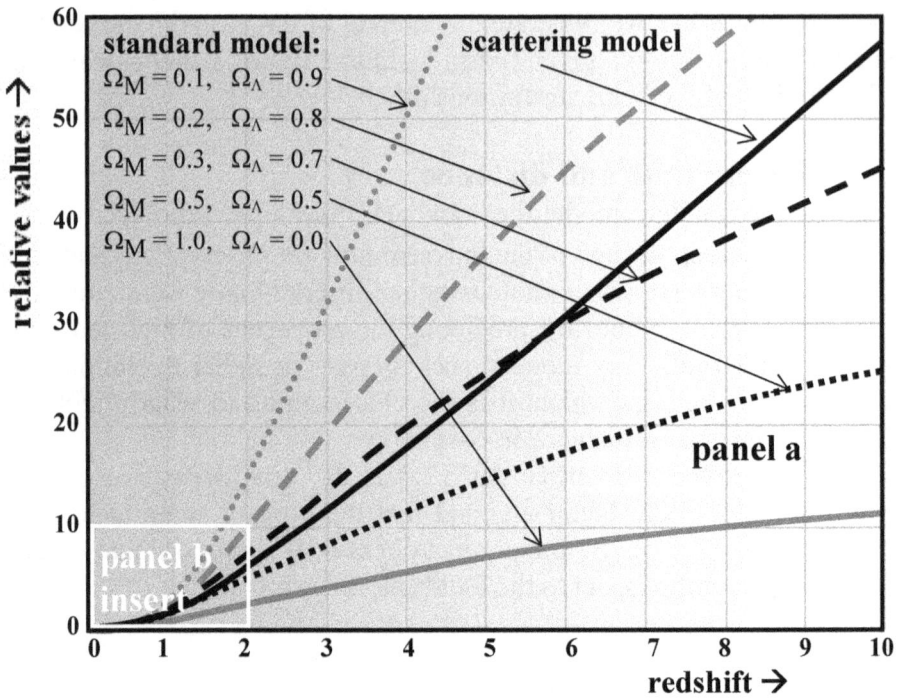

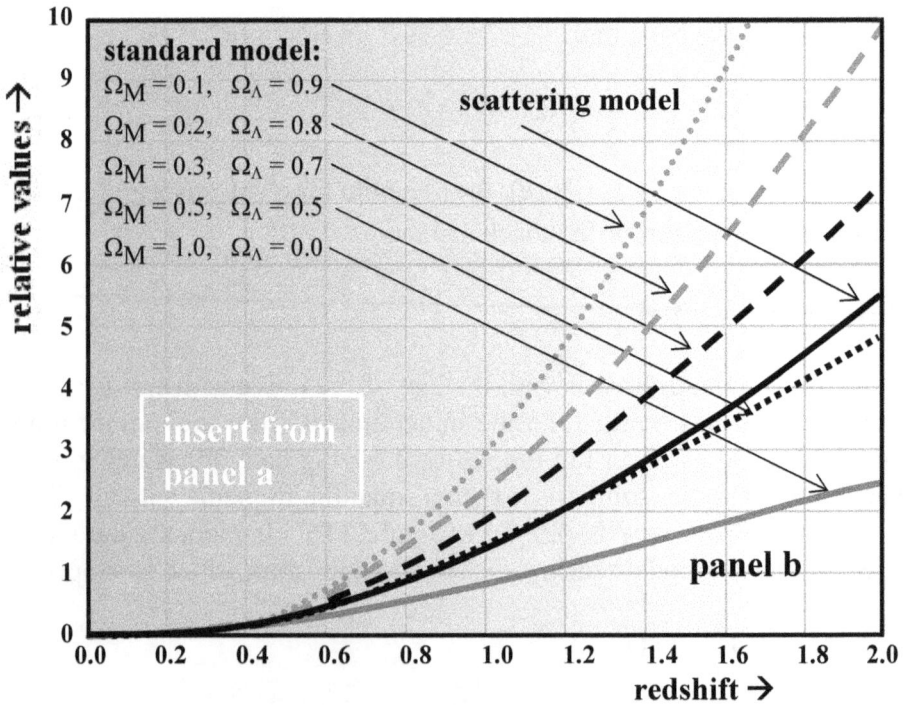

Figure 115:   Circumscribed volume out to a given redshift

j. confusions on assignment of distance

The distance-redshift relation can not be observed directly, and so we are left to infer what we can based on the assumed mechanisms involved in redshift phenomena. A Doppler interpretation of redshift naturally has ramifications with regard to compatible 'causes' so that the standard model is committed to various effects that would result from that interpretation.   It is obvious that although all of the various 'distances' that we have discussed above are necessary consequences of the expanding universe assumption, they do not share common values except at short distances. At cosmological distances, i. e., distances greater that a billion light years or so, they differ awkwardly. Line-of-sight comoving distance, transverse comoving distance, and luminosity distance all exceed the supposed radius of the universe itself by a redshift of about three for all versions. Only Lookback distance is constrained to be less than this value for reasonable density assignments for the standard model.

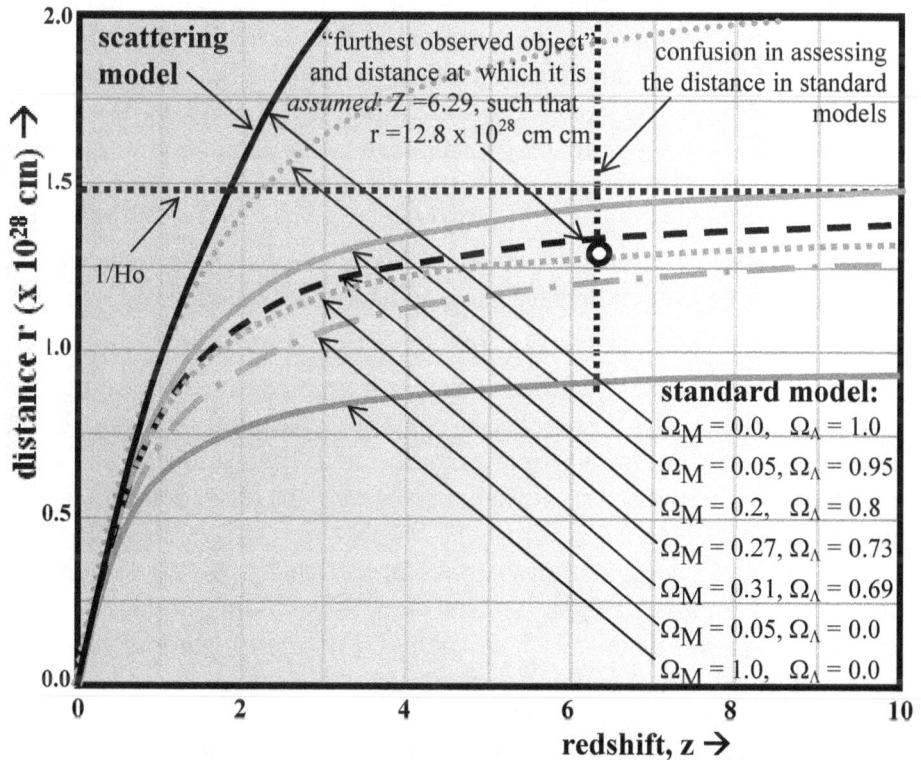

Figure 116:   **Lookback time/distance predictions for the *standard model* with various density parameter values as well as the *scattering model***

Lookback distance is a parameter whose values are envisioned as something that might appropriately be considered the 'actual' distance compatible with both the standard model and conventional parlance. For example, the source of a recent gamma ray burst has been cited as the most distant object ever observed. It was concluded by Cowen (2006) and others that the distance to this object was 12.8 billion light years. The galaxy in which this event seems to have taken place has a redshift of 6.29. The event is indicated as the open circle in figure 116. This distance was obviously merely *assigned* based on the acceptability of a particular world model. The vertical line in the figure indicates the diversity of possible assignments of distance that would be given by proponents of other models. Clearly, the scattering model would assign a much greater value to its distance, as indeed many of the other alternatives of the standard model would as well. Some would assign a smaller value.

Determination of distance in the standard model depends quite intimately on the particulars of the subset model one decides to accept. Clearly, the assignment of distance to the gamma ray burst cited above required many assumptions to be made even beyond the applicability of the standard big bang model. For one thing, the appropriateness of the 'lookback distance' parameter, was chosen rather than another measure accepted for other purposes by the standard model, such as say, comoving 'distances' shown in figures 108 and 109 or luminosity distance shown in figure 112. In addition a baryon density that is about one third of the critical density has also had to be assigned to Ω_M along with values of lambda for Ω_Λ and spacetime curvature to Ω_R. This assignment has been justified based on implications including the extreme scatter in supercluster galaxy redshifts interpreted in accordance with a Doppler interpretation and application of the virial theorem. We will discuss this network of hypotheses in detail in a later chapter.

Another combination of the three density parameters might have been (and yet might be) selected that would produce a very different inferred distance. The picture can easily become muddled when all of the alternatives for all three 'densities' (a vast three-dimensional space of possibilities) are considered in the mix of the standard model. Needless to say, there is a reason that press releases usually just refer to redshift with no specific mention of distance per se.

As Ellis (1997) appropriately states, "The cosmological model is a crucial, but often overlooked, variable in linking time and redshift. For $H_o = 70$ and $\Lambda = 0$, the redshift corresponding to a look-back time of, say, 7 Gyr [7 x 10^9 years], varies from $z = 1$ to 3 depending on Ω."

Lookback distance may be as free of theoretical accoutrements as it gets since it derives merely from the time it took light to arrive (in accordance with the various versions of the theory) multiplied by the speed of light along its path. This must be integrated along the spacetime path it took to get here of course. This is simplified even in relativity theories because the space portion of the spacetime interval Δr^2 applicable at each step along the line of sight, i. e., the path of a photon, is equal to the time interval $\Delta \tau^2$. Since along a line of sight, $\Delta \tau^2 = \Delta r^2$, where $d\tau$ is defined as $-c$ dt by scaling preference. The value, $-ct$, is negative by virtue of the observed photon emission events having occurred in the past. The distance is, of course, positive. We assume the observation occurs at time zero, i. e., 'now'. The simplest solutions assume $\Omega_R = 0$, i. e., that the curvature is minimal along the way, so that one can employ the more traditional treatment of time.

To understand the complexity of the issues and assumptions involved has required some understanding of the theoretical basis of the standard model, which is why we have dedicated a chapter primarily to that discussion. This has been necessary to support any comparisons of observed 'distance' and other more reliable cosmological measures with the predictions of the scattering model.

k. the Tolman test of surface brightness

Nearly eighty years ago Richard Tolman determined that the surface brightness of "standard candles" should decrease as $(1 + z)^{-4}$ if indeed there is an expanding universe as presumed by the standard cosmological model (1930). This prediction applies for all of the geometries included under that umbrella. So he phrased his finding as a refutability test for expansion.

His derivation involved recognition that luminous flux of a distant source will be reduced as a function of redshift as identified in chapter 4. This involves two factors of 1+z in the denominator, one of which is attributed to reduced energy of each photon, the other to time dilation in the standard model but to absorption in the scattering model. Neither of these factors are specifically distance-related. There is also, of course, an inverse square of distance factor no matter how it is characterized as a function of redshift. However, since surface brightness involves *total* flux from a given observed solid angular area, there is a cancellation of this factor by the square of distance in Euclidean geometry. So distance to the object should not affect measurements of surface brightness in Euclidean geometry. Therefore, one would expect the redshift functionality of this parameter to diminish as $(1+z)^{-2}$. Although since Tolman did not anticipate another factor of $(1+z)^{-1}$ attributable to absorption by an intergalactic plasma medium that we have seen applicable to the scattering model, he and his successors suggested that predicted functionality in Euclidean geometry would involve merely a single factor of 1+z in the denominator.

As discussed above with regard to cosmographics, in the standard model there are several defined 'distance' parameters such that at cosmological distances the geometrical effects due to the Hubble flow and differences in space curvature come into play. Tolman noticed, however, that since all versions of this cosmological model involve expansion, independent of the associated cosmological parameter values, an aberration functionality will be involved in any prediction of surface brightness. This functionality that is associated with the square of angular diameter distance, which as we saw, is predicted to increase with increasing redshift, introduces two additional factors of 1+z in the denominator for the standard model. The resulting, clearly-refutable difference in the average surface brightness predictions <SB> of the standard cosmological model and the scattering model are the following:

$$<SB>_{SM} \sim 2.5 \log (1 + z)^{-4}$$

for the standard model, whereas, the scattering model predicts:

$$<SB>_{SC} \sim 2.5 \log (1 + z)^{-2}$$

However, there is additionally the consideration of determining objects that can be legitimately employed as 'standard candles' for this test. We have already noted and will see in many specific instances where standard model apologists must assume that galaxies, for

256

example, are not the same now as they were in the distant past. To the extent that this involves their radii, additional compensation will be required that will involve model dependencies. In the next chapter we will assess the degree to which the two models satisfy Tolman's test.

l. issues of model flexibility versus scientific refutability

One cannot doubt that the standard model provides flexibility. Solutions included under that aegis span a vast range of conceivable cosmologies unified primarily by the presumption of an infinitesimal origin in some sort of big bang and the current value of Hubble's constant that determines the redshift per unit distance in our immediate environs assuming a Doppler interpretation. Alternatives include universes that expand forever and ones that collapse back into oblivion, disappearing into singularities. Some even exhibit variability claimed by Reiss et al. (1998). Figure 117 shows the *brief histories* of time that are endorsed by one version or another of this model.

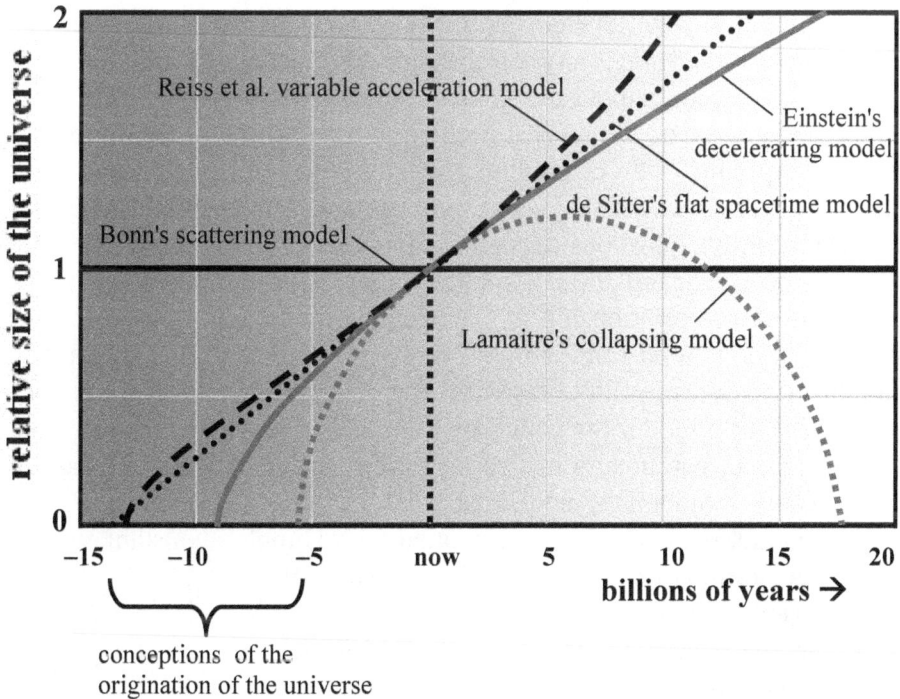

Figure 117:   Standard model flexibility versus scattering model stability

The scattering model is represented on the this plot as a flat line. The reason, of course, is that there are no presumed changes to the size of the universe or any other of its intrinsic properties over time. There is no "history of time" in this model. Redshift is used as a metric for distance in this model, but it does not presume thereby to be assessing an associated size of the universe or how long its processes whose progress in individual cases is measured using time have persisted.

Perhaps the extreme flexibility of the standard model is not such a 'good thing' in a scientific theory. The objective of refutability is certainly thereby frustrated. Indeed, as the theory is treated by many of its staunchest proponents the standard model repeatedly demonstrates inherent irrefutability. With its adjustable parameters it seems that virtually any redshift phenomena might be matched by one setting or another. Unfortunately for proponents, this has sometimes required the use of different parameter values to match one set of phenomena and yet another set of values to match another observation. Resolution is further obfuscated by recent tendencies to change parameters 'on the fly' as in Reiss's model. In even these cases that require disparate values of the key parameters, confirmation of the theory is typically claimed. Furthermore, any observation that is not in accord with prediction is accompanied by an associated "evolutionary profile" that would accommodate the missed prediction being right on the button. It is not surprising that cosmology has turned into disputations with regard to parameter settings and evolutionary tendencies rather than more substantive questions concerning the viability of the model itself.

The scattering model, on the other hand, employs the same redshift-distance relationship in each metric prediction. The extent to which this is a 'good thing' might certainly be debated, but the author cannot conceive of a scientist not attributing value to the inherent refutability of a model for which no quarter is clamed or taken. There are, of course, parameters in the scattering model as well. These are the electron temperature and density of the plasma medium. They effect a value, which we have associated with Hubble's constant. There is no acceleration or deceleration of expansion, of course, because there is no assumed expansion in the first place. There is in addition, however, the possibility of some variation in the average intergalactic plasma absorption coefficient.

Interestingly, the standard model that precisely matches the scattering model in line-of-sight distance is the de Sitter empty universe model, whereas with regard to lookback distance it is the high lambda version. However, these agreements are in reference to distances are not subject to direct observation. With regard to reliable observations of luminosity dependence on redshift, it is the 'concordance' model propounded by Riess et al. (1998) that is a remarkably close fit. However, the scattering model certainly doesn't require their suggested acceleration, deceleration and an intervening 'jerk' to match the data by expansion. That 'concordance' model is characterized by the triad of density parameter values, 0.31, 0.69, and 0.0 without the suggested variation in H_0. Heavy dashed lines were used for the curves in figures 105 through 112 that most closely represented its predictions.   In figures 112 and 113, it is clearly the model that most closely matches the scattering model predictions.

Perceived behavior of the universe such as Riess et al. (2004) have suggested is just an artifact of an associated model. The 'given' is basically only observations (typically angular separation and luminosity of 'standard candles') that we interpret one way or another. Distance and velocity (expansion rate, if you will), as well as the further inferences of acceleration and jerk are, of course, *not* perceived by anyone as being directly observable.   We do not feel the universe accelerating beneath us, although a scientist might certainly wonder, why not?

Whereas there does seem to be one value of Hubble's constant that characterizes the observed cosmological redshift, it is observed that there are dense conglomerations of galaxies at similar redshifts. This fact is certainly in part due to an actual clumpiness of galaxy clusters with a scarcity between – a clumpiness that would have taken ten times the currently accepted age of the universe to have been so effected by gravitation. But the clumpiness seems to involve even more than that.

We will see in a later chapter that observed dense plasma within rich galaxy clusters contributes substantially to an increase in the overall 'cosmological' redshift along the direction of such clusters according to the scattering model. So to the extent that the universe itself, including an associated plasma strata, is 'clumpy', the redshifting process defined for the scattering model will reflect that.   Separate effective values of Hubble's constant will pertain in those intervals where light propagates through the hot dense interior of rich galaxy clusters.   Since such rich clusters exist more or less uniformly throughout the universe, the contribution of an associated redshift boost to overall cosmological redshift will be appreciable. This will reduce otherwise-required higher densities and/or temperatures that one might have thought would be required by the model. Thus, as was derived in accordance with the scattering model, the dynamic pressure involving the average of the product of temperature and density must be on the order of 4,000 K gm cm^{-3}. This is only an average in the same sense that H_o is an average as we will see. As demonstrated for the scattering model with the illustration of figure 103 on page 225, a redshift will occur in light propagating through *any* hot plasma. We'll discuss the ramifications that eliminate the need for 'dark matter' in more depth in a later chapter.

Another parametric 'flexibility' of the scattering model involves the average absorption coefficient of the intergalactic medium as illustrated in figure 55.b on page 117. The extent to which it differs from the nominal value of $\gamma_C = 1.48 \times 10^{27} \rho_e \sec^{-1}$ as illustrated in the figure could effect slight alteration to the scattering model plots in figures 112 and 113 above.

Chapter 13

Comparing Predictions of the Scattering and Standard Models against Observations

In the previous chapter we determined ostensible cosmographic predictions of the various versions of the standard model and also for the scattering model. Now we must determine whether the predictions tentatively confirm or refute their respective models.

a. measurements of angular separation

In the previous chapter we saw that the standard model predicts rather strange phenomena with regard to the angular separations of objects in the field of view of a telescope, for example. This is a bold prediction that should be fairly easy to test as a discriminator.

In figure 118 observations are included that seem to refute the strange predictions of the standard model, seeming to confirm a Euclidean relationship with redshift itself. This is somewhat different than predicted by the scattering model in a Euclidean space, but is much closer than for any of the alternatives of the standard model.

The data provided in the figure derive from, Kellermann (1993, p. 663) where median angular sizes of sample galaxies and quasars are plotted against redshift in better agreement with the predictions of the scattering model than for the standard model curves. A Euclidean relationship of angle and redshift is shown to be compatible with observation, but no credible model insists on Hubble's initial prediction.

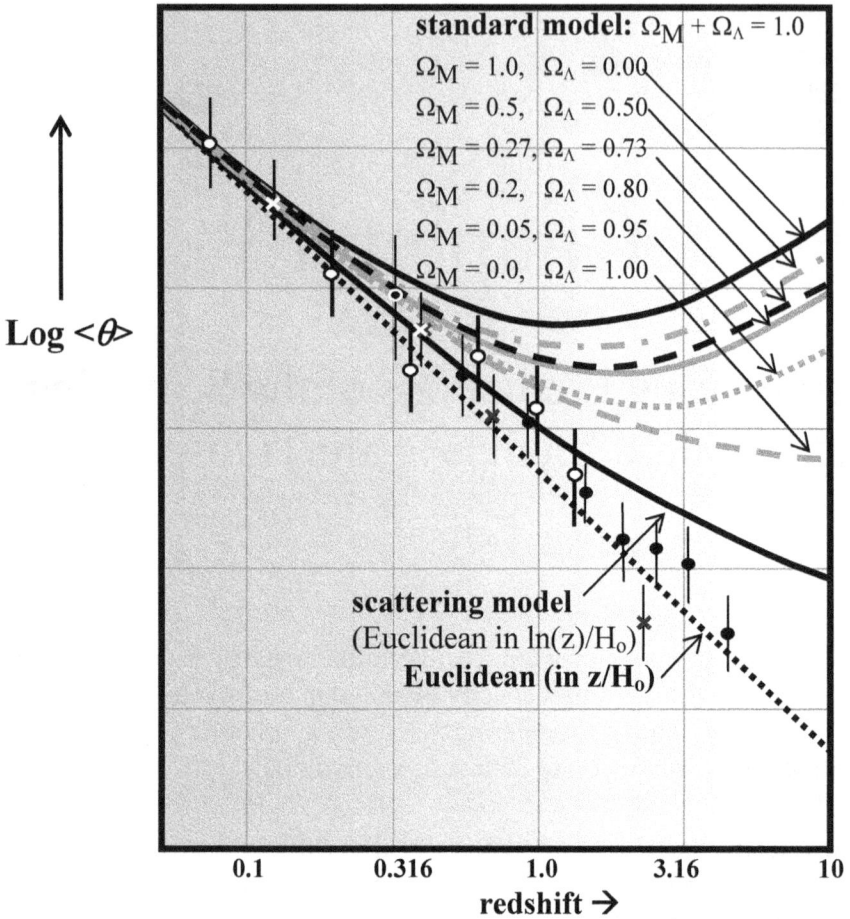

Figure 118: The angular size data predictions for the various models with plots of median angular sizes of samples of galaxies (open circles), galaxies in narrow luminosity range (crosses), and quasars (filled circles) data derived from, Kellermann (1993, p. 663)

Peebles (1993, p. 326) states that, "a galaxy at redshift z ~ 1 is expected to have an angular size on the order of one arc second, which coincidentally is comparable to the angular resolution, or seeing, permitted by our atmosphere. At z ≥ 2 the angular size at given linear size is predicted to increase with increasing redshift. Observations above the atmosphere may show that there are classes of elliptical or spiral galaxies at z ≥ 1 that look similar enough to their low redshift cousins so that there is a reasonable case for an estimate of their physical sizes relative to nearer galaxies. If so, the angular sizes will provide a very useful cosmological test." The author is aware of no data to refute a Euclidean relationship as of the date of publication.

The fact that nearly linear relationships are measured to persist out to significant values of z has been used on occasion, rather than to call these major theories into question, merely to eliminate angular diameter from consideration by most cosmologists as a legitimate test of their distance metrics. Kellermann mentions angle and frequency selection effects that may reduce apparent sized for the more remote measurements.

Also, the Sunyaev-Zel'dovich effect has been used in an attempt to obtain such angular data with regard to entire clusters of galaxies. See figure 119 taken from Carlstrom, et al (2002), with a plot appropriate to the scattering model added. But this data – despite its large uncertainties – does little to alter the obvious conclusion to be drawn from figure 118. See also figure 110 and the associated text where D_A is derived and plotted for the various models.

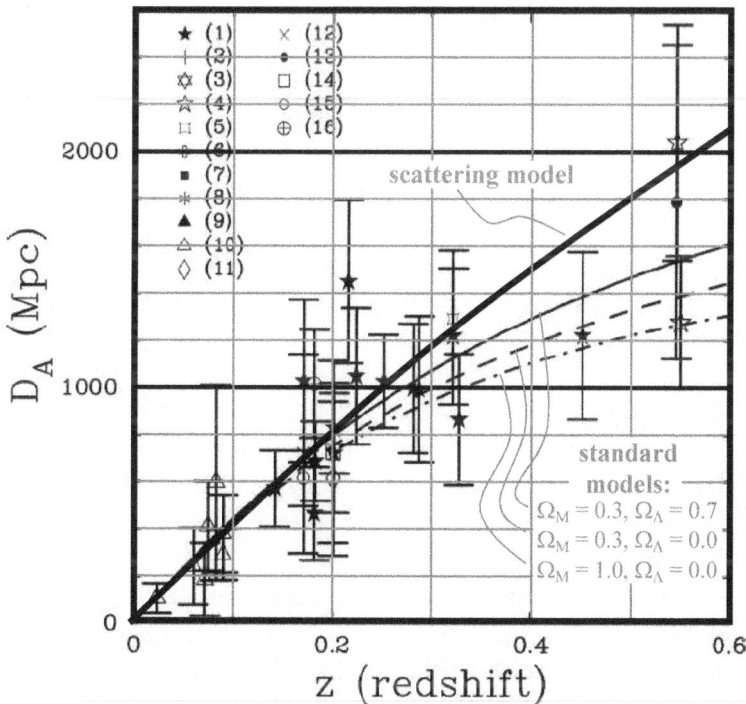

Figure 119: **The angular distance relation is plotted for three different standard model cosmologies, all assuming H_0 = 60 km per sec per Mpc; scattering model added.** [Plot from Carlstrom, et. al., 2002.]

It seems however to have been presumed that 'the appearance of' Euclidean relationships persist because of unspecified evolutionary effects associated with expansion. These would have to be precisely (and quite perversely one must conclude) those which would conspire to make the sizes and associated physical characteristics of such remote objects evolve so as to give the universe a Euclidean facade when such a direct metric is used. Kellermann says, for example, "The relation suggests size evolution in Friedmann models. "It is instructive here to recall Gallileo's arguments concerning giants necessarily having had to have had distorted proportions if they were to maintain the same functionality as normal human beings. He argued rationally that they could, therefore, not manifest human form and so would be incapable of aping a 'standard human' that appeared to be closer than the giant actually was. Similar arguments would seem to defy the evolution of standard candles whose functional characteristics, other than a coincidental down-scaled size, remain unchanged. Physics has, typically defied such conspiracies.

More recently, it has been proposed that, rather than galaxy diameters and the easily measured separation of the lobes of double radio sources, that much less precisely measurable compact radio sources might be employed as angular standard candles. This quantity has been somewhat more in agreement with the accepted models although the uncertainty of the measurements is considerable, notwithstanding which, compact radio sources now seem to be perceived by utilitarian cosmologists as being perhaps immune to evolutionary effects.

b. luminosity vs. redshift of standard candles

The most typical measurement of cosmological significance other than possibly the microwave background, is the luminosity of standard candles at cosmological distances. But even this measure is not without its major uncertainties.

Clearly the preparation of the data requires extreme expertise. Wavelength filtering and corrections for specific galaxy type spectral distributions makes this a difficult field, whose practitioners provide the data to be employed in confirming cosmological theories. We will not question the data itself although as Hogg (2002) points out, the uncertainties in 'K correction' is at least a few percent. We will, therefore perform those alterations to functional form of predictions to accommodate these redshift-dependent observational artifact changes in the same way that it is done for any other cosmological model. Since in particular very close agreement is realized for the scattering model with the standard model luminosity predictions we assume this closeness of fit will be preserved through such effects. This reasoning does not, of course, apply to what is referred to as "evolutionary changes" in the observed structures that sometimes find their way into the data.

There is a dearth of what could be considered complete data for observations in the optical region of the spectrum beyond $z = 1$, but there is data for radio wavelength emitting galaxies out to a considerable distance. In figure 120 we present data provided by Peebles (1993, p. 84) that was in turn obtained from McCarthy as noted. This data is plotted on a logarithmic scale along with predictions for several of the standard models as well as the scattering model. The agreement is good for the scattering model and for the concordance standard model, and the fact that the "best correlation" linear fit differs, is of little significance since the data clearly shows an upward bend.

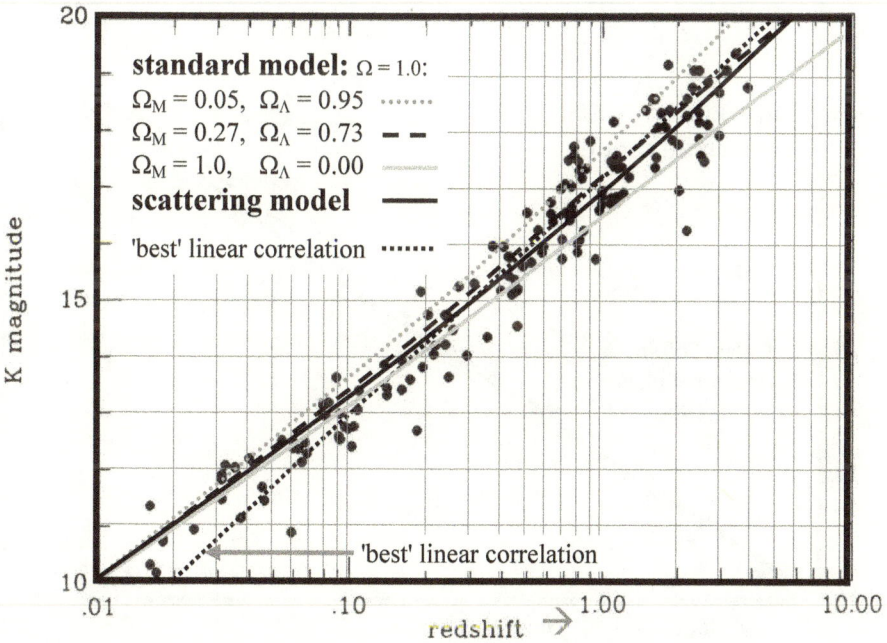

Figure 120: Redshift-magnitude relation for radio galaxies (McCarthy 1992)

Fortunately SN1A type supernova data is not vulnerable to the observational vagaries of galaxy redshift analyses. Figure 121 provides data originally presented by Perelmuter et al. (1998) which included error bars and discussion of models. The curves plotted in figure 121 are not from that source. The author has added them from the analyses presented in the previous chapter and plotted as figure 112. In that plot it is clear that the upward trend in the scattering model predictions is evident in observed data. This feature of high redshift data is being attributed by current cosmologists to "acceleration". See also Gondoin (2006). Clearly the form of this magnitude-redshift plot is extremely similar to that for the very different object types whose data is plotted in figure 120. The same upward trend is present in both sets of data.

Figure 122 provides Riess et al.'s (1998) data set that goes out to a redshift of 1.8 with a straight redshift scale. This plot also includes the author's predictions from the formulas that were included for figure 113 in the previous chapter of the variously parameterized standard model variations as well as the scattering model.

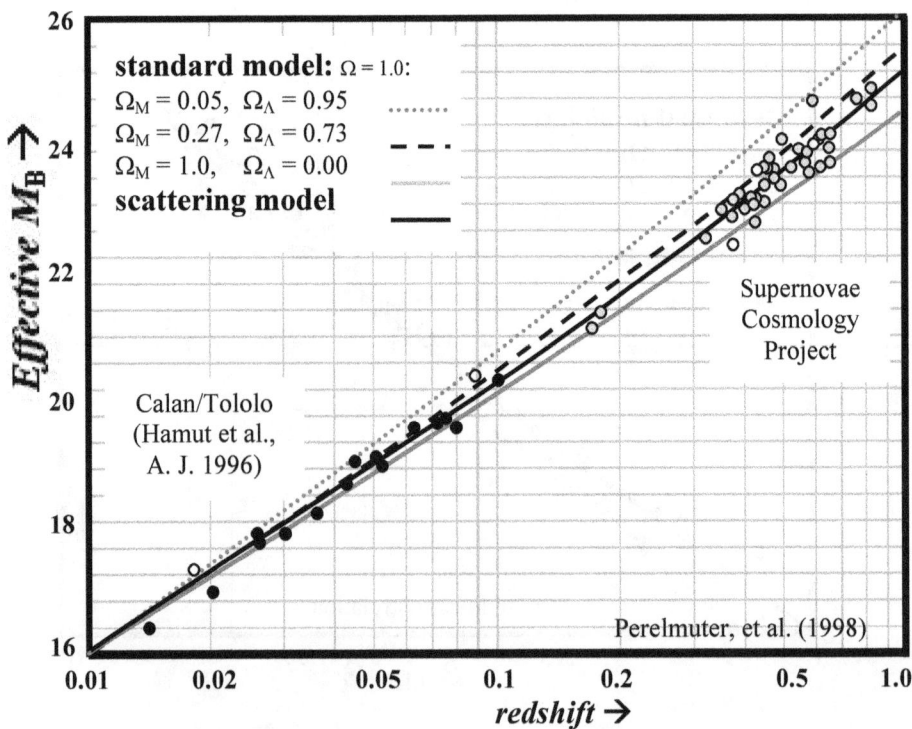

Figure 121:   The SNIA supernovae data (log plot)

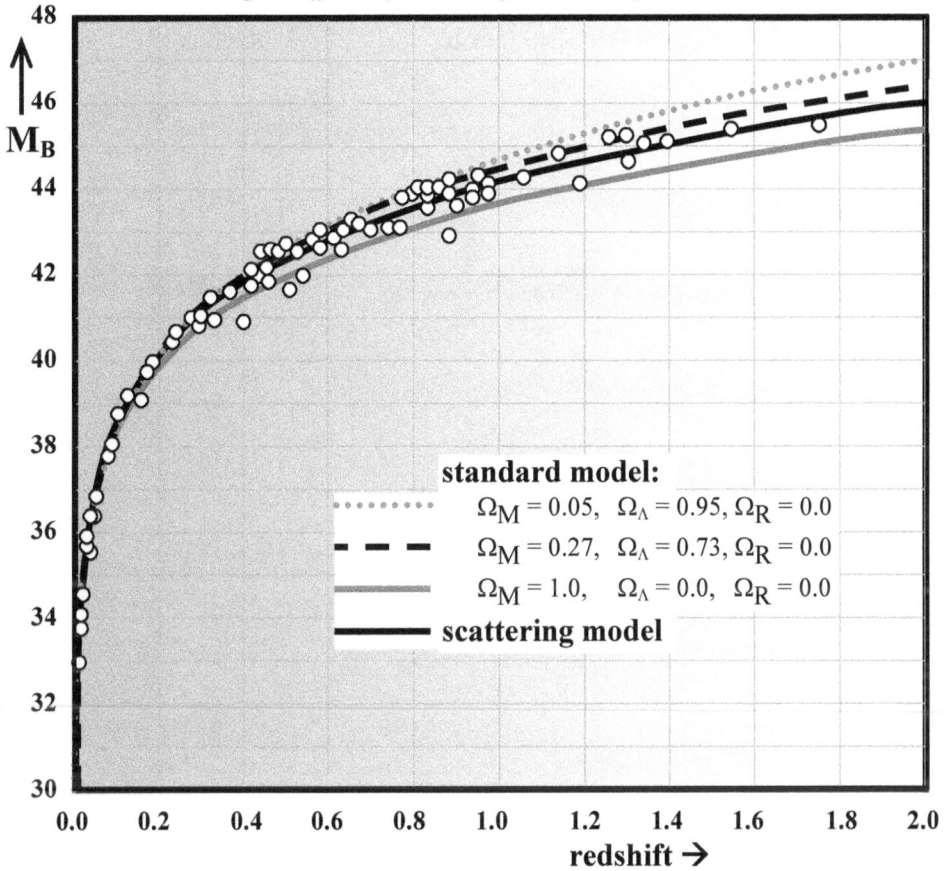

Figure 122: The SNIA supernovae data superimposed on model predictions

c. details of SN1A data

Kowalski et al. (2008) state that, "The SN Ia measurements remain a key ingredient in all current determinations of cosmological parameters". To support the preeminence of this data and the extreme inferences that are being drawn from it, they provide a comprehensive tabulation of data sets taken from various researchers. Refer to figure 123, which provides data sets from various researchers. For each data set they provide residual of variations from the best-fit curve to all the data. This additional residual data is provided in figure 123 to give a feel for the integrity of the various data sets. See figure 122 that provides this best fit and extremes of the standard models, but without the error bars that are available in figure 123.

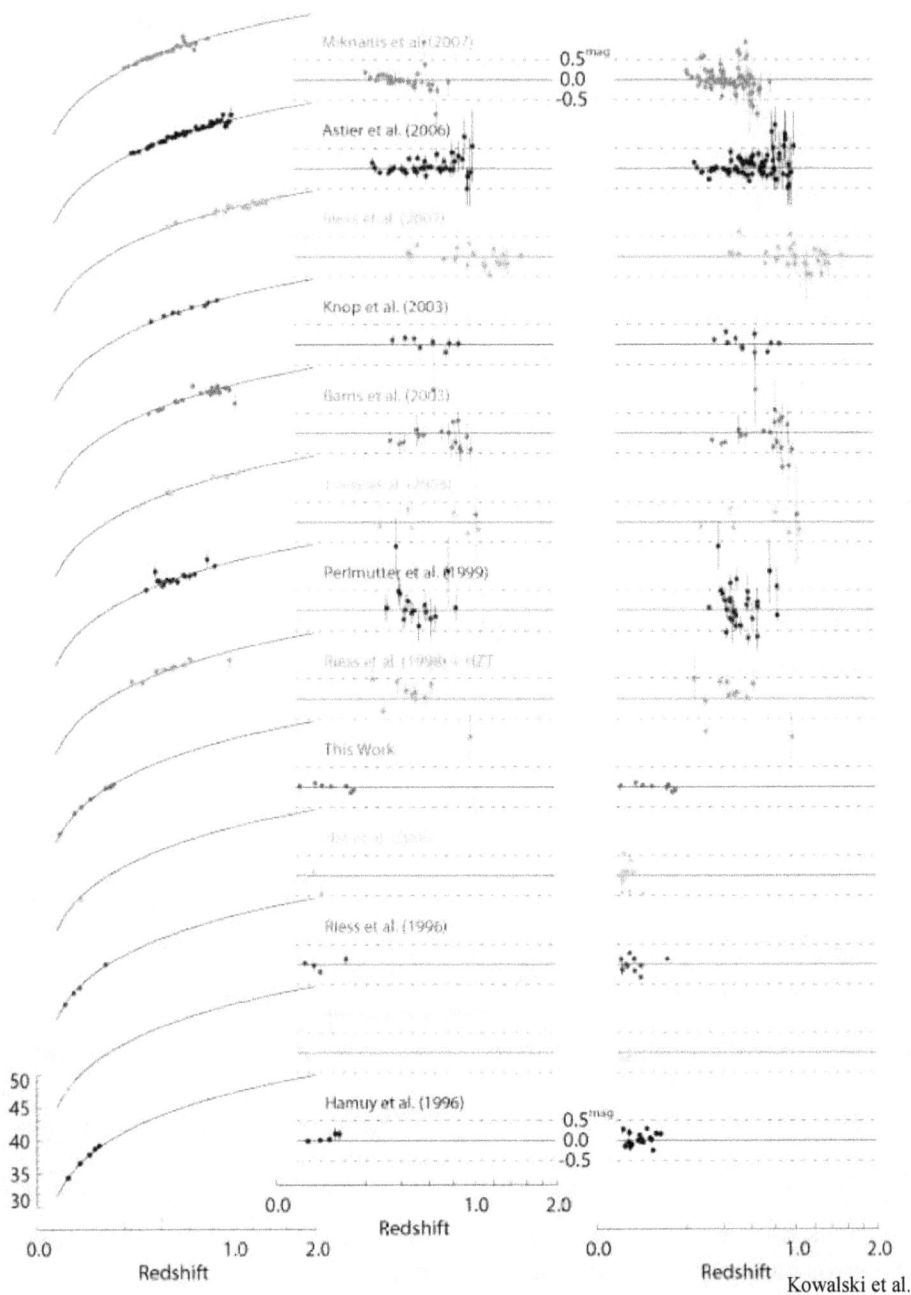

Binned Hubble Plots **Binned Residuals** **Residuals**

Miknaitis et al. (2007)

0.5^{mag}
0.0
-0.5

Astier et al. (2006)

Riess et al. (2007)

Knop et al. (2003)

Barris et al. (2003)

Tonry et al. (2003)

Perlmutter et al. (1999)

Riess et al. (1998) + HZT

This Work

Jha et al. (2006)

Riess et al. (1996)

Hamuy et al. (1996)

0.5^{mag}
0.0
-0.5

Redshift

Redshift

Redshift

Kowalski et al.

Figure 123: The SNIA supernova individual data sets and set residuals

It seems clear that the consensus, or as it is now known, the "concordance" view of standard big bang cosmology associated with the density parameters, $\Omega_M = 0.31$, $\Omega_\Lambda = 0.69$, and $\Omega_R = 0.0$, has reached that consensus in large part because of the close agreement those values provide with the actual SN1A luminosity data.

As shown in figures 120, 121, and in earlier figures, differences in predicted values between those of the concordance model and those for the scattering model are minimal. At a redshift of 2.0, that difference is, 0.190 magnitude. At a redshift of 4 the difference reduces to 0.063; at z=10 it would only be 0.22. Notwithstanding the obvious fact of the tremendous difference between the rationale for these two models, SN1A data cannot discriminate between them although clearly the scattering model is the closest fit.

Only the angular data presented earlier discriminates predictions of the two models.

d. gamma ray burst data

Gamma ray bursts (GRBs) have made their entry on this stage with bright promise of extending the range of accurate redshift/ luminosity data that should ultimately clear up the picture. New methods of using these high luminosity events as standard candles are emerging. See for example, Ghirlanda et al. (2006) who have developed a relationship by which one can relate luminosity and redshift in traditional ways. Their data is shown in figure 124.

Predictions made by the scattering model and several standard cosmological model versions have been included as plots along with data for the observed phenomena. Once again we find that the scattering model fares at least as well and better than any of the many standard model versions.

One would like to say more about GRB data, but there is a dearth of data so far. The future will no doubt provide a wealth of information; but it is not yet available.

As shown in the discussions and illustrations in this chapter, the cosmographic predictions for the various standard cosmological model versions and the scattering model derived and discussed in the previous chapter have tended to favor the scattering model wherever there is a discriminating test. Time dilation analyses of Blondin et al. might seem to present a serious challenge, however, that requires the additional data and analyses we will give that warranted concern in the next chapter.

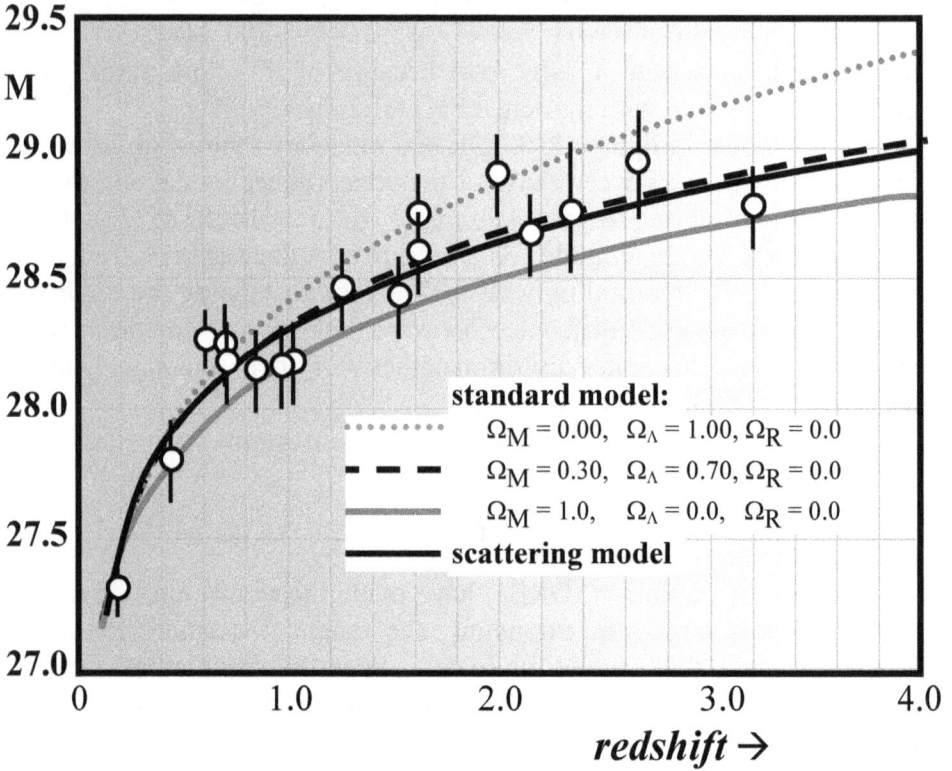

Figure 124: GRB luminosity equivalent [from Ghirlanda et al, 2006]

e. surface brightness data

Of course the technology was not available when Tolman (1930) first defined his refutation test of the expansion predicted by standard cosmological models using surface brightness measures. That prediction is that:

$$<SB>_{SM} \sim 2.5 \log (1 + z)^{-4}$$

Whereas, for the scattering model the prediction is:

$$<SB>_{SC} \sim 2.5 \log (1 + z)^{-2}$$

Until very recently, observations have been largely inconclusive.

Although the test still has not been performed to everyone's satisfaction, with Lubin and Sandage's (2001) preparations for "testing the reality of expansion" and their follow through in papers II through IV, one can say that a definitive test, if not yet completed satisfactorily according to everyone, is at least marginally feasible at this time. Actually, of course, there was much more than just technology that stood in the way of proceeding directly to the testing phase following Tolman's suggestion in 1930.

Sandage and Lubin identified four major obstacles: The first was the need for an operationally robust definition of the average surface brightness that had to be obtained for a galaxy class that could then act as a 'standard candle' for this test. Even for the most regular of galaxies the surface brightness is strongly dependent on where in an image of the galaxy one defines this *standard* for surface brightness? The value varies by a factor of as much as a thousand across the face of a galaxy image. So just where on such a *surface* can one standardize a region in which to measure a class-typical mean of surface brightness? The angular metric size must be independent of cosmological geometry (which is a problem) and depend only on observed photometric data.

Additionally, of course, at large redshifts systematic errors tend to creep in due to the resolution of the instrument used. Sandage and Lubin indicate that angular resolutions on the order of 0.1" (a tenth of an arc second) are required for observations at redshifts in excess of $z = 0.5$. They cite earlier efforts for which test results were significantly marred by insufficient angular resolution of ground-based instruments that were on the order of 1.0" with photographic data used exclusively. A proposal for a cosmological project to perform the Tolman test was made during the early planning phase of the Hubble telescope because at least 10 times better resolution would be obtained by that instrument. This test was finally undertaken 25 years later.

Three high redshift galaxy clusters were studied in this series of papers using the Keck and the Kitt Peak ground based telescopes in addition to the data obtained from the Hubble space telescope. See Oke et al. (1998) and Postman et al. (1998 and 2001) for clarification. The redshifts of the three clusters that were used are 0.76, 0.90, and 0.92.

Test preparations proceeded by obtaining first-ranked cluster elliptical and S0 galaxies in low redshift clusters and analyzing them to obtain the pertinent correlations between average surface brightness, linear radius, and absolute magnitude. This effort provided a calibration of correlations between parameters in the zero redshift limit. Extensive photometry had been done by Postman and Lauer (1995) of cluster galaxies at low redshift such that with minor modifications that sample could be used to define this zero-redshift calibration. In this sample fiducial data set the averages of surface brightness for elliptical (E) and S0 galaxies were calculated by measurements made at each of multiple radii from the center of each galaxy.

To define an appropriate metric Petrosian's (1976) analyses were used. He had defined functions of luminosity magnitude, m_η to be evaluated at various radii r_η that are particularly relevant to the Tolman test because, as he proved, his parametric definition provided equivalence of the difference in magnitude between the mean surface brightness averaged over the interior area and the surface brightness at that particular radius. With this definition, the mean surface brightness magnitude per square arc second can be calculated as a function of radius as follows,

$$<\text{Sb}>_\eta = 2.5 \log (\pi \, r_\eta 2) + m_\eta$$

where m_η is the apparent magnitude of the light encompassed within the respective radii and r_η is measured in arc seconds. Sandage and Lubin make a special note of this being "independent of all cosmology as to 'proper distance.' It is a directly observed quantity, no matter what its interpretation." See also Menanteau et al. (2006) who state, "The advantage of using a Petrosian radius over traditional surface brightness limit radii, … is that it only depends on the galaxy light profile, and it is therefore independent of the redshift of observation."

So surface brightness values were determined for 118 first-ranked cluster galaxies at Petrosian radii characterized by $m_\eta = 1.0$, 1.3, 1.5, 1.7, 2.0, and 2.5 magnitude. The authors determined intensity profiles as functions of radii from the Hubble telescope images where there was sufficient angular resolution to permit reliable angular radii to be measured at Petrosian values from $m_\eta = 1.0$ to 2.0 magnitude. Observed surface brightnesses at each of the Petrosian radii were then obtained and corrected to the rest wavelengths using K correction. In this way they could compare commensurable surface brightnesses of the high-redshift cluster galaxies against similar data for the calibrated fiducial local galaxies of known absolute magnitudes.

Sandage and Lubin argue that, although the formulas of the standard cosmological model are employed in rendering the linear radii and absolute magnitudes from observed data, "the test is nevertheless free from the hermeneutical circularity dilemma occasionally claimed in the literature." This more-typically theological phraseology seems strange in a scientific paper. Nonetheless they provide logical reasons for observed mean surface brightness being independent of the assumptions of cosmological models because their calculations proceed directly from the measured data using only angular radii and apparent magnitudes of that data. However, there is an implicit rest frame galaxy *size* obligation in the formulation that cannot avoid model dependence. Tolman's assumptions assume identical rest frame size in a 'standard candle'. As we will quote later Conselice (2004) concludes that there must be size evolution within galaxy type such that early (high redshift) galaxies would have to have been smaller than they are now – so much for standard candles in an evolving universe.

Lubin and Sandage do (as they must) discuss presumed effects of luminosity evolution and the requirement that absolute magnitudes and linear radii distances must come from cosmographic predictions at high redshift, but the extent to which their 'evolution' conjectures have been separated from 'observed data' is not completely clear. They define n as a luminosity evolution exponent to quantify what is required to force observed data to match standard model predictions as follows:

$$\Delta M_{evol} = 2.5 \log (1+z)^{4-n}$$

SUMMARY OF THE TOLMAN SIGNAL AND THE INFERRED LUMINOSITY EVOLUTION FOR Cl 1604 + 4321 at z = 0.9243 IN THE R BAND USING $q_o = 1/2$

m_η	$\langle Sb \rangle_\eta$ (mag)	n	ΔM_{evol} (mag)	4−n	Number of Galaxies
1.0	2.44±0.16	3.43±0.23	0.40±0.16	0.57±0.23	14
1.3	2.24±0.16	3.15±0.23	0.60±0.16	0.85±0.23	14
1.5	2.21±0.16	3.11±0.23	0.63±0.16	0.89±0.23	14
1.7	1.96±0.17	2.76±0.24	0.88±0.17	1.24±0.24	14
2.0	1.76±0.18	2.48±0.25	1.08±0.18	1.52±0.25	13

Significantly, they state that, "we disregard the data at m_η values of less than 1.7 as undoubtedly unreliable". They then average the values for

$m_\eta = 1.7$ and 2.0, seemingly without regard to the trend in the "reliable" data toward a power of 2 rather than 4. What they come up with is:

$$\langle SB \rangle_{SM} = 2.5 \log (1+z)^{2.59 \pm 0.17} \quad \text{magnitude for R band, and}$$

$$\langle SB \rangle_{SM} = 2.5 \log (1+z)^{3.37 \pm 0.13} \quad \text{magnitude for I band.}$$

where the same procedure of disregarding data for low m_η values is required and used for the I band data as well.

Lubin and Sandage also calculate results for cases with standard model parameter values $q_o = 0$ and 1. In the former case the "required evolutionary correction" is increased by 0.18 mag at $z = 0.7565$, by 0.23 mag at $z = 0.8967$, and by 0.22 at $z = 0.9243$. They translate this difference into changing the exponents in the previous equations to:

$$\langle SB \rangle_{SM} = 2.5 \log (1+z)^{2.28 \pm 0.17} \quad \text{magnitude for R band, and}$$

$$\langle SB \rangle_{SM} = 2.5 \log (1+z)^{3.06 \pm 0.13} \quad \text{magnitude for I band.}$$

Clearly, the standard model does not score well on the Tolman test without significant help from ad hoc assumptions of evolution.

Since the requisite equations for their analyses presume the standard cosmological model, those same authors provided a separate section in their paper IV in which they modify analyses appropriate to their perception of requirements for *any* tired light model. This backs out the strange angular distortion metrics of the standard model that we encountered in the previous chapter. They obtained:

$$\langle SB \rangle_{TL} = 2.5 \log (1+z)^{1.61 \pm 0.13} \quad \text{magnitude for R band, and}$$

$$\langle SB \rangle_{TL} = 2.5 \log (1+z)^{2.27 \pm 0.12} \quad \text{magnitude for I band.}$$

These results are in excellent agreement with exponent predictions of the scattering model of 2.0. This is certainly much better than was obtained for any of the various versions of the standard model. However, since Lubin and Sandage did not take the scattering model's broadband absorption factor into account, they disparage what is in actuality excellent agreement with predictions of the scattering model.

Chapter 14

Claims of Acceleration of Expansion and Time Dilation in SN1A Data

The observational data that has been inspected in the previous chapter provides clear choices between models in some cases and less obvious ones in others. However, in all these cases the scattering model fares at least as well as any version of the standard model. If these were the only tests for model viability, the scattering model would seem to warrant acceptance as the cosmological model of choice. But there is more data to be evaluated, including the distribution of matter in the universe, the background radiation, and the relative abundances of the light elements. We will get to those additional predictions and observations presently, but first there are a couple of inferences that have been made by standard model cosmologists with regard to SN1A data in particular that must be addressed.

The SN1A data has been taken to imply a general acceleration of the expansion predicted by the standard model as well as providing evidence of time dilation effects in the proceedings of the supernovae at extreme distances. These two inferences differ significantly with regard to the logic of their construction. The only recently cited acceleration is not a phenomena that was predicted by any version of the standard model but has been put forward to salvage the version that is most viable with regard to other cosmographic predictions. The most distant SN1A observations tend to deny even this alternative unless some alteration of theory is accepted. There is no viable explanation for the hypothesis except that it salvages the standard model.

On the other hand, time dilation is a prediction intimately associated with the expansion of the standard model in accordance with the special theory of relativity. It has been assumed as a factor in luminous flux predictions in the standard model and unless there is some evidence to support its actually occurring in observed processes that develop over time, the concept of expansion comes into question.

a. rationale for inferring an acceleration of expansion

As explained, inferences of acceleration of our entire universe derive from the conviction that the 'concordance' version of the standard cosmological model is correct. This belief that is now firmly held by a consensus of cosmologists is bedeviled by the fact that at redshifts in excess of unity its predictions do not match observations. The data seems to suggest that rather than any universal expansion being a mere "coasting" as characterized by a constant value of Hubble's 'constant', since its inflationary phase, it has suddenly undergone acceleration with a "jerk" (a rate of change of acceleration). Proclamations have resulted concerning the factuality of the entire universe having had to accelerate to effect a fit with the consensus view, and that it currently 'must' begin a phase of deceleration. See figure 125 provided by Reiss et al. (2004) where the expectations of the alternatives are illustrated. The figure shows the implication to the concordance model. Notice however that the scattering model provides an ideal fit even at the extremity of available data without requiring the extravagance of having the universe perform extraordinary maneuvers to fit its predictions. Reiss's conjecture – typically denominated a "discovery" – of the acceleration of the universe – was heralded by _Science Magazine_ as 'Breakthrough of the Year' in 1998.

Despite the reorganization of the data in figure 125, the points to the right in the top panel are easily identified with the same points in figure 121. The ostensible rationale for the several curves as attributed by Riess et al. are labeled in the bottom panel where the data have been binned. Clearly there is a complicated splining of functionality involved in fitting this data by standard models. With the scattering model there is no such manipulation of metaphysical baggage required to fit the data. It just *fits*. See the line segment added to the top panel of the figure for this purpose.

The complex of explanations of the presumed acceleration continues to bedazzle cosmologists and conflate the scene of cosmological conferences and journals. See for example *Science Daily* (Jan. 12, 2005), where the following appeared:

"The 2dFGRS [galaxy survey] has shown that baryons are a small component of our universe, making up a mere 18% of the total mass, with the remaining 82% appearing as dark matter. For the first time, the 2dFGRS team have broken the 10 percent accuracy barrier in measuring the total mass of the Universe.

"As if this picture weren't strange enough, the 2dFGRS team also showed that all the mass in the universe (both luminous and dark) is outweighed 4:1 by an even more exotic component called 'vacuum energy'. This has antigravity properties, causing the expansion of the universe to speed up. This conclusion arises when combining 2dFGRS results with data on the microwave background radiation, which is left over from the time when baryon features were created. The origin and identity of the dark energy remains one of the deepest mysteries of modern science."

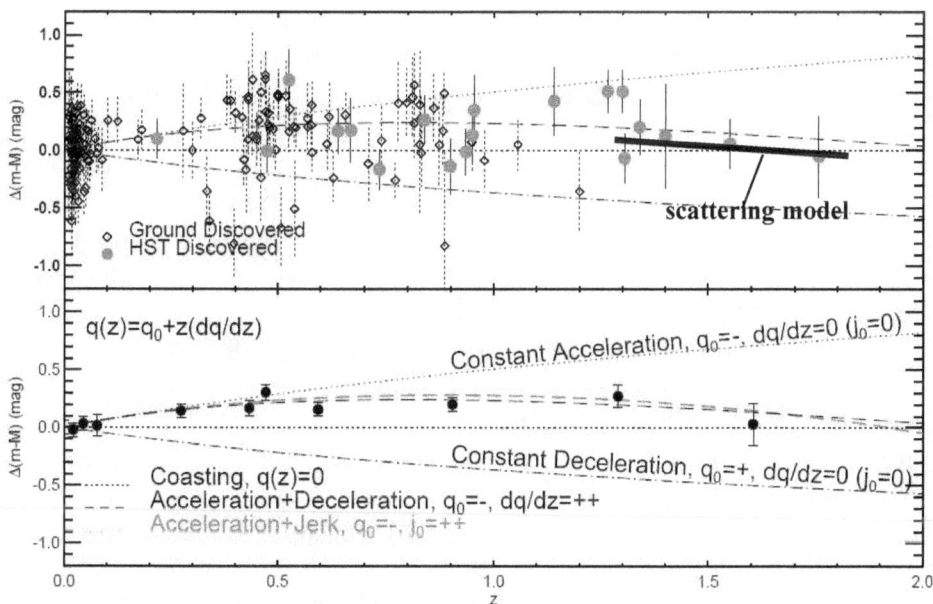

Figure 125: The basis for proclamations of acceleration of the universe

277

It is difficult indeed to know how to deal objectively with attitudes expressed in this piece. Certainly one must acknowledge that inferences made from this confusion are astounding. Clifton and Ferreira (2009) explore the continuing ramifications of Riess's conjecture. After ten years of failure to identify anything that might remotely suffice to account for this obscure dark energy, cosmologists are trying to find anything that might account for the data. They report an increasing willingness of cosmologists to reject the only remaining aspect of Copernicus's principle of humility with regard to our position in the universe that the standard model does not seem to accommodate. The temporal aspect of this principle was sacrificed with the big bang; now the spatial aspect is in jeopardy. All this because to right-most bubbles in figure 121 don't lie right on the dashed line identified as the "concordance model".

Needless to say, there is no problem with the scattering model's accounting of the phenomena – it is right on the path.

b. background on time dilation claims

In reference to figures 13 and 14 in chapter 2 that depict SN1A supernovae data, we discussed the protraction of the decay period for the more luminous of these *inherently* very luminous events. We mentioned that the associated elongated decay process suggests the possibility of tests for time dilation. We were somewhat skeptical and in proposing a model for which there are no exceptional recessional velocities of distant cosmological objects, reticence to accept such claims might seem suspect. In fact this author has on other occasions expressed a deep-seated skepticism of the very notion of time dilation that has been accepted as central to most interpretations of formalities and associated experimental results of the special theory of relativity. So he is doubly suspect in that regard.

There was a rush to embrace the concept of time dilation after the many other early successes of the special theory. Max Born cited, for example, the facts associated with muons created in the upper atmosphere being able to survive the travel time all the way down to sea level even though their usual decay rate would not have allowed them to survive that long. He suggested that high-speed mu mesons experience longer lifetimes than relatively stationary ones encountered in the laboratory or they would have decayed long before arrival.

But muon decay and other claims of time dilation invariably involve processes and reactions, the cause of whose transition rates remain ambiguous because of a commensurable energy dependence (Bonn, 2008). The ambiguity involves processes taking longer to accomplish when an object has been accelerated to, or is bound by, a higher energy. In other words, these processes invariably are intimately tied to the energy of the particle itself. Typically cited examples where time dilation would come into play include particle decay rates and cesium clock energy transition effects with experimental data indicating that the decay rates are increased commensurably with the time dilation factor, $\gamma = (1 - \beta^2)^{-\frac{1}{2}}$. However, muons that are tightly bound in potential energy wells also decay less rapidly because precisely the same special relativistic 'gamma factor' affects the mass-to-energy conversion. So inevitably one is left with somewhat of a quandary – do muons decay more slowly because of clock time dilation or just because they happen to have been infused with additional energy by acceleration? The effects would be the same. If it's clock time dilation, where is the clock? Are there really clocks and random number generators that control particle decay? This author doubts it. Can time dilation and relativistic effects of increased energy *both* be operative, and if so why is the decay rate not affected by the square of the gamma factor? That would clearly be refuted by the data itself.

That is the author's baggage that he carries into this discussion, notwithstanding his obvious acceptance of the formalities of the theory.

c. SN1A decay rate considerations

There has long been a presumption that the predicted extreme recessional velocities of distant cosmological objects might provide a testable instance where time dilation, unencumbered by environmental energy differences, would present a situation appropriate for refutation testing. Where better for it to appear than in a demonstrable aging process of luminosity profiles of standard candles? Indeed, as we noted earlier, distant SN1A events for which more extreme velocities would be associated according to the standard model *do* take longer to decay.

However – and yes, this is a major *caveat emptor* – in this case too the familiar time-dilation/energy equivalence alternative duality question cannot be avoided. Again convincing evidence suggests it is the associated energy and not some obscure 'clock' that is involved.

Initially it was presumed that all SN1A events had identical profiles to justify their emerging status as very powerful 'standard candles'. However, the problem with observations of any very distant object is that there is an unavoidable observational filtering that takes place in such observations known as the 'Malmquist bias' whereby only the brightest objects and events can be observed at great distances. Any differences in the objects based on inherent luminosity will inevitably appear to exhibit redshift dependence. This conflates arguments pertaining to the 'evolution' of objects as well. So SN1A data is inevitably skewed to extremely energetic objects/events of a given type, where again we find that same conflation of issues involved with the more energetic of SN1A events taking longer to decay – even where there is no appreciable redshift at all.

As we will see, there are still advocates who maintain that time dilation is *the* explanation for longer decay periods at high redshift, although Jensen (2008) and others continue to doubt the legitimacy of such interpretations because of the now well-known longer decay period for high-energy, low-redshift supernova events. There is also some confusion about whether extremely powerful events such as supernova 2006gy are even properly classified as SN1A rather than Type II or other category supernovae that do not even qualify as a 'standard candles' to be used in such analyses. In any case, the data clearly indicate the occurrence of a more powerful supernova; it certainly does *not ipso facto* verify the occurrence of time dilation. A longer decay period associated with more luminous supernovae is a general characteristic of SN1A supernova variations independent of redshift and its supposed cause.

Ruiz-LaPuente (2004) illustrates the correlation between the brightness at maximum and the rate of decline of the light curve at various wavelengths in figure 126 obtained from that source. It was the mid-nineties before corrections were incorporated into data that legitimized the status of SN1A events as 'standard candles'. The problems were illustrated earlier in panel a of figure 13, page 29, also from Ruiz-LaPuente (2004) with the correction applied in panel b. This corrected data provided the invaluable contributions to observational cosmology that we saw in figures 120 through 123. The data give us the most accurate value of H_0 that matches similar plots obtained for other standard candles. They also provide the estimates of overall baryonic mass density Ω_M and cosmological constant Ω_Λ

applicable to the standard cosmological models whose predictions were determined previously. The necessary corrections to the SN1A data also fit the scattering model predictions. The degree to which this data has been warranted is apparent in the willingness of the community to accept Riess's claim of an unaccounted acceleration of the entire universe based upon it.

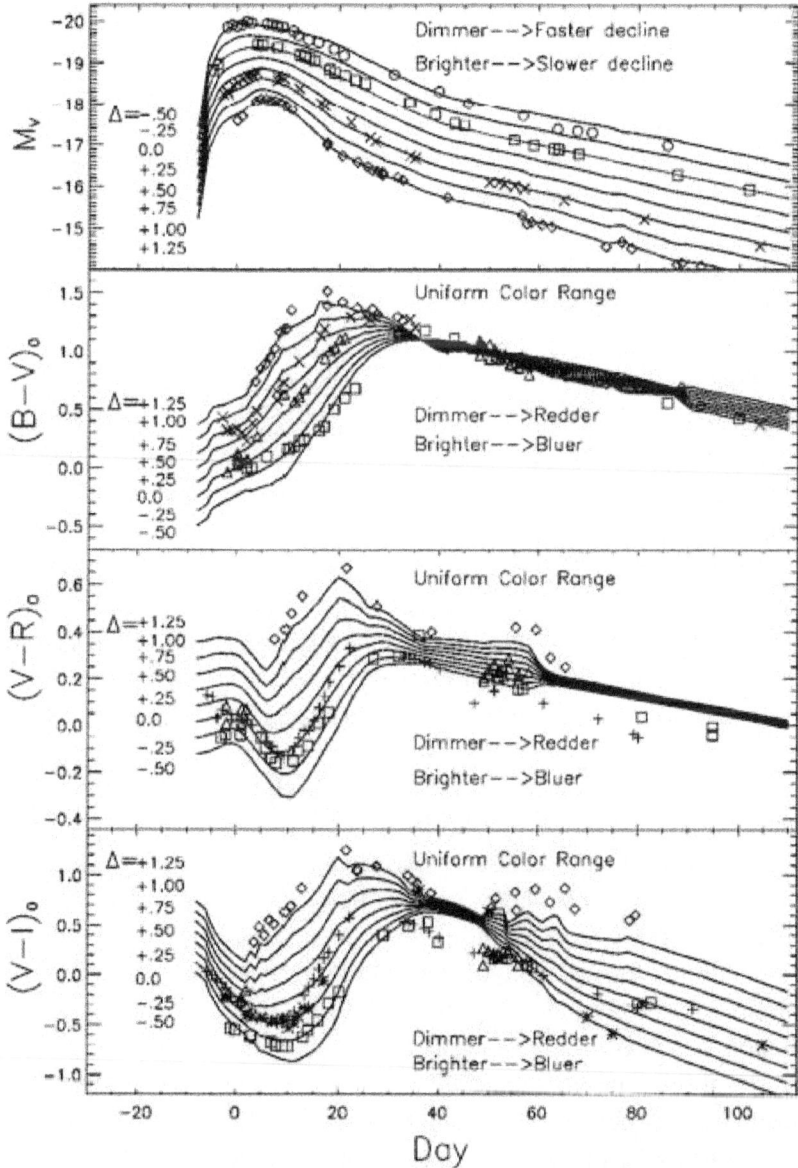

Figure 126: **SN1A decay data for events of differing brightness**

d. mimicking of time dilation by a Malmquist bias

None of these corrections involve time dilation, although of course, the standard model depends on a factor in the luminous flux projection that depends upon it. So time dilation per se still needs to be addressed directly. Blondin, et al. (2008) have presented a unique case for time dilation based on intrinsic characteristics independent of direct comparisons of the problematical luminous flux decay intervals for which they state: "Comparison with the observed elapsed time yields an apparent aging rate consistent with the $1 = 1/(1+z)$ factor (where z is the redshift) expected in a homogeneous, isotropic, expanding universe. These measurements thus confirm the expansion hypothesis, while unambiguously excluding models that predict no time dilation, such as Zwicky's 'tired light' hypothesis."

This perception that time dilation can be confirmed and that an associated refutation of all "tired light" models can be effected in a single stroke using SN1A data is a claim with which we must take issue. The scattering model is, of course, in that class of *tired light* cosmologies Blondin, et al. perceive themselves as disqualifying in one fell swoop. In any case, they have formulated and presented their case in a straight-forward manner. The issue we have discussed concerning the usual conflation between attribution of the decreased decay rate and inherent brightness rather than time dilation is not an interpretational difference that Blondin, et al. have ignored mind you. In fact, they began their discussion by broaching just such reservations in reference to data shown in figure 127, which was presented in their paper.

They addressed these counter arguments head on as follows: "...one might argue that at high redshift we are preferentially finding the brighter events (akin to a Malmquist bias). Such a selection effect would produce a spurious relation in which there would be broader light curves at higher redshifts, without any time dilation." Apparently they believe that they have been able to entirely circumvent this criticism by innovating another aspect of supernovae spectra that avoids a direct comparison of temporal decay rate data with all the arguments and counter arguments that would accompany any conclusion based on such an approach. We will discuss their findings, but it should be noted first that despite a minor concession to the Malmquist bias, they have not shown how it affects the *appearance* of a time dilation when there is, in fact, none. Let us do that now.

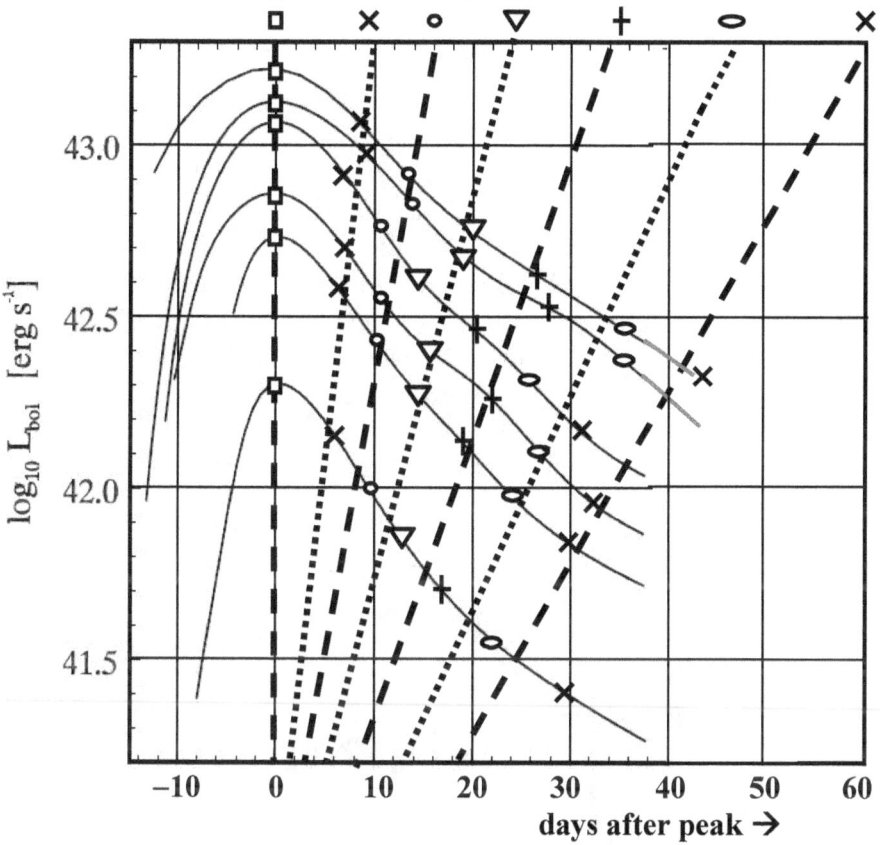

Figure 127: Rest frame SN1A brightness profiles [augmenting Blondin et al.]

Trend lines have been added to the original plot in figure 127 to show amounts of time required for magnitude to drop from peak values in quarter magnitude increments. It is important to determine to what extent such trend lines can be said to mimic time dilation effects. Consider the luminous flux formula elaborated diagrammatically on page 66 and discussed elsewhere. In all those cases we considered the luminous flux *f*, that is being referred to in the Blondin plots as L_{bol}, to have been a function of the inherent luminosity L_0 and redshift Z. Now, however, consider the inverse case in which the bolometric maximum is the given and inherent luminosity of the event in its 'rest frame' must be determined as a function of redshift as follows:

$$L_0[Z] = (4 \pi / \Omega) L_{bol} (Z+1)^2 r^2[Z]$$

283

$$= (4\pi / \Omega H_o^2) L_{bol} (Z+1)^2 \ln^2(Z+1)$$

Here the final expression pertains to the case of the scattering model.

From the peak of the bottom curve in figure 127 to that of the top one is very nearly a factor of 10. Thus, if the event associated with the top curve had been at a redshift such that $L_{bol} = L_o / 10$, then the two events would be characterized by the same observed bolometric maximums. Without presuming time dilation there would be a clear distinction between the redshifted curve and that of a higher luminosity SN1A event as shown in figure 128, since the narrower profile would be retained despite the effect of redshift on luminous flux. Compare the lower solid and upper dotted curves illustrated in the figure. However, if we were also to infer a time dilation effect, we would have:

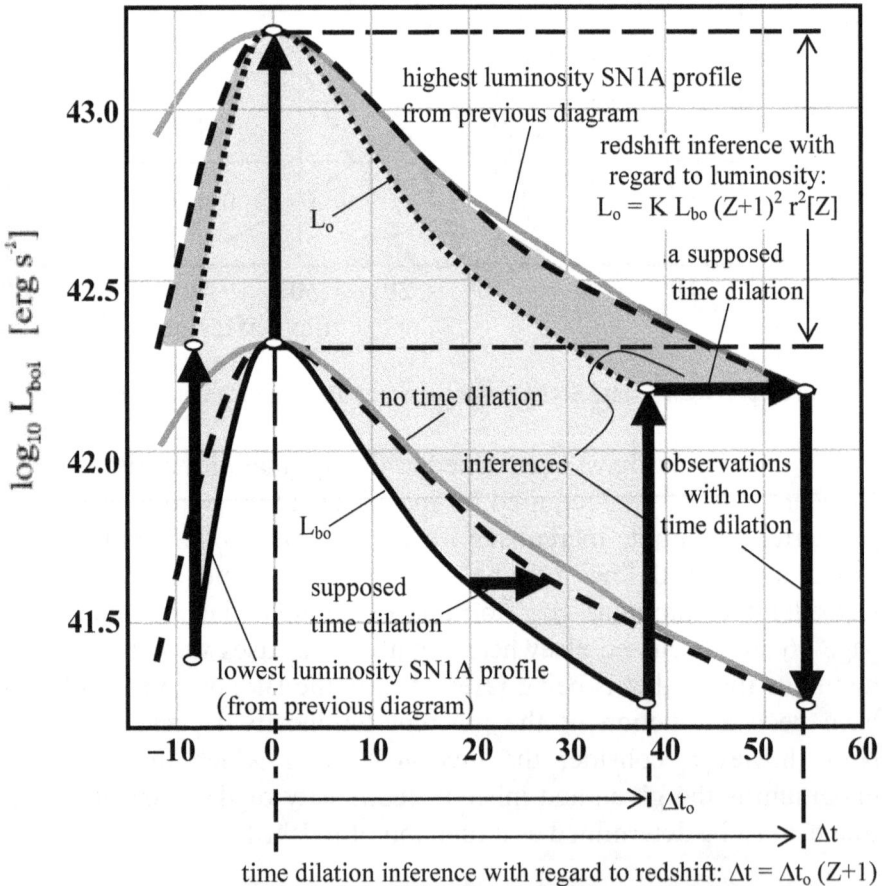

Figure 128: Confused inferences of time dilation in SN1A profiles

284

$\Delta t = \Delta t_o \, (Z+1)$

Here the Δt terms refer to the amounts of time after the peak luminosity is realized. We see that the profile is stretched to very much the same profile as the more luminous event (solid gray curve at top) with no redshift, nor therefore, time dilation. There is certainly little more than what could be considered 'individual differences' other than a slower rise to peak. At any rate, to argue that time dilation actually occurs in SN1A profiles would seem to require one to argue that more luminous SN1A events in the distant past had narrower profiles than they do in the current epoch. Otherwise we are not observing time dilation but merely the fact that more luminous SN1A events have broader profiles.

e. more recent claim of time dilation in SN1A data

Nonetheless, Blondin, et al. (2008) argue based on their accumulating archive of the day-by-day spectral changes in individual SN1A events. They find that spectral changes from one day to the next in SN1A emissions vary in uniform ways that seems to apply without regard to the peak luminosity of the individual supernova. Their initial analyses were done appropriate to the rest frames of the events. Figure 129 is the result of superimposing the spectra of 20 supernovae onto the same plot for four particular days-since-peak-luminosity.

According to their analyses cross spectral correlations of the spectra of individual SN1A against these archived day-to-day profiles provide an indication of how far along in the decay profile (in the rest frame) the measurement is made. They presume this intrinsic variation in spectra is more or less equivalent to a programmed (rest frame) clock time value that they also presume to have little if anything to do with the peak luminosity. It is this data – each datum obtained at a single point in time – that they claim to indicate how far along in some internal process the supernova has progressed. This they have plotted in what we have reproduced as figure 130. So far their data is limited to 13 high redshift samples, but if the day-to-day spectral profiles prove indeed to be independent of maximum brightness, then their analyses would seem to legitimately suggest the possibility of time dilation.

However, this conclusion does not seem likely at this point. There is a fairly high degree of similarity in the daily profiles of figure 129, giving rise to the large uncertainties in figure 130; that is an issue that their error analyses must surely have handled. But upon inspection of the magnitude data in the different bands presented by Ruiz-LaPuente (2004) presented in figure 126 and other similar profiles, one is left to wonder how data illustrated in figure 129 can possibly result. These two data sets seem entirely incompatible. Inspecting the profiles for the color bands in figure 126 on a day-to-day basis, it is difficult to rectify that with the composite profiles of Blondin et al.. If all the bands shown by Ruiz-LaPuente (2004) are weighted equally, there is an interval between about day 10 and 18 for which the relationships might hold, but quite clearly outside that interval it would not.

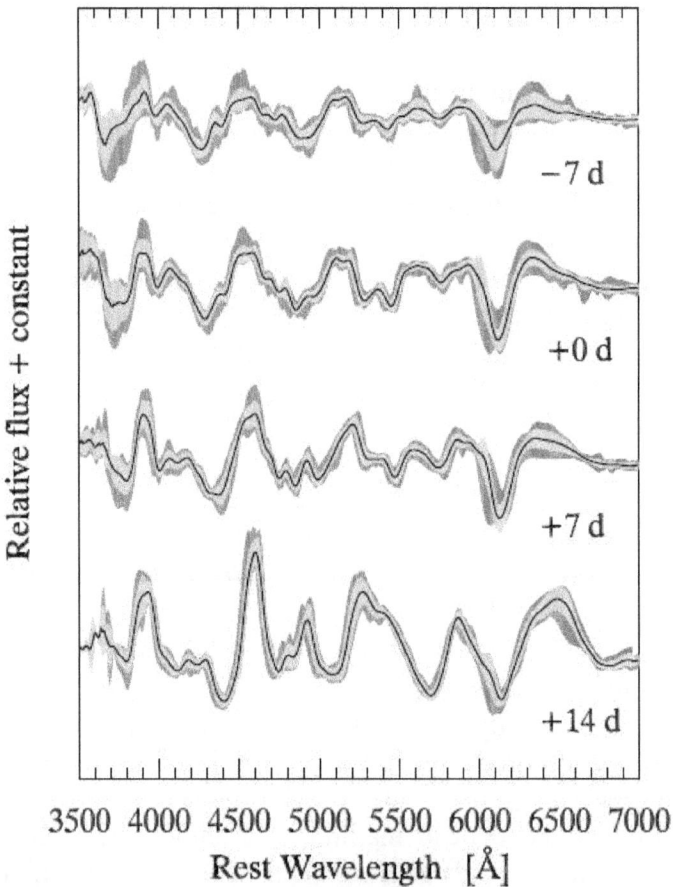

Figure 129: Day-by-day spectra of 13 SN1A supernova. The standard and maximum deviations are shaded light and dark respectively with the mean indicated by the dark line. [from Blondin et al. (2008)]

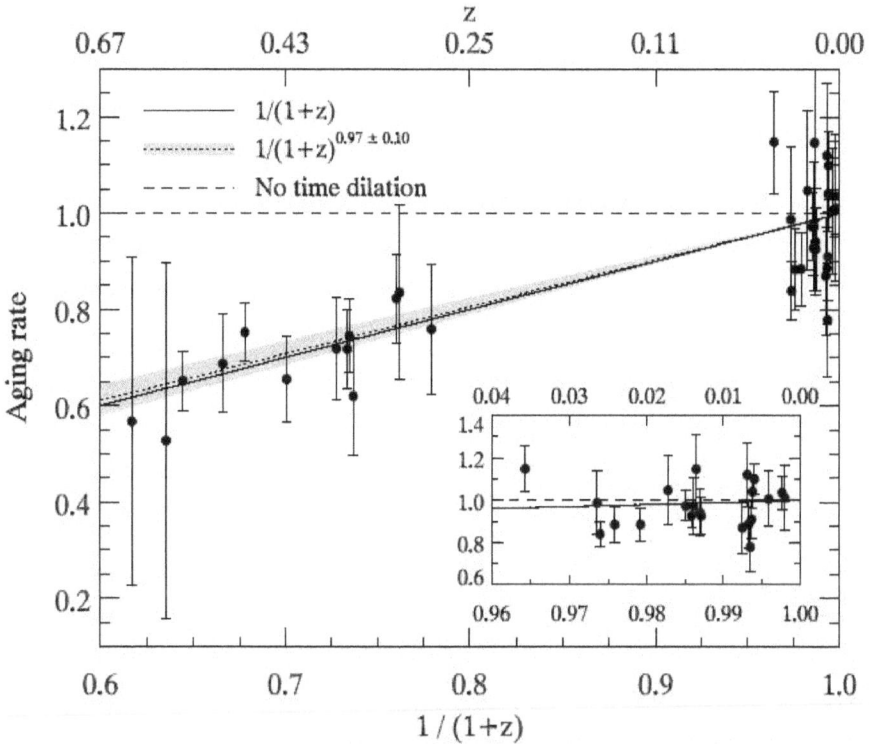

Figure 130: SN1A aging data interpreted as time dilation [Blondin et al. (2008)]

More significantly, it is difficult to rectify obvious luminosity selection inherent in the inevitable Malmquist bias with the longer decay periods *not* having a major impact on the comparison of decay rate, which they do not analyze. Whatever the internal workings of an SN1A event, the processes must necessarily proceed on a time scale determined by the involved energetics. One must certainly suspect that the (however few) slowly aging profiles at low redshift derive from the more luminous objects. At high redshift which involves roughly a cubed spatial volume metric out to the redshift with which they concern themselves, there would be many more of these (and even brighter) types of profiles and virtually none of the lower luminosity types. Inspection of the potential for a Malmquist bias mimicking the effect of time dilation in figure 128 suggests that it is that functionality that has established itself in their data. One would like to see this comparison made and discussed more quantitatively rather than summarily dismissed as they have done.

Chapter 15

Comoving Number Densities of Galaxies

Observed *comoving number densities* of various populations of galaxy types, quasars, and other phenomena provide essential data from which to assess validity of cosmological models and model-dependent metrics. This data helps to discriminate between models and ultimately may determine which of the various cosmological models is correct.

a. cautions in predicting comoving number densities

In a homogeneous flat Euclidean universe the actual population of objects of a given class within any solid angle Ω of observation would increase as the cube of the distance, in particular as:

$$N = N_0 \, \Omega \, r^3 \, / \, 3$$

where N is the number of galaxies within the solid angle (measured in steradians) out to a distance r, and N_0 is the number density of such galaxies per unit volume of space. In the vicinity of our Milky Way galaxy N_0 is on the order of 5 x 10^{-20} per cubic light-year ~ 1.73 Mpc^{-3}. If this local density were to be applied throughout the extent of a spherical volume in Euclidean space of the generally accepted 'radius of the universe', or Hubble distance, the universe would contain an astronomical number – certainly many trillions of galaxies. Well...

But what constitutes the immediate vicinity of the Milky Way to which this density is known to apply? Related to this is a second question: *Why is that an issue?*

As we saw earlier, redshift is not considered a reliable indicator for distances that are less than hundreds of millions of light years, and are not all that reliably calibrated beyond that. One major reason is that random motions of galaxies relative to that of the Milky Way produce red (as well as blue) shifts that dwarf any cosmological effect.

The scattering model in particular, involves related issues of the logarithmic form of a distance-redshift relation not being established until at least thousands of extinction intervals have occurred. That process requires on the order of 10^{26} cm for visible light propagation, a distance that coincides with the observed reliable correlation. At any rate, it makes little sense to refer to "the redshift of Andromeda" other than to indicate its specific velocity relative to the Milky Way.

But it makes sense to characterize distance by association with 'cosmological' redshift for convenience rather than changing units mid-stream. This requires addressing several additional issues. Consider,

$$r(z) = H_0^{-1} \ln(z + 1)$$

that tends to zero as z goes to zero, but in a rather non-intuitive way. First, $H_0^{-1} \cong 1.4$ x 10^{28} cm is a huge multiplier. For another, for a redshift much less than unity, we have Hubble's law plain and simple.

$$r(z) \rightarrow H_0^{-1} z.$$

'Immediate vicinity' comes to mean $z_{local} \leq$ ~ 0.01 where $\ln(z + 1) \cong z$ to a high degree of accuracy. For such purposes Andromeda is in the immediate vicinity of the Milky Way since the distance is on the order of two million light years for which cosmological redshift, z < 0.00015.

In what follows we will use redshift in our discussion of galaxy count even in the immediate vicinity of the Milky Way. It should not be misconstrued to imply that the author believes cosmological redshift to be pertinent in assessing such small distances, but only as a means of describing all distances self-consistently.

The differential number count, i. e., the number within some unit volume in the interval at r in the simplistic Euclidean case is:

$$dN(r) = N_o \, d\Omega \, r^2 \, dr$$

But when one extends this to take into account the functionality of distance with its only measurable concomitant we obtain:

$$dN(z) = N_o \, dV_C(z)$$

where $dV_C(z)$ is the 'comoving volume element' whose functional dependence on redshift was derived earlier for various of the standard models and also uniquely for the scattering model. It was illustrated for the various models in figure 114. In figure 115 the integral of $dV_C(z)$ with respect to redshift was illustrated with curves for the total volume encompassed out to a given redshift. The two panels in figure 115 show the same functional dependence, but with panel b providing a clearer view of the behavior for redshifts below $z = 2$.

Astronomical instruments are limited with regard to resolution (currently to on the order of an arc second or so), apparent luminosity (to about magnitude 27 in the optical range for the best land-based telescopes and 30 for the Hubble space telescope). Therefore, as one nears these limitations of the employed instrumentation, the observed galaxy population counts must necessarily diminish more and more significantly as numerous small, faint, or otherwise-obscured, galaxies defy detection.

A specification of the instrument capabilities is essential to accurate prediction of the number of objects one should expect to observe in looking back to various distances/redshifts. In Euclidean three-space without redshifting, an instrument would be capable of detecting luminous flux f for all galaxies with an inherent luminosity of L_o for which $f \geq L_o / (4 \pi \, r^2)$. However, with redshift this becomes:

$$f \geq L_o / (4 \pi \, (z+1)^2 \, r(z)^2)$$

where one of the factors of z+1 is only coincidentally the same for the standard and scattering models with unique rationales. This has already been explained as dependent on the functional form of absorption in a redshifting plasma. That the inferred distance r(z) as a function of redshift differs for each of the models has also been explained.

The distance r(z) at which a given instrument will no longer be capable of detecting a galaxy of inherent luminosity L_o is a function of redshift according to any of the models. The following formula gives its limit value derived directly from the previous inequality:

$$r_{Lo} = (z+1)^{-1} \sqrt{L_o / 4 \pi f}$$

If, for example, all galaxies had an identical luminosity L_o, then no galaxies would be observed for which $r(z) > r_{Lo}$. So, suppose that an instrument's flux detection capability, f were geared to just be capable of detecting galaxies of inherent luminosity 0.5, 1.0, 1.5, 2.0 or 2.5 times L_o within a redshift of 2 of the Milky Way. Then the density of galaxies observed through the instrument would drop to zero at the redshifts corresponding to the crossover of the breakpoint curves labeled Lf in figure 131 according to the built-in capabilities of the instrument. Breakpoint curves are superimposed on the total encompassed volume curves from figure 115 panel b.

Thus the slope of any number-of-galaxies curve obtained from a deep space galaxy survey will decrease continually as fainter and fainter galaxies can no longer be seen with the given instrument beyond the limiting redshift. The predicted curves according to the most usual density parameter versions of the standard model as well as that for the scattering model are shown in figure 132. Clearly, they are indistinguishable at low redshift. The observed phenomena is illustrated in figure 133 taken from Brown et al. (2001) where it is manifestly clear that galaxies for which $L < L_o$ cannot be seen. In this figure the ordinate value for the constraint curve is galaxy magnitude beneath which no observation is possible.

However, even without there being a drop in observed galaxy counts due to inherent limitations of instruments, there would still be a predicted flattening out of the differential galaxy count implicit in the flattening of the comoving differential volume as a function of redshift for all models that was shown in figure 114. This does not *ipso facto* imply an associated change in *spatial* differential volume. It is inferred rather from cosmographic metrics based on redshift resulting from the distance-to-redshift relationship – at least in the scattering model that is the entire significance of the difference.

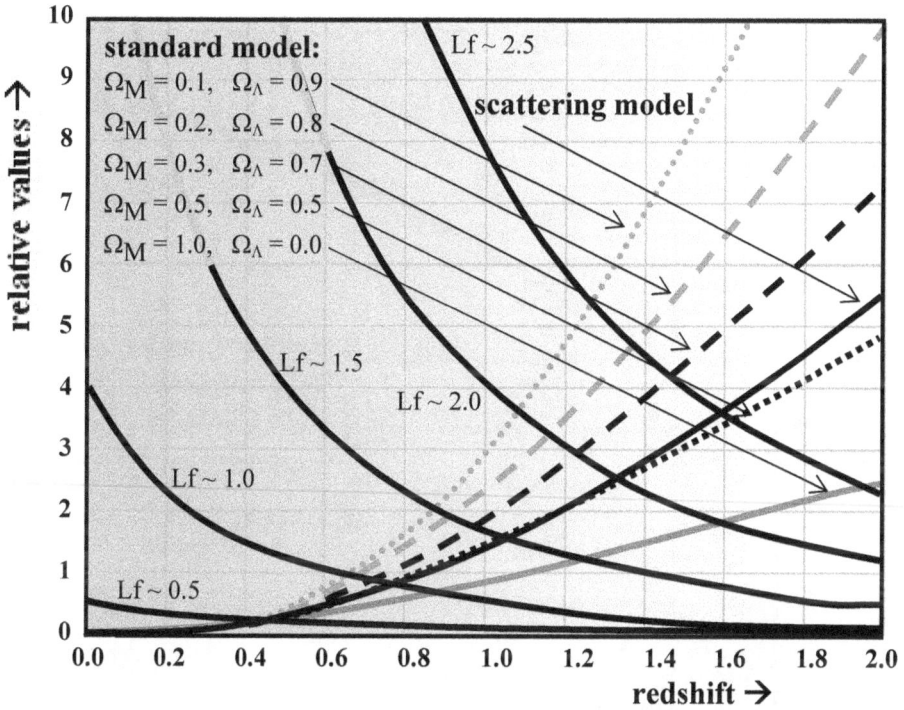

Figure 131: Breakpoints labeled (Lf) of number of galaxies of a given inherent luminosity that will be observed with a given instrument

The non-Euclidean differential volume will significantly alter the form of any predicted comoving number density as a function of redshift. In any case, when the volume differential is taken into account, even in a flat spacetime Euclidean universe uniformly filled with galaxies, the differential count with respect to redshift will level off. For standard models it will actually begin to diminish. There is, of course, a strong correlation between redshift and magnitude of observed galaxies such that if we plot the differential volume versus the luminosity modulus under the simplifying assumption that all galaxies have the same luminosity, we obtain the plots illustrated for the various models in figure 134.

Data taken from Kochanek et al (2001) is provided in figure 135 that illustrates a somewhat similar pattern in observed data. This pattern is similar to the scattering model prediction except that there are clearly more faint galaxies than the simplifying assumption allows. We will clarify that further on taking into account the Schecter function.

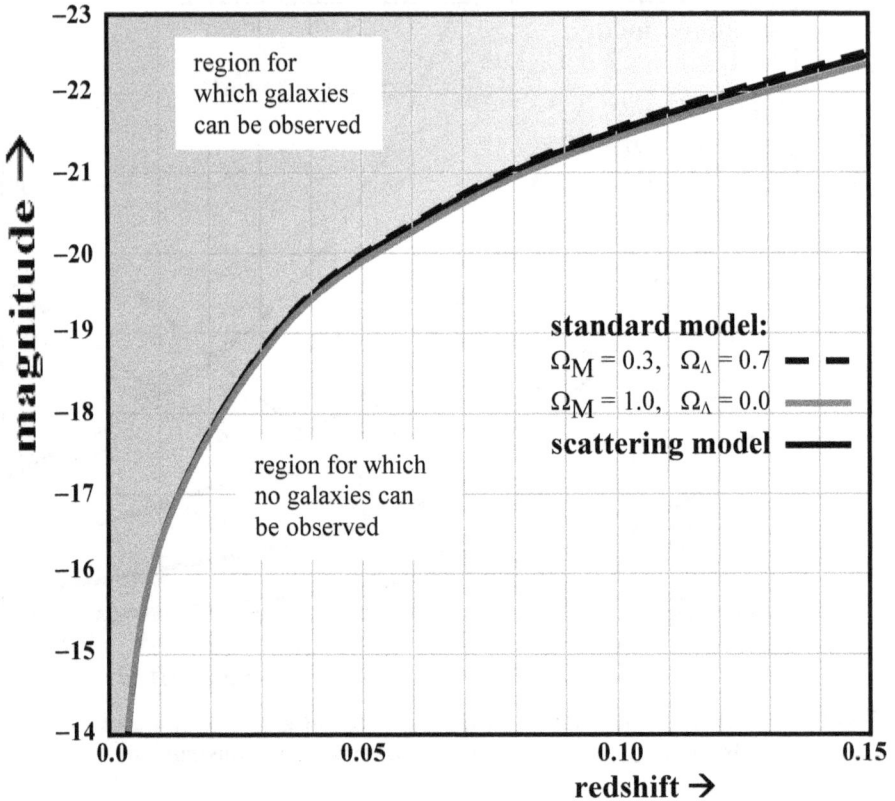

Figure 132: Predicted constraint from the various models

The predicted total number count of galaxies that should be observed in a given survey depends intimately on the model, including the associated luminosity and differential volume functionality that is appropriate to that model. Even the essentially Euclidean scattering model defies the simplified Euclidean three halves power relationship with distance because of the conflated relationship of luminosity and differential volume with factors involving redshift. So a naively considered uniform distribution of galaxies in a flat Euclidean spacetime universe, all with the same absolute luminosity, would meet

294

with frustration. We will discuss the Schecter luminosity function that characterizes the distribution of numbers of galaxies as a function of absolute magnitude in the next section. That distribution must certainly be taken into account to obtain accurate predictions.

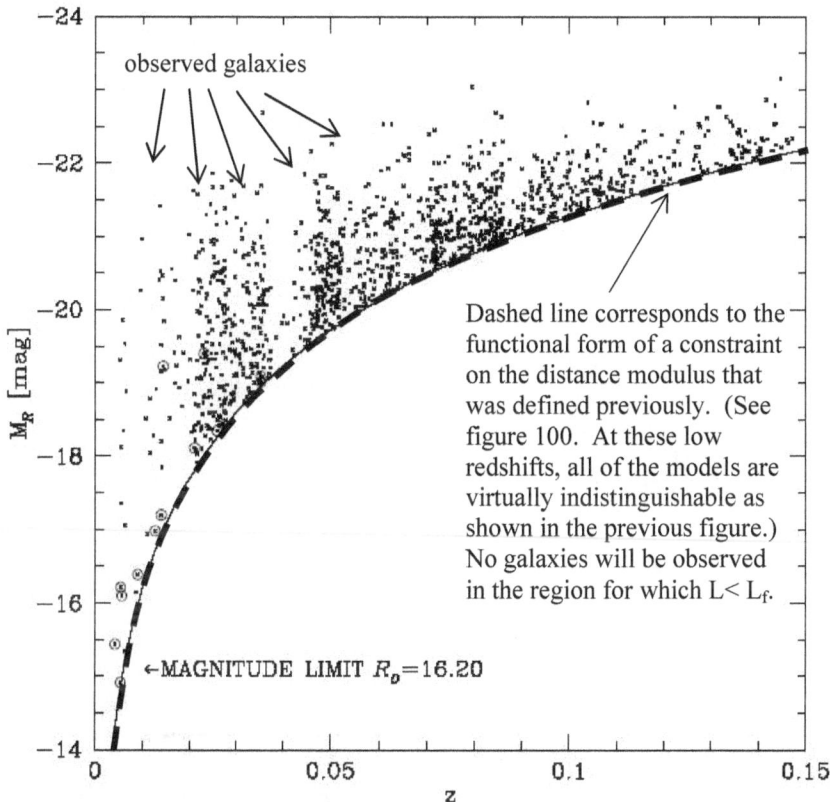

Figure 133: Distribution of absolute (inherent galaxy luminosity) with redshift. The line shows the limiting absolute magnitude resolving power vs. redshift. (Circles are low surface brightness galaxies.) [Taken from Brown et al (2001).]

If one just takes into account those galaxies that appear within a given slice of an angular window of the sky, one obtains a pattern like that shown in figure 135. However, before attempting to get an accurate predicted functional dependence applicable to such galaxy surveys, we must take into account the luminosity characteristics of the galaxies being observed. Although this issue is debated, the scarcity of galaxies at the outer edge of any survey seems certainly to have been caused more from faint galaxies escaping detection as discussed above than from an actual dearth in the galaxy population at extreme distances. In figure 136 taken from Drory, et al. (2001) the number of observed faint galaxies clearly drops off rapidly as redshift increases.

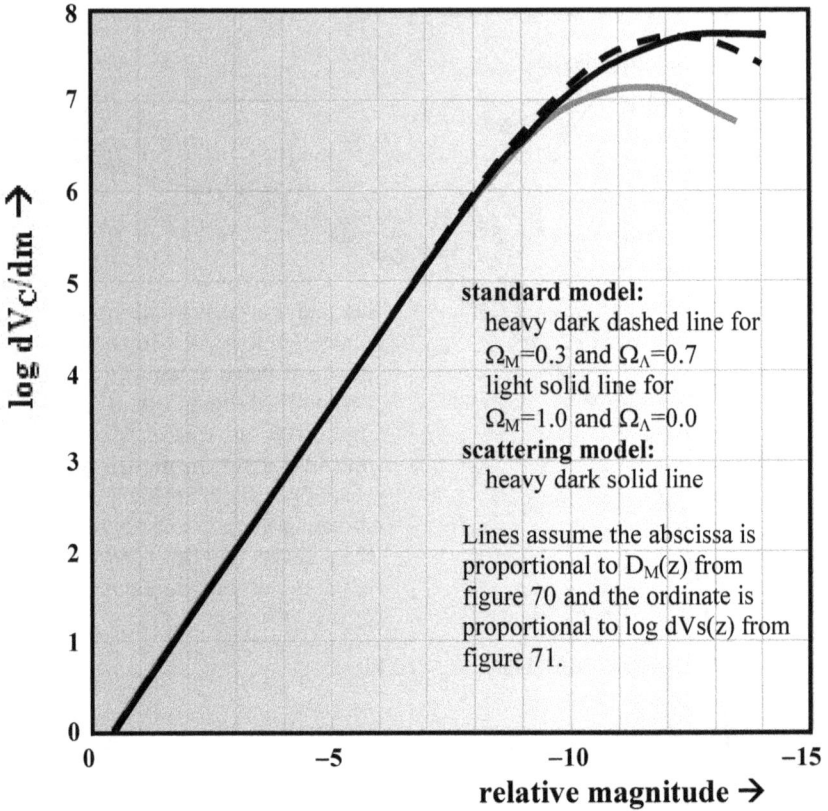

Figure 134: Simplified predictions of comoving density

This observation obviously results because faint galaxies at large distances increasingly escape detection. That the shape of the Schecter luminosity function remains essentially unchanged for the inherently brighter galaxies seems obvious in the figures as well. If there were some fundamental difference in galaxy makeup in the past, why would only the faint end of the spectrum be affected? The unobserved galaxies in turn affect the scarcity noted in surveys at extreme redshifts like that which appears in figure 135 and in the survey made by Brown et al. (2001), with regard to which, refer back to figure 133.

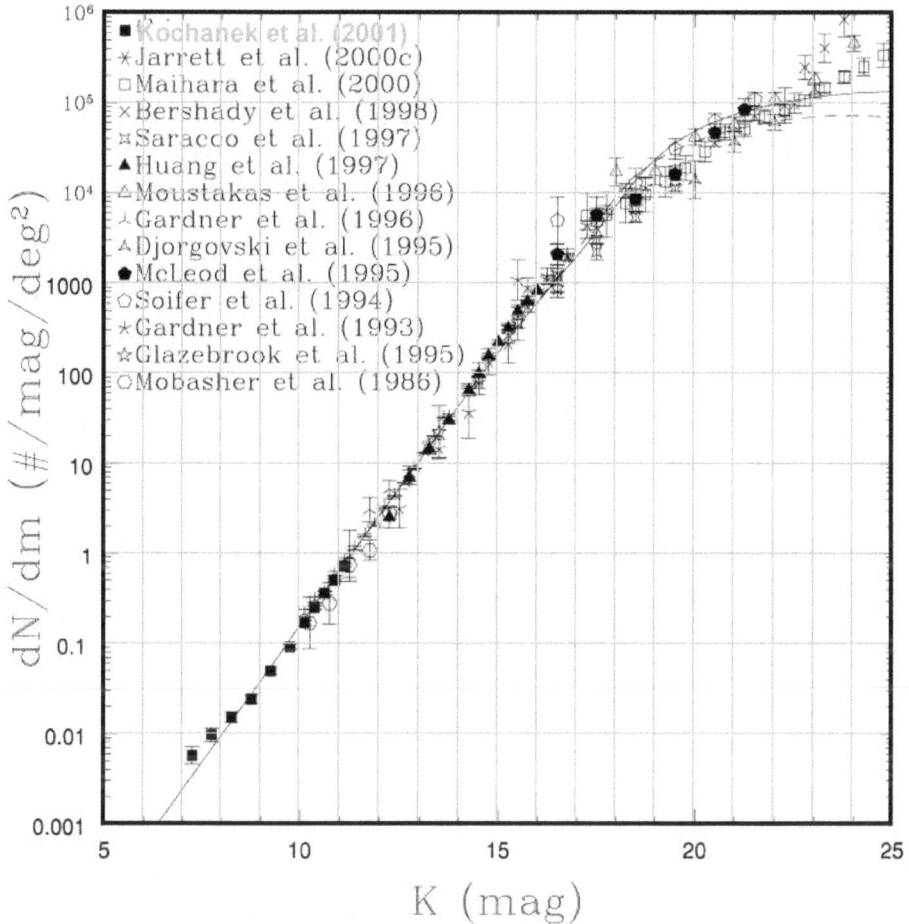

Figure 135: Surveys indicating the dramatic increases in numbers of observed galaxies as one increases the resolving power of instrumentation [Taken from Kochanek et al. (2001).]

b. the Schecter distribution of galaxies

There is a tremendous range of inherent galaxy luminosities; giant ellipticals are on the order of 10^7 times more luminous than dwarf galaxies, for example. There is a similar extreme range in frequency of occurrence of each of the various galaxy types by luminosity. Even within classes of galaxies such as spiral and elliptical galaxies one sees a similar distribution favoring faint galaxies. So to accurately survey galaxies at various distances, as indicated by redshift, in addition to instrumentation limitations that affect the apparent form of this distribution, the number count must emphasize galaxy types, all of whose luminosities are sufficient to be observed at the limit of a survey. A 'galaxy luminosity' function that supports such endeavors is shown in figure 137. Its functional form indicates that there are fewer extremely luminous galaxies to be expected than faint ones. 'Brightest galaxy' in a cluster is a very select class – shown as being randomly distributed at the bright end of the galaxy luminosity spectrum.

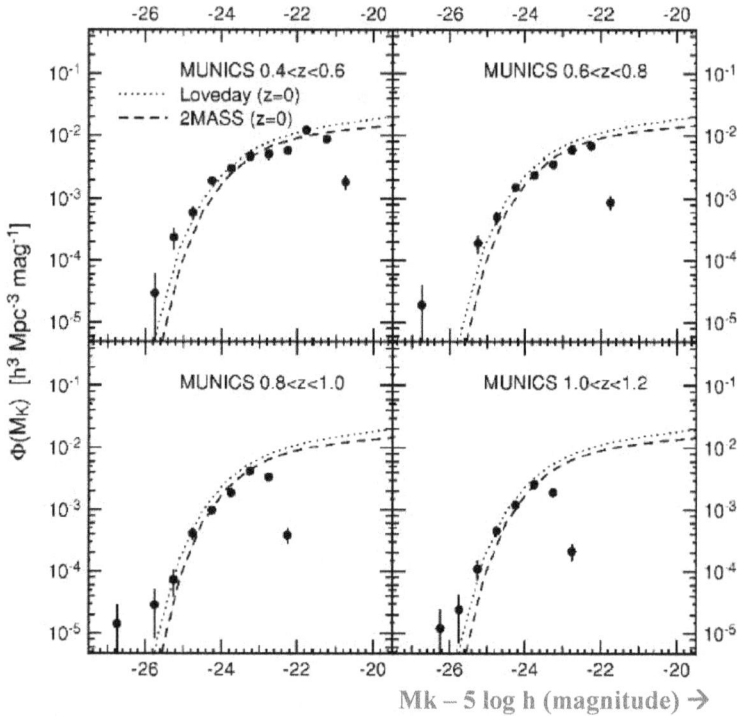

Figure 136:   Rest-frame K-band luminosity function in four redshift bins spanning 0.4 < z < 1.2. The dotted and the dashed curves are the z = 0 LFs by Loveday (2000) and Kochanek et al. (2001), respectively. [MUNICS data from Drory et. al. (2001)]

This generic galaxy luminosity function is readily demonstrated as an approximate match to actual data provided in complete surveys of galaxies. See for example, the data provided by Loveday (2000), Kochanek et al. (2001), and Blanton et al. (2001) in figure 138. The form of this data in all such complete surveys is invariably the same and its form is characterized by the empirical formula known as the Schecter galaxy luminosity function whose given by:

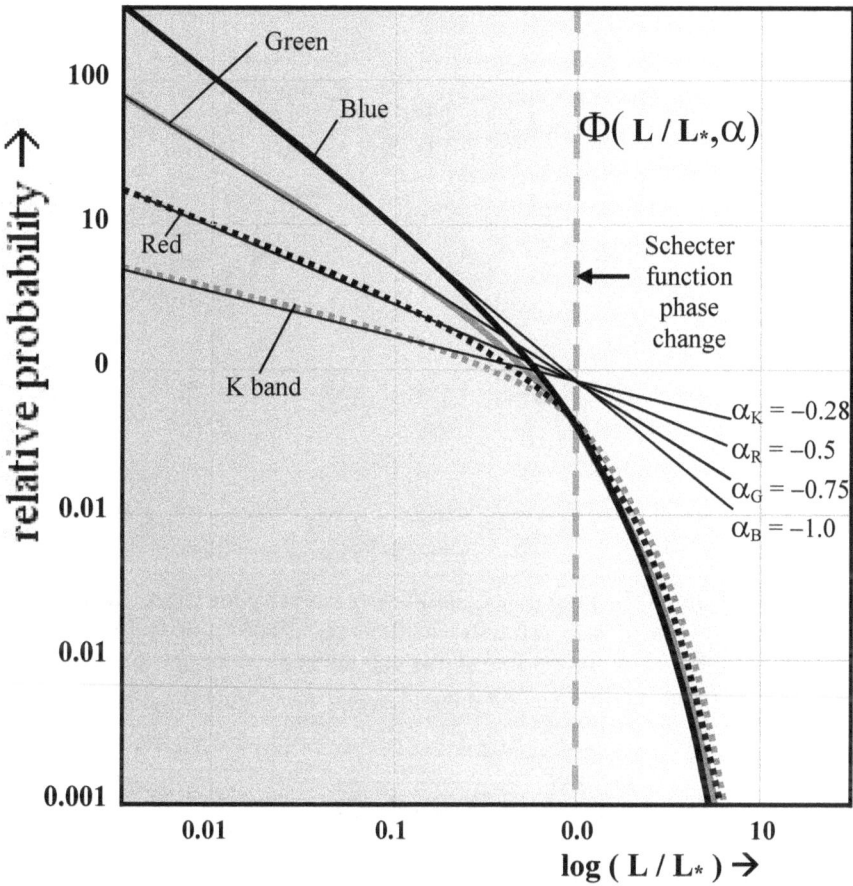

Figure 137: Schecter luminosity distribution function and its color dependence

$$\Phi(L)\,dL = N\,(L/L_*)^\alpha\,e^{-(L/L_*)}\,dL/L_*$$

where $\Phi(L)$ is the relative number of galaxies of luminosity L in the interval dL. All the parameters α, L_*, and N depend on the band of the electromagnetic spectrum being used to make the observations.

N *characterizes* the density of galaxies per unit volume of space for normalization purposes. It is a nominal value whose purpose is to obtain the best fit to the function. L_* characterizes a nominally bright galaxy luminosity on the order of say the brightness of the Milky Way galaxy relative to which a dwarf galaxy would be but a tiny fraction. $L_* = 5.46 \times 10^{43}$ ergs sec^{-1}, for example.

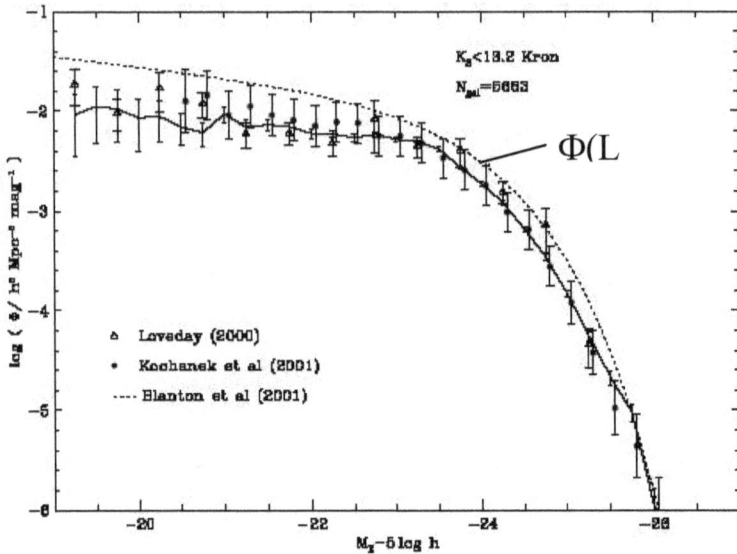

Figure 138: Observed K-band galaxy number-luminosity function. Symbols and error bars show estimates of Loveday (2000) and Kochanek et al. (2001). The dotted line shows the Schecter function estimate of the K-band luminosity function. [Taken from Cole et al. (2001).]

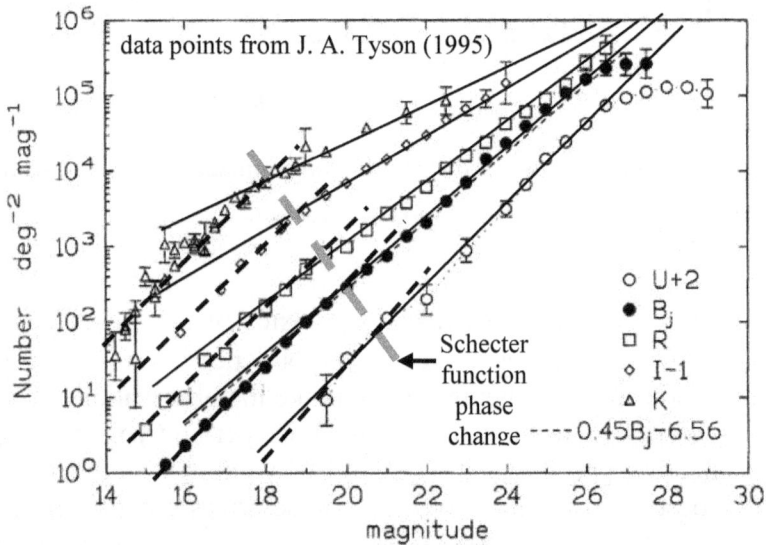

Figure 139: Galaxy differential number counts dN(m) from a number of surveys are plotted to completeness limits in five color bands. (Notice trend lines that have been added by the current author to the original data.)

The negative slope of the nearly linear phase of the curve at the faint end is α. This slope depends on the wavelength band (color) employed by the instruments in making observations. The probability of finding a galaxy of a given relative luminosity $\Phi(L/L_*,\alpha)$ in a given volume of space is plotted for the different wavelength bands. The slopes of these linear portion of the curve for the various bands are given by: $\alpha_{blue} = -1.0$, $\alpha_{green} = -0.75$, $\alpha_{red} = -0.5$, and $\alpha_{k\ band} = -0.28$. Thus, more sensitive instrumentation would naturally reveal increasing percentage of 'blue' galaxies – the slope being much steeper for blue galaxies than for longer wavelengths. A different spectral mix is observed as fainter and fainter galaxy counts are obtained. This is illustrated in figure 139 where the lines superimposed on the data can be obtained by integrating the two respective phases of the Schecter distribution function over a range of redshifts that we will investigate in more detail in a later chapter. Needless to say, this conflates disputes with regard to the issue of color evolution of galaxies.

Redshift dependence is also sometimes assumed to apply to the galaxy luminosity function as well. But here, as in many other cases, there tends to be a certain amount of ambiguity in claims suggesting evolutionary effects. Of course, this will be also conflated by different predicted luminosity diminution effects for different cosmological models, etc.. We will discuss in details further on. Suffice it to say that all of the various models make somewhat different predictions in this regard.

In 1997 it was found (by Lowenthal, et al.) that the comoving number densities at high redshift are 34 times higher than a just previous estimate by Steidel, et al. (1996). The revised estimates were based on ground-based photometry at the W. M. Keck Observatory with slightly fainter limits and the Hubble telescope *deep field*. These estimates place the number density at a level comparable with more accurate estimates of the local volume density of galaxies brighter than a specified luminosity. Besides this fact, one must note that their estimate is still only a lower limit. They indicated that numbers might even be as much as 3 times higher than what they had found. These galaxies appeared to be small with most of their light usually contained within a 1 arc second disc but were still quite luminous.

Nonetheless, Kochanek et al. (2001) claim redshift dependence, where they inspected the differences shown in the panels of figure 140, for example. But the superimposed identical dotted Schecter function curves that have been added by the current author to both panels would seem to account equally well for the two data sets with the last count (faintest reliable observation) in each case dropping down considerably as happens in redshift surveys like that shown in figure 141. So the claim seems questionable to this author. In any case much of the discussion of galaxy number count predictions involves the issue of whether there is 'evolution' of the Schecter function with redshift.

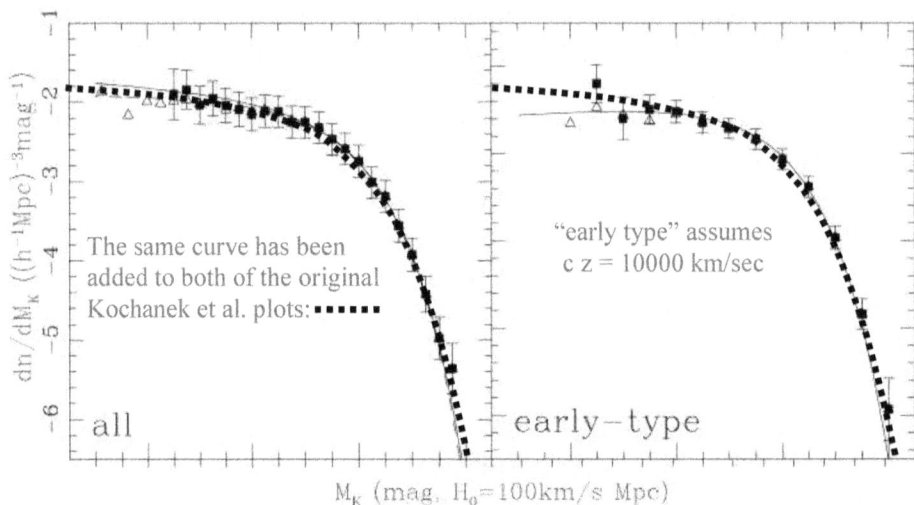

Figure 140: Rest-frame K-band luminosity function – current and redshifted

If we consider the distribution of galaxies at a given redshift *as observed* in the rest frame of the observer prior to adjustments necessary to render what is shown in figure 137, the distribution would be quite different. It would slide considerably to the left due to a luminosity diminution factor of,

$$1 / (4 \pi (z+1)^2 r(z)^2)$$

Here, $r(z) = H_0^{-1} \ln (z+1)$ for the scattering model with more complicated but similar expressions applying to the various versions of the standard model. We have discussed the rationale for luminosity dependence for the scattering model in chapter 4.

We will defer more detailed discussion of the impact of redshift on apparent galaxy distributions, but will come back to it with regard to

302

discussions of actual data from galaxy redshift surveys in the next chapter.

c. estimating galaxy counts versus redshift

In evaluating the number of galaxies to be expected in a given survey, of course, the flux detection capability of instrumentation must be taken into account as discussed above. The Schecter luminosity function could then be used directly to estimate the number of galaxies that would exist within the scope of observation in a uniformly filled Euclidean space. However, in a universe that experiences cosmological redshift one must evaluate the luminosity flux that would be observed uniquely at each redshift and integrate that altered Schecter function using the redshift-dependent differential 'comoving' volume. In figure 142 luminosity distributions are given for galaxy types in the 2df redshift survey. (Notice that magnitude signs are reversed relative to the Schecter diagrams and in figures presented earlier. Thus the abscissa will appear reversed from Schecter function presentations.)

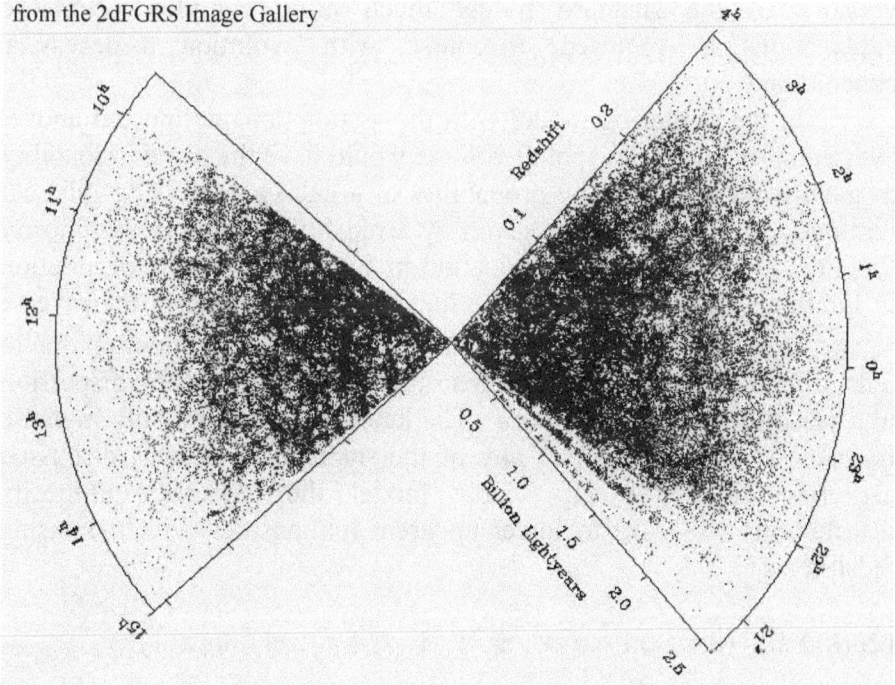

from the 2dFGRS Image Gallery

Figure 141: 2dF galaxy redshift survey (2dFGRS) data

The curves in figure 137 are at least reminiscent of similar data that are unique to each of the galaxy types cited by Colless (1999) earlier. The plots in figure 142 that show a decline in counts at the faint end of the distributions do not explicitly involve redshifting effects, but perhaps there is some category selection involved in what shows up both here and in figure 140 as well.

d. assessing luminosity effects on observations

Since the scattering model embraces Euclidean geometry, albeit as a logarithmic function of redshift, the assumption of uniformly distributed galaxies throughout space (as against redshift) seems reasonable. Under these assumptions inherent galaxy luminosities at all redshifts will be represented by the very same (rest frame) Schecter luminosity function described in the previous section. Nonetheless, even with this assumption, *apparent* luminosities and galaxy counts as functions of redshift will diminish for the scattering model in a somewhat similar way to what would be expected of all of the various versions of standard model as we have already discussed. For these versions of the standard model much more complex geometric explanations are required, of course, with 'evolution' a necessary concomitant.

In the scattering model with the assumptions delineated above, every unit of Euclidean spatial volume would have the same probability of galaxy occupancy. The probability of a galaxy possessing inherent luminosity L would then be given by a redshifted Schecter luminosity function. That involves modification in accordance with a diminution in luminous flux of the reference luminosity, $L_*(z)$ due to the inverse square of distance to each galaxy, the quantum energy reduction due to redshifting of wavelengths, and absorption losses implied by dispersion in a redshifting plasma medium. The latter factor differs only in cause from the predicted effect of any of the standard models as discussed earlier. Thus, for any viable model the detected luminosity distributions will shift to lower apparent luminosities with increasing redshift as:

$$L_*(z) \rightarrow L_* / (4 \pi (z+1)^2 r(z)^2 \rightarrow H_o^2 L_* / (4 \pi (z+1)^2 \ln(z+1)^2)$$

Only the final expression is limited specifically to the scattering model.

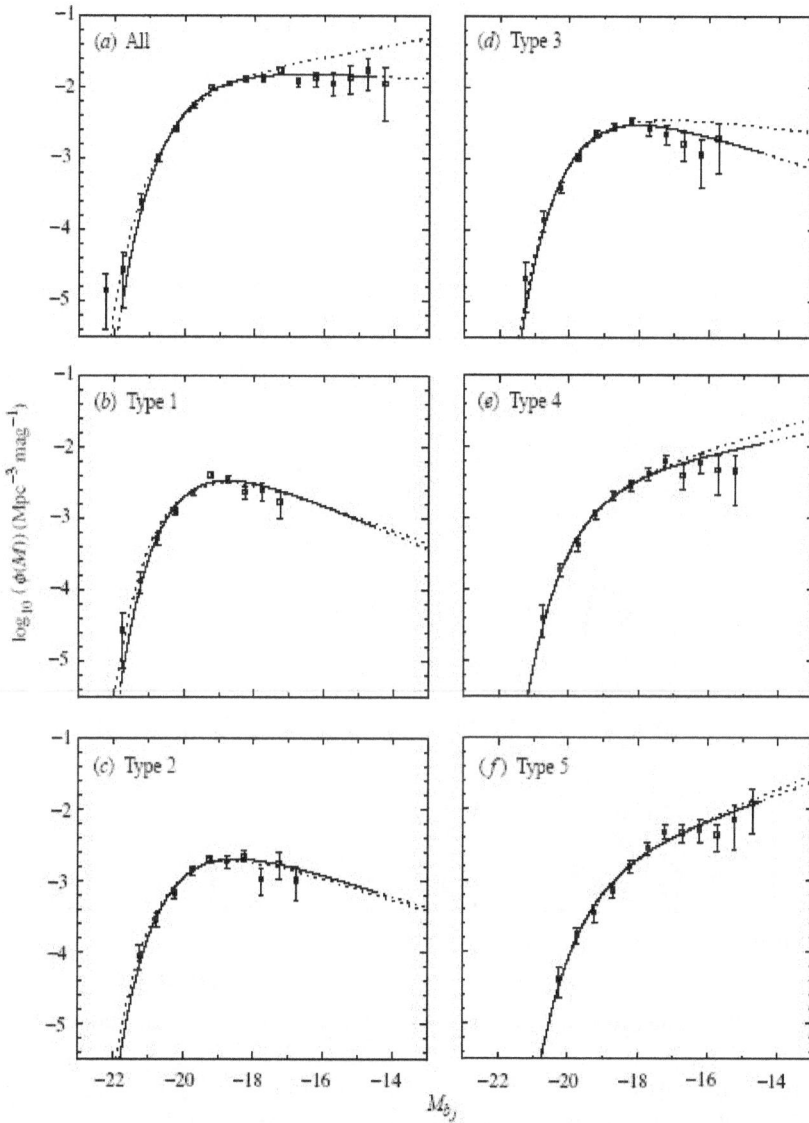

Figure 142: Luminosity functions of 3000 galaxies partitioned into the five identified 2dF types

The actual number of galaxies in the 'rest frame' at a given redshift will be unaffected by this shift except in regard to instrument visibility for an observer at z = 0. So 'evolution' is definitely not an issue (at least directly) with regard to the Schecter formula in this case. In figure 143 the expected galaxy count vs. luminosity distribution of the Schecter function is illustrated at a number of redshifts for $\alpha = 0.75$. For curves to the left of any visibility cutoff line, L_{inst} as shown, no galaxies could be observed at that redshift due to instrument limitations illustrated by the vertical line. The curves are embodied in the following formula:

$$\Phi(L, z) = N_0 \, (4\pi \, (z+1)^2 \, \ln(z+1)^2 / \, H_0^2 \, L_*)^{1-\alpha} \, L^{-\alpha} \, e^{-(4\pi \, (z+1)^2 \, \ln(z+1)^2 \cdot L/ \, H_0^2 L_*)}$$

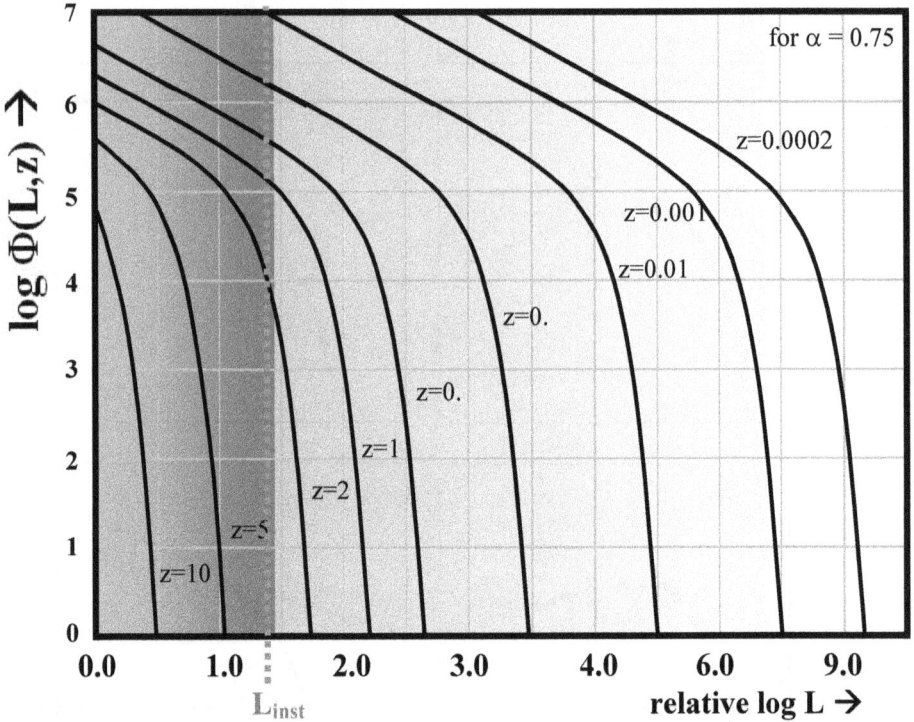

Figure 143: The effect of redshift on the galaxy luminosity function $\Phi(L,z)$

This relation defines a family of Schecter curves, one for each unique redshift. Clearly, dropping the vertical line at say $L = L_{inst}$ to represent a faintest galaxy that can be observed by an instrument of capability f_{inst} will be associated with a masking effect as depicted previously in figures 132 and 133.

The expected relative number of galaxies to be observed at each redshift for a given relative luminosity is obtained as:

$$N(L,z) \, d\Omega \, dz = \Phi(L, z) \, d\Omega \, \frac{\partial}{\partial z} V_C(z) \, dz$$

Refer to figure 144 for plots of this function using the scattering model formulation as well as for formulas applicable to a couple of typically promoted versions of the standard model.

In these plots it is again evident that the scattering model makes predictions that are well within the range of those predicted using realistic parameter values for the standard model. The plotted curves will be quite different for other values of α, but the sense in which the scattering model prediction is in the midst of those for the standard model remains the same. Of course observation of absolute luminosity slices would be difficult if not impossible to disentangle from any actual deep space survey. Nonetheless it should be obvious from the figure that although there is no presumption that the number of galaxies in each luminosity band diminishes with redshift, there will nonetheless be the perception of a significant decrease with increasing redshift as indicated by the plots in figures 143 and 144.

Let us consider the expected total numbers of galaxies in a restricted luminosity window at given redshifts according to the various models. To plot this we need to integrate the previous equation over the width of the window of luminosities. This has been done to obtain predictions for the several models. Results are shown in figure 145 for the formulation:

$$N_{ob}(L_{inst}, z) \, d\Omega \, dz = \int_{L_{inst}}^{L_{max}} \Phi(L, z) \, dL \, d\Omega \, \frac{\partial V_C(z)}{\partial z} dz$$

In the figure the integrals have each been taken over only four orders of luminosity magnitude.

However, if we look at the total number of galaxies of a given luminosity out to a given redshift, Z_{max}, we must intergate as indicated in the following formula:

$$dN(L, Z_{max}) \, d\Omega = \int_{Z_{min}}^{Z_{max}} \Phi(L, z) \, \frac{\partial V_C(z)}{\partial z} \, dz \, L \, d\Omega$$

The results of this equation are plotted in figure 146.

If we look now at various nonzero levels of L_{inst} restrictions, obtaining complete number counts for all greater luminosities, we obtain for each model a set of curves similar to those provided in figure 147 for the scattering model. Observations typifying differential galaxy count data are shown in figure 148 taken from Drory (2001).

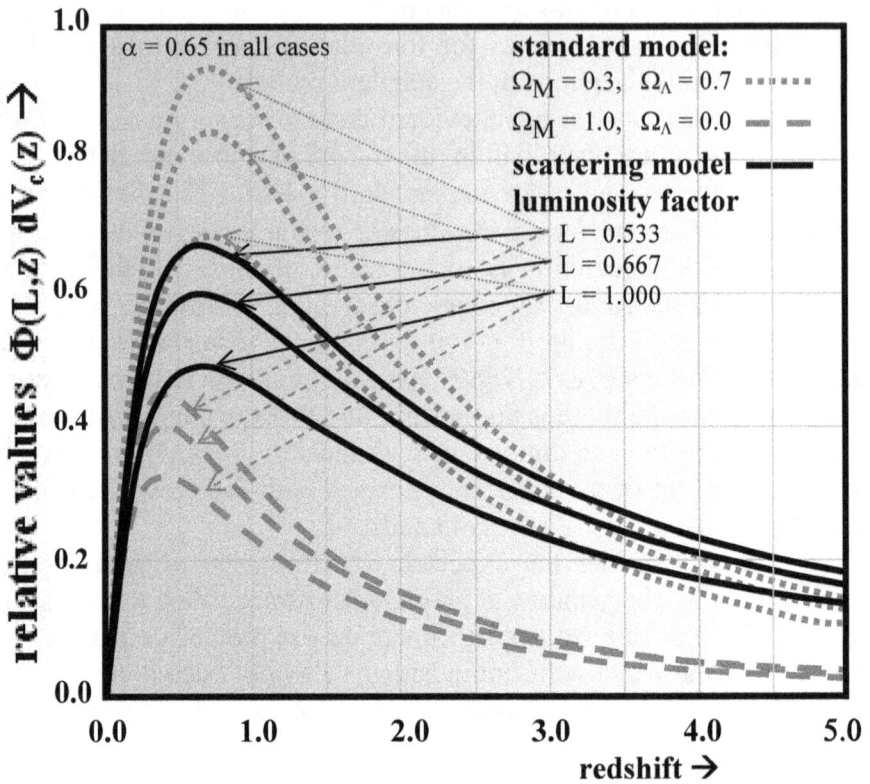

Figure 144: The differential number of galaxies to be encountered of a given inherent luminosity in a solid angle and redshift interval

In figure 149 the relationship between the photometric and spectroscopic redshift data presented in figure 148 is shown. Clearly there are uncertainty issues with the data that is presented in figure 148, but within those limitations, the resulting differential number count distributions of figure 146 and 147 seems reasonably well-established.

There is a fair degree of uncertainty in the assignments of the redshift in the diagram of figure 148 as shown in panels of figure 149.

The inability to observe more of the increasingly numerous fainter galaxies is a severe limitation on such tests of theory. Clearly, even for modest redshift surveys, the disparity of faint galaxies is substantial. It is clear that the scattering model predicts comoving number densities that fit available data as well as any of the standard models and that this model, therefore, warrants equal consideration and currently no model can claim confirmation based on number densities.

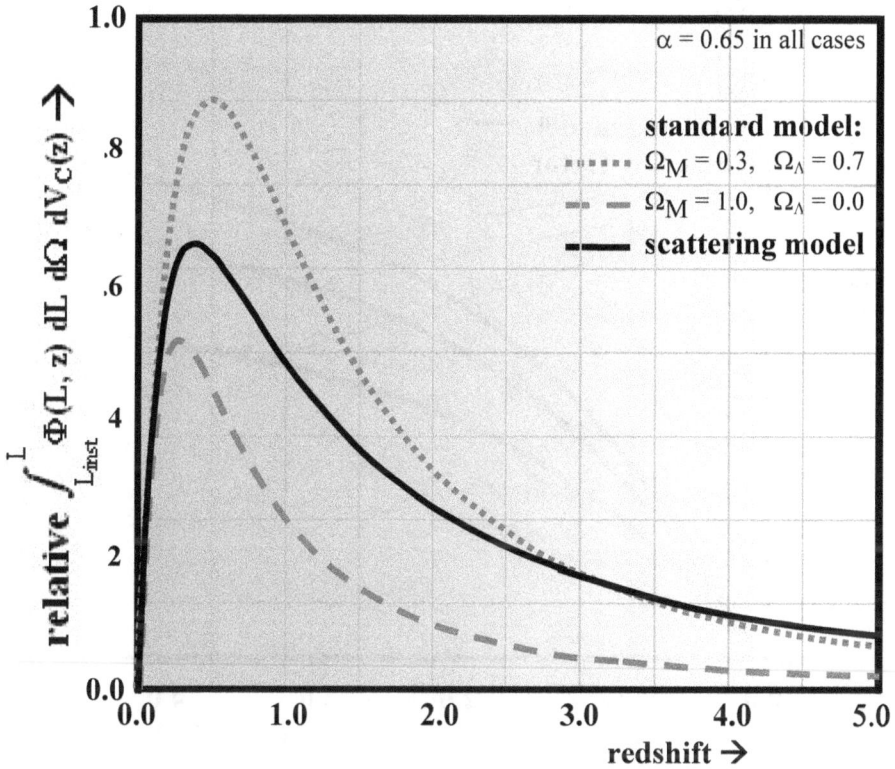

Figure 145: **The total galaxy count in a three order of magnitude interval of inherent luminosity as a function of redshift for various models**

Nonetheless we will address the appropriate formulation for possible later application. To calculate a predicted total number of observed galaxies out to a given redshift of all inherent luminosities limited only by L_{inst} (as further limited by redshift) within a solid angle, one must integrate over the comoving volume increment out to that redshift as follows:

$$N_{total}(L_{inst}, z) \, d\Omega \;=\; \int_{o}^{z} \int_{L_{inst}(z+1)^2 \ln(z+1)^2 / H_o^2}^{L_{max}} \Phi(L, \zeta) \, dL \frac{\partial}{\partial \zeta} V_C(\zeta) \, d\zeta \, d\Omega$$

The redshift increment limited total number of galaxies was plotted for the various models in figure 146 with curves differing predictably. The instrument luminosity limitation effect was shown for the scattering model in figure 147. Figure 148 showed the observations of analogous photometric and spectroscopic redshift data from Drory et al. (2001).

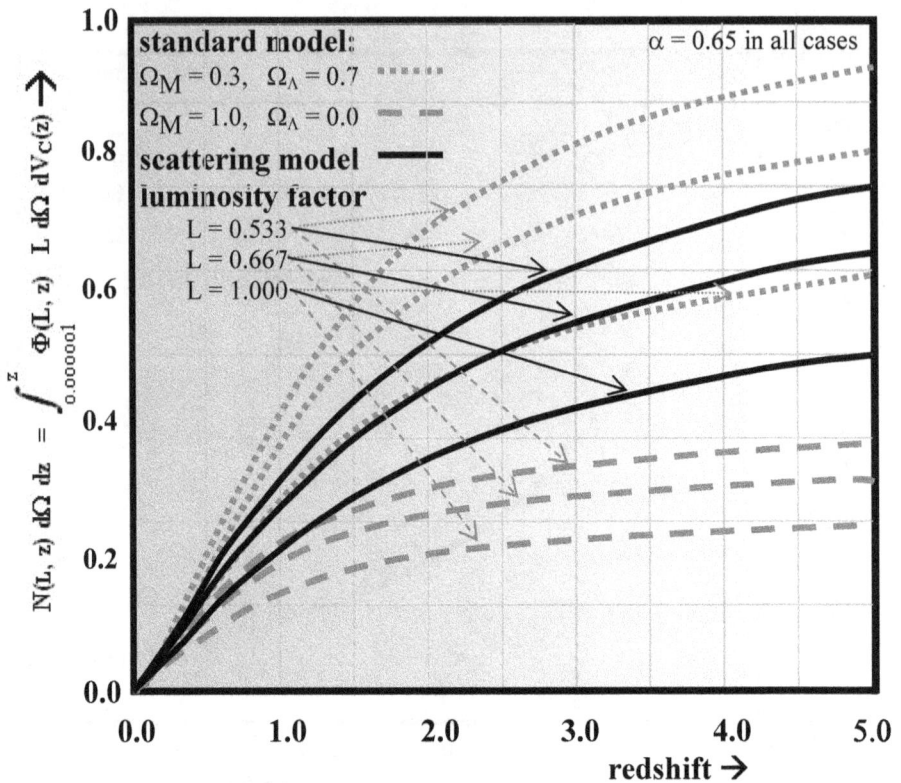

Figure 146: The differential galaxy count as a function of redshift for the standard and scattering models

After all this preparatory work with the Schecter function, we are now at a place where we can predict the form of the total number of galaxies to be observed by a given instrument as a function of observed magnitude which is the more usual observation. The previous equation is used to obtain the total numbers of galaxies in the various magnitude ranges. With these plots in figures 150 and 151, we begin to understand the single bend in such curves. This was shown earlier in figures 135 and 139. The expected total numbers of galaxies is plotted for the three principle models . The behavior is seen also in the data provided in figure 152 from Ellis (1997).

e. wrap up on comoving number density analyses

It is clear that comoving number density predictions of the scattering model are about midway between those of the two standard

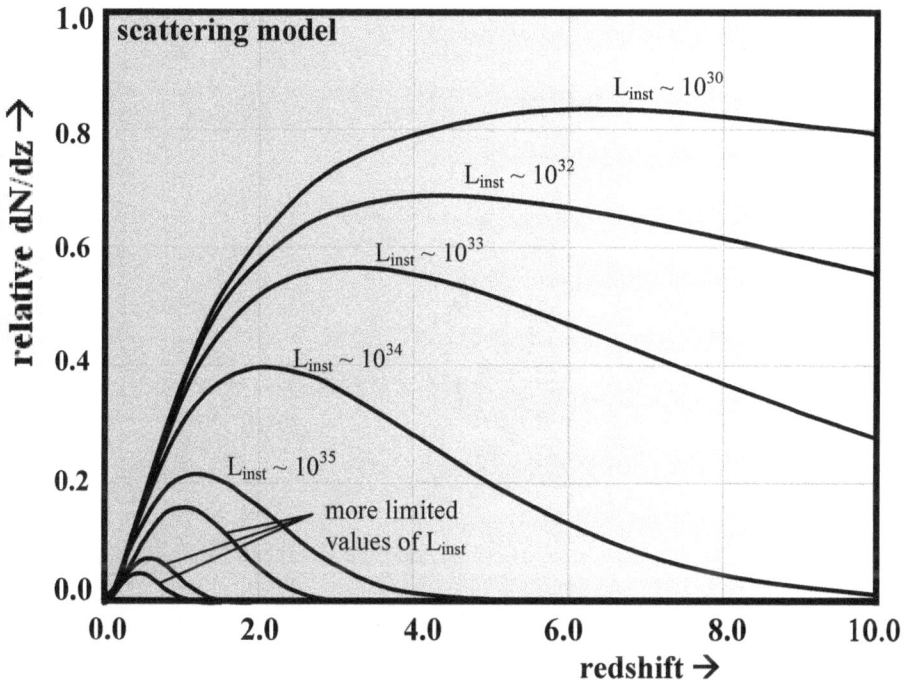

Figure 147: The differential galaxy count for various instrument capabilities as predicted by the scattering model

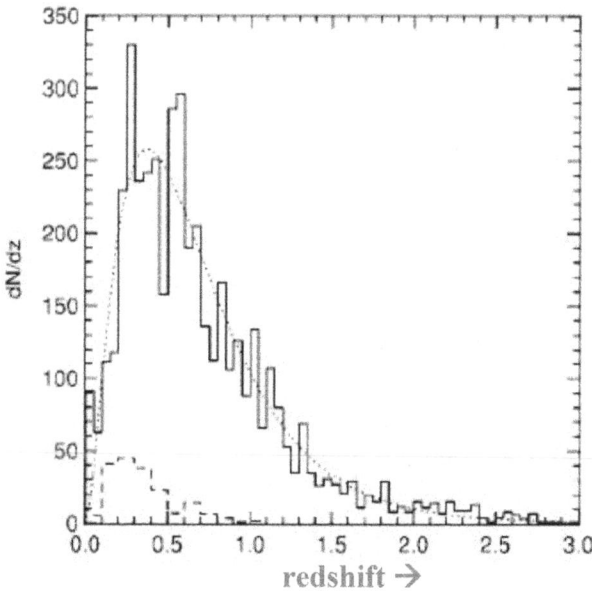

Figure 148: Distribution of photometric redshifts (*solid histogram*) and a best-fit analytic description (*dotted line*) as well as the distribution of those entries for which spectroscopic redshifts (*dashed line*) were available. [taken from Drory (2001)]

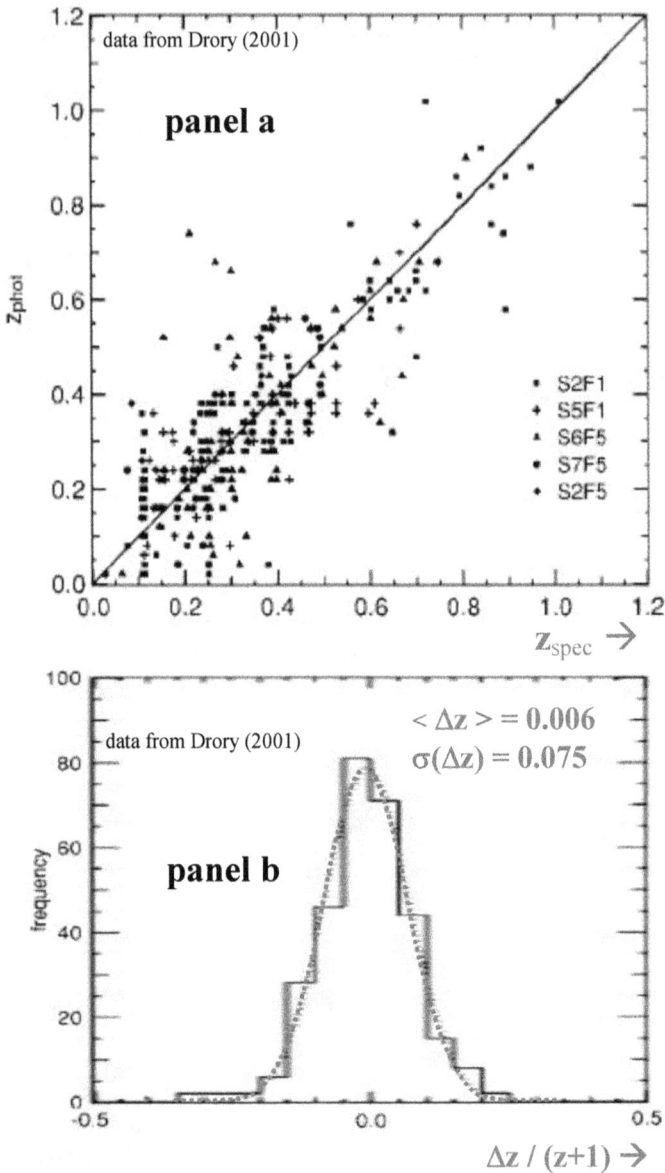

Figure 149: The uncertainty in photometric versus spectroscopic redshifts

model versions for which $\Omega_M = 1.0$, $\Omega_\Lambda = 0.0$ and are $\Omega_M = 0.3$, $\Omega_\Lambda = 0.7$. Any resolution of which models might be refuted by comoving number densities of observed galaxies will have to wait until more complete observations are available for galaxies out to redshifts in excess of unity. For data out to a redshift of 0.2 there is virtually no way to definitively distinguish predictions as shown in figures 150 and 151. The bending in these otherwise straight curves is an observed phenomenon, however, as shown in figure 152 from Ellis (1997). Clearly none of this data discriminates between viable cosmological models or discriminates the scattering model from the major versions of the standard model.

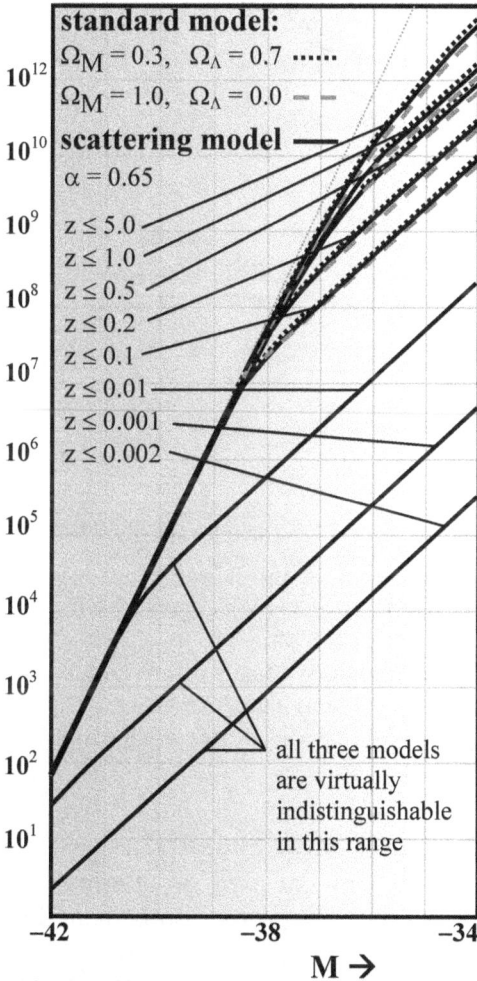

Figure 150:
Number of galaxies versus observed luminosity out to the various redshift limits ($\alpha = 0.65$)

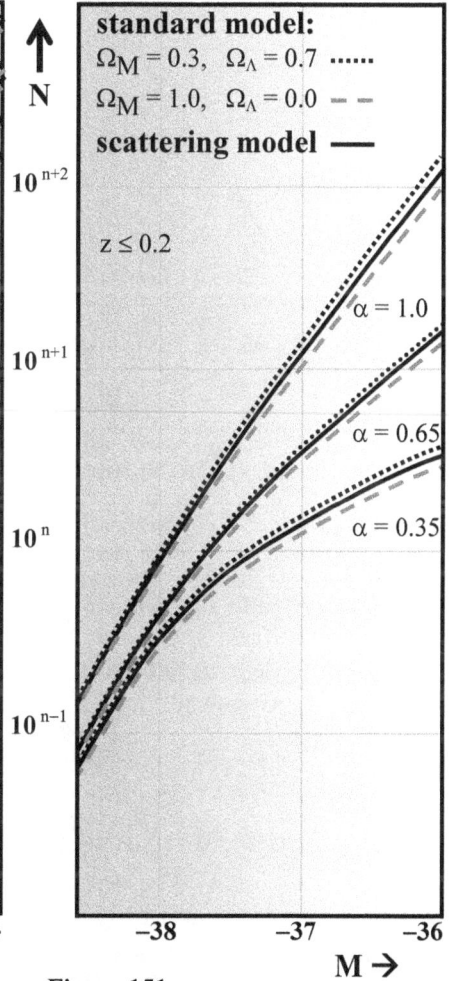

Figure 151:
Elbow in the number of galaxies versus observed luminosity curves for various values of α

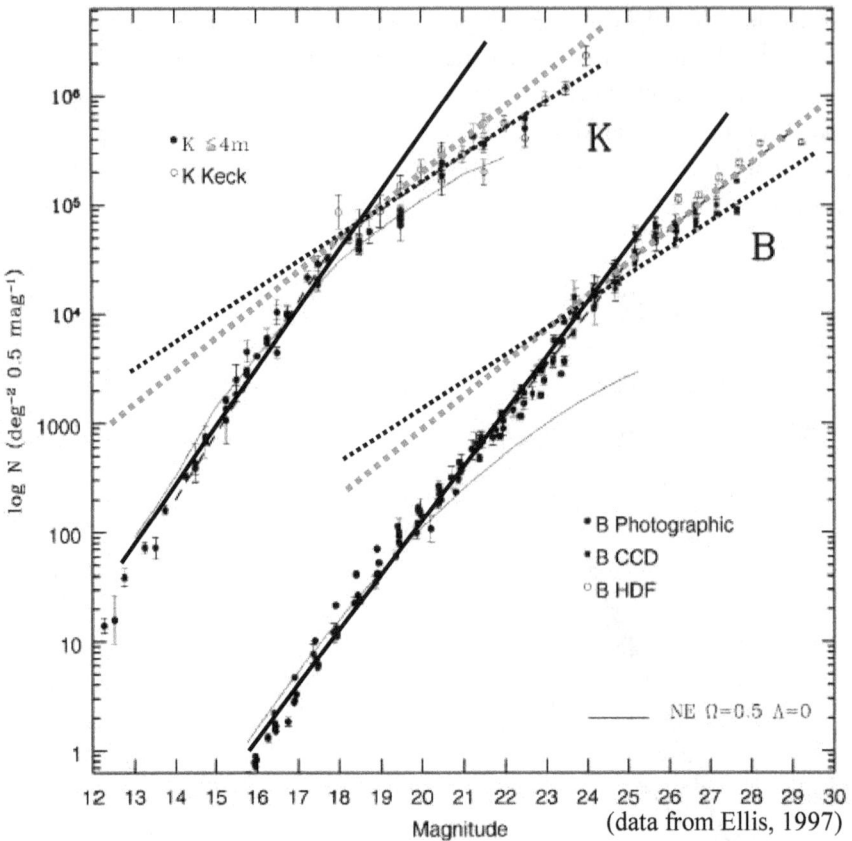

Figure 152: Magnitude of galaxy number counts in the B and K passbands from the compilation of Metcalfe et al (1996) augmented by the Keck K counts of Moustakas et al (1997). The two power law slopes (solid lines) drawn have gamma (= d log N / dm) = 0.47, 0.30 around B = 25 (Metcalfe et al 1995b) and 0.60, 0.25 around K = 18 (Gardner et al 1993). The solid curve indicates a no-evolution prediction for an Einstein-de Sitter universe.

f. claimed evolution of galaxy types with redshift

Much research is being expended in the area of attempts to identify early phase protogalaxies as a predominant portion of galaxies at high redshift that would indicate that galaxy evolution is taking place. Conselice et al. (2004), for example, have detected large concentrations of star-forming luminous diffuse disc-shaped as well as luminous asymmetric objects at red shifts from z = 0.5 to z = 2.0, with the numbers decreasing beyond that.

These images of some of these objects are shown in Conselice (2004) from which it is apparent that the observations are at the very edge of observability. Referring back to figure 141 and additional redshift survey data to be presented in the next chapter, it becomes obvious that the observations to which Conselice refers are well beyond reasonable capabilities to distinguish the morphology of the objects and that classifying them in that way involves a considerable vulnerability. Nonetheless, his abstract states:

"There exists a gradual, but persistent, evolutionary effect in the galaxy population such that galaxy structure and morphology change with redshift. This galaxy structure-redshift relationship is such that an increasingly large fraction of all bright and massive galaxies at redshifts $2 < z < 3$ are morphologically peculiar at wavelengths from rest-frame ultraviolet to rest-frame optical. There are however examples of morphologically selected spirals and ellipticals at all redshifts up to $z \sim 3$. At lower redshift, the bright galaxy population smoothly transforms into normal ellipticals and spirals. The rate of this transformation strongly depends on redshift, with the swiftest evolution occurring between $1 < z < 2$.

First of all, he is citing observations with fifteen to twenty times greater redshift than appear in redshift surveys. One must assume that this data is not included in such surveys because it is unreliable. The broadband absorption characteristics that we identified with the intergalactic medium suggests an optical depth of $z < 2$ so that even with sufficiently powerful telescopes images would be blurred so considerably that attempts to identify these objects by morphological type would be futile. By $z = 3$ identification of galaxies would become extremely difficult. The transition of objects in his analyses from definite ellipticals and spirals into what he refers to as "peculiars" and those that are "too faint" to specifically identify is exactly what one would expect when searching data that is so far beyond current capabilities and the optical depth of the medium.

He states however that, "High redshift galaxies also tend to be small with likely small stellar masses", but this of course, is conflated by the presumption of angular measures in the standard model that he accepts, which if not (as it most likely *is not*) realized would have to imply the galaxies themselves were smaller. And this, like the rest of his conclusions, must be considered very doubtful.

Significantly, Conselice cites no changes in the overall numbers of bright and massive galaxies out to a redshift of on the order of $z = 3$.

Chapter 16

Techniques and Estimates Used in Evaluating the Density of the Universe

The determination of the mass density of the universe is key to a determination of credibility of cosmological models compatible with unadorned facts. As with Hubble's constant, much excellent scientific work has gone into trying to assess the value of the associated baryonic mass density. Much of the early work in assessing the density of the universe involved the luminosity of various standard candles, but in addition there are techniques involving the apparent motions of astronomical objects and, to a lesser extent, gravitational lensing. We will review these techniques that have been used, explain how they work, and discuss problems with their application. In the next couple of chapters we will explore results of application of these techniques.

Most of these techniques are outgrowths of the theories of classical physics including electrodynamics and celestial mechanics. But at lease with regard to gravitational lensing, the application of the general theory of relativity is required. The results of these techniques will not in general be in agreement. That is a source of consternation.

a. mass-luminosity relationships

Ultimately our knowledge of the far reaches of the universe derive from what we have learned of our own solar system. From Keplerian analyses the masses of the sun and other members of the solar system have been assessed. And of course, we have been able to directly measure its luminosity. These parameters are denoted as $M_\odot$

and $L_\odot$, respectively. The values of these parameters have been determined quite accurately as:

$M_\odot$ $= 2 \times 10^{33}$ gm

$L_\odot$ $= 4 \times 10^{33}$ ergs/sec.

Extending our reach with Hertzsprung-Russell diagrams, such as that presented in figure 1, page 18, has provided the relationship of luminosity vs. temperature for the various types of stars we observe in our galaxy. As was shown, our sun is a rather average member of the main sequence of stars on such diagrams.

From Keplerian analyses of multiple star systems we have luminosity vs. mass relationships for them as well. The results are as follows:

$$\frac{L}{L_\odot} \sim \left(\frac{M}{M_\odot}\right)^{3.5}$$

In actuality the exponent has been determined to be somewhat larger by some observers. A value of 3.9 has been cited, for example.

These parameters and the mass-luminosity relationship has been extended to galaxies. Our Milky Way galaxy with its approximately 500 billion stars exhibits the following characteristics:

$M_{MW} \sim 6 \times 10^{11} M_\odot = 1.2 \times 10^{45}$ gm

$L_{MW} \sim 2 \times 10^{10} L_\odot = 8 \times 10^{43}$ ergs/sec.

Thus we have that,

$$\frac{M_{MW}}{L_{MW}} \sim 30 \frac{M_\odot}{L_\odot}$$

Elliptical and SO galaxies exhibit mass-to-luminosity ratios that are 3 to 4 times greater than for other galaxies. Kochanek (1996) provides a ratio of as small as on the order of 16 for some galaxies.

Similar parameters and relationships are known for clusters of galaxies as well, for which mysteriously the ratio considerably exceeds that for galaxies and stellar systems generally. Thus, for 'rich clusters' of galaxies, to a rough approximation,

$$\frac{M}{L_B} \sim 300 \frac{M_{\odot}}{L_{\odot}}$$

Here L_B is the total luminosity of the cluster within the half-luminosity radius in the blue region of the electromagnetic spectrum. That this ratio greatly exceeds that of the constituent stars is deemed evidence that 'dark matter' is somehow involved. The same basic relationship holds for smaller clusters as well as huge superclusters of galaxies.

There are considerable changes to the acknowledged ratios that have occurred over the last decades. Loewenstein (2003) provides recent data on these ratios in the table below taken from his article. 'Dark matter' and, more recently the separately accounted exotic 'dark energy' has dominated discussion during these last couple of decades, accordingly in these discussions leaving only a small portion of the mass of the universe is attributed to baryons. Baryonic mass is, of course, the mundane stuff that pervades our everyday world and has been ordered into a Periodic Table of the elements appropriate for discussion in university chemistry and physics classes. There is not much of a similar definitive nature to state concerning the 'dark' substances at this point in time, and since we will be showing why they are uneccessary, we will concentrate, as Lowenstein does, on baryonic matter in the objects that can actually be seen both by day and by night.

Mass-to-Light Ratios and Mass Fractions

parameters	universe as a whole	galaxy clusters
$< M_{total} / L_B >$	270	300
$< M_{stars} / L_B >$	3.5	4
$< M_{gas} / L_B >$	41	35
mass fraction in baryons	0.17	0.13
mass fraction in stars	0.013	0.013
mass fraction in gas	0.15	0.12
stars/gas ratio	1/12	1/9

It is now evident that most of the baryonic mass contained in the universe at large is in actuality primarily involved as plasma gases in and around galactic structures – primarily in galaxy clusters that emit intense X-ray radiation.

Parameters including luminosity within a core radius, R_c and half-luminosity radii, R_A that are on the order of two megaparsecs, apply to the analyses of rich clusters.

Rather than luminosity – that had traditionally been considered the observable of astronomical matter – occupying the seat of honor in the pantheon of mass measuring techniques, in current cosmological circles, it plays the roll of 'little brother'. Other techniques we will explore are considered more reliable.

b. Newton's laws and Keplerian mechanics

Newton developed his laws of motion derivative to the principle that it requires force to alter momentum of massive bodies as follows:

$$\mathbf{F} = d\mathbf{p}/dt$$

Here $\mathbf{F}$ is the force, $\mathbf{p}$ the momentum, both vector quantities. This is the familiar formula $\mathbf{F} = m\mathbf{a}$ that applies where relativistic analysis is not required; it will be adequate for the analyses used in determining galactic motions. Primary among forces with which Newton dealt was gravitation, which originally he investigated within the context of those phenomena taking place on earth.

The legendary apple incident suggests an epiphany associated with recognition that the same force that causes an apple to fall also forces the planets to fall toward the sun, but in so doing, to fall along paths whose formulas had been determined earlier by Johannes Kepler. These are the elliptical paths of orbits of the planets with the sun at one foci and parabolic trajectories of the comets. In the process, Kepler had established how an orbiting object sweeps out equal areas in equal times, how periods are related to radii, etc.. See figure 153.

Newton discovered a generalized law of gravity that involves a force acting to bring the objects together in the amount of:

$$F = - \, G \, M \, m \, / \, r^2$$

where r is the distance of the center of the object of mass m from the center of the object of mass M and G is Newton's gravitational constant,

320

$G \equiv 6.7 \times 10^{-8}$ erg-cm/gm^2.

Combining Newton's laws of motion and gravitation allows us to determine the mass M of the material substance inside an orbiting object of mass m. The formula, relating to Kepler's efforts, is:

$$M = 4\pi^2 r^3 / G P^2$$

where r is the orbit radius and P is its period to make a complete orbit. Thus we can calculate the mass of the sun knowing the periods of, and distances to, its planets. Similarly, we can determine the mass of our own and other planets from the orbits of their moons, the mass of galaxies by their periods of rotation, and the mass of galactic clusters from the motions of the gravitationally bound galaxies.

Of course most of what is involved in these calculations is only as complicated as typical home work in a freshman physics class. But its significance to cosmology warrants our being perfectly clear about it. To this end figure 153 is provided, which illustrates Kepler's discovery with Newton's contribution of gravity being the driving force behind the phenomena, for which Einstein would later find a different basis. The figure also illustrates several orbital mechanics principles to be discussed in the next section.

Of course, as we observe objects at great distances from our solar system, we more typically measure a Doppler redshift from which we infer an associated 'spectroscopic' velocity, from which we infer a period, etc.. So what might seem to be a cut-and-dried observation of a required parameter is often times in actuality an inference based on the measurement of something totally other than what we infer from it. This is particularly the case with regard to efforts to disambiguate radial motions and cosmological redshift appropriate to our current endeavor.

c. the virial theorem

The virial theorem is derived directly from Newton's laws of motion. It relates more generally to the energy contained within the constituent parts of a system of gravitationally bound objects in this particular application of inverse square law forces. The energy of a system characterizes what has been, and/or *can be*, done by a force acting on a body. In mechanical systems there are two ostensible forms of energy: A velocity-dependent 'kinetic' energy, T and a position-dependent 'potential' energy, U. When a force acts upon an object, it transforms the potential energy of that object into kinetic energy. Throughout whatever processes take place, however, the *total* amount of energy will be 'conserved'.

panel a. Kepler's law

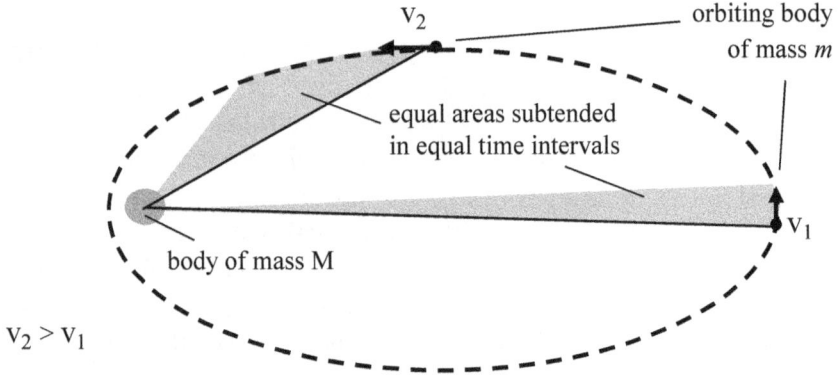

equal areas subtended in equal time intervals

orbiting body of mass m

body of mass M

$v_2 > v_1$

panel b. Newton's laws

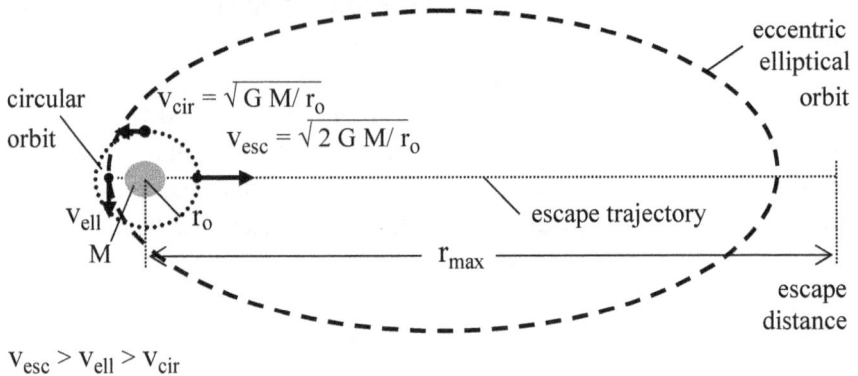

circular orbit

$v_{cir} = \sqrt{G M / r_o}$

$v_{esc} = \sqrt{2 G M / r_o}$

eccentric elliptical orbit

escape trajectory

v_{ell}

M

r_o

r_{max}

escape distance

$v_{esc} > v_{ell} > v_{cir}$

Figure 153: Kepler's and Newton's contributions to orbital mechanics

The virial theorem involves the relationship between a velocity-dependent kinetic energy, T and the gravitational potential energy, U of a self-gravitating system. The relationship is as follows:

$2T + U = 0$

We can make some simple assumptions about T and U for observed galaxies under consideration. First of all, the kinetic energy of an object moving relative to us at non-relativistic speeds is just:

$$T = \tfrac{1}{2} m v^2,$$

where m is the mass of the object and v its velocity. The total kinetic energy of an ensemble of I objects is just:

$$T = \sum_{i=1}^{I} \tfrac{1}{2} m_i v_i^2, \text{ where } M = \sum_{i=1}^{I} m_i$$

The gravitational potential energy is given by:

$$U = - G M m / r$$

Typically one can reason either from the perspective of the acting force or from energy conservation considerations. In circular centripetal motion the force holding an object in its orbit – the Newtonian gravitational force – is:

$$F = - G M m / r^2 = m v^2 / r = m a$$

where v^2 / r is just the centripetal acceleration of the object. But one could also reason concerning the energy considerations:

$$U = - G M m / r = -m v^2 = - 2T$$

In either case, from such considerations for an ensemble of objects one can obtain a composite relationship for the entire system.

$$M = V^2 R / G$$

where, for example, M is the total mass of the galaxy, V is the mean velocity (combining the rotation and velocity dispersion) of stars in the galaxy, and R is the effective radius (size) of the galaxy. This equation relates observable properties of galaxies and clusters of galaxies – the velocity dispersion and effective radius – to the ever important property of associated mass that is not directly observable. The virial theorem is the primary approach to the determination of the mass of distant astronomical systems such as galaxies and clusters of galaxies.

This powerful technique allows the average total kinetic energy to be calculated even for very complicated systems for which it would be impossible to obtain an exact solution. However, although this technique is solidly based in the classical theory of mechanics, it too depends intimately on the validity of inferences from observed Doppler shifts in wavelength to assess the velocities from which kinetic energy can be determined. Conclusions based on this technique are maintained in spite of equally obvious objections to associated conclusions that imply the objects within the evaluated system should have escaped.

It is by the comparison of mass estimates based on this virial theorem to estimates based on the luminosity relationships of galaxies and galactic clusters that cosmologists have become increasingly convinced of the existence of *dark matter* in galaxies and clusters of galaxies. Fritz Zwicky first used the virial theorem to deduce the existence of matter that was not otherwise observed, what is now called "dark matter". This additional matter is absolutely required if the measured redshifts (or 'spectroscopic velocities') are indeed of Doppler origin because the inferred velocities could not otherwise occur in gravitationally bound systems. The two situations are mutually exclusive, and since the clusters are clearly bound together, there must be additional mass to bind them. Refer to figures 153 and 154 which illustrate the escape criteria for gravitational systems. The included diagonal lines in figure 154 each reflect the following virial theorem threshold formula applicable to maximum velocities of bound objects:

$$r_{max} < 2\,G\,M\,/\,v_{max}^{2}$$

Here M is the total mass of the cluster. A galaxy possessing the maximum velocity will escape from the cluster if the criterion expressed by the inequality is not met. Clearly a large mass is required to justify large velocities in sizable clusters. Notice that this is basically the same criterion as that which establishes the Schwartzchild radius of a black hole, where escape can be precluded even at the velocity of light by merely limiting the size of the gravitationally bound unit.

So all estimates of the mass of galactic clusters have been retrofitted with sufficient 'dark matter' to accommodate the observed stability of the clusters. This has involved doubling, tripling, and even adding ten times the mass of these objects in order to make the physics work, and then with the presumption of the added mass in place, efforts proceed to come up with conjectures concerning what in heavens name the matter could be. Of course that is not how physics, and science in general, is supposed to work. In short, observations seem to refute the Doppler conjecture, but the refutations are being ignored.

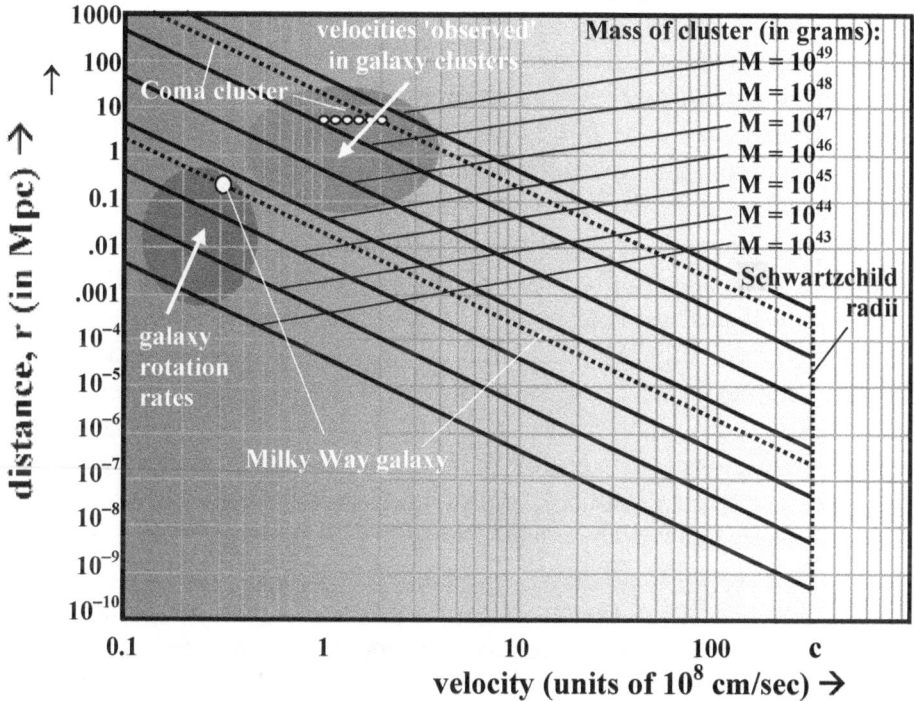

Figure 154: Escape criterion for gravitationally bound systems

So… that is the approach that has been taken by the established cosmology community, with efforts primarily directed at determining whether the added mass is more compatible with it being 'cold' or 'hot', etc.. A more productive approach would seem to be to have a more critical look at the ostensible implications of the additional matter conjecture. One such implication is that adding mass must effectively increase the size of the cluster to preclude escape no matter how the added material is distributed. The problem that is being solved in this way is to assure that the 'observed' velocities are not allowed to propel galaxies beyond the confines of the cluster. So the geometrical dimensions of the cluster must increase accordingly as shown in figure 154, and this size parameter is a directly observable quantity.

Let us consider the Coma cluster, for example, which is the closest rich cluster with on the order of ten thousand galaxies. It is situated at between 90 and 100 Mpc from us and subtends about 4 degrees of the sky, giving it a physical diameter of about 6 Mpc. The velocity distribution of galaxies is included as dots in figure 154. The virial mass within 5 Mpc of the center is estimated at 2×10^{15} $M_\square$, with a mass-to-luminosity ratio of 160 $M_\square/L_\square$. Biviano (1997) provides a plot of velocities of galaxies within the Coma cluster which we have reproduced in figure 155. This range of velocities is what has been superimposed on figure 154 as well, where it should be clear that even with the huge augmentation of mass, it is insufficient to keep galaxies with velocities at 2×10^8 cm/sec from escaping the cluster.

This problem is aggravated by the fact that even those galaxies with less than the required escape velocity could escape by the sling shot effect of passing near another galaxy that could propel it outward with greater velocity. This is similar to how NASA augmented thrust by sling-shot effect in going around the moon to help escape terrestrial and solar gravitational potentials. Certainly galaxies must occasionally escape from clusters to account for the infrequent but ubiquitous 'field galaxies' that occur throughout the vast regions between clusters.

Observed Doppler-inferred radial velocity components provide major clues to what is wrong with the conjecture. Skewing of the distribution toward recessional velocities obviates this interpretation of redshift as in error. We address the incorrectness in the next chapter.

d. gravitational lensing

It is thought by many that the gravitational deflection of light is an exclusive claim of the general theory of relativity. This is not the case. Ironically, Einstein's initial prediction made in 1914 was precisely what could have been predicted using Newtonian mechanics.

Newton had considered this possibility as early as 1704. There were several subsequent attempts using Newton's laws of motion that anticipated gravity acting upon light. They include a prediction by the British polymath John Mitchell, who first conceived of black holes based on a similar interaction between light and matter as referenced in Appendix C. In 1799 Laplace broached the same topic. Perhaps more definitively the German mathematician and astronomer Johann Soldner (1804) wrote "On the Deflection of a Light Ray from Its Straight

Motion Due to the Attraction of a World Body Which it Passes Closely". These inferences from Newtonian mechanics all predict a deflection angle, θ in light passing such a massive object as follows:

$$\theta = 2 (1 + \gamma) \, G \, M_\square / b \, c^2$$

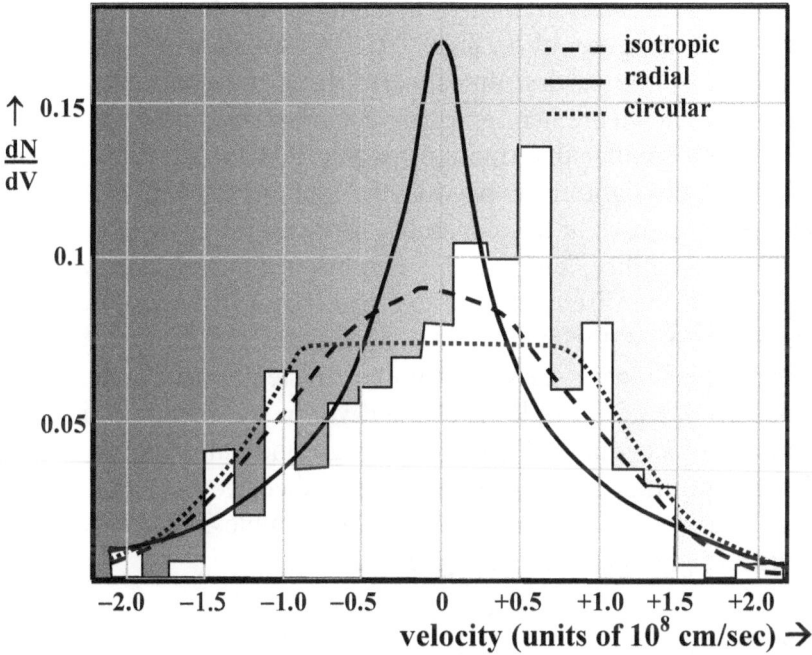

Figure 155: Velocity dispersion in the Coma cluster

where b is called the 'impact parameter'; it is the minimum distance between the light path and the object. For observations of deflection at the limb of the sun, $b = R_\square$ where, $R_\square = 695,500$ km is the radius of the sun. For calculation purposes, the expression, $G \, M_\square / b \, c^2 = 2.120$ km at the limb of the sun. The parameter γ, whose value is zero according to the Newtonian theory is replaced by unity in the proper application of Einstein's general relativity. Clearly, it is the parameter that discriminates predictions made using the two theories. Iess et al (1999) state that its present observationally-determined empirical value is,

$$\gamma = 1.000 \pm 0.001,$$

where they claim that "the quoted uncertainty is a rough estimate of the results of different measurements."

Because the effect is through an angle that is proportional to the inverse of the distance of closest approach, the angle of deflection is similarly affected as illustrated in figure 156. At the limb of the Sun the predicted deflection is 1.75 arc seconds. It results in a displacement of a star's apparent position through that angle away from the Sun. This was illustrated in figure 49 on page 103. As we showed in Chapter 5 for refraction by the earth's atmosphere, if one were further removed from the sun than we are here on earth, such that the angular size of the sun were much smaller than the half degree it is for us, the maximum deflection of star light from beyond the sun would still be 1.75 arc seconds. Several uses of this capability recommend themselves.

This capability, enabled by Einstein's general relativity, has been applied to the discovery and deciphering of otherwise confusing phenomena. As Wilhelm de Sitter once used the *non*-existence of 'ghost images' of binary stars to promote Einstein's special theory, the actual occurrence of multiple 'ghost' and variously distorted images of astronomical objects beyond massive foreground galaxies and clusters of galaxies confirms aspects of his general theory. See, for example figure 157 described in the inset. Here the double image of a quasar resulting from gravitational lensing is exhibited.

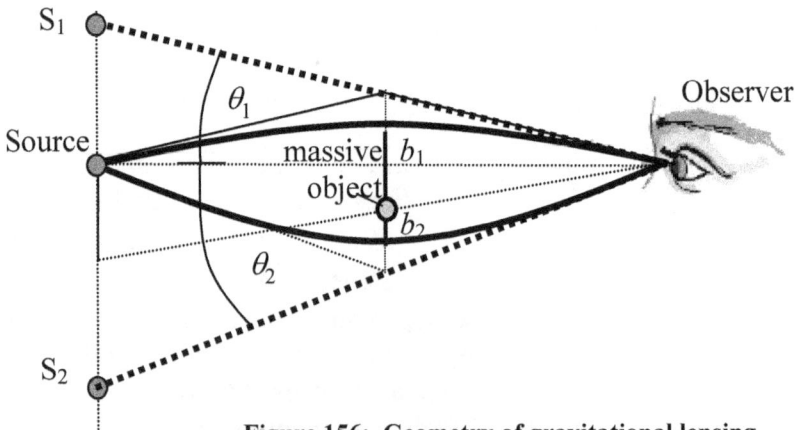

Figure 156: Geometry of gravitational lensing

These phenomena produce somewhat similar effects to refraction in the atmosphere of foreground objects described in chapter 5. However, there are a couple of essential differences including the fact of there being no wavelength-dependence with gravitational lensing and a dependence on impact distance rather than variations in the electron density of atmospheric media results in an extremely different angular dependence. For more information refer to Kolb (2007), Fischer et. al. (2000), and Sutton and Bridle (2006).

Kochanek (1996) has applied gravitational lensing to assess the maximum allowed values of Einstein's rejected Λ. The rationale pertains to the availability of multiple images. Although there are cases of multiple clear images of the same object as demonstrated in figure 157., it is primarily merely the distortion of background galaxies tangentially stretched by foreground clusters that are encountered.

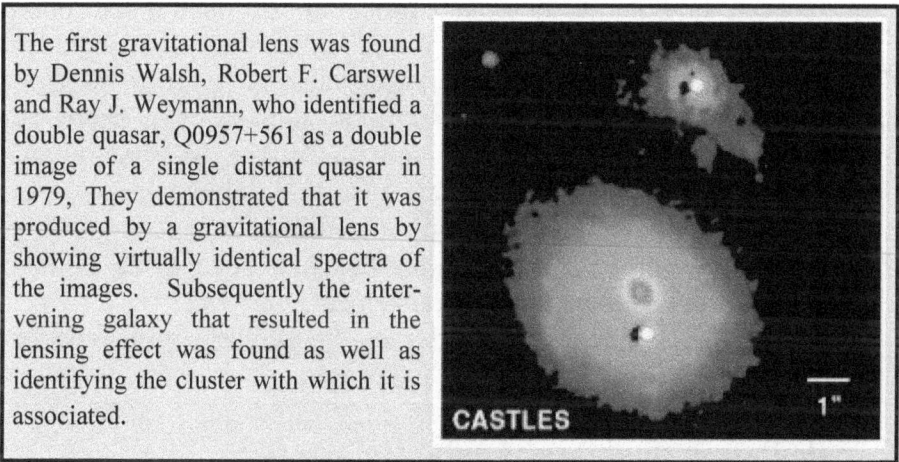

The first gravitational lens was found by Dennis Walsh, Robert F. Carswell and Ray J. Weymann, who identified a double quasar, Q0957+561 as a double image of a single distant quasar in 1979, They demonstrated that it was produced by a gravitational lens by showing virtually identical spectra of the images. Subsequently the inter-vening galaxy that resulted in the lensing effect was found as well as identifying the cluster with which it is associated.

Figure 157: Observation using a gravitational lens

Importantly, gravitational lensing is not envisioned as a viable replacement for use of the virial theorem and classical mechanical methods of assessing mass. The measures obtained from gravitational lensing where it is viable are consistently less than those obtained using more traditional methods.

e. mass measurement results

Spurred on by theoretical anticipations, the assessment of the actual mass density of the universe as a whole has been a priority among investigators. The obvious place to have started such an endeavor was with surveys of the numbers of galaxies within typical volumes of space along the lines of investigation concerning comoving number density observations discussed in the previous chapter.

The CFA Redshift Survey and other projects have provided a tremendous amount of data on which to perform analyses for assessing the overall mass density. Figure 158 is just a token example of the data that has been provided, here for but one small angular slice out to a maximum redshift of $z = 0.05$, which if the recessional (expansion) interpretation is correct, would be five percent of the radius of the entire universe, implying that case that spectroscopic 'velocities' of up to five percent of the speed of light have been measured.

There are some obvious features of data displayed in figure 158: Galaxies are *not* uniformly distributed. They do, of course, in some cases appear as isolated 'field galaxies', but this is the exception. Primarily they occur in 'groups' of up to 50, in 'clusters' that have from fifty to several thousand, and in 'super clusters' with millions of galaxies spanning several hundred million light-years. There are millions of these superclusters in the observable universe as we illustrated in chapter 2. And as we saw in section a above, the assessed mass-to-luminosity ratio increases with increasing numbers involved in these units. It, therefore, has seemed reasonable to begin an assessment at the top level in this hierarchical arrangement.

Then, of course, we will address the similarly problematical issue of the mass of the individual galaxies.

f. calculating the mass of galaxy clusters

Let us apply the virial theorem to see how the accounting would go for a cluster with uniform mass density out to a radius, $R_c = 1.5$ Mpc with a velocity dispersion of $\sigma_c = 750$ km/sec. We will use the formula, $M_c = \sigma_c^2 R_c / G$ from section c above. With $M_c = 4/3 \, \pi \, \rho_c \, R_c^3$, in accordance with the simplifying assumption of uniform density of galaxies within this region, we obtain:

$$\sigma_c^2 = 4/3 \, \pi \, G \, \rho_c \, R_c^2$$

From this application of the virial theorem we can calculate,

$$7.5 \times 10^7 = (\rho_c \times 2.81 \times 10^{-7})^{\frac{1}{2}} \, 4.63 \times 10^{24} \, \text{cm/sec}$$

$$\cong 2.45 \times 10^{20} \, \sqrt{\rho_c}$$

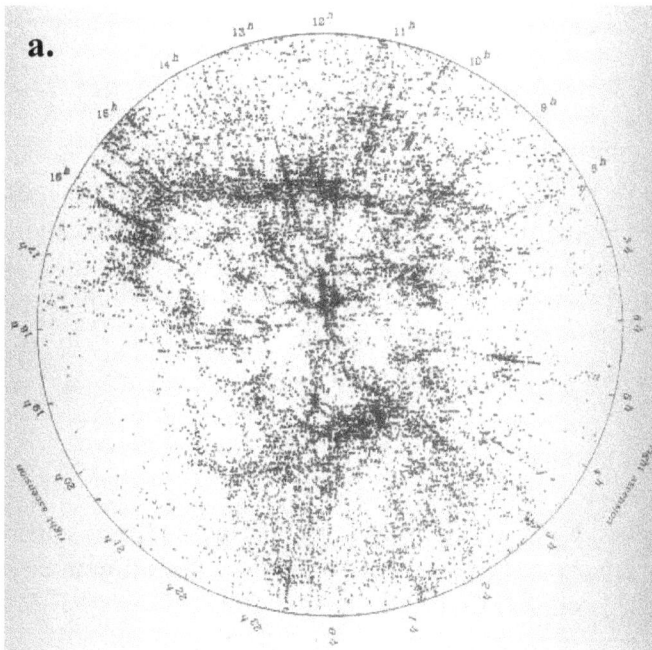

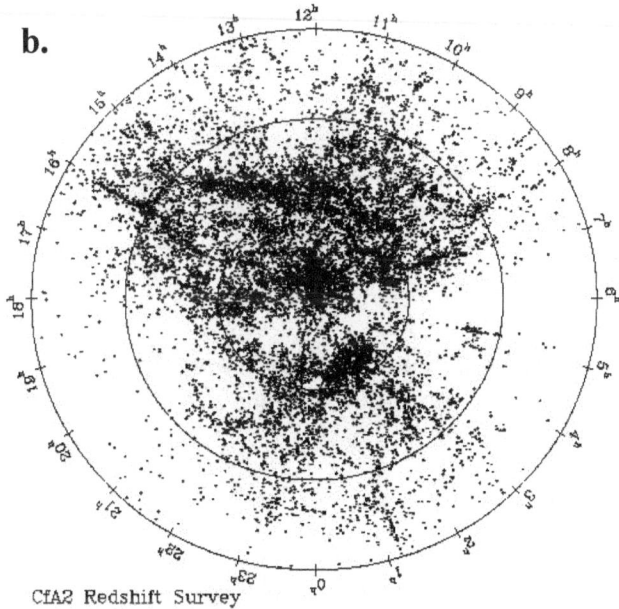

CfA2 Redshift Survey

Max Radius 15000
$0 \le h < 18000$ (km/s)
$m_B \le 15.5$

Puck

Copyright SAO 2001
Smithsonian Astrophysical Observatory

Figure 158: CFA Redshift Survey data

$$\rho_c \cong 9.4 \times 10^{-26} \text{ gm/cm}^3$$

$$M_c(r \le 1.5 \text{ Mpc}) \cong 3.9 \times 10^{49} \text{ gm} \sim 2.0 \times 10^{16} \text{ M}_\odot$$

This is somewhat larger than the average cluster mass because of our assumption of uniformity out to R_c, but we have used the median value of σ_c in this assessment. We are certainly within the range of what is observed   So let us proceed to allocate mass associated with these characteristics among luminous (observed) galaxies whose mass density we will call ρ_{gal}, intracluster plasma gas of density ρ_{gas}, and the mysterious 'dark' component whose density we identify as ρ_{dark} so that:

$$\rho_c = \rho_{gal} + \rho_{gas} + \rho_{dark}$$

Then, if indeed 82% of this total mass is non-baryonic 'dark' matter as is currently envisioned in order to account for the extreme mass that has not otherwise been accounted, we have:

$$\rho_{dark} / (\rho_{gal} + \rho_{gas} + \rho_d) = 0.82, \text{ and}$$

$$\rho_{dark} = (0.82 / 0.18)(\rho_{gal} + \rho_{gas}) \cong 4.56 (\rho_{gal} + \rho_{gas})$$

Although, clusters have been found that contain nearly as much mass in hot core gases as in all their galaxies, Bahcall (1999) indicates that typically $\rho_{gas} \sim 0.07 \rho_{gal}$. In this case, the *dark matter* density would be:
$\rho_{dark} \sim 4.55 \times 1.07 \rho_{gal} \cong 4.9 \rho_{gal}$

By this analysis the overall mass of the core of a cluster of galaxies would contain nearly six times that associated with the galaxies that are observed.

$$\rho_c = 5.62 \rho_{gal}$$

This result implies that only a little more than four tenths of velocity dispersion can be attributed to observed galaxy mass. Clearly that is why dark matter is thought to contribute so substantially to the high mass-to-luminosity ratio of galaxy clusters.

g. characteristics of 'rich' galaxy clusters

Cowie and Perrenod (1978), Bahcall (1999), Reiprich (2006), and others have summarized much of what has been learned about 'rich clusters' of galaxies. These rich clusters typically have many thousands of galaxies caught in a common gravitational milieu. The numbers and types of galaxies within about 1.5 Mpc of the center of these structures seem to be distributed in accordance with the Schecter function discussed in the previous chapter. The following list provides some of the more ostensible features of these rich clusters:

core radius, R_C	$0.15 \rightarrow 0.3$ Mpc
half luminosity radius, R_A	$1.5 \rightarrow 3$ Mpc
mass, M_{cl} ($r \leq 1.5$ Mpc)	$1.5 \times 10^{14} \rightarrow 3 \times 10^{15}$ $M_\odot$
luminosity, L_B ($r \leq 1.5$ Mpc)	$10^{12} \rightarrow 10^{13}$ $L_\odot$
core number density	$3 \times 10^5 \rightarrow 3 \times 10^6$ Mpc^{-3}
range of velocity (scatter), σ_r	$400 \rightarrow 1400$ km/sec
median scatter velocity	750 km/sec
intracluster gas:	
mass, M_{gas} ($r \leq 1.5$ Mpc)	$10^{13} \rightarrow 10^{14}$ $M_\odot$
M_{gas}/M_{cl} ($r \leq 1.5$ Mpc)	$0.03 \rightarrow 0.15$
X-ray temperature	$2 \rightarrow 14$ keV
electron density	$\sim 10^{-3}$ cm^{-3} $\rightarrow 10^3$ cm^{-3}

There are several things of note in this list. One is that the scatter or dispersion of galaxy velocities, which is really just a redshift scatter, seems underestimated in this summary. Becker et al. (2007) determined the mean and scatter of this parameter for the maxBCG galaxy cluster catalogue, and they obtained on the order of 200 km/sec for small groups of galaxies and 850 km/sec for larger clusters. But again the numbers seem to this author always to be underestimated. In the previous section we saw how this phenomena shows up on redshift galaxy maps as readily detectable straight lines, comprised of an extreme number of galaxies, whose length, and therefore velocity dispersion, can easily be measured. There are several obvious examples that seem at least twice as large as the upper ends of the ranges in the list above, as cited by Becker et al., and by others.

All rich clusters produce extended X-ray emission from hot intracluster plasma gas. One of the most studied clusters in recent years is the cluster PKS 0745-191 for which a large amount of X-ray data is available. Figure 159 borrowed from Hicks et al. (2002) displays a map of X-ray intensity (counts) in bins about the center of the cluster. The highest temperatures are realized at some distance away from the center, with significant counts registered at distances of 3.5 Mpc and as shown in other sources. Clearly temperatures approach 10^9 K. In other papers it is clear that X-rays with energies approaching 10 keV persist at least to radii of 3.5 Mpc. In figure 160 derived from Hicks et al. (2002) this is also apparent, with intensity levels settling to the uniform background level beyond that. The drop in intensity has more to do with a steep gradient in plasma density than temperature since there is X-ray emission at a considerable distance from the center of the cluster to where it merges into the uniform X-ray background, which it seems clear, arise in similar circumstances throughout the cosmos.

Of course, plasma gas density and temperature profiles are difficult to assess other than by their X-ray emissions that diminish considerably with decreasing density. Although density does directly affects the dynamics of the involved galaxies one must disentangle this involvement from conjectures of 'dark matter'. We present an educated guess at a simplified galaxy cluster temperature profile out to larger radial distances in figure 161. It is based on data provided by Hicks et al. (2002), Bahcall (1999), plots by Cowrie and Perrenod (1978) to be discussed later, and others. This formula is the following.

$$T(r) = k_1 \, e^{-k_2 \, (r/R_c)^2}$$

Interestingly, galaxy velocity dispersion correlates directly with intracluster gas temperature as shown in figure 162. This provides key evidence with regard to redshift *scatter* predicted by the scattering model. The scatter is predicted to be linear with peak temperature, k_1 in figure 161. A linear relationship has been added to figure 162 from Lubin and Bahcall (1993) who investigated isothermal models with gases and galaxies in hydrostatic equilibrium in a cluster potential that suggested to them a squared relationship, $\sigma_r^2 \propto \beta kT$.

The densities of intracluster plasma gas and individual galaxies within the cluster seem to track each other – at least within central cores where plasma gases are measured. The mass density of this gas as a function of distance from the center of the cluster varies as follows:

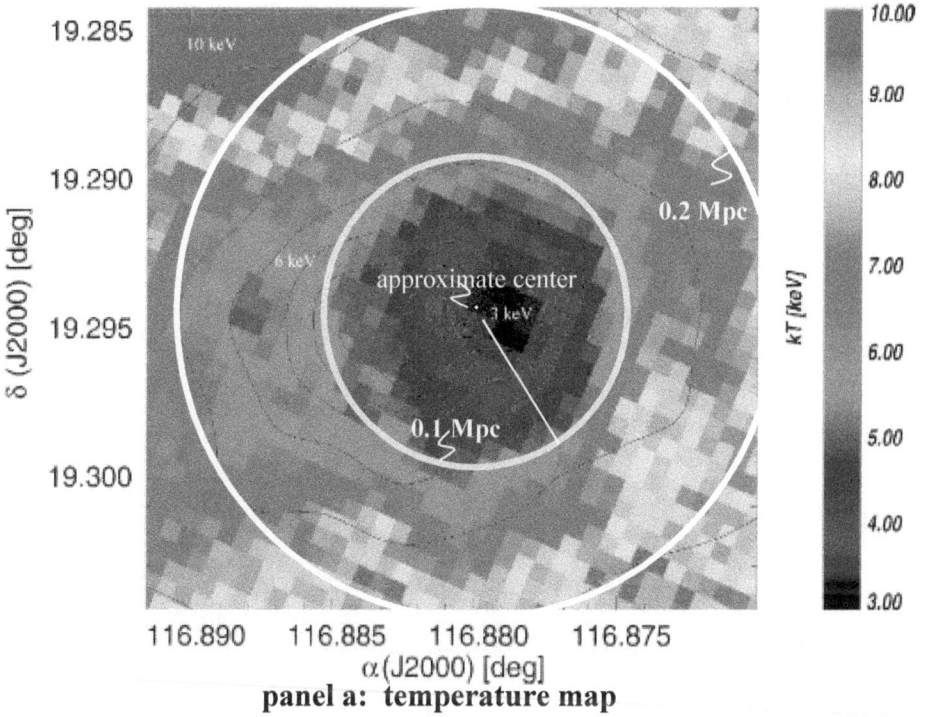

panel a: temperature map

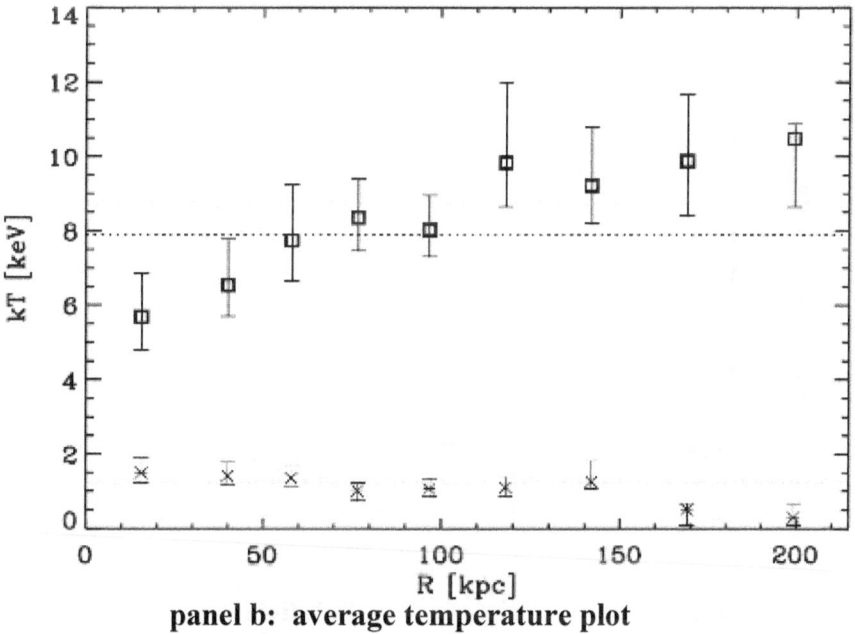

panel b: average temperature plot

Figure 159: Temperature in intracluster core regions of PKS 0745-191

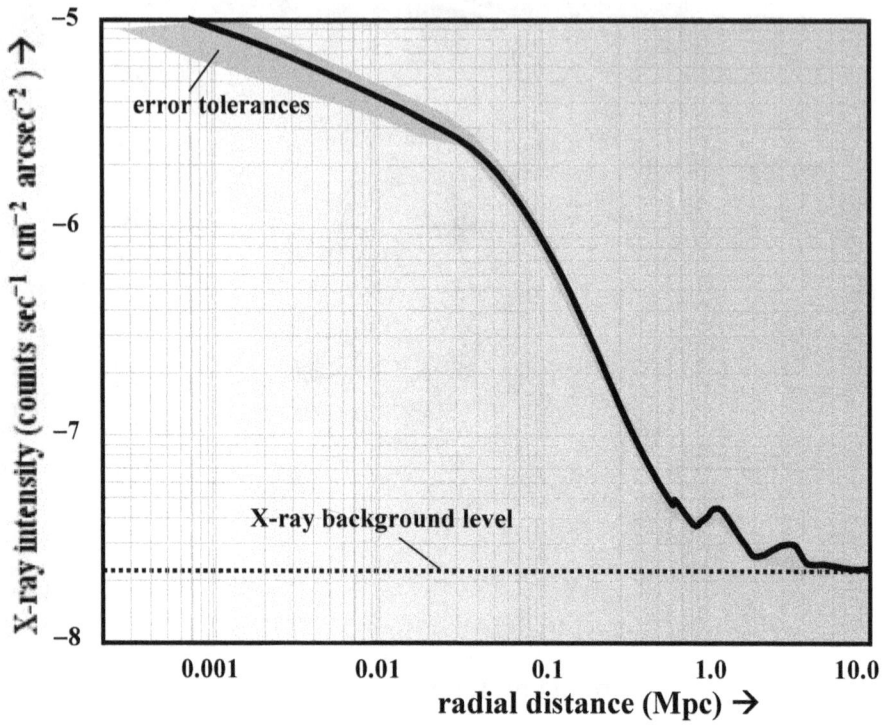

Figure 160: X-ray intensity of cluster gases in PKS 0745-191

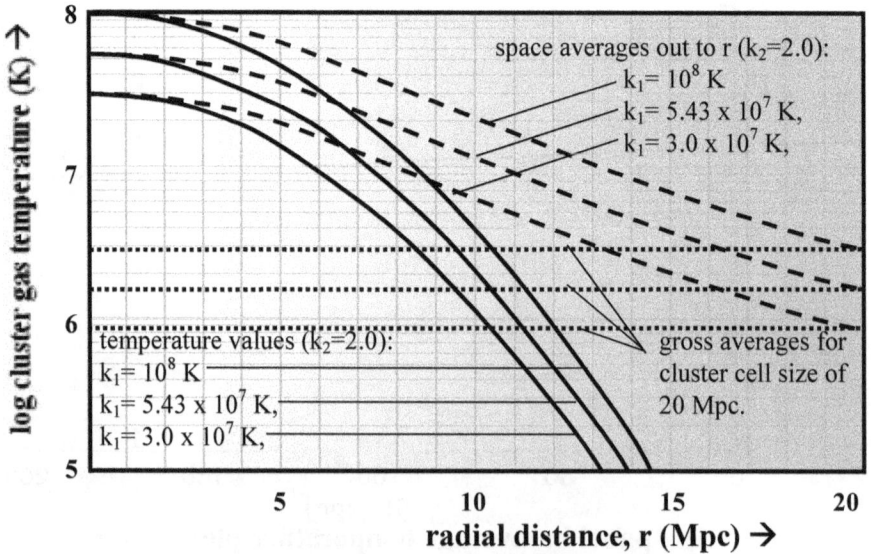

Figure 161: Estimated temperature of cluster gases in PKS 0745-191

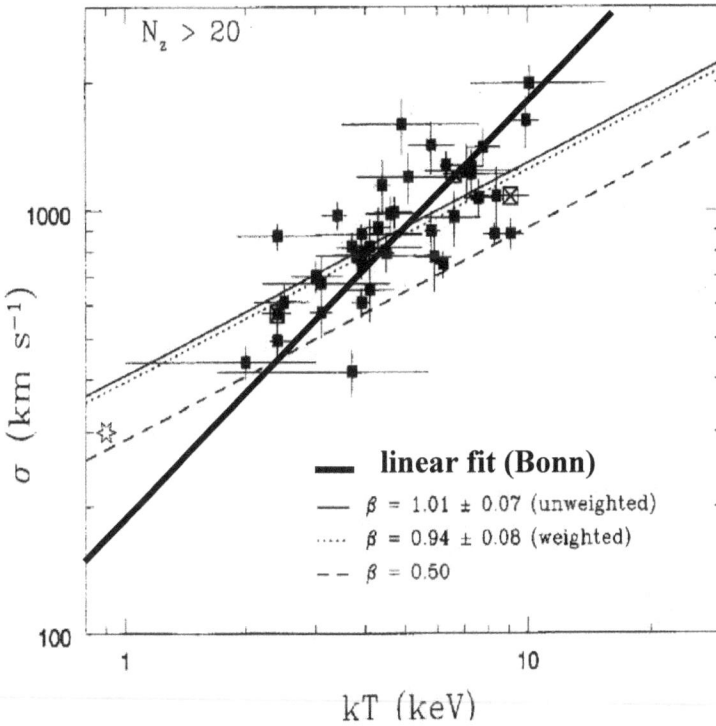

Figure 162: Cluster radial velocity dispersion (sigma-r) vs. gas temperature (kT) for 41 clusters (Lubin and Bahcall, 1993)

$$\rho_{gas}(r) = n_o [\, 1 + (r/R_c)^2 \,]^{-3\beta/2}, \text{ where } \beta \cong 0.47.$$

This formula presented by Hicks et al. (2002) for cluster PKS 0745-191 is similar to Bahcall's (1999); although she used −1 as the exponent in a simplified form. The formula assumes spherical symmetry with both isothermal and hydrostatic equilibrium applying to the cluster core region. The parameter $n_o \cong 0.069$ cm^{-3} is the plasma gas density at the center of that particular cluster. X-ray spectra is used to determine both the central plasma temperature and density in this case. Within the central core region of 0.5 Mpc, Hicks et al. (2002) infer a total mass of 29.1 x 10^{13} M$_\odot$ = 5.82 x 10^{47} gm for PKS 0745-191; 19% of this is plasma gas, consistent with the constraint on average mass density of the cluster core region of,

$$< \rho_{baryon}(r<R_c) > \;\geq\; (1300 \,/\, h^2) \,\rho_o$$

Here they have assumed $h \cong 0.5$ rather than 0.714 as we have elsewhere based on more recent values of H_o, and $\rho_o = 5 \times 10^{-30}$ gm cm^{-3}. This value is huge with regard to the observed overall density of the universe. Clearly density drops off considerably beyond a core radius.

The mass density of galactic structures correlates quite directly to plasma density, so we will characterize the overall mass density in a similar way. The previous formula for plasma density has modified to accommodate mass density as follows:

$$\rho_{baryon}(r) \cong 7.580 \times 10^3 \, \rho_o \, [\, 1 + (\, r / R_c \,)^2 \,]^{-3\beta/2}$$

This formula is plotted as the top-most dark solid line in figure 163.

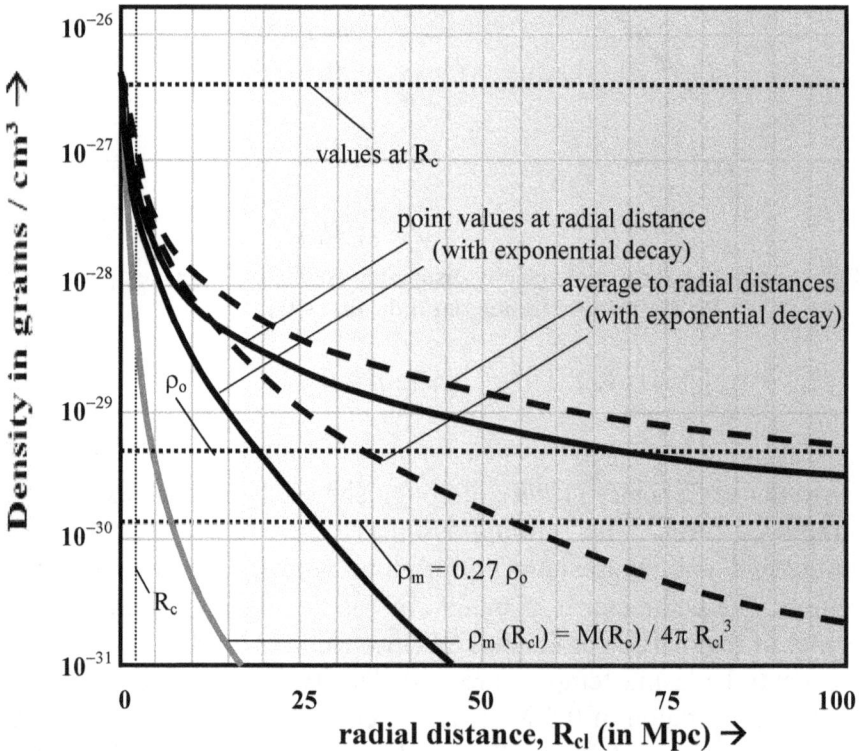

Figure 163: Implied baryonic mass density for various clustering assumptions

Thus, the total mass contained out to a radius R_{cl} from the cluster center can be characterized as,

$$M(R_{cl}, g) \cong 9.5 \times 10^4 \, \rho_o \int_0^{R_{cl}} [\, 1 + (x/R_c)^2 \,]^{-3\beta/2} \, x^2 \, dx$$

From this there is an implied mass associated with the cluster if it is separated from other clusters so as to effect the average density ρ_m of the universe. In this case, if Hicks' formula persists unabated to its gravitational limits – to where if the average density were realized, there would be an associated baryonic mass of over 5×10^{50} grams. It seems clear that there must be a steeper gradient at radii further from the center. Suppose an exponential decrease that becomes appreciable only at large distances as expressed in the following formulation:

$$\rho_{baryon}(r) \cong 7.580 \times 10^3 \, e^{-(r/R_d)} \, \rho_o [\, 1 + (\, r / R_c \,)^2 \,]^{-3\beta/2}$$

Then this result would be ameliorated as shown in the lower two plots in figure 163.

h. determining the mass of individual galaxies

Similar analyses are used to determine the mass of the galaxy themselves. But again the estimates of the masses of spiral galaxies based exclusively on their visible luminous matter are insufficient to account for how rapidly they are spinning using Keplerian celestial mechanics as described previously. In many cases they are, in fact, spinning faster than the escape velocity at all regions beyond their central bulges.

Under such conditions, how can galaxies retain their structure rather than dissipating as matter escapes? Also, there is a strange characteristic that does not accord well with traditional analyses of such structures. Galactic disc regions rotate at approximately the same speed from just outside a central bulge all the way out to their visible edge. The outer extremes seem to rotate almost as though the galaxy were a solid object rather than gravitationally bound individual stars and dust. By Newton's laws and Keplerian analysis the rates of rotation should decrease rapidly as the inverse square root of distance. It doesn't.

The data is clear. Figure 164 shows the kind of emission line redshifts, from carbon monoxide and hydrogen, that have been used to measure rotation rates. In figure 165 from the same source, Sofue and Rubin (2001), the rotation curves for many galaxies are shown.

See also the curves for the Milky Way, NGC4258, and M31 in figure 166. The centrally directed gravitational force that drives the rotation of such massive objects results in behavior of orbiting masses as shown in this figure. The effect of a condensed central bulge is an inverse proportionality. The masses in the disk itself accumulate to further reaches of the structure, with rotation rates declining much less rapidly. The rotation effects of what has been thought to be unseen 'dark matter halos' accumulate linearly with distance from the center. These various effects are illustrated specifically in figure 167.

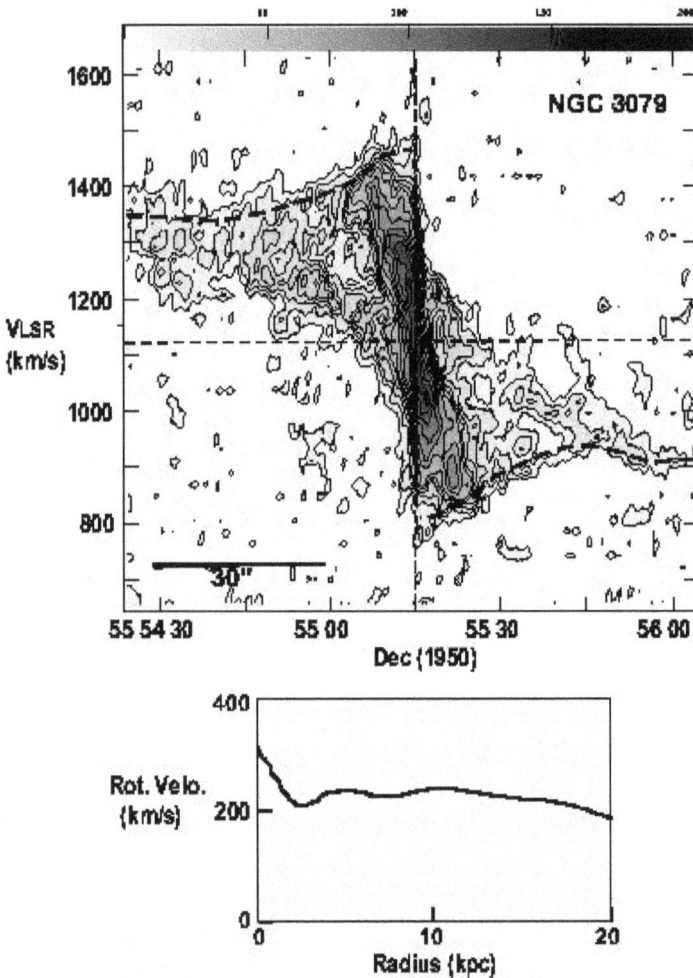

Figure 164: **The lower panel shows a composite rotation curve produced by combining the CO result and HI data (Irwin and Seaquist, 1991) for the outer regions from the top panel.** From Sofue and Rubin. (2001)

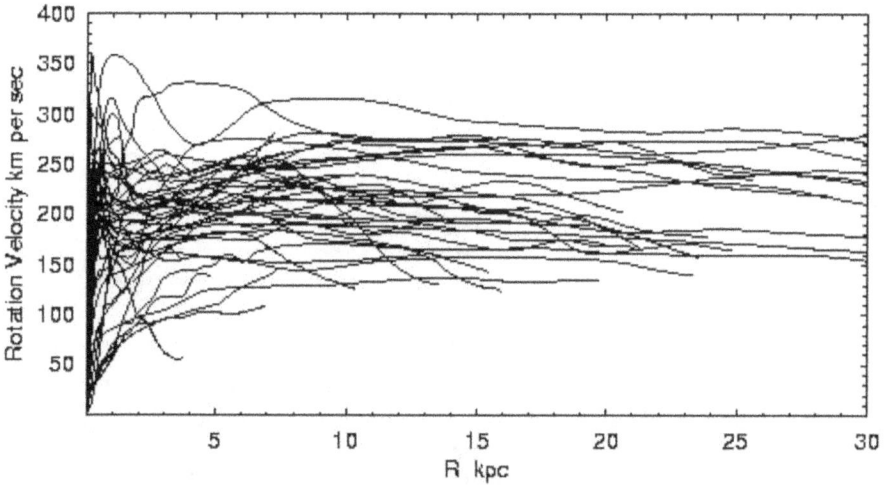

Figure 165: Rotation curves of spiral galaxies obtained by combining CO data for the central regions, optical for disks, and HI for outer disk and halo. Taken from Sofue and Rubin. (2001)

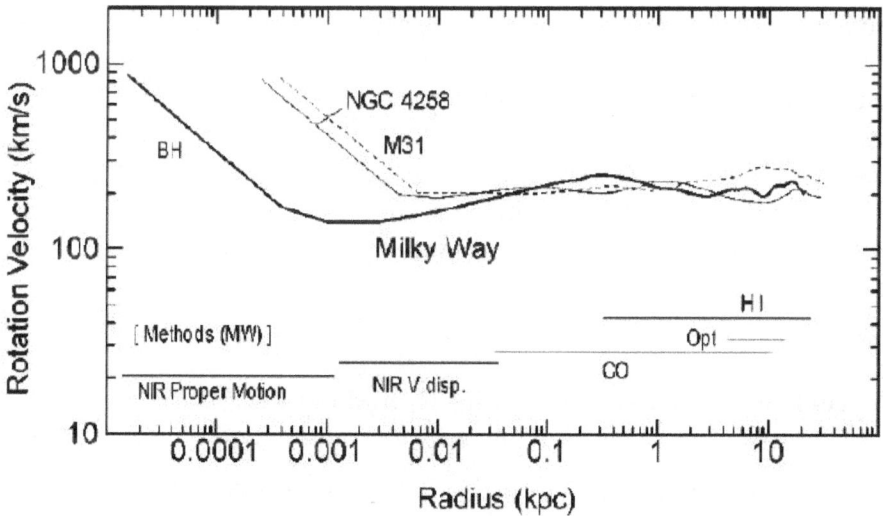

Figure 166: Rotation rates plotted on a log scale show the bulge rotation is Keplerian, in the case of the Milky Way galaxy, from the center out to 0.001 of its total radius. Beyond this, there is a clear flattening of the rotation curve. Taken from Haramein, et. al., (2008) who cite the original in Sofue and Rubin (2001)

341

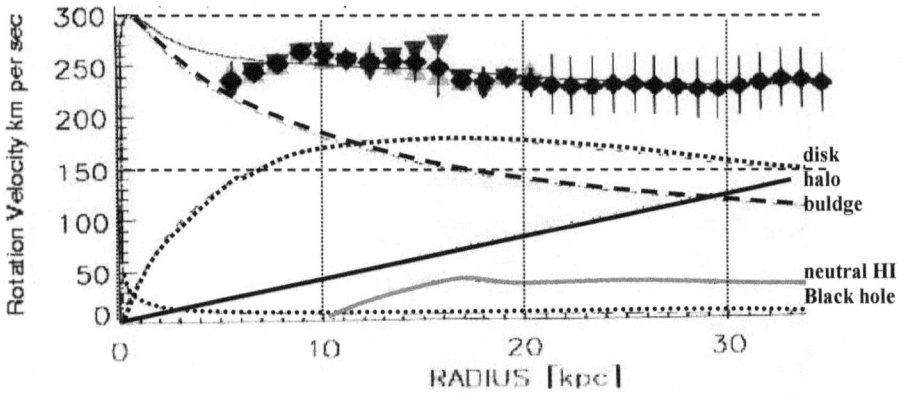

Figure 167: **Rotation rates vs. distance from the center of the galaxy M31. The individual curves denote the supposed contribution to the total rotation of the central black hole, hydrogen, the bulge, the dark matter halo, and the disk masses.** Haramein, et. al. (2008)

i. calculating contributions to galaxy rotation

In figure 167 taken from Haramein et al. (2008) the actual rotation profile of the galaxy M31 is shown with various superimposed curves. These show individual contributions to the total rotation curve from a central black hole region, dispersed hydrogen, the bulge, the various known luminous disc masses, and the generally-assumed dark matter halo. Again we encounter a perceived necessity for the illusive *dark matter* – in this case as a 'halo' surrounding the galaxy. Unique mass distributions M(r) each produce a unique velocity profile. We will consider each of the various possibilities illustrated in figure 167 using the formula derived in earlier sections of this chapter:

$$v_{rot}^2 / r = M(r) \, G / r^2$$

1. central massive bulge – radius R_b and density ρ_b

For a central massive bulge, $M(r) = (4/3) \, \pi \, \rho_b \, r^3$, if $r < R_b$, and $M(r) = (4/3) \, \pi \, \rho_b \, R_b^3$, if $r \geq R_b$, In these two situations we have:

$$v_{1b}(r) = \sqrt{(4/3) \, \pi \, \rho_b \, G} \; r, \text{ if } r < R_b - \text{linear increase, and}$$

$$v_{2b}(r) = \sqrt{(4/3) \, \pi \, \rho_b \, G \, R_b^3 / r}, \text{ if } r \geq R_b - \text{square root decrease.}$$

342

This is essentially compatible with what is observed as is readily apparent in figure 167.

2. **uniform thickness disk – thickness Δ_d and density ρ_d**

For mass distributed uniformly throughout the main disc, $M(r) = \pi \rho_d \Delta_d r^2$. So that in this situation we have:

$$v_d(r) = \sqrt{\pi G \Delta_d \rho_d(r) \, r} \quad - \text{square root increase.}$$

Clearly the density $\rho_d(r)$ of the disc must diminish with distance from the center in this case to account for the trailing component .

3. **uniform density spherical halo – density ρ_h**

For a surrounding halo of uniform density, the rotation velocity $v_h(r)$ formula is the same as for $v_{1b}(r)$ in situation 1 above except that the density ρ_h is considerably less than ρ_b, of course. Also, the uniform density continues out to a considerably larger distance in this case. Using the data from figure 167, let us see what that density turns out to be.

$$v_h(35 \text{ Mpc}) = \sqrt{(4/3) \pi \rho_h G} \; (35 \text{ Mpc}) = 1.5 \times 10^7 \text{ cm sec}^{-1}$$

So we obtain:

$$\rho_h = 6.88 \times 10^{-30} \text{ gm / cm}^3$$

This is the only mysterious contribution to the galaxy rotation curve. It is this linear contribution that compensates for the otherwise diminishing effects of the other velocity components. This required halo density value is a very conservative estimate of the density of intracluster plasma gas known to populate such regions in and around galaxies in any case. Hicks et al. (2002) indicate that 19 % of the cluster mass of the cluster PKS 0745-191 is in plasma gas. Refer back to figure 163.

This cluster plasma is primarily hydrogenous for which the mass density is related to the electron density as:

$$\rho_h = 1.67 \times 10^{-24} \, \rho_e$$

So that,

$$\rho_e = 4.11 \times 10^{-6} \text{ electrons / cm}^3$$

In the next chapter we will discuss the expected electron densities in and around clusters as shown in figure 168. Clearly densities of this magnitude persist out past five times the cluster core radius, and would certainly gravitate to surround large galaxies. In short, this particular 'dark matter' has already been accounted for.

In this chapter we have addressed the issues of how the mass of the universe is being assessed, the facts as they have been determined, and examples of galaxy clusters and individual galaxies where apparent incompatibilities have been noted. We have demonstrated how the mass determination techniques are applied to reach those conclusions that have been embraced by cosmologists. We have also hinted at the alternative solutions to the associated dilemmas.

But many issues have been left open with regard to how this data can be properly explained. What has seemed to demand that there be more mass than that associated with what has been observed seems almost inescapable. It is the exclusive use of the Doppler interpretation of redshift that attempts to account for all of observed 'spectroscopic velocities' as associated with the motions of orbiting objects. We will question these issues in much more detail in the next couple of chapters.

Chapter 17

Profound Implications of the Plasma Redshifting in Rich Cluster Cores That Produces Apparent Velocity Scatter

A major thesis propounded in this volume is that cosmological redshift is an artifact of forward scattering in a hot plasma. We have seen that the equivalent of Hubble's constant results if the average of the product of the kinetic temperature and density of plasma electrons is equal to 4.13×10^3 K cm^{-3}. As was noted in chapter 3, this exceeds the expected value of the product of these two parameters averaged separately. However, since a volumetric average of the product and not a product of volumetric averages is at issue, it is understandable that in an intergalactic medium that accommodates large variations in these two parameters, such an average dynamic pressure is actually realized.

In this chapter we will demonstrate how various characteristics of plasma gases within and between galactic clusters produce observed redshifting effects. The result is equivalent to what would be the case if it were all to have been effected by recessional Doppler, but with very major exceptions including resolution of the 'dark matter' dilemma.

a. partitioning properties of the intergalactic medium

In looking out into the cosmos, observations inevitably involve light paths that intersect clusters of galaxies along the way. The density and kinetic temperature of interior regions of these colossal structures are orders of magnitude more extreme than for the exterior regions. To quantify this difference we provide diagrams from Hicks et al. (2002) and Cowie and Perrenod (1978). Figures 168 and 169, show electron density and kinetic temperature data applicable to galaxy clusters. Peak

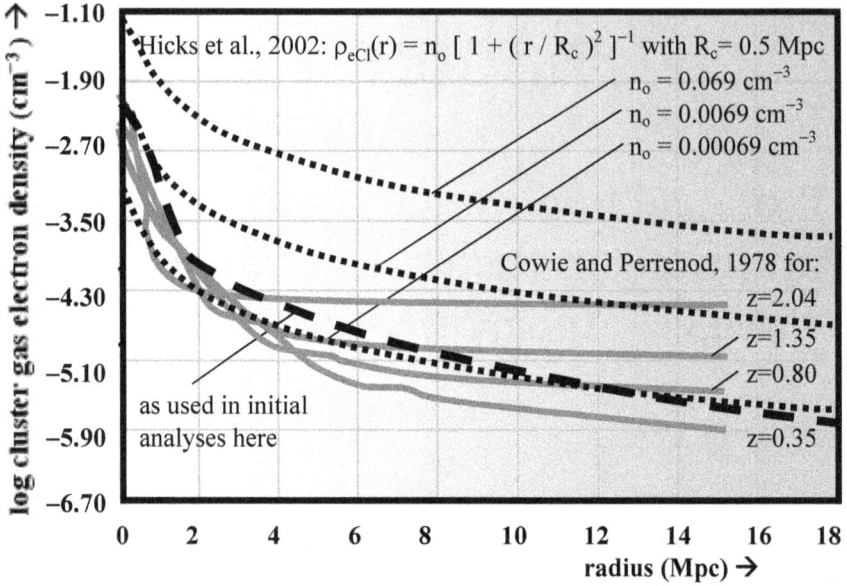

Figure 168: Electron density measures in intracluster plasma

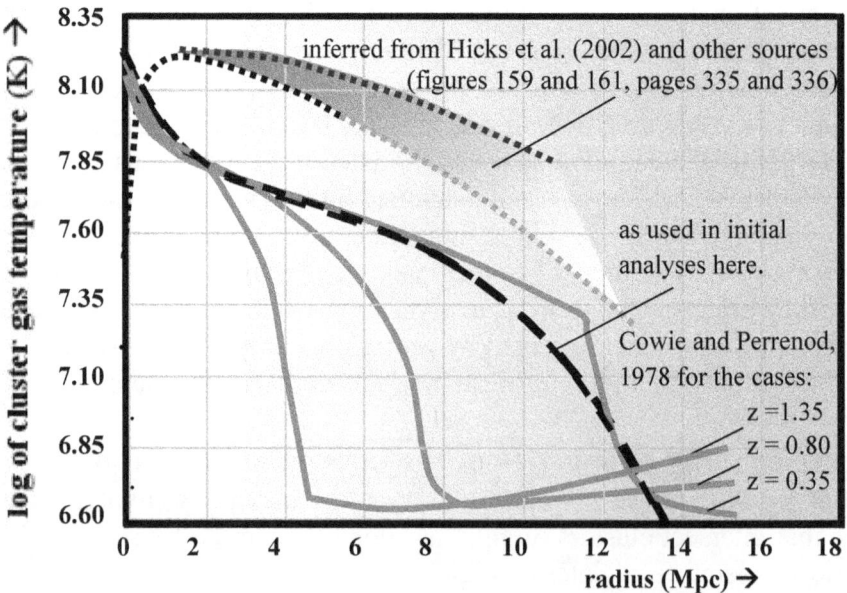

Figure 169: Electron temperature measures in intracluster plasma

values are quite dramatic at the centers of these regions and remain at much higher than average levels to many Mpc from centers of clusters.

In the vast regions between clusters these quantities are much smaller, however. In chapter 23 we will provide evidence based on microwave background radiation considerations for a universal average of related parameter values that just sustain ionization, whereas, as illustrated above, the following values (and in illustrated cases much more extreme values) are realized toward the centers of rich clusters:

$$\rho_{eCl}(0) = 0.027 \text{ cm}^{-3} \text{ and } T_{eCl}(0) = 1.71 \times 10^8 \text{ K}$$

In addition to plots provided by Cowie and Perrenod (1978), the approximations provided in the previous chapter taken from Bahcall (1999), Hicks et al. (2002), and Lowenstein (2003) are plotted as dotted lines in figures 168 and 169 from more recent observations of cluster parameters. We explore such situations in more depth further on.

The heavy dashed lines in the figures 168 and 169 were drawn using the following expressions as functions of radial distance from the center of such an (assumed spherical) cluster:

$$\rho_{eCl}(r) = 0.013 \times 10^{-1.615 \, r^{0.3}} + 0.0056 \times 10^{-.8 \, r^2} + 2.88 \times 10^{-7}$$

$$T_{eCl}(r) = [111 \times 10^{-.5 \, r} + 25 \times 10^{-0.01 \, r^2} + 35 \times 10^{-0.0003 \, r^4}] \times 10^6$$

where, in keeping with the premises of this volume, we assume no distance (i. e., no 'evolution' with cosmological redshift) dependence. We do not presume forms of these formulas have any significance with regard to modeling the underlying behavior of intracluster gases. The selection of these expressions was based entirely on conservative empirical considerations to obtain a reasonable fit from which to calculate associated effects. We use these formulas here to characterize electron density and kinetic electron temperatures of galactic cluster cores in our initial analyses.

Significantly, it is the product of these parameters that is of immediate interest to us with regard to effects on plasma redshift, whose derivation was provided earlier, as follows:

$$P(r) = T_{eCl}(r) \, \rho_{eCl}(r)$$

This product of the above expressions is plotted in figure 170. The parameters k_1 an k_2 are defined on page 334. Included in the figure is an indication of the average that must be realized to produce redshift equivalent to Hubble's constant. Clearly significant redshift will accrue through the core of such a cluster.

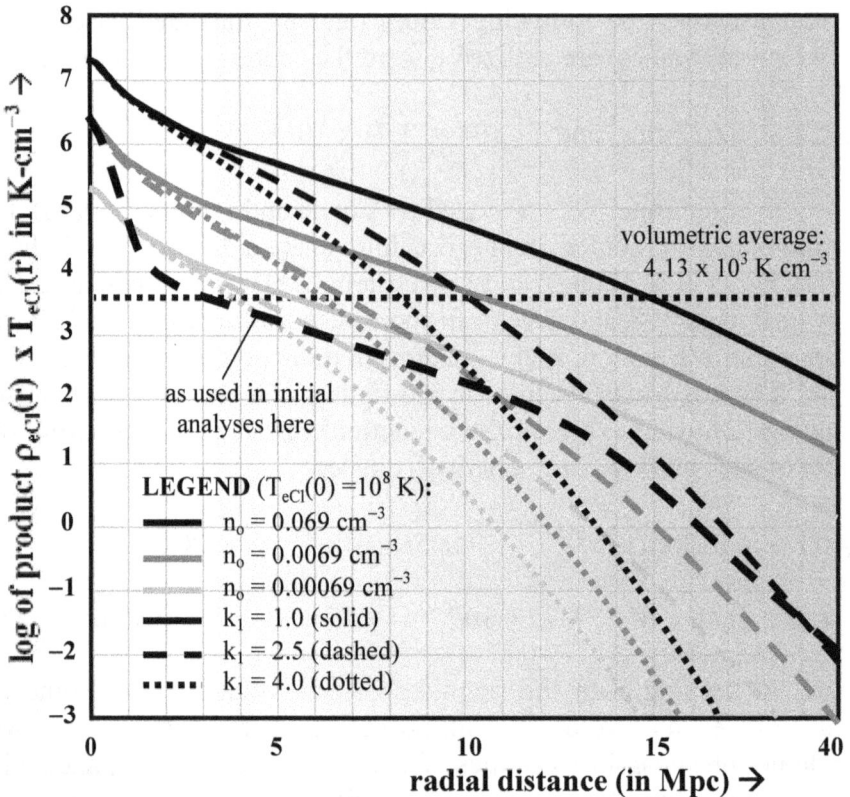

Figure 170: Dynamic pressure product of intracluster plasma parameters

b. determining static redshift scatter of cluster galaxies

With reasonably good data now in hand for plasma properties within galaxy clusters, we will proceed directly to show that these properties produce the appearance of extreme velocity scatter in galaxy clusters. Later whatever the distribution of actual virial imposed by gravitational energy of the cluster velocities can be added directly into this picture, but initially we will investigate only the impact of the plasma redshift on apparent velocity distributions exclusive of such added complexity.

In our analyses in chapter 11, beginning on page 211, we demonstrated the integration approach to determining the net redshift at a given distance. Now, however, instead of a constant product of density and temperature that could produce a uniform medium constant analogous to Hubble's, we have a variable value $H_{cl}(x,y,z)$ with which to deal. Recall that observed cosmological redshift attributed as a Doppler effect by the standard model was attributed by the scattering model to incremental energy losses incurred by propagation of light through a hot plasma that emulated the Doppler effect as follows:

$$\Delta v_x(x,y,y) = c\ H_{cl}(x,y,z)\ \Delta x, \text{ where } H_{cl}(x,y,z) = 2.42 \times 10^{-4}\ H_o\ P(x,y,z)$$

The incremental illusory effects must be integrated along the path that the light takes in getting to the observer as follows:

$$V_x(x_g,y_g,z_g) = 2.42 \times 10^{-4}\ H_o \int_0^{x_g} P(x,y_g,z_g)\ dx$$

where $x_g = X_{Cl} + r_g \cos \theta_g$ as illustrated in figure 171. Notice that $P(x,y_g,z_g) \rightarrow P(r,\theta)$ must be evaluated at each point along the path (r_x,θ_x) in order to perform the integration. Thus, we must obtain the radial distance from the center of the cluster to the position on the path.

$$r(x) = \sqrt{(x - X_{Cl})^2 + y_g^2}$$

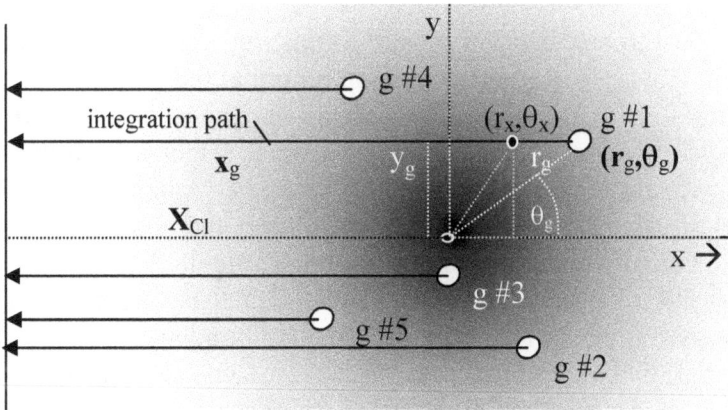

Figure 171: Determination of redshift 'velocity' of galaxy #1 in cluster plasma

Since we are assuming a symmetric plasma density and temperature distribution, θ_x does not explicitly enter into the determination of P(r). Then if the galaxy cluster is centered at a distance X_{Cl}, characterized by the plasma product parameter, P(r), let galaxy #1 be situated at the location (r_g, θ_g) relative to the center of the cluster, so that $y_g = r_g \sin \theta_g$. and we integrate from $x = x_g$ to 0 to evaluate the apparent 'redshift velocity' imposed on observations of any object situated at that position. Later we will incorporate the 'actual' velocity distribution appropriate to the symmetries of the situation, but it is important to separate out the extent to which the plasma redshift alters the inferred distribution.

In the integral equation above, we choose velocity in km sec^{-1} for the units of Hubble's constant, which is usual among cosmologists, although a tradition we have typically not honored for obvious reasons:

$$H_o \cong 7.14 \times 10^{-29} \text{ cm}^{-1} \cong 67 \text{ km sec}^{-1} \text{ Mpc}^{-1}$$

We will take X_{Cl} as 100 Mpc for this analysis, which is appropriate to the distance of the Coma cluster from earth.

$$V_x(x_g, y_g, z_g) = 0.016013 \int_o^{X_g} P(x, y_g, z_g) \, dx$$

In figures 172 and 173 we see the amount of 'redshift velocity' imposed as a function of angle and radial distance from the center of the cluster. The next figure provides detail for the insert indicated at the top right.

Just to be clear with regard to this tremendous additional (quote) 'velocity' that is being added by assuming that redshift is all caused by a Doppler shift in wavelength, an additional figure is provided showing the redshift 'velocity' as a function of radial distance from the center of the cluster at the various angles as figure 174. Again, it is apparent that within the angular dimensions of the cluster there is a huge amount of pseudo velocity being accounted as actual velocity.

The obvious fact to be taken from these figures is that even without virial velocities of gravitationally bound systems, spectroscopic 'velocities' will take on the appearance of speeds that are significant relative to the speed of light based on these redshifts. This extreme extraneous contribution is not limited to galaxies at specially selected points within the cluster, but apply generally to positions close to the center and those at extreme distances from the center. So it is key to determine what this does to the expected scatter of velocities.

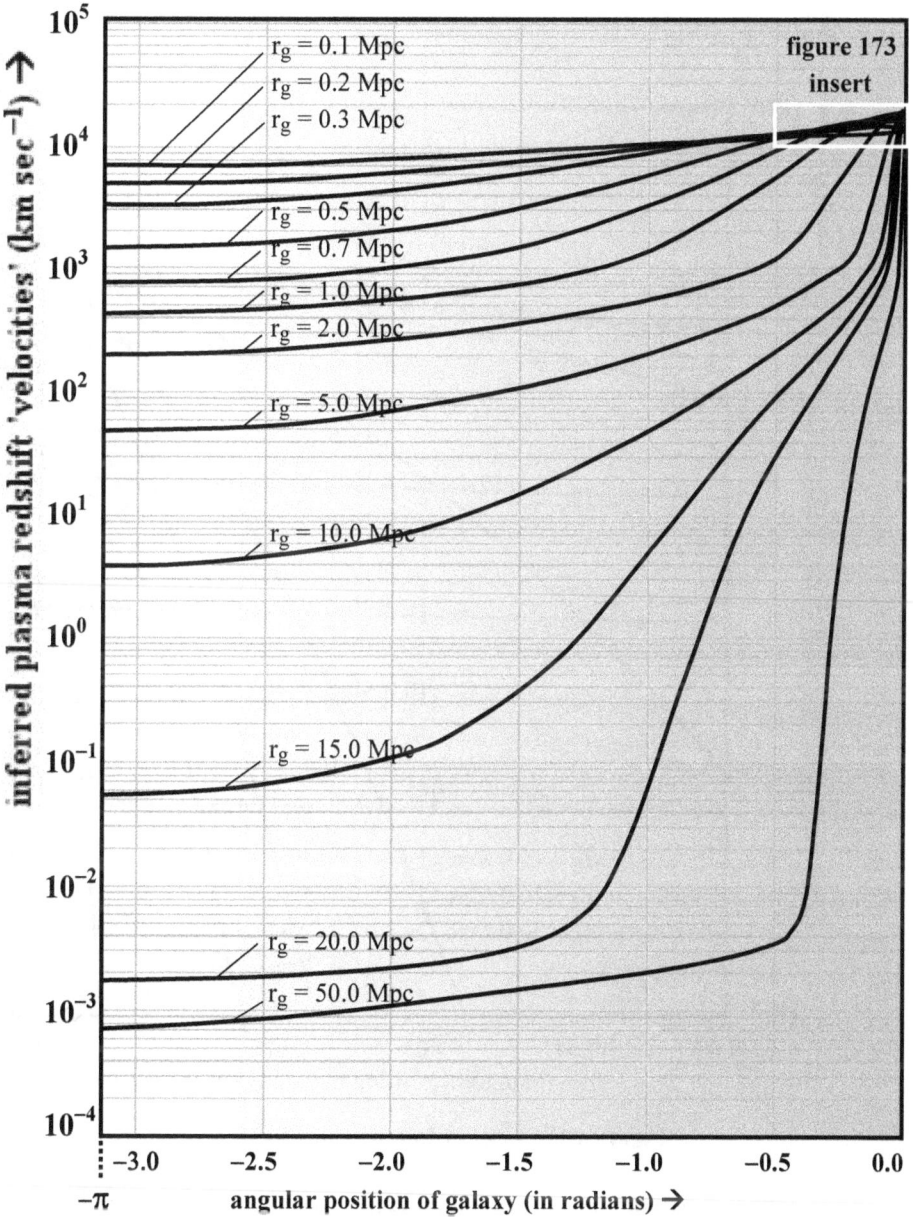

Figure 172: Redshift 'velocity' of galaxies at various angles and radii in cluster

For a cluster centered along the x axis at a distance of 100 Mpc, the angle α_g to the galaxy identified as g will be approximately:

$$\alpha_g \cong \sin^{-1}(y_g / X_{Cl}) \cong 0.01 \, y_g \text{ radians.}$$

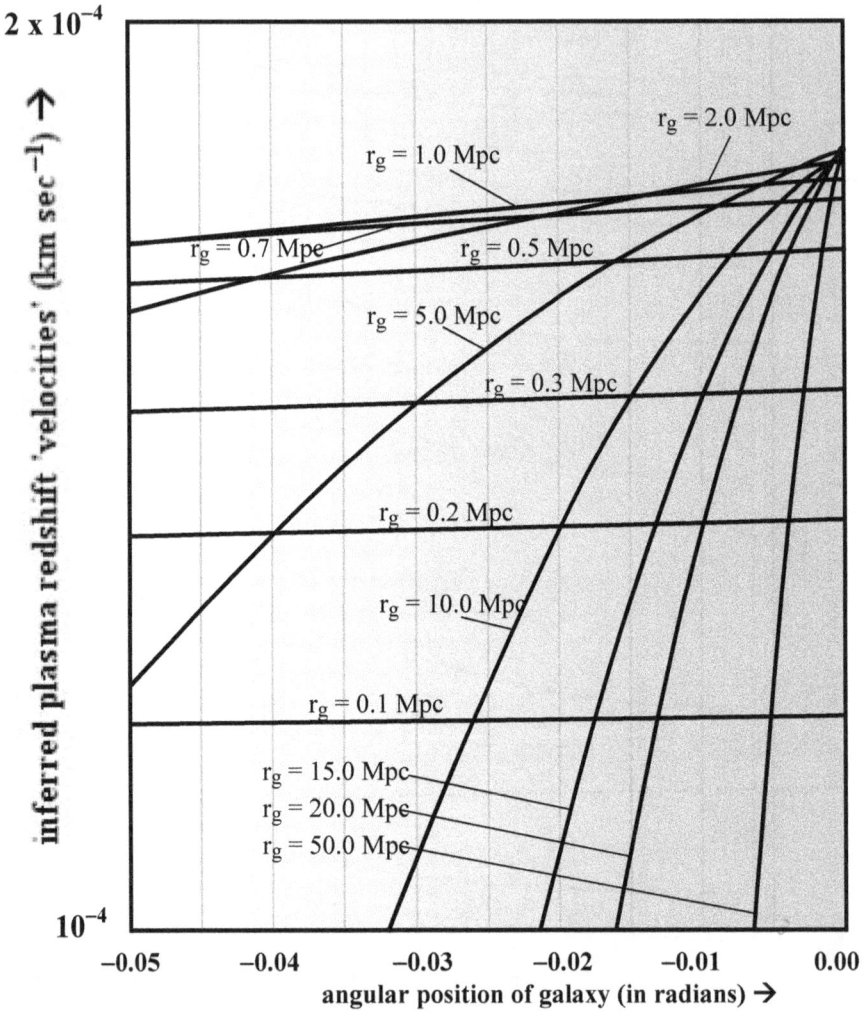

Figure 173: Insert from previous figure

The dense concentration of galaxies within a couple of Mpc of the center recommends this approximation. By far the largest percentage of the 'velocity' scatter is concentrated at the center of this domain, the validity of the approximation is certainly apparent in figure

175. In this figure it can readily be seen that the extreme extent of the rays of galaxies denominated 'fingers of god' in redshift surveys derive from this rich plasma core.

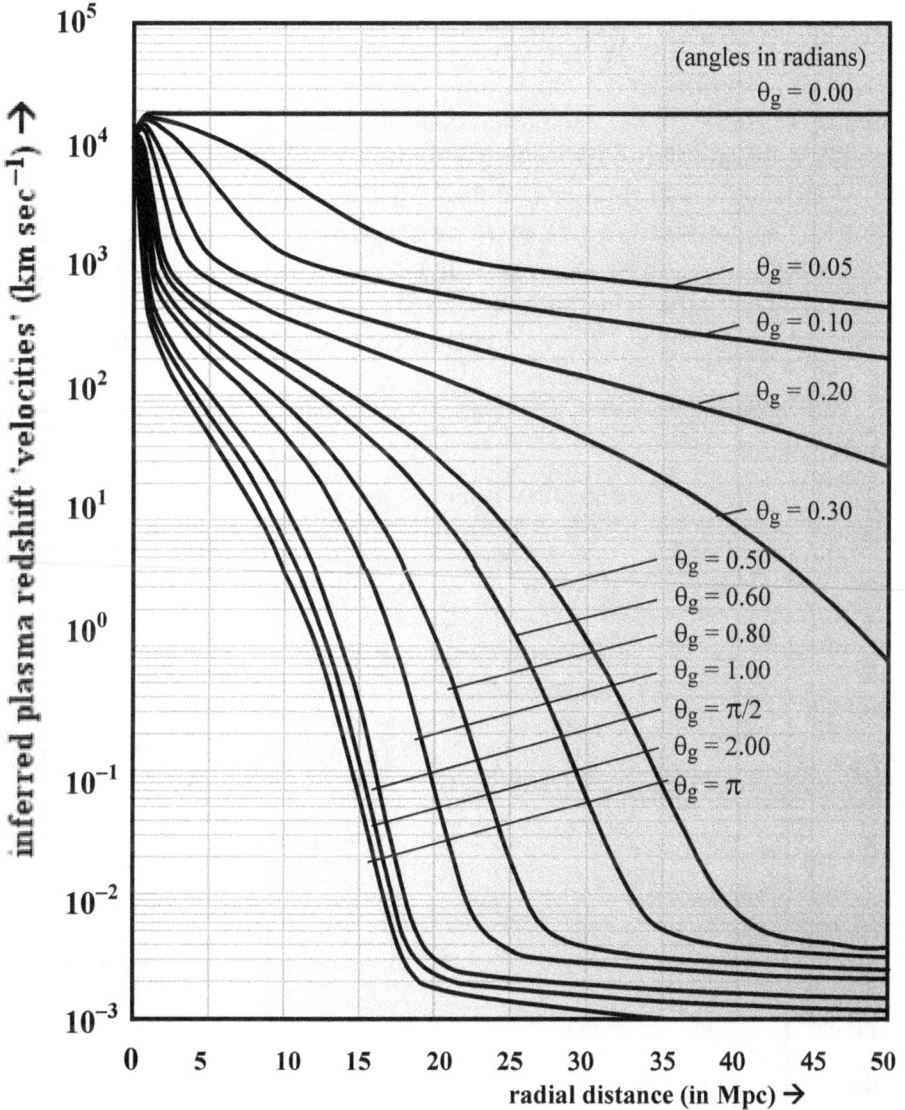

Figure 174: Redshift 'velocity' of galaxies at various angles and radii in cluster

The significance of a punctuated plasma redshift through intergalactic space, in contrast to what has been considered a uniform 'Hubble flow' associated with the exclusive interpretation of Doppler recessional expansion is apparent. Clearly clusters encountered along every line of sight into deep cosmic regions support the same effectual redshifting at large scales, which Hubble noted as the domain to which it applies as a distance indicator. It is significant to note that although we have included only one cluster situated at a distance of 100 Mpc more or less similar to the Coma cluster, that there are lesser and greater clusters to be encountered at intervals and that each cluster includes a penumbra of plasma gas with redshifting capabilities similar to what is shown in figures 172 through 173, and slightly differently in figure 175. So that the less extreme spectroscopic velocity increments at the bottom of these figures will be encountered at more frequent intervals, with all effects along a line of sight being cumulative. So the first 95 Mpc in figure 175 would also experience some redshifting.

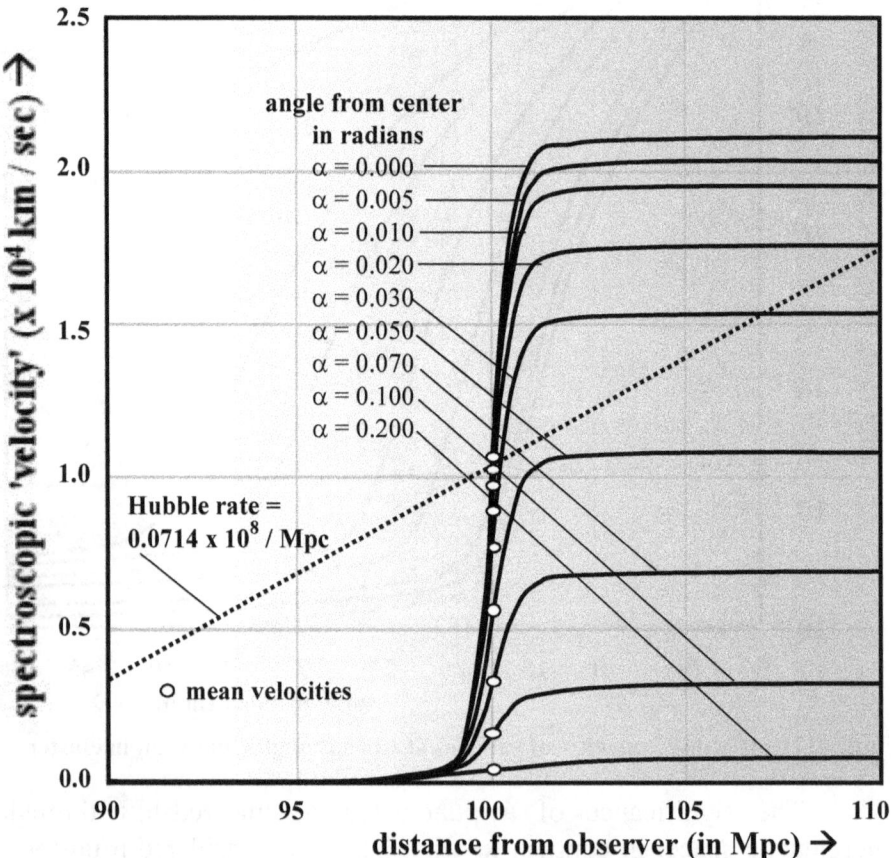

Figure 175: Plasma redshift scatter at various angles through a galactic cluster

In figure 175 a velocity spread at each value of α_g does not take into account the distribution of galaxies from which the redshifted light is being received. It merely represents the amount of redshifting to be expected if there were a 'stationary' galaxy as a source of radiation at that distance and angle from the center through the cluster

In all of these preceding plots we merely assumed a uniform distribution of a set of identical galaxies. We will now assume gaussian distribution of galaxies in clusters that are characterized by a mean radius of concentration as follows:

$$\rho_g(r)\, dV(r,\theta,\phi) = (N_{Total} / \sqrt{2\,\pi\,\sigma^2}\,)\, e^{-r^2/2\sigma^2}\, r^2 \sin^2\theta\, d\theta\, d\phi\, dr,$$

where σ is the standard deviation of density concentration, with other parameters as defined in figure 176 below. Defining the distribution of position and velocity of galaxies in rich clusters is used by researchers in attempts to understand what produces the 'fingers of god'. Jarrett (2006), as example, modeled spectroscopic 'velocity' scatter of cluster galaxies with gaussian distributions of position and velocity of galaxies primarily out to one or two Mpc from the center of clusters. Of course Jarrett assumed that 100% of 'spectroscopic' velocities were virial effects rather than a major portion of these being caused by plasma pressure. So his conclusions differ substantively from ours.

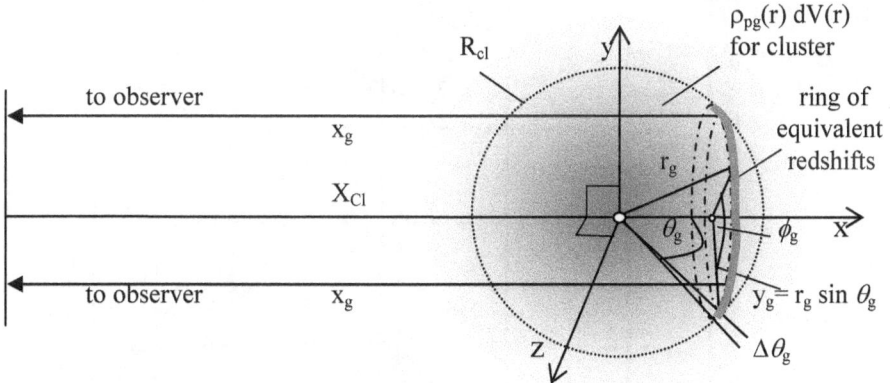

Figure 176: Simplifying geometrical symmetry of galaxy cluster distribution

In our case, we addressed merely components of spectroscopic velocities attributable to location-dependent properties of the plasma parameters in cluster gases. In such an assessment baryonic mass of the cluster is all that is relevant since dark matter is part of the counter conclusion unnecessary in the scattering model. We will consider here only aspects of the probability density that pertain to the angular effect of plasma redshift on what is observed. This involves $2\pi \rho_g(\theta_g) \Delta\theta_g$, which is the distribution of cluster galaxies at various observed angles relative to the center of the cluster. We assume $\theta_g \cong y_g / X_{Cl}$. This is a valid simplification as long as $y_g << X_{Cl}$. That is the case in situations in which we will be primarily interested, as for example, the Coma cluster. In that particular case $y_g < 3$ Mpc and $X_{Cl} \cong 100$ Mpc. Although the distance of a planar cross section at X_{Cl} for which 'stationary' galaxies at each angle θ_g exhibit their mean spectroscopic velocity, these mean velocities in each annulus differ appreciably as illustrated by open circles in figures 175 and 177.

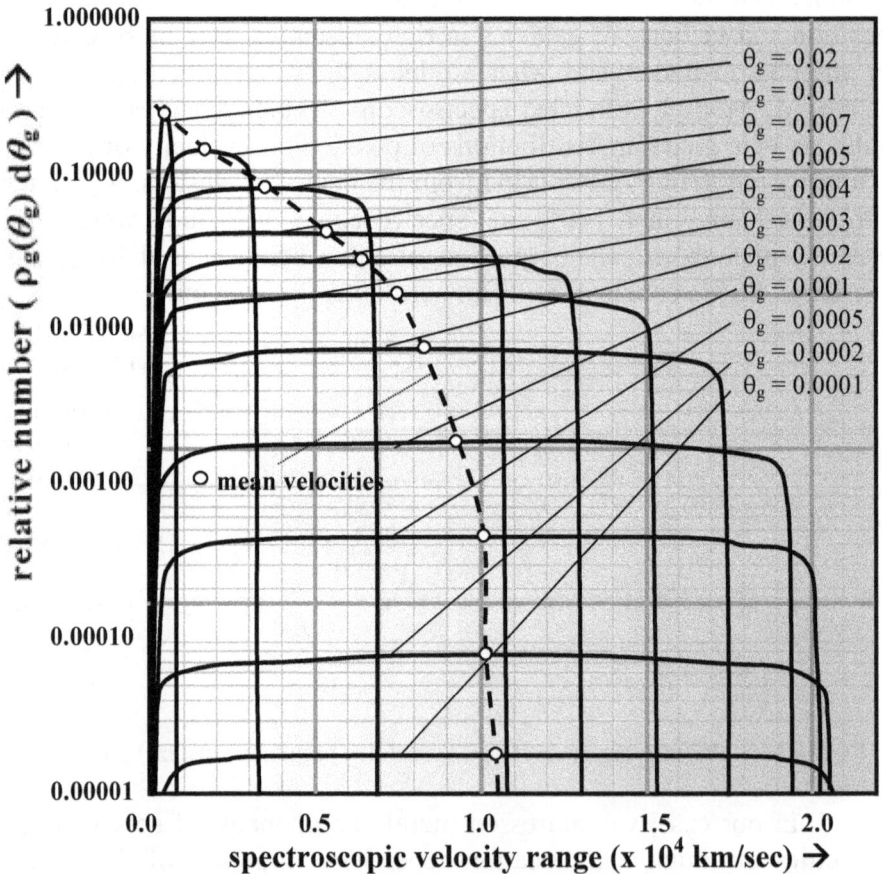

Figure 177: Plasma redshift scatter at various angles through a galactic cluster

A significant aspect of observed distributions of 'spectroscopic' velocities is that they are asymmetric about the value of the Hubble flow at the cluster center. In the previous chapter (refer to figure 155 on page 327) we noted the observed asymmetric skew. This is very unlike what would be observed if only virial velocities were involved. Thus arises a significant difference between appearances of the plasma and virial redshifts in this regard. This disambiguating characteristic provides another refutation test that can be used to distinguish the standard and scattering cosmological models.

Because of assumed cluster symmetry, observation from a great distance as shown in figure 176, we were able to reduce the dimensionality so that representation requires only the two parameters, still providing integrity to the analysis used to provide the illustrations. All galaxies at the angle θ and distance r share observational properties, which depend primarily on distance in the x direction; only the numbers in each category are increased in varying degrees by this consideration.

$$\rho_{pg}(r)\, dV(r,\theta) \;=\; N_{Total}\, \sqrt{2\,\pi/\sigma^2}\; e^{-r^2/2\sigma^2}\, r^2\, \sin^2\theta\, d\theta\; dr.$$

In the above probability distribution equations N_{Total} is the total number of galaxies in the cluster, determined by volume integration as:

$$N_{Total} \;=\; \int_0^{R_{Cl}} \rho_g(r)\, dV$$

Here R_{Cl} is taken as the effective escape radius of the galaxy cluster, beyond which no galaxy will be considered associated with the cluster. This escape radius discussed in the previous chapter is determined by the maximum value of the orbital velocity scatter. This was addressed in the discussion of the virial theorem.

The obvious fact to be taken from the figures in the previous section is that even without addressing orbital velocities of the galaxies there would be a broad distribution of 'spectroscopic velocities' as a function of spatial distance relative to the center of the cluster given by:

$$\rho_{vg}(r)\, dV(r,\theta) = V_x(r_g,\theta_g)\; \rho_{pg}(r_g)\, dV(r,\theta)$$

In Cartesian coordinates (x_g, y_g) we determined the distribution in terms of the transverse measure that is an observable, y_g, where

$$\rho_{vg}(y_g)\Delta y_g = 0.016 \sqrt{2\pi/\sigma^2}\, N_{Total} \int_0^{R_{Cl}+110} e^{-(x_g^2+y_g^2)/2\sigma^2} \int_0^{R_{Cl}} P(x,y_g)dx\; dx_g\, \Delta y_g$$

Of course the actual virial distribution of velocities must be superimposed on this. Clearly what is left after what is shown in figure 175 will be well within expectations of the baryonic – and therefore luminous – mass of the cluster. It certainly does not require additional mysterious 'dark' mass.

What this does is reduce the contribution to inferred virial 'velocity' to one half the value indicated in the figure, i. e., to about 1.1 x 10^4 km/sec in keeping with the range of velocity scatter identified by Bahcall (1999) as detailed in the previous chapter.

c. follow-on analyses with regard to galaxy clusters

Then one should proceed as Jarrett did to allocate that velocity remainder among galaxies. In any attempt at a totally realistic model that process would employ the Schecter distribution rather than assume the same mass for all galaxies in the cluster, etc..

These analyses should be provided as a follow-on to the basic treatment of the mechanism of plasma redshifting provided in earlier chapters in order to clarify its expanded appropriateness to a universe with variations in the properties of this intergalactic – and particularly intracluster – medium. In the next chapter we will employ these and other analyses to do a fuller accounting of observations that have lead most cosmologists to acquiesce to the concept of dark matter.

It does strike one as somewhat non-intuitive that the intense redshifting encountered across the relatively short diameters of rich galaxy cluster cores (on the order of one or at most a few Mpc), which are separated by much more vast distances, could somehow average out to effect Hubble's constant of cosmological redshift. So let us look at how this plays out by actually performing the averaging procedures.

The first step in this process is to note that clusters of galaxies are the primary contributors to the baryonic mass of the universe – ignoring for the moment what has been attributed to 'dark' matter, which also, of course, is perceived as primarily a cluster phenomenon.

There are three aspects to this averaging that must be taken into account. First of all, the distribution of 'average size' clusters must be what determines the ultimate average baryonic mass density of the universe as a whole. This we will later assess as on the order of, ρ_u = 7.6 × 10^{-31} gm cm^{-3}. From such an estimate we can constrain the mass and cell size of the average cluster:

$$3\, M_{Cl} / 4\, \pi\, R_{Cl}{}^3 = 7.613 \times 10^{-31} \text{ gm cm}^{-3}$$

If $M_{Cl} \approx 10^{47} - 10^{48}$ gm is a reasonable upper limit of the average *baryonic* mass of a galactic cluster as we suppose along with Bahcall (1999) and others. Then for its average radial dimension R_{Cl} we have:

$$R_{Cl} \cong \sqrt[3]{M_{Cl} / (4\pi / 3) \text{ x } 7.6 \text{ x } 10^{-31}} \,] \,/ (3.26 \text{ x } 10^6 \text{ x } 9.46 \text{ x } 10^{17})$$

$$\cong 11 - 24 \text{ Mpc}$$

A sphere of a radius in this range would contain one cluster on average. So to obtain a volumetric average of dynamic pressure, we compute:

$$P_{ave}(R_{Cl}) = 3\, R_{Cl}{}^{-3} \int_0^{R_{Cl}} P(r)\, dr = 4.13 \text{ x } 10^3 \rightarrow H \text{ , where}$$

$$P(r) = \rho_{eCl}(r) \text{ x } T_{eCl}(r)$$

The other average of interest is a line-of-sight distance between regions of high dynamic pressure of radii R_{cc} that contribute most substantially to cosmological redshift in viewing the distant cosmos.

$$d_{l\text{-}o\text{-}s} = (4\pi / 3)\, R_{Cl}{}^3 / R_{cc}{}^2 \cong 150 - 300 \text{ Mpc}$$

where R_{cc} is the radius of the core region of interest with regard to the most intense redshifting shown in figure 170 as ranging from about 3 upwards to 15 Mpc. The determination of this line-of-sight distance is illustrated in figure 178, where it can be seen that a line-of-sight separation between intense redshifting regions is on the order of a hundred to a couple of hundred Mpc. This accords well with the degree to which spectroscopic redshift is added in passing through these regions as shown in figures 175 and 177.

The reader will remember that originally Hubble noted that redshift could only reliably be used as an indicator of distance for observations of objects that are well in excess of several hundred Mpc. Refer to figure 11 on page 27 where it is manifestly obvious that the dimensions for such an averaging are indeed appropriate.

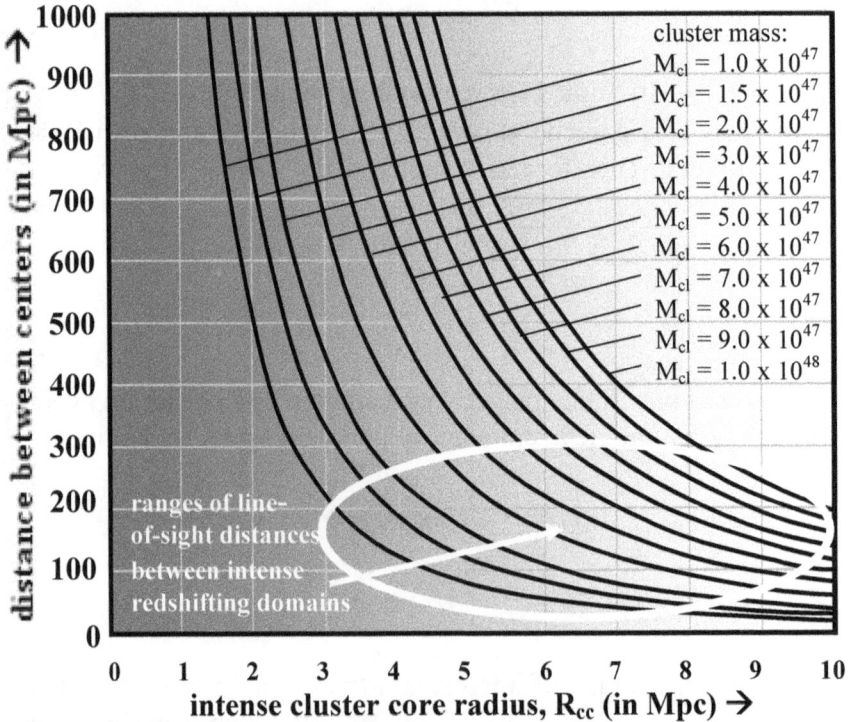

Figure 178: Determination of average line-of-sight separation of galaxy clusters

We will discuss a ripple effect in redshift surveys in the next chapter, where it is obvious that an observation artifact is involved that manifests itself as a density wave emanating outwards from our observation location here in the Milky Way. This results from the spacing of the most intense redshifting centers along any line of sight.

As a result, we have an expression for the redshifting capability of the plasma at each point within it as follows:

$$H_{igm} = H_o \, P(r) \, / \, 4.13 \times 10^3$$

H_{igm} is not a constant, of course. Redshifting will not be a uniform phenomenon throughout intergalactic space. In fact, as we will show, this gives rise to very strange observation artifacts in virial analyses and redshift surveys. Thus, we arrive at an allocation of electron density and kinetic electron temperature of critical regions of intergalactic space, that although matching the effect of Hubble's 'constant' as characterizing cosmological redshift on average, does not result in an identical constant value at every point in space.

Chapter 18

Assessing Requirements for, and an Alternative Explanation of 'Dark Matter'

We have seen that when one comes to assess just how much matter there is in the universe, one is quite literally face-to-face with the dilemma of whether there may be more to this issue than meets the eye. That has ultimately come to characterize the final conclusion of the matter rather than constituting the rationale for reopening the discussion of why the Doppler interpretation has not been subjected to much more scrutiny. In this chapter we will provide that much needed scrutiny.

a. current rationale for 'dark matter'

To the question, "What is dark matter?" the only answer that can be given is that it is any material substance that doesn't involve itself in electromagnetic interactions. In contrast, the matter with which we are familiar is primarily 'baryonic', i. e., the protons and neutrons that comprise the nucleus of the known elements. Together with much lighter electrons, these nuclei are intimately involved in the emission and absorption of radiation. They are also involved in what we will investigate further on as the "thermalization" process. In short, every substantive 'thing' that laborious research has brought into the realm of what is *known* is comprised of what can be directly measured and observed. Physics, chemistry, biology, and anything pretending to be a 'hard science' concerns itself with what can thus be *observed*.

With this admittedly inadequate understanding, let us try to comprehend how 'science' is proceeding without the advantage of this precedent. Let us consider, for example, comments like the following from *Science Daily* (Jan. 12, 2005),

"The 2dFGRS [galaxy survey] has shown that baryons are a small component of our universe, making up a mere 18% of the total mass, with the remaining 82% appearing as dark matter."

"Shown", "appearing"? That's what the statement says, is it not? Science seems not to understand its roots in *refutation*. The reference is to a survey shown in figure 141 taken from Matthew Colless's 2dFGRS Image Gallery. Notice that the pattern exhibited there is claimed to have collaborated the microwave background distribution with regard to revealing the structure to the universe. It doesn't. It is certainly not uniform, but to suggest that it might only be *non*uniform to one part in 10^4 or 10^5 is absurd. Such extremely minor variations *are* what are found in the background radiation as we will see in another chapter. Structure in the universe is clearly much more pronounced than that.

Secondly, it is demonstrated quite clearly in CFA survey data provided in figure 158, that galaxies tend to be organized in rays like spokes in a bicycle wheel spreading out from our observation point in the Milky Way, as for example the rays extending out at approximately right ascension 5^h, 13^h, and 15^h to 16^h, etc. in figure 158.a. When many slices are mapped together as in figure 141 or 158.b, this predominant feature becomes muted by the additional slices of data so that what might seem to be a mere artifact of viewing no longer obscures these patterns in the data. But are these really just artifacts of limited data?

The CfA Redshift Survey was started in 1977 by Marc Davis, John Huchra, Dave Latham, John Tonry, and others including Margaret Geller was completed in 1982. The early thin slice of data obtained (similar to panel a of figure 158) clearly showed these rays of galaxies. As data was added by juxtaposing additional slices of observations, these radial lines became obscured in more comprehensive surveys. This is shown in panel b of figure 158 and in the 2dFGRS data in figure 141, and increasingly on even more comprehensive recent data published by the project. But on each slice the rays still predominate.

These 'spokes' were apparently termed "fingers of god" by Brent Tully. The assumption has been that they are caused by Doppler shifts associated with the peculiar scatter velocities of galaxies within a cluster. These velocities are in turn assumed to have all been produced by the gravity of the cluster and therefore subject to the virial theorem. Such 'real' velocities do, in fact, alter the observed redshifts of galaxies in the cluster making major deviations from any cosmological distance-redshift relation. Redshift surveys map each observed galaxy to a location at a distance from the center of the map that is proportional to its redshift. Redshift on such diagrams is – first and foremost – presumed to have been caused by the radial Doppler effect and cannot distinguish between individual motions and presumed cosmological expansion or 'Hubble flow'. There is a somewhat related 'Kaiser effect' (Kaiser, 1987) and Hartnett and Hirano (2008) have identified ripples (actually more like major waves) in these redshift surveys also centered at our peculiar position in the universe that we will discuss further on.

In panel b. of figure 158, at about the right ascension 16^h direction from our position in the Milky Way, there is a 'spoke', if one might call it that in preference to *finger of god*, whose length extends 5,000 km/sec. There are many more such clusters with nearly as long an extent as for example at 13^h and at about 5^h. Our 'local' supercluster (Virgo) appears at $12^h\ 27^m$; it has about 1500 galaxies. The extent of these anomalies on survey images such as figure 158 a. seem even greater to the naked eye than the velocity dispersion characterized by Bahcall (1999) and Becker et al. (2007) as discussed in summary above. If we interpret the lines in figure 158 as actual velocity dispersion within a single rich cluster, it would imply that some of these massive galaxies are moving about each other at nearly 5 percent of the speed of light. Since at least two-thirds of the mass responsible for such motions has been attributed to 'dark matter' as we saw in our calculations in chapter 16, the 'actual' velocity extent in all these cases would be less than half as large if only luminous baryonic matter were involved, with the remainder accounted for in other ways as discussed in the earlier chapter. Clearly, it is still a large velocity for so massive an object as a galaxy in a cluster, but would be a much more reasonable value for such large structures.

Jarrett (2006) has graphically addressed how this "spectroscopic velocity" issue is being handled academically. A Gaussian position and velocity distribution of galaxies within 1 to 2 Mpc of the center of the galaxy cluster is assumed. His effort with regard to the Coma cluster involved 390 sources. In the Hercules cluster he says, "Even after removing the *finger*, it is clear that another finger exists undoubtedly due to a nearby cluster."

b. What acts as 'dark matter' and where does it occur?

Since dark matter has taken center stage in cosmology, it has naturally received broad attention. Investigations to determine what it is has developed into a major field of astrophysics. Conjectures to account for it range in scale from subatomic to astronomical objects.

Particle physicists and speculative cosmologists have developed multifarious theories concerning its nature. Theories include 'cold' (CDM) explanations and 'hot' ones (HDM), incorporating baryonic or weirder particles. Theorists develop predictions for their theories that hopefully can then be tested against observation. The currently favored theory is that it consists of virtually invisible, slow-moving, collision-less particles that possess only gravitational attractions. Currently the CDM approach seems to be more compatible with observations than HDM theories involving excess neutrinos or hypothetical axions, for example. Theories of particle physics that predict Weakly Interactive Massive particles (WIMPs) tend to corroborate the CDM approach. New particle accelerators like the Large Hadron Collider are designed to find them. The Massive Astrophysical Compact Halo Objects (MACHOs) explanation involves only 'normal' baryonic matter in a wide variety of objects that might remain unobserved. Remnants of dead stars including black holes, neutron stars, white dwarfs, brown dwarfs, and massive planets are among the proposals. All these objects remain unobserved because they emit relatively little radiation. So, many dark matter candidates have been proposed, each having inherent strengths and weaknesses – none are convincingly real.

An explanation proposed by Milgrom (1983) challenged the very physics by which one performs mass assessments, proposing to MOdify Newton's laws of Dynamics (MOND). MOND theories thereby successfully explain the dynamics of galaxies and clusters but are phenomenological, strictly speaking *irrefutable*, with no theoretical basis. They are incompatible with relativity (and all other) theories.

The assessed demand for dark matter to accommodate observed dynamics varies from a factor of unity to a hundred times the observed mass of associated gravitational structures. There is definitely a trend, although it too seems to have escaped observation, in which those objects involving the hottest and densest plasmas tend to exhibit the largest factor of *dark* with regard to *baryonic* matter. Galaxy clusters may exhibit a factor of 5, super cluster cores may involve a factor of 10 or more. Dwarf galaxies exhibit a factor of ten or more (Milgrom, 2009) .

c. a different explanation of the velocity dispersion

In chapter 16 we performed calculations that illustrated the established view of how much dark matter is present in galaxy clusters and spiral galaxies. Now we will perform a similar calculation for what we consider to be a more viable alternative approach.

In earlier chapters we discussed the rationale for redshifts resulting from forward scattering in a hot plasma (i. e., one containing electrons possessing relativistic speeds).   Initially, this thesis was developed initially under the simplifying assumption of a uniform intergalactic medium with a dynamic pressure requiring the product of electron density and kinetic temperature of 4.13 x 10^3 K cm^{-3}. In the previous chapter we extended this approach to incorporate a variable dynamic pressure that averages to that same value over all of space, but produces redshift in passing through the medium that is commensurable with the value at each segment of the light path. We provided detailed information with regard to the implications of redshifting in plasmas known to exist in the galaxy cluster cores and found these redshifts to be roughly equivalent to the 'spectroscopic velocities' that have given rise to the dark matter conjectures.

Now let us examine specifically what intra-cluster plasma characteristics would be required to produce the redshift dispersion that is the observable aspect on which dark matter theories are based. In particular the data provided by Bahcall (1999) and others with regard to the cluster core plasmas justifies the claim that it would exhibit itself as equivalent to an actual velocity-induced Doppler redshift scatter.

Densities of the plasma medium in intracluster domains are significantly greater than in intergalactic regions as we have discussed. Electron densities in such plasma have been observed to be as high as $\rho_{ic} = 10^3$ cm^{-3} according to Bahcall, but we have used considerably lesser numbers that seem more typical to produce the scatter effect.

Intracluster plasma temperatures, according to the table provided in chapter 16, section g, are in the range of 3.2 x 10^7 to 2.25 x 10^8 K as shown also in figure 159 and 161. However, Marshall et al. (1980) and others more recently have observed that X-ray background associated with such clusters exhibits temperatures as high as 40 keV (i.

367

e., approaching 10^9 K). So that a volumetric average of 4.13 x 10^3 K cm^{-3} as illustrated in figure 170 on page 348 is inherently reasonable.

As we showed in an earlier chapter, the effective redshift *constant* predicted by the scattering model for the intergalactic medium is directly analogous to Hubble's constant. However, it is the average of a product of the temperature and density of the plasma medium that determine its value. So, for the intracluster plasma we would also have:

$$H_{cl} \cong 1.146 \times 10^{-32} \, T_{cl} \, \rho_{cl} \, cm^{-1}$$

For the intergalactic medium as a whole, a much lesser product of temperature and electron density was required in order to match Hubble's constant. Estimates discussed in chapter 3 above suggested that ρ_e would be less than 10^{-5} cm^{-3}. But the intracluster electron density exceeds this by at least three orders of magnitude with the electron temperature in excess of 10^8 K in many cases. So for observations through the central core of rich clusters the effective redshift constant could be as much as a hundred times larger than suggested for the intergalactic medium in our earlier analyses:

$$H_{cl} \sim 100 \, H_o.$$

In observing cluster galaxies across the diameter D_{cl} of the central region of a rich cluster as illustrated in earlier diagrams and figure 179, there would necessarily be a considerable augmentation to the otherwise applicable cosmological redshift just because of the dense cluster plasma: This additional contribution would be on the order of:

$$Z_{cl} \sim 100 \, H_o \, D_{cl} \cong 100 \times 1 \, Mpc \times 71.4 \, km \, sec^{-1} \, Mpc^{-1}$$

$$= 7,140 \, km \, sec^{-1}$$

Since the Doppler interpretation of redshift assumes that, $Z_{cl} = \sigma_{cl} \, / \, c$, this predicted redshift is indeed equivalent to spectroscopic 'velocities' that show up in the spokes of the CFA survey data as shown in figure 158. The results appropriate to observed clusters was plotted in figure 175 in the previous chapter.

This approach has not required the introduction of mysterious 'dark matter', nor does it impose the otherwise-required excessive mass-to-luminosity ratios for clusters of galaxies. It assumes nothing about the plasma other than the observed plasma.

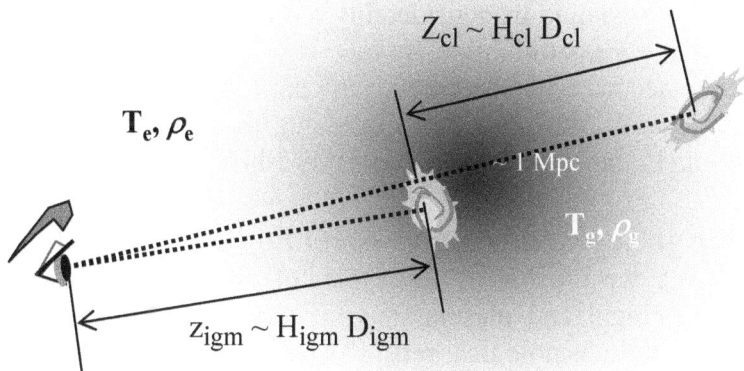

Figure 179: Increased plasma redshift effect through the core of galaxy clusters

Let us check this another way using simplifying assumptions concerning the plasma distribution: If, as in our example employing the virial theorem in chapter 16, we accept $\rho_{gas} = 0.07 \, \rho_{gal}$, a uniform distribution through the core we would exhibit an electron density of,

$$\rho_{cl} \cong 7.0 \times 10^{-4} \text{ cm}^{-3}$$

The velocity scatter of the galaxies across the diameter of the cluster core would be,

$$\sigma_{cl} = 7.5 \times 10^{7} = 1.146 \times 10^{-32} \, T_{cl} \, \rho_{cl} \, D_{cl} \, c$$

To satisfy this equation, we would have to have,

$$T_{cl} \cong 6.7 \times 10^{7} \text{ K}$$

This is completely in accord with what is observed. If the density were greater, the required temperature could be lower. Thus, using only the plasma gas known to exist in rich clusters accommodates the otherwise perceived need for five to six times as much mass as is observed. William of Ockham would certainly have preferred this approach.

Since the redshift induced by this plasma scattering is not in actuality associated with a velocity at all, it necessitates only that the 'observed' mass be used in the virial formulas – all but less than ten percent of the rest of the 'dark matter' is merely an artifact. Thus, mass attributed to the cluster should be considerably reduced. The otherwise presumed value of the mass-to-luminosity ratio drops to well below 50 for rich clusters in accord with the amount of 'dark matter' in galaxy 'halos'. We will discuss this after first discussing the contribution of intragalactic plasma to the overall cosmological redshift.

d. intracluster gas contribution to cosmological redshift

In figure 175 we illustrated the aspects of observing galaxies embedded in the intragalactic plasma, but in what way would this contribute to the overall cosmological redshift? If an isolated cluster is observed for which there is a 'field galaxy' on the observer's side of the cluster with a cosmological redshift of z, how do the redshifts of the clustered galaxies map in comparison? And what about *field galaxies* on the other side of the cluster?

With the 'dark matter' assumption, the cosmological redshift of neither of the two cited ranges of field galaxies would be affected by an observed dispersion of redshift within the cluster. Other than totally confusing the picture, dark matter does not directly affect the redshifts of galaxies outside the cluster whether in front or beyond it. In short, this redshift effect is independent of cosmological redshift. This is *not* the case for the scattering model of cosmological (and other plasma induced) redshifts, of course, as shown in figure 180 as earlier in 175.

So however major the distortion to redshift survey mappings, the impact of clusters on cosmological redshift would merely have a transient effect with absolutely no impact on overall 'cosmological' redshift according to the established view, very like the coins pasted to the balloon analogy to Einstein's expanding universe. Notwithstanding the fact that the established approach has no consensus opinion of what the dark matter *is*. On the other hand, with the plasma scattering model, there would indeed be a persisting impact of observing more distant galaxies through intermediate clusters of galaxies. In fact the balloon analogy does not apply at all – the apparent 'stretching' takes place primarily in the 'coins'. Considers that the galaxies that are observed and registered to positions within the universe according to their observed redshift are predominantly cluster galaxies with vast distances between clusters. That means that in the scattering model, rather than a smooth redshift-distance relation that we have been led to believe pertains, it would predict a rather bumpy relationship. And in fact, since the bumps are indeed *very* major as is easily seen in any redshift survey, this 'average value' of 'Hubble's constant' would be very significantly affected by the amount of clustering.

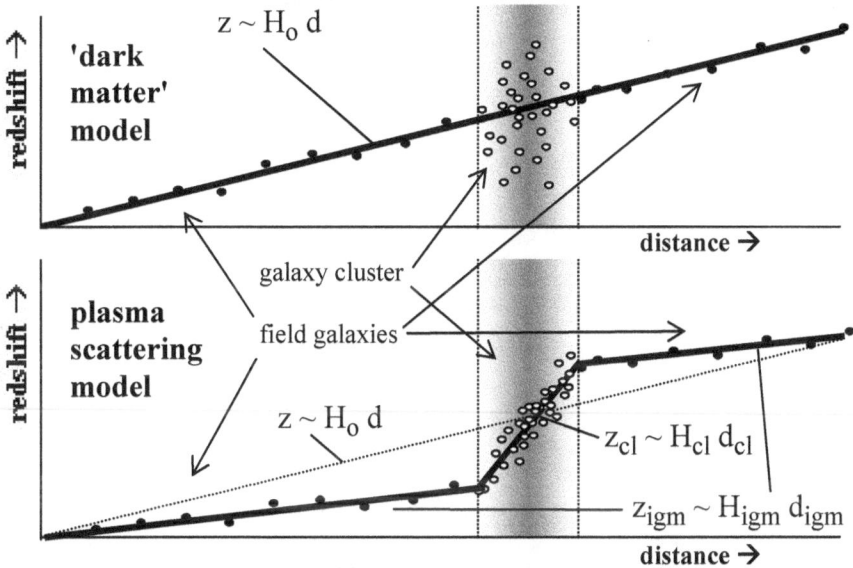

Figure 180: Cosmological impact of redshift through the core of galaxy clusters

Although the redshift augmentation across a galaxy cluster is significant as illustrated in figure 175, the line of sight distance between cluster cores is also very significant. Thus, in any final analysis Hubble's constant would be an average of values that is in large part determined by the clustering in our universe as illustrated in figure 181.

This conjecture that perhaps Hubble's constant in not just an average of *individual* galaxy redshift contributions but much more significantly the accumulation of the effects of the more intense plasma scattering that occurs within galaxy clusters with lesser contributions from intergalactic plasma needs to be considered in more detail. Is it reasonable, and more importantly, is it refuted by observation?

In consideration of figure 180 it should be noted that in neither diagram (top or bottom) are galaxies plotted where they would appear on a redshift survey. To do that would require a knowledge (that we do not have) of distance *per se*. At extreme distances all we really have to go on is redshift. So let us plot in duplicate where each galaxy shown in figure 180 would appear on such a redshift survey according to the two models. For the model that attributes increased cluster redshifts to dark matter by augmenting the mass parameter in the virial theorem, only the cluster galaxies are affected. They will be spread out as shown in the top diagram of figure 182. The field galaxies (identified as black circles) will be unaffected since their redshifts are assumed to be exclusively 'cosmological' in origin.

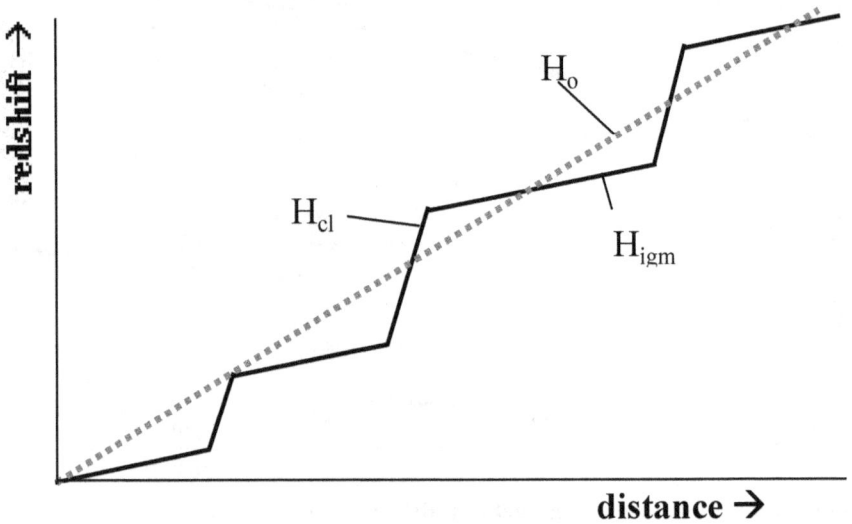

Figure 181: Punctuated expression of Hubble's cosmological redshift constant

However, in the plasma scattering model, there are two different redshift slopes to be taken into account, but both have a very similar cause – plasma scattering – so that even the field galaxies need to be shifted to the Hubble curve which is after all just a composite effect.

In figure 180 and also 182 all the galaxies (the same number for each model) have been partitioned into 8 separate redshift bins for each model. It should be clear that the clustered galaxies would all fall into the same redshift bins for each model. The different redshifting slopes for field galaxies embedded in intergalactic space with its considerably lesser dynamic pressure would produce a shrinkage in the separations of galaxy redshifts in these domains relative to the standard model hypothesis. Therefore, in the scattering model one would expect an apparent clumping of field galaxies even between clusters.

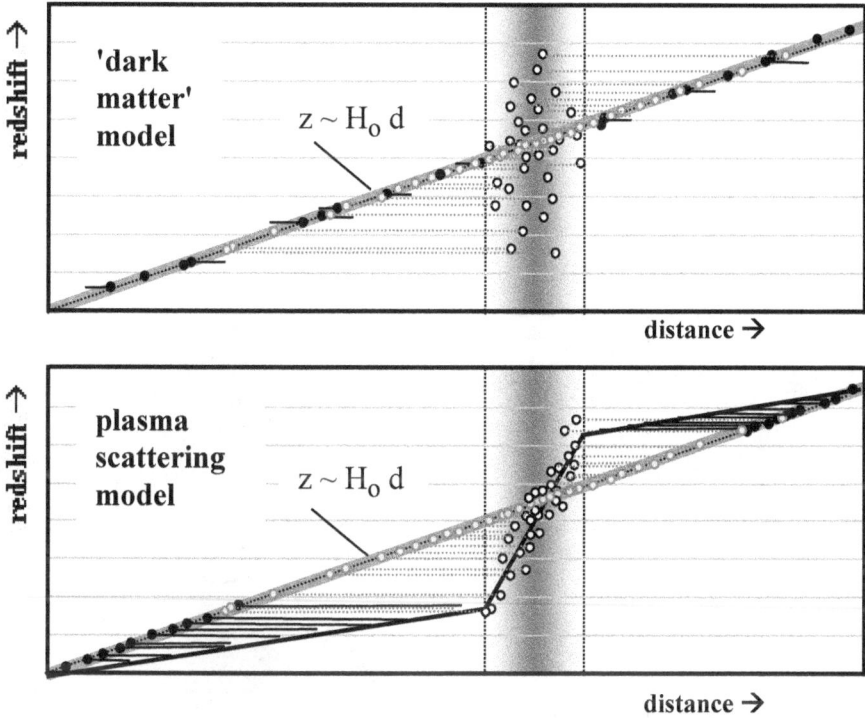

Figure 182: Mapping process using Hubble's cosmological constant

For the dark matter of the standard model, as the distance between clusters increases to say ten times the radii of the cores of these huge conglomerations, the redshift of the field galaxies would form a smooth background for the over densities of clusters smeared out over that background. This is illustrated at the bottom left and upper right of figure 182. However, for the scattering model, the larger the distance of separation and the more dense the clusters, the more bunching there will be in the redshift distributions of the field galaxies.

Figure 179 was set up to illustrate the smearing effect of clustering on the overall redshift pattern of cluster galaxies. In figure 183 we emphasize the predicted counter development that takes place for field galaxies where compression occurs according to the scattering model when they are plotted on a redshift scale employing Hubble's constant. The separation between clusters in this diagram is taken as ten times the diameter of the cluster cores. Predicted lumpiness associated with the plasma scattering model is twofold. Galaxies within clusters will appear smeared over large redshifts. In contrast, the field galaxies that exist in those vast regions between clusters will appear to be compressed when plotted using the Hubble redshift constant.

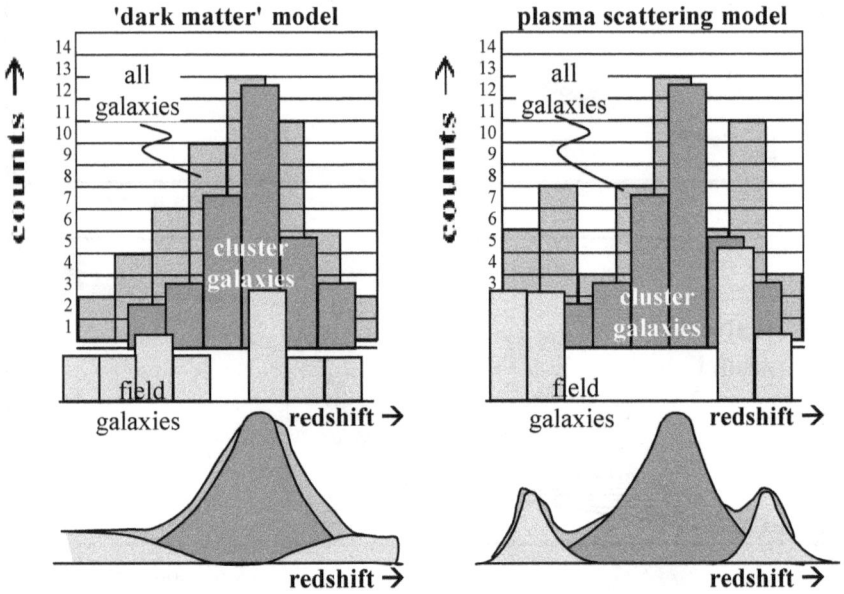

Figure 183: Grouping of galaxies that results using Hubble mapping

Guzzo describes observed structure at the highest levels in the universe. Clearly the distinction between 'cold dark matter' or any other version of the standard model that assumes dark matter and the scattering model is a major refutable difference in this regard.

e. observations of redshift clumping

The obvious observation is that the galaxies map differently for the two models. This is a refutation discriminator between the models that should be easy to verify with the wealth of data currently available. Is there redshift clumping or isn't there? There is.

Hartnett and Hirano (2008) have analyzed an obvious lumpiness structure in the SDSS redshift data which is shown again in figure 185 taken from their paper. The 'Great Wall' super clusters of galaxies at about 200 Mpc from the Milky Way is an obvious feature in the data.

But it is not alone. There seem to be ripples of extreme magnitude as one proceeds deeper into in the data even as the amount of data dissipates. This waffling does seem in some sense centered on our observation point. This prompted Hartnett and Hirano to perform a Fourier analysis of the data to determine any evident periodicity. Their results are shown in part in figure 186 taken from that source.

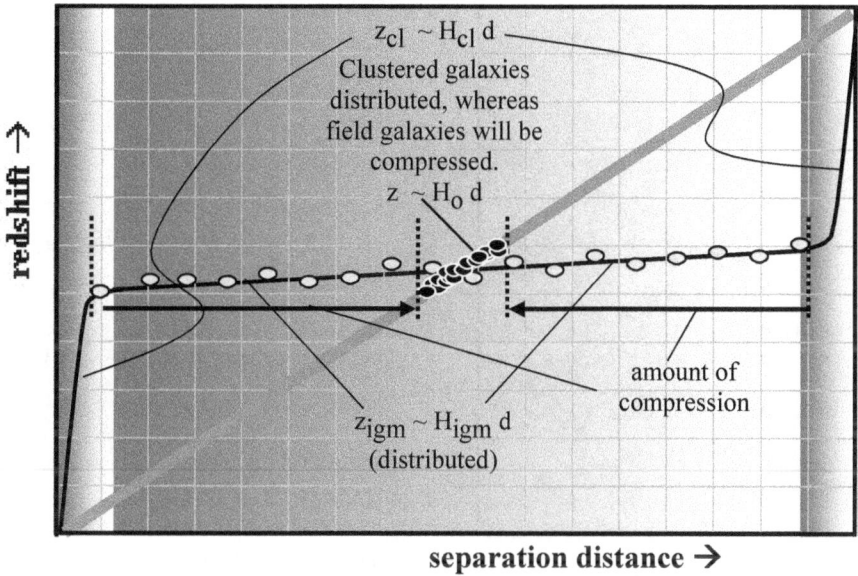

Figure 184: Apparent compression of field galaxy redshift spacing

However, on inspection of the various slices whose profiles are provided in figure 186, we see that the major humps and valleys are unique to each slice through space. For example, in the image slice for declinations 52 to 58 degrees, the humps and valleys are completely out of phase relative to the other slices, but exhibit the same periodicity. So the phenomenon is not spacetime related, but clearly is a function of what is encountered in observations made in that particular direction. It is clearly an artifact of observation employing a redshifting mechanism as a means of interpreting data. Hartnett and Hirano found that the redshift data is characterized by extreme peaks and valleys just as it appears to be. The peaks and valleys in the different slices of the data don't correspond in any very direct way although they were able to obtain a rough periodicity applicable to multiple slices. Clearly there is a similar type of pattern that applies to each slice. That pattern entails a huge number of galaxies grouped around a particular redshift with a dearth of galaxies before and aft. With this phenomena repeated throughout the data.

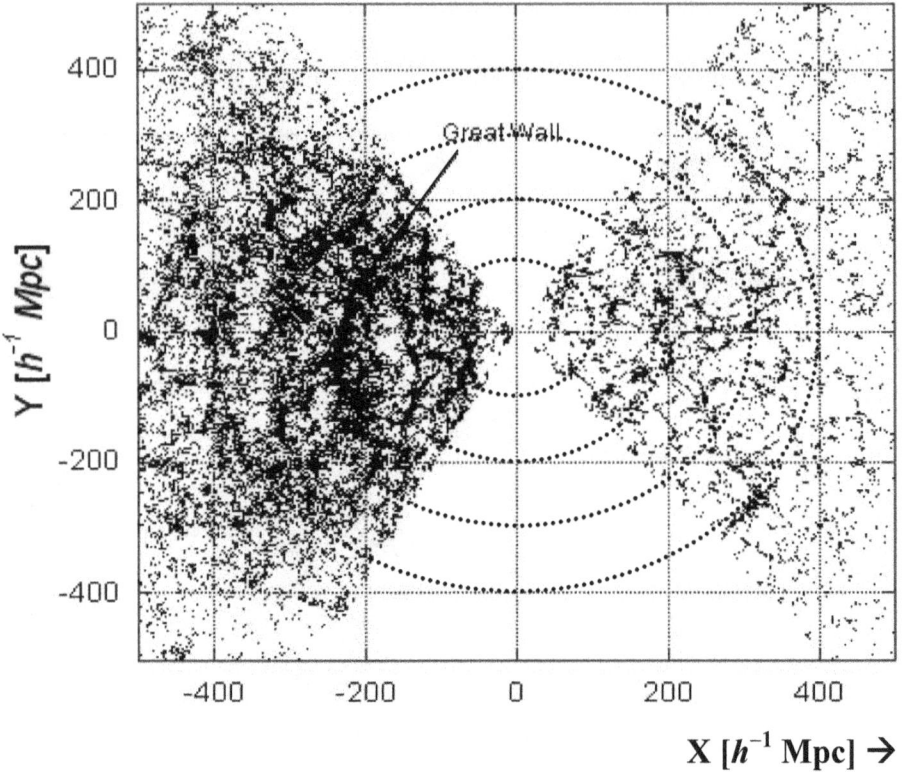

Figure 185: Density ripples in the SDSS redshift survey from 0 < Dec < 6

This is clearly in agreement with arguments presented for the plasma scattering model shown in figures 179 through 180 and earlier figures in the previous chapter. So it is certainly not embarrassing to say that the model predicts a punctuated rather than continuous trend in the redshift-distance relationship even at cosmological distances.

For the established approach there is no such agreement with the observed clumpiness of redshift. The arrangement of clusters would have no impact on the redshifts of clusters behind them; so they could easily overlap. There would be no periodic clumping of redshift. Cluster galaxies would be 'smeared' in with field galaxies obfuscating their differences. In the scattering model, the cluster would be smeared almost exactly the same as for the 'dark matter' model, but in this case the field galaxy redshifts would be compressed away from the clusters. Refer to figures 179 and 180. Furthermore, the line-of-sight distances between cluster cores discussed in the previous chapter produce a rough periodicity of redshift density not accounted for in the standard model.

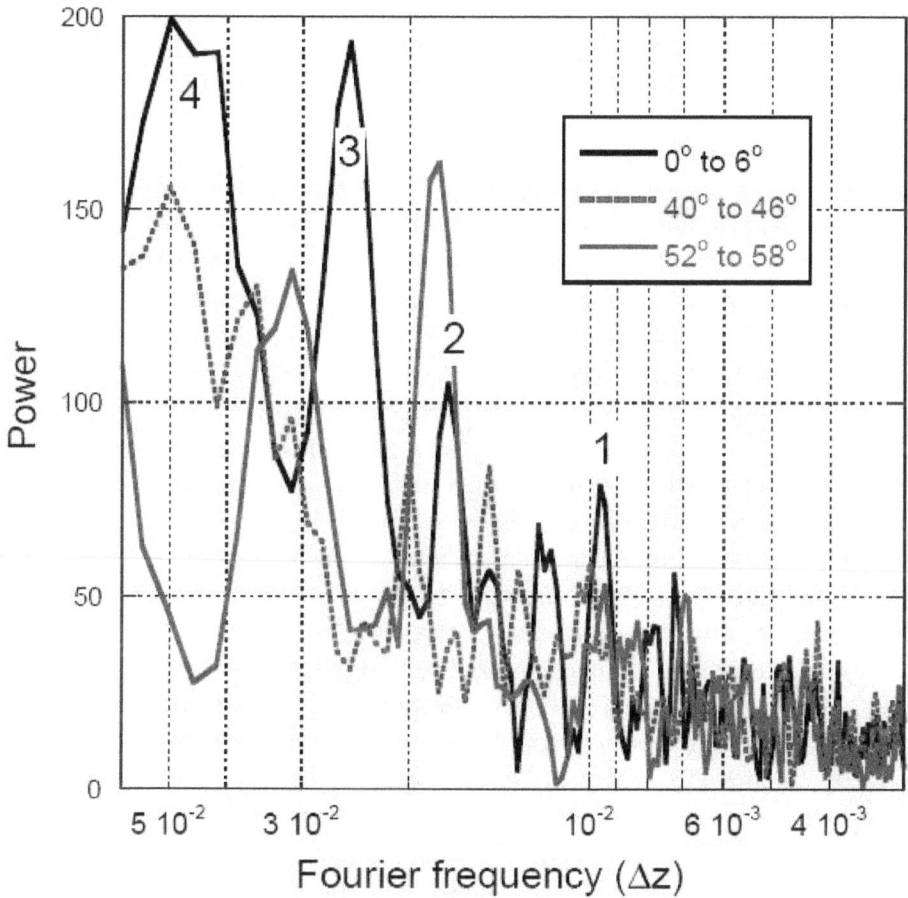

Figure 186: Peaks and valleys in the SDSS redshift profile

What had seemed to be an awkward aspect of the plasma scattering model derived in earlier chapters, was that it required a dynamic pressure of the intergalactic medium that was considerably larger than had been anticipated for intergalactic regions. The density had to remain small to comply with observations and that resulted in a very lengthy implied extinction interval. Compensating for that to obtain the requisite dynamic pressure by enforcing higher temperatures – besides stretching credibility for that parameter in the open space between field galaxies and clusters of galaxies where neutral hydrogen clouds appear – would have necessitated wider aberration angles over extreme distances to converge. Together these constraints would have involved a tremendous swath of space to effect each step in the forward scattering process.

However, in intracluster plasma the extinction interval is many hundreds of times shorter. So in retrospect, we see that H_{igm} need *not* correspond directly to H_0 but that a weighted combination of H_{cl} and H_{igm} is what effects the value of H_0. Certainly the distances between clusters is significant, but the redshift effect is so much more effective in denser intracluster regions that very little additional effect will be required in the sparse intergalactic regions to match Hubble's constant.

f. the rationale for 'dark matter' in rotating galaxies

Another area in which dark matter has been cited is in the mass determination of individual galaxies as we discussed in chapter 16. In accounting for rotation rates, galaxies have seemed to have insufficient mass. Here too gravitational aspects of unseen matter is demanded to supplement gravitational effects associated with the masses of luminous sources of radiation. Accounting for the excessive rotation rates at distances far removed from the central nucleus of the galaxies has required assumptions concerning unobserved galaxy 'halos'.

On pages 343 and 344 we computed the uniform density of a spherical halo that would account for the required additional mass. It proved a necessary and sufficient amount to produce the observed rotation velocity profile that was shown in figure 167 for the galaxy M31. The observed rotation profile was thus obtained by assuming the existence of a spherical halo for which,

$$\rho_{H\text{-}M31} = 6.9 \times 10^{-30} \text{ gm / cm}^3$$

This 'requirement' is easily met by a plasma electron density of 4×10^{-5} cm^{-3}. It is considerably less than the observed plasma density that occurs in all galaxy clusters and must probably surround all galaxies. It exceeds the universal baryonic mass density by very little and, as we saw in chapter 3, in this form at a temperature above ionization levels but beneath that for which significant X-ray emissions result, it is virtually invisible, and that seems to be the operative demand.

g. additional data on the 'dark matter halos' of galaxies

Sofue and Rubin (2001) indicate that the widely adopted custom of drawing a rotation curve by linking positive and negative velocities from the opposite sides across the nucleus along the major axis is incorrect, at least in massive galaxies. (See figure 164 on page 340.) What this asymmetry seems to imply is that there is plasma redshifting involved here that is similar what was discussed for intracluster plasma gas that produces the similar asymmetry shown in figure 155 on page 327. Kochanek (1996) observes that there is on the order of a 250 km/sec excess dispersion of the central stars in some galaxies over and above what would be expected from the luminous properties. This, he concludes, must be attributed to dark matter. However, of course, this can much more easily be interpreted as due to plasma redshifting in the hot interiors of these galaxies using the scattering model to the same effect as we have shown elsewhere.

In addition Rubin has noted that 43 of 81 galaxies studied exhibit rotation curves describes as "disturbed" and "abnormal". These abnormalities include asymmetric curves. This too is expected according to the scattering model with no dark matter required.

The Milky Way halo has been said to extend at least 200 kpc – about halfway to the neighboring galaxy M31, which is 350 kpc away. If halos are as large as those suggested by the gravitational distortion of background galaxies seen in the vicinity of foreground galaxies (Fischer, et al., 2000), the cause of which we would attribute quite differently to his accounting, then the halo of our Galaxy may brush an equivalently large halo of M31. In fact, this same plasma overdensity merges into the intracluster medium such that there is very little distinction to be made. Bahcall (1999) indicates that "Unlike previous expectations, this suggests that most of the dark matter is associated with the dark halos, and not intracluster medium."

"Regularly rotating spiral disks" are observed at a redshift of 1, when the universe is said to have been less than half of its present age. A correlation of their rotation velocities with their blue magnitude matches to within 0.5 magnitudes that of nearby spiral galaxies. Evolution, if that had been involved over what is considered the last half of the age of the universe, has certainly not dramatically altered this correlation according to Sofue and Rubin (2001).

h. resolution of 'dark matter' in the scattering model

As detailed in the previous chapters, and sections above, 'dark matter' has typically been hypothesized to resolve extreme spectroscopic redshift dispersion anomalies in galaxies and galaxy clusters. These extremes that have precipitated claims of dark matter have been interpreted exclusively as Doppler phenomena derivative to the virial theorem, or in the case of MOND theories to the failure of Newtonian dynamics.

In contradistinction to this unilaterally accepted approach, according to the scattering model the increased redshifting in the galactic halos and rich clusters that biases redshift survey data is due to forward scattering in plasma. The effect is to substantially increase the redshifts of objects seen through clusters of galaxies. As our outward investigations proceed further and further, they inevitably encompass observations made through regions with rich clusters and super clusters whose rich plasma cores substantially increase redshift. In fact the Milky Way is itself imbedded in such an environment. So statistically Hubble's redshift constant becomes an average of more intense redshifting through intracluster regions and lesser shifts that take place in the sparser intergalactic regions.

Collaborating evidence for the favored dark matter theories have not been sufficient to confirm that the associated effects are, in fact, due to increased mass. Gravitational lensing that we discussed in passing in the chapter 16, has provided results that correlate with the mass assessed by other means, but the agreement is not all that good, typically providing much smaller estimates. So there really is no collaborating evidence for dark matter other than its supposed affect on galaxy motions, which are only 'confirmed' by spectroscopic, i. e., redshift, measurements.

Chapter 19

The Measured Background Radiation Spectrum

As we saw in chapter 3, an intergalactic plasma must exhibit some sort of stable state – both *currently* and in epochs past – with cluster core hotspots evenly distributed throughout. Such a medium – in fact, any medium – emits thermal radiation. Beyond that, because of the redshifting of such radiation, it is a matter of degree, i. e., at what temperature will it radiate? That the intergalactic medium must radiate blackbody radiation at some temperature just as our atmosphere does; it is after all just thermodynamics, not cosmology *per se*. Naturally from the perspective of standard models the current intergalactic medium would radiate predominantly *foreground* radiation. However, hypotheses put forward in this volume demand that this plasma medium must also in some hitherto unknown way account for observed *background* radiation that differs substantially from the extremely high temperature spectrum one might naively expect from such a fully ionized plasma. There would be no other way to account for the ubiquitous microwave background radiation according to this theory. But first, what, if any, spectral evidence is there for the existence of such a uniform high temperature intergalactic plasma medium?

a. high temperature thermal spectrum

Since the temperature of intergalactic and/or intracluster plasma would have to be extreme to produce the observed redshift via scattering effects described earlier, the *classical impact cross section* or *Landau parameter* would be less than the *de Broglie wavelength*. Such a plasma would emit 'free-free' collision-induced radiation known as *bremsstrahlung* radiation with telltale photon energies hc/λ on the order of $k\,T_e$, with T_e the kinetic temperature of the electrons. (See Haines, 1993 for example, for further explanation of this and other thermal properties of an ionized plasma.) Electromagnetic characteristics of the intergalactic plasma constrained as described elsewhere to account for measurements of Hubble's constant according to the scattering model, would imply intracluster temperatures as high as $T_e \approx 10^8$ K over an appreciable percentage of space. Previous considerations reduce the average intergalactic temperature requirement.

Such an unresolved high temperature thermal background X-radiation distribution was indeed observed quite accidentally in 1956. Although Parmar, et al. (1999, p. 611) seem to confuse this in stating,

"The diffuse Cosmic X-ray Background (CXB) was discovered in 1962 predating the discovery of the Cosmic Microwave Background (CMB) by several years. The spectral characteristics and the spatial distribution of the CXB have been measured by many X-ray missions, but its origin is still not fully understood. See Wu et al. (1991) who also provide a discussion of this diffuse X-ray background. The 3–60 keV spectrum is well described by a thermal bremsstrahlung model with a temperature, kT, of 40 keV (Marshall et al. 1980)."

In fact, Marshall, et al. noted that the spectrum detected by HEAO-A2 in the 3 to 50 keV range was remarkably well fitted by a 40 keV isothermal Bremsstrahlung model with a temperature of 40 keV as shown in figure 187 below. They note that no other special features are present and that the spectrum matches no other known sources.

Ricker and Mézáros (1993) showed that starburst galaxies or other active galactic nuclei from distant sources that had been thought to be responsible for this radiation could not be culpable in the range above 3 keV. With all efforts to date 60% of the intensity of this radiation is still unaccounted. Ricker (2007) provides data from Gruber et al. (1999) shown in figure 188. Notice that the intensity is even greater than modeled by Marshall, et al..

However, with regard to such measures Verschuur and Kellermann state (p. 637) that "the main problem is that the temperature of the gas is uncertain, as the X-rays normally provide only a lower limit." It is for this reason, among others, that the temperature of the intergalactic medium has been so easily underestimated. In fact, Post (pp. 1052-1053) states that, "the radiation rates from a plasma are much less than the Planck or blackbody value. For example, at a radiation temperature of 10^8 K, the Planck radiation, being proportional to T^4, would amount to the almost inconceivable value of 6 x 10^{24} W/m^2 [6 x 10^{27} erg/cm^2 in CGS units]. But the tenuous plasma is optically very 'thin' over most of its emission spectrum...so that a plasma with a kinetic temperature of 10^8 K might radiate at the radiation rate from a blackbody at radiation temperatures of only a few hundred kelvins."

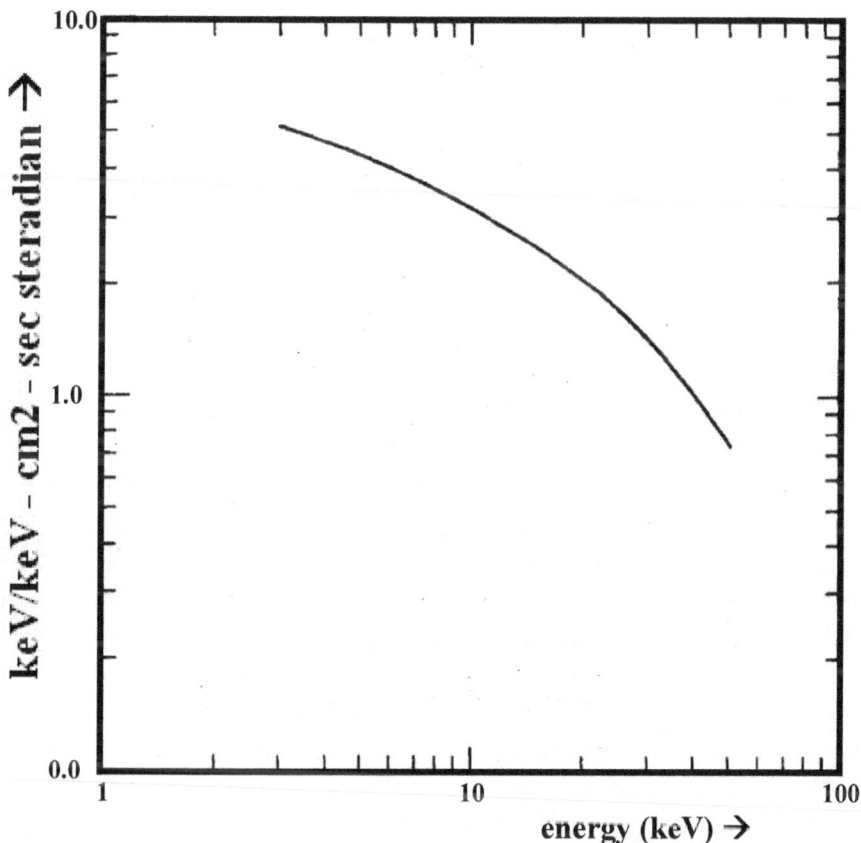

Figure 187: The background spectrum modeled at 40 keV with approximately 10 % accuracy. [From Marshall, et al, 1980]

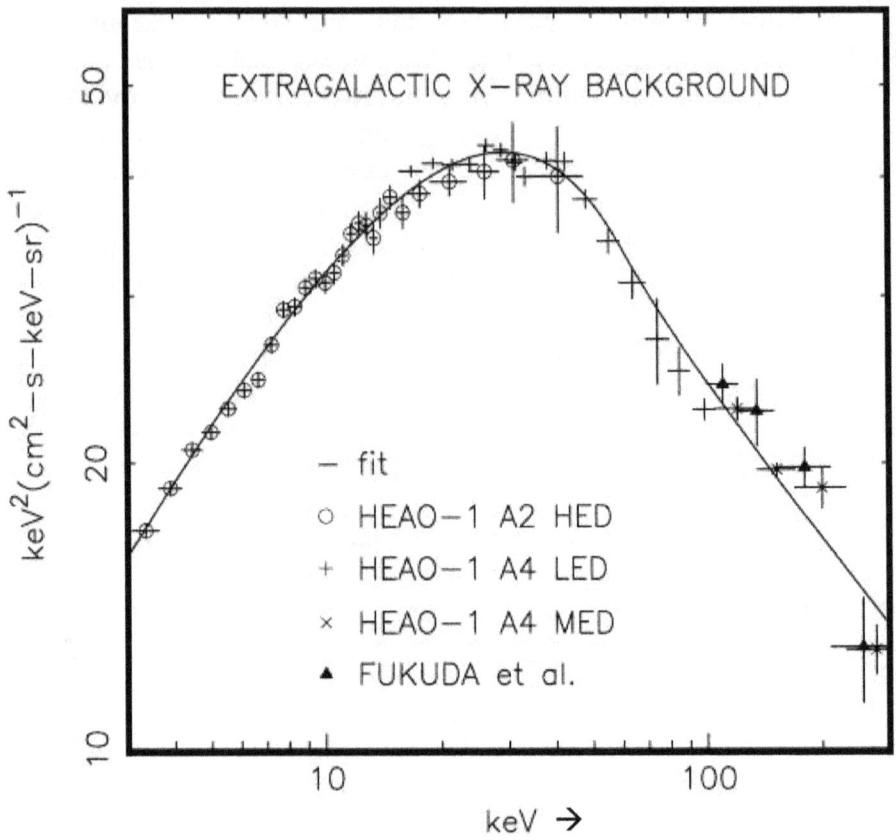

Figure 188: The measured x-ray background spectrum from Ricker (2007)

Low levels of impurities in a hydrogenous plasma produce tell tale 'excitation radiation' that is significantly (several orders of magnitude, in fact) greater than the associated thermal radiation levels except at temperatures up around 10^8 K or greater. So intracluster plasma with its higher percentage of heavier ions is much more readily observed than the virtually invisible intergalactic plasma medium. Verschuur and Kellermann (p. 638) also cite data from several sources of astronomical observations indicating temperatures in excess of 10 keV (corresponding to thermal radiation wavelengths less than 10^{-9} cm such as indicated here) in the Abell and Coma galactic clusters. Within galactic clusters there are, as we have seen, considerably higher densities with commensurably higher excitation radiation due to large ion 'impurities' that makes it easier to observe. Silk (1980, p. 185) also indicated that there is an X-ray spectrum in the Perseus cluster corresponding to a *tenuous* hot thermal gas with temperatures in excess of 10^8 K. But significantly the broad background X-ray distribution is not limited to any (or all) such specific sources.

This X-ray background has long been a source of problems for standard cosmological models since all observations to date find it to be independent of individual sources and *incompatible* even with any models of individual sources to have been proposed. When all known sources are subtracted it is uniform across the sky to one part in 10^3. Its spectrum differs substantially from any known individual sources including active galactic nuclei and quasars. That it fits so perfectly the thermal radiation distribution of a background hot gas preceding 'decoupling' such that the microwave background would necessarily have had to have been scattered by it, is indeed a source of on-going investigation. David Burrows (1996) has stated that: "Although this discrepancy may not seem serious when compared to other uncertainties often encountered in astrophysics, it is currently considered to be substantially beyond the uncertainties in the precision of our knowledge concerning the known classes of X-ray emitting objects and how they behave at large cosmic distances."

In defending the standard model, however, Burrows states: "that the remaining emission is contributed by a hot intergalactic medium, seems difficult to reconcile with COBE results (Mather, et al. 1990), which do not find any distortion of the blackbody spectrum of the microwave background due to scattering of the microwave photons by this hot gas." But of course the perceived problems have to do with assumptions of the radiation passing through a limited distance of such a gas. In the model proposed here, the "hot gas" is not an obstacle, but in a very real sense the vehicle of the thermalization of the energy embodied in the microwave background radiation.

Peebles (1993, p. 580) also brings up the incompatible pressure gradient that should accrue at the interface with Lyman-α forest clouds. But as proposed here, there would be no substantial pressure gradient.

With the launch of the Chandra X-ray Observatory in 1999, NASA began high quality X-ray astronomy. It was determined that there are hot plasma sources within our own galaxy, but 10% remained unexplained. By 2008 two independent teams of astronomers had pointed Chandra toward a couple of apparently empty areas of sky for deep exposures lasting 140 hours in one case and 280 in the other. This data is known as the Chandra Deep Fields North and South. These studies suggest that the uniform background might derive from many extremely distant and uniformly distributed individual quasar sources of x-rays, identified by their signature energy spectra.

The facts remain that diffuse emission of radiation in the X-ray band of the electromagnetic spectrum from about 1 keV to over 100 keV (i. e., from over 10^7 K to well over 10^9 K) is observed. Although this spectrum is dominated by sources at very great distances, it retains a high degree of isotropy. See for example, Jaffe (2009) for recent clarifications as well as added mystery. It is a diffuse flux with respect to local sources of emission such as our own Milky Way galaxy or the local supercluster of galaxies. Only in the microwave band of the spectrum does a similar condition exist.

Even the best attempts to account for radiation as deriving from discrete sources of known luminosities such as clusters of galaxies, Seyfert galaxies, N galaxies, BL Lac objects, and quasars have all failed by nearly a factor of three.

Indeed, this short wavelength radiation is second only to the microwave background in overall intensity, being only a few orders of magnitude less. Figure 189 is from Fabian and Barcons (1992). The smooth curve in the X-ray domain is from Marshall, et al. (1980), other sources provided isolated observations. See Tyson (1995) for information concerning the optical range. A horizontal line would represent equal energy at every frequency. This brings us to the most significant feature of the figure – the microwave and submillimeter background spectrum (MWB) that derives from Mather, et al. (1990).

b. the microwave background spectrum

Considerable histrionics surround the few facts concerning how radiation may have 'evolved' into an observed spectral distribution of 2.725 K radiation. It would be easy to explain if one could justify that there had been but one slice in time at a redshift of say $Ƶ \approx 10^9$ when the emission took place, as vaguely suggested by the standard models. Although, in that case, it is mysterious how radiation on the order of 3 x 10^8 K could be 'tax deferrable', or "frozen in", or other *ad hoc* terminology of choice used from time to time to describe the suspended animation of that single aspect of the universe. This supposed thermal suspension is explained as lasting until a cooler time variously denominated "time of last scattering", or "recombination" (or other terminology) occurred, at which time radiation and matter are said to have become 'decoupled' at a redshift of anywhere from 10^4 to 10^3 K.

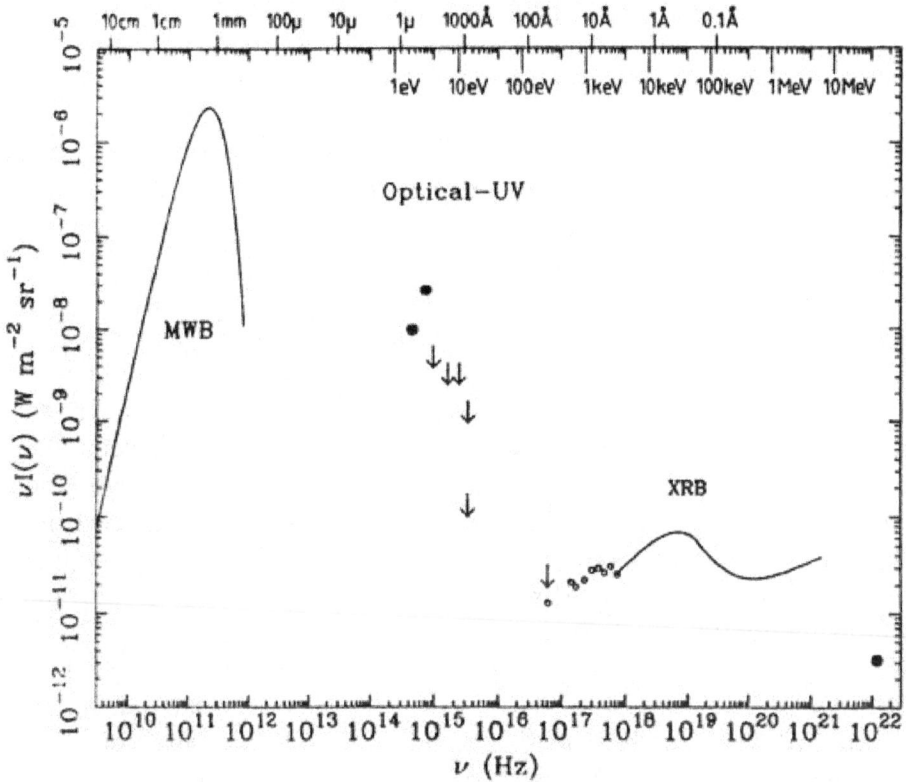

Figure 189: Spectrum of the extragalactic sky – energy from the radio band to gamma-rays (arrows denote upper limits)

Robert Dicke's team predicted microwave radiation from "cosmic matter" (Dicke, 1946). Cosmic microwave background as such was predicted more specifically based on the assumption of a big bang in several articles in 1948 by George Gamow and by Ralph Alpher and Robert Herman (Alpher, 1948). Alpher and Herman estimated the temperature of the cosmic microwave background to be 5 K. Two years later, they revised their estimate to 28 K. Gamow would later predict 6 K and a few years after that re-evaluate it at 7 K. It would not be until the 1970s that a consensus would be established that the, by now well-observed, microwave background radiation was a remnant of the big bang. To say it was 'predicted' by the standard model is shaky at best.

c. the precise determination of the spectrum

In figure 190 the Far-Infrared Absolute Spectrophotometer (FIRAS) instrumentation data obtained from NASA's Cosmic Background Explorer (COBE) satellite is shown. (For more detailed information, see for example Mather, 1994; and Fixsen, 1996.) This data is plotted with the abscissa given in waves per centimeter, i. e., one over wavelength, and the intensity plotted as the ordinate is given in units of ergs-per-centimeter-squared-steradian-seconds-per-centimeter. Figure 190 shows the theoretical curve for 2.725 K blackbody data plotted in units of 'spectral photon sterance'.

Notice that "THEORY AND OBSERVATION AGREE" has been prominently displayed on the plot in figure 190. This was done prior to dissemination by NASA. Why? The answer can only be that the observed data very accurately represents the theoretical blackbody curve shown in figure 191 for the temperature 2.725 K. It is accurate, in fact, to within the width of the line used to draw the curve, and therefore, no error bars need to be shown. See figure 192 taken from Mather (1994) where actual tolerances *are* shown. To demonstrate that tolerances are indeed less than the width of the lines used to plot the data in figure 190, we have used data taken from Ned Wright (2007) where the error tolerances have been multiplied by 400 and plotted on the NASA data. See figure 193. Figure 194 shows the same data as a function of wavelength. Clearly the microwave background radiation corresponds to the 2.725 K blackbody spectrum.

However, as the paragraph above indicates, that is *all* that the widely disseminated comment legitimately implies. Yet, of course, it is *meant* to imply much more than that. The obvious intention would seem to be that this chart has fully confirmed the aspirations of a generation of cosmologists who had tentatively accepted the veracity of the standard cosmological model. But, however exceptional the technology that went into obtaining this data, it is *only*, or perhaps more appropriately one should say that it is *superlatively*, information that is pertinent to assessing whatever theory can fully explain it.

The COBE satellite was also used to detect the variations in this intensity of the microwave background data. Instruments called Differential Microwave Radiometer (DMR) gathered signals from three antennas and compared the measurements against each other very precisely. By measuring differences in the intensity of radiation from different areas of the sky, variations (other than an offset for the Milky Way velocity) were determined to be about 1 part in 100,000 across the entire sky. That data is shown in figure 195. The dark band across the center of the image indicates more intense radio emission including that from material in the disk of our Milky Way galaxy. This is quite understandable, as is the next step in the data processing of the associated data image. It involves modeling of the radio sources in our galaxy in an attempt to estimate the uniformity of the background if we were not observing from deep within a galaxy. In figure 196 the resulting data is shown. NASA's Wilkinson Microwave Anisotropy Probe (WMAP) data is provided in figure 197.

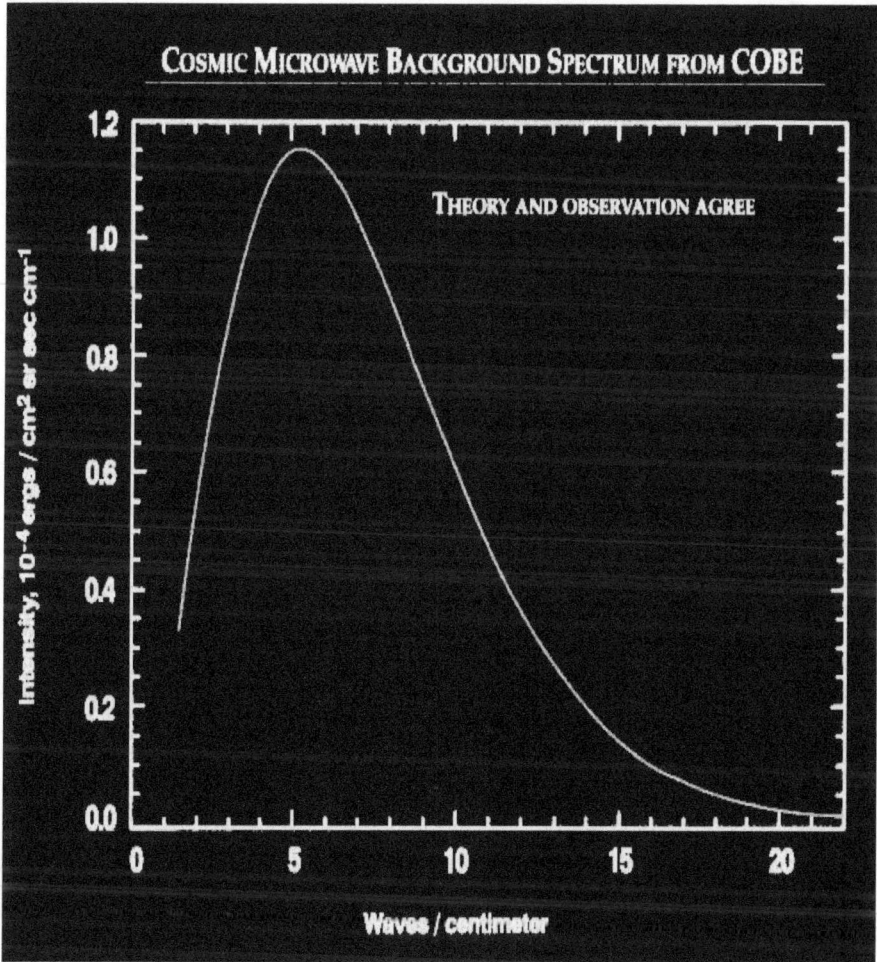

Figure 190: Measured microwave background radiation

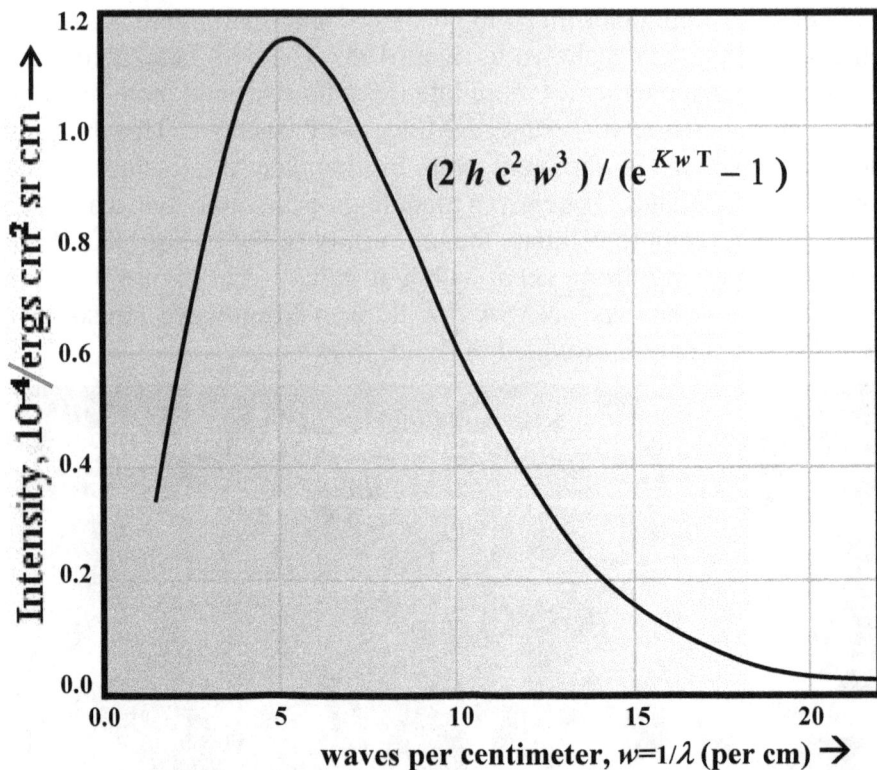

$$(2\,h\,c^2\,w^3\,)\,/\,(e^{K\,w\,T} - 1\,)$$

Figure 191: "Theoretical" blackbody radiation for 2.725 K temperature

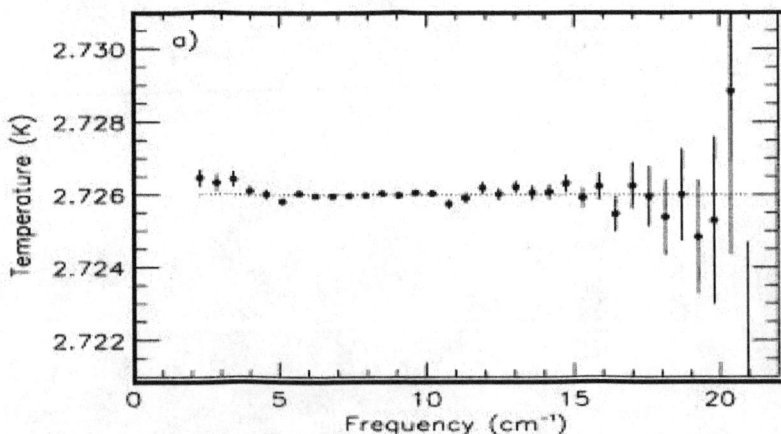

Figure 192: Tolerances on NASA Microwave background data

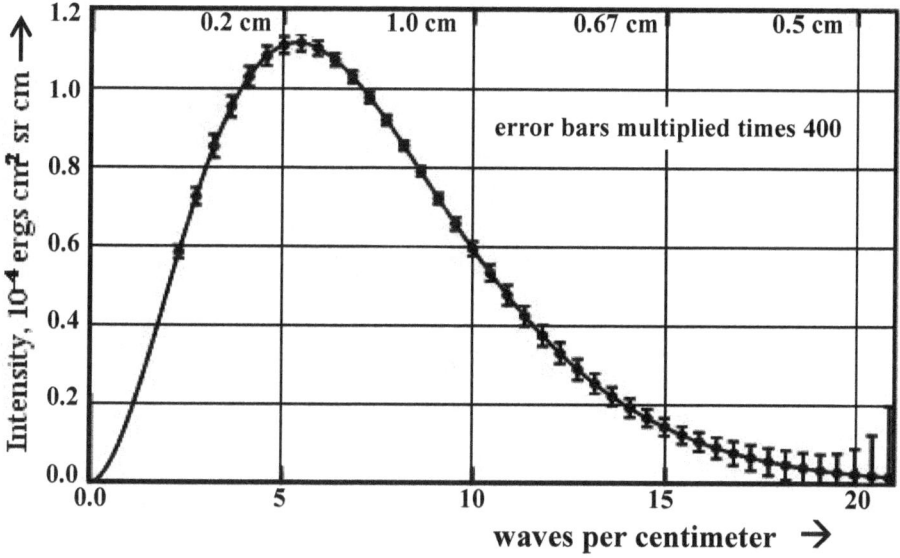

Figure 193: FIRAS data with exaggerated error bars included

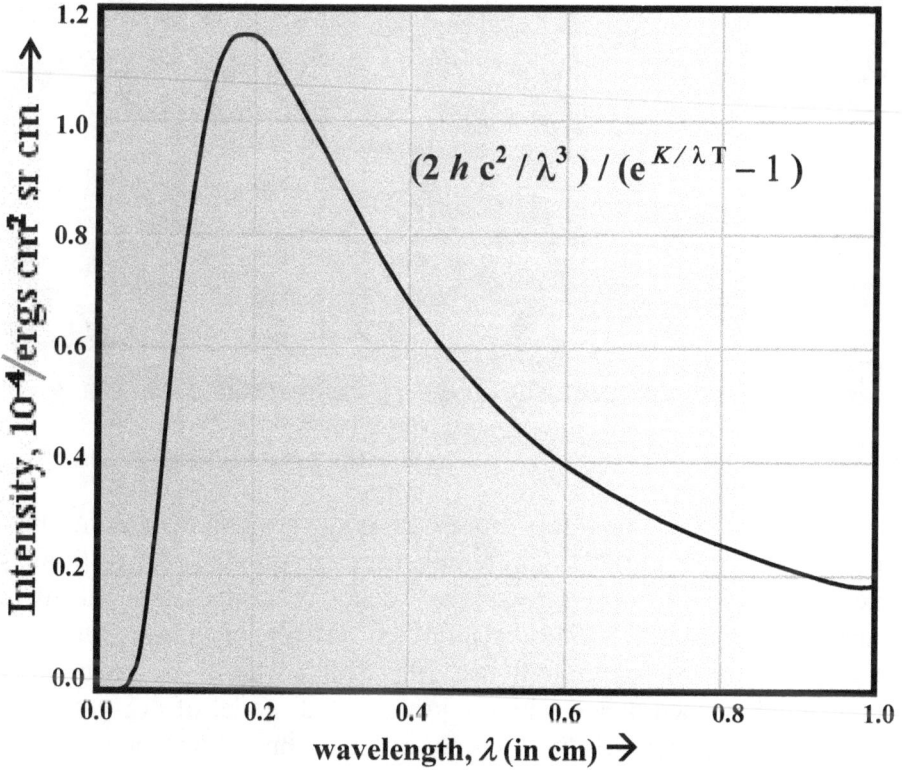

Figure 194: Blackbody radiation for 2.725 K temperature

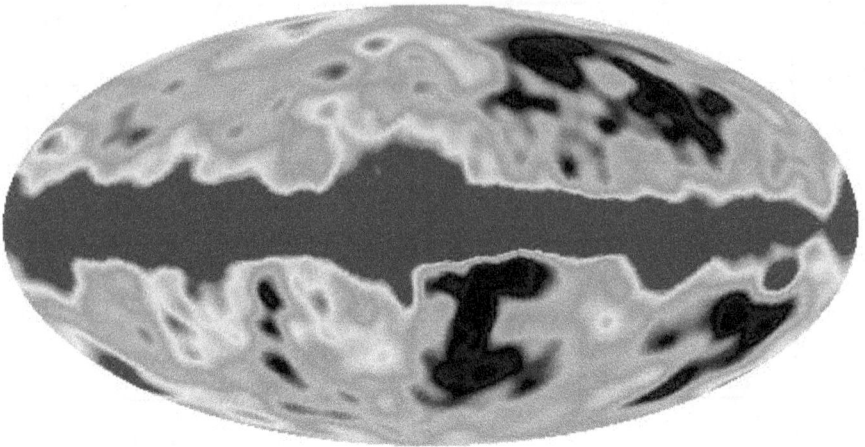

Figure 195: Measured variations in microwave background radiation intensity

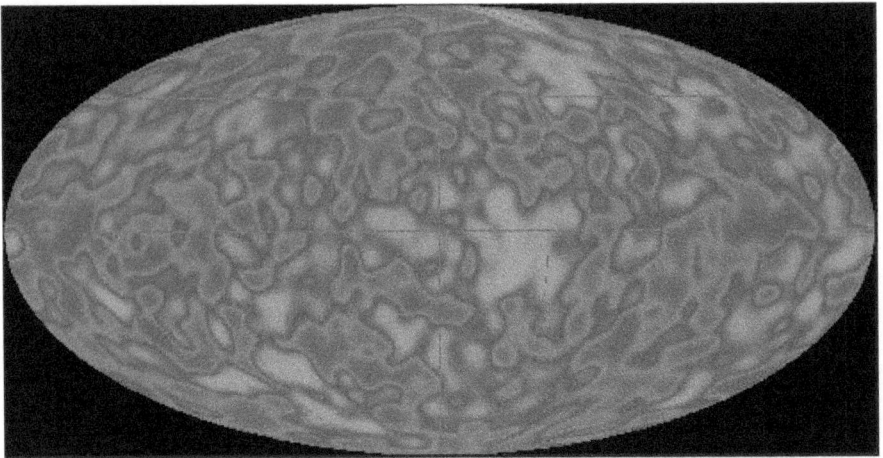

Figure 196: "Processed" variations in background radiation intensity

The variations in this processed image are less than about 0.001 percent, i. e., the variations in temperature are on the order of 0.00001 degree Kelvin. One might very well ask, "What does this add to the unretouched figure?" The disenthralled answer, of course, would be that it is just the data in figure 195 with portions of the intensity subtracted in accordance with a 'programmed model of our galaxy', a nominal indication of the degree of smoothness obscured by our galaxy.

Figure 197: WMAP cosmic microwave background data [from Hu (2007)]

This mapping does not attempt to correlate tiny bumps of microscopic intensity with the 'great wall' or other ostensible features so readily apparent to the naked eye on such galaxy surveys. Although Hu (2007) and many others have addressed models to do just that. Claims made for this data, however, suggested that that had, in fact, been successfully accomplished.

It would definitely seem that all such hype is intended to tie this instrumental success in measuring background radiation to theoretical success in predicting it, suggesting that this was somehow precisely predicted from theory. It wasn't. The temperature was not predicted. The form of the distribution has nothing to do with the model for a thermodynamically stable universe. Any theoretical discussion of structure should address first and foremost, its very obvious appearance, i. e., that the observed structures would have taken tens of billions of years to develop and move into the locations they occupy. And, of course, they do not appear to have been following any pattern laid down at creation.

The following comment is from Yost and Daunt (2008),

"Any theory of the origin of large scale structure must account for the formation of voids, which occur on scales of approximately 100 Mpc. We can immediately place one constraint on their formation. If we take as a reasonable estimate of a peculiar velocity for a galaxy ~ 600 km/s, it would take 160 billion years for such a galaxy to travel 100 Mpc and cross a void – far longer than the age of the Universe. Therefore, it is extremely unlikely that the voids were formed by galaxies moving out of a region the size of a void after their formation. The galaxies must (on the scale set by voids) have formed near where they are today, so voids must reflect the distribution of our present galaxies at the time when they were created."

That is apparently the party line. However, since appearance is what science is all about, why do we not give credence to that so obvious appearance of having taken much longer to develop. The observed large-scale structure of the universe, as shown in the diagrams of the previous chapter, defies association with variations of one part in ten to the fifth. It is this author's opinion that these structures did, in fact, take very many billions of years to form – not just 14.8. We should have moved past pronouncements such as Bishop Ussher's. Our eyes and neural capacities have evolved; we have designed instruments that exceed their specific abilities; we must use all these advantages to decipher the universe and not limit our understanding to initial pronouncements no matter how profound they may have seemed at the time.

Of course, supporting arguments for *any* viable cosmological explanation of background radiation will necessarily be complex. Although the author does not consider efforts to date to be extremely credible, he is keenly aware that the development of a viable solution to the origin and thermalization of the background radiation that is compatible with the scattering model is not an inconsiderable task. But neither is it insurmountable. The solution has been around for some time. The resolution of this issue will be addressed in some depth after first discussing more general theoretical considerations of blackbody radiation.

Chapter 20

Theoretical Considerations Involving the Redshifting of a Blackbody Spectrum

Early experimental analyses of blackbody radiation were performed using experimental apparatuses called 'cavities' like that shown in figure 198. For obvious reasons what was seen down the hole in the apparatus was called "blackbody" radiation. The device would be surrounded by what is called a heat bath to maintain the cavity device at a fixed temperature. The detected radiation invariably has the form given by the Planck radiation distribution, which theoretical form was derived following a notorious failure of its predecessor that was based on classical physics called 'the Rayleigh-Jeans distribution'.

a. characterization of the blackbody spectrum

In thermal equilibrium, the energy density of emitted radiation from a plasma or any other medium will inevitably be distributed by wavelength $\rho(\lambda_e)d\lambda_e$ with a 'blackbody' spectrum whose functional form is illustrated graphically in figure 199. Figure 200 presents the same distribution on a log scale. Its formula is:

$$\rho(\lambda_e, T_e)\, d\lambda_e = (2\pi\, h\, c\, /\, \lambda_e^5)\, (e^{K/\lambda_e T_e} - 1)^{-1}\, d\lambda_e$$

This particular parametrical representation is denominated, *'spectral radiant exitance'* and is expressed per unit wavelength. The units of ρ are ergs/cm^2 sec. The constant factor K in the exponent is given as:

$$K \equiv h\,c\,/\,k = 1.441 \text{ cm K},$$

where the individual factors have all been defined earlier with regard to other discussions. The wavelength associated with the peak of this distribution function is:

$$[\lambda_e]_{peak} \equiv \lambda_p = 0.2\,(\,h\,c\,/\,k\,T_e\,) \cong 0.2898\,/\,T_e.$$

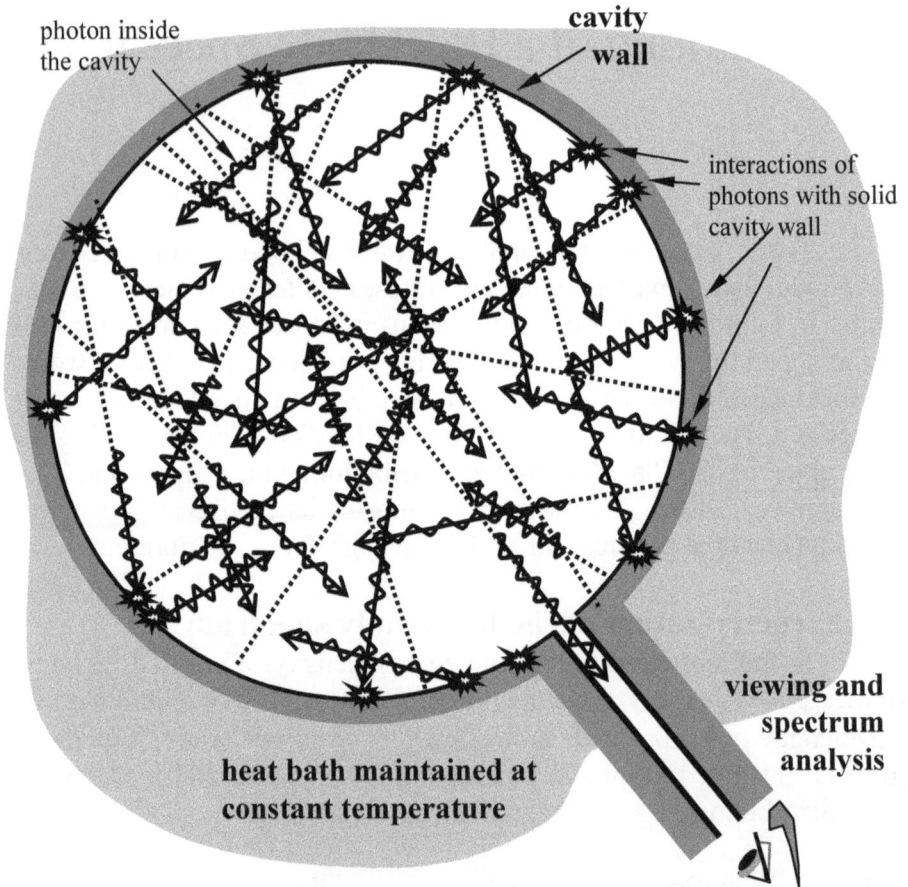

Figure 198: Apparatus for observing 'cavity' radiation

When divided by the speed of light, the distribution equation above provides the emission energy per cubic centimeter in the wavelength interval λ_e to $\lambda_e + d\lambda_e$, which with the assignment of the value $T_e = 3.1 \times 10^8$ K, would result in the peak radiation wavelength of $\lambda_p \approx 10^{-9}$ cm.

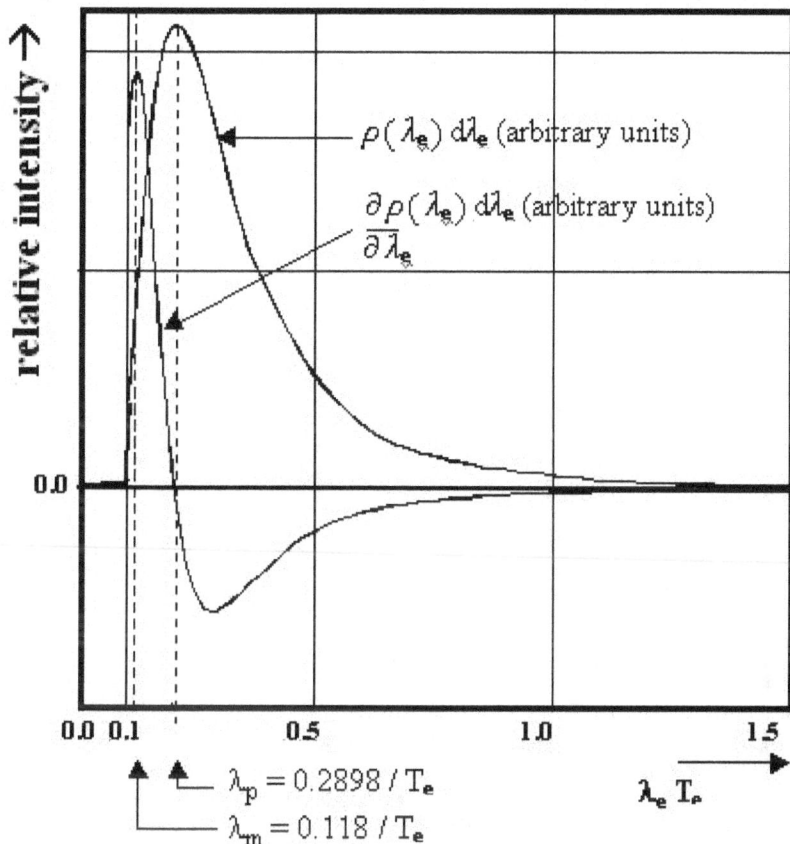

Figure 199: Blackbody spectrum and its first derivative

The total energy radiated in one second through a square centimeter of surface area of any substance in thermal equilibrium is given by Stefan's empirical formula:

$$I_T = \sigma \varepsilon \, T_e^{\,4}$$

Here $\sigma = 2.268 \times 10^{-4}$ erg-cm^{-2}-deg^{-4} is the Stefan-Boltzmann constant and ε is the *emissivity*, i. e., the efficiency of emission of the medium relative to that of a theoretically perfect blackbody. (The *emissivity* and *absorptivity* are equal for a given substance, which is quite understandable with regard to the cavity apparatus which is the appropriate representation of a 'blackbody'.) Since I_T is defined as the energy transported across a one square centimeter area in one second, the radiant energy density in one cubic centimeter is, $E_T = I_T / c$.

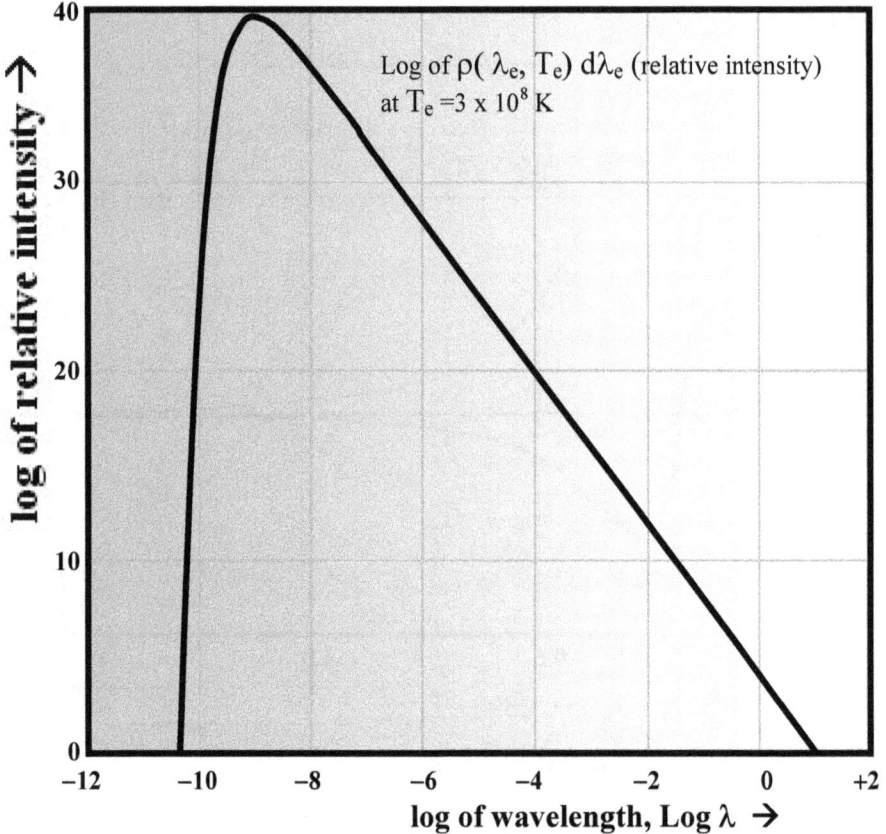

Figure 200: Blackbody spectrum plotted on log scale

Indeed, this is what we obtain by integrating the Planck energy density profile provided above over all wavelengths from λ_e equal zero to infinity. This results in the theoretical energy density for a 'blackbody' as follows:

$$E_T = (2 \pi^5 k^4 / 15 h^3 c^3) \, T_e^4 \cong 7.56 \times 10^{-15} \, T_e^4$$

b. temperature dependence of the blackbody spectrum

The curves of overall intensity of blackbody radiation plotted in figures 199 and 200 involve an extreme dependence on temperature. It is a fourth order effect that is readily apparent in plots made for several temperatures that show corresponding peak wavelengths, λ_{peak}. A blackbody at twice the number of degrees Kelvin will radiate at sixteen times the intensity as shown in figure 201 and at left in figure 202.

From these facts one might expect radiated energy, E_T to be greater than 10^{18} ergs/cm^3 for emissions of an intergalactic medium that accommodates the plasma redshift mechanism. This phenomenal rate of emission of X-radiation is not observed. Explanations for this will be forthcoming further on in this chapter and in subsequent chapters.

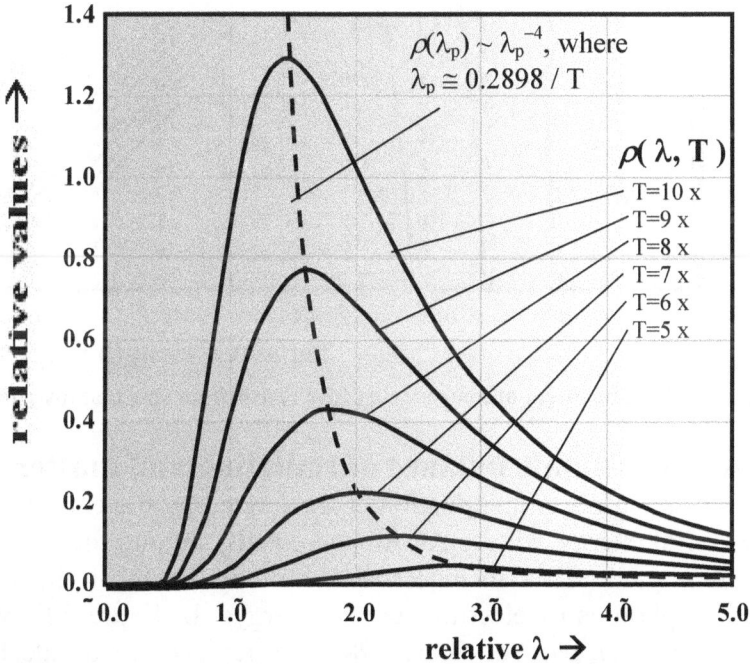

Figure 201: The effect of temperature on spectral distribution

Intriguing questions for a scattering model are: "*How could high-energy radiation frequencies emitted throughout an indefinitely extended medium be reduced so dramatically in accounting for observed microwave background radiation?*" "*How could a resultant sum of all such radiation attain a microwave background blackbody form and exhibit a thermal X-ray spectrum as well?*"

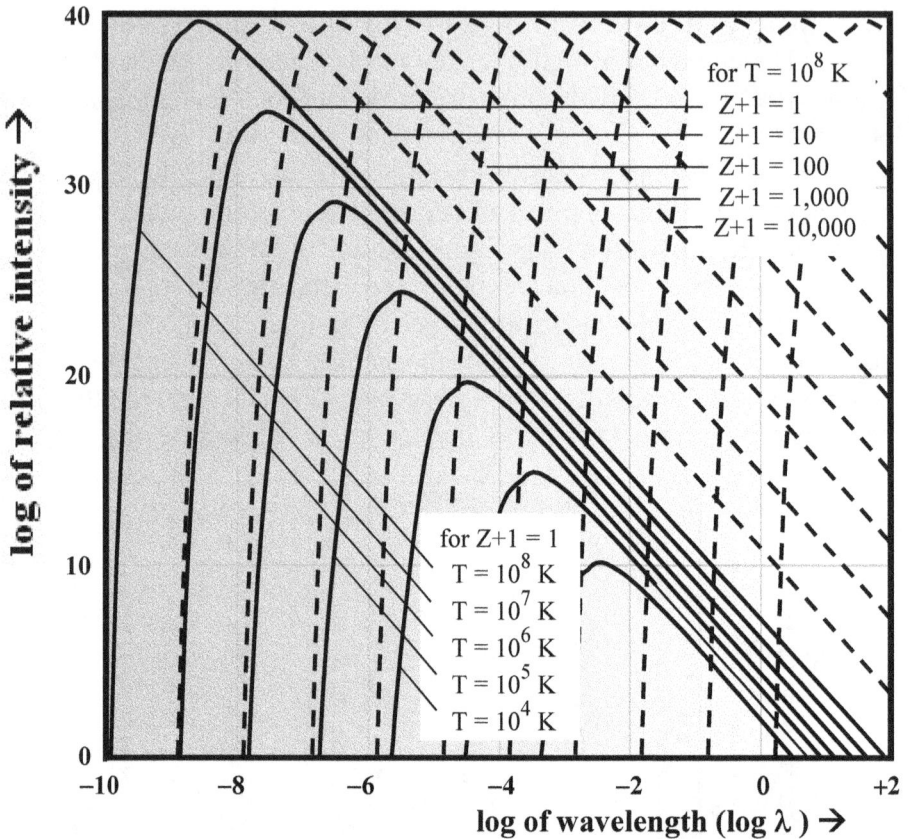

Figure 202: The effect of temperature and redshift on spectral distribution

c. thermodynamic balance of radiation and matter

In an equilibrium situation characterized by the conservation of energy, total energy will be partitioned equally among the constituents of such a gas. This includes all the components of the gas, *not* excluding photons of electromagnetic energy. In figure 203 we have depicted the 'heat bath' of figure 198 being replaced by an ideal gas of indefinite extension so as to have no fringe effects. The gas is assumed to be maintained at the given temperature.

But for a stationary state ideal gas this situation is similar to taking away the cavity altogether with photons scattering off of the particles in the gas to similar effect. In figure 204 we show only those photons shown in the previous figures within the conceptual cavity radius. Although some photons originate outside the spherical region where the cavity was drawn, collisions all occur with a material particle at some distance. Importantly, the distribution within the cavity region will be essentially the same whether the solid cavity is there or not.

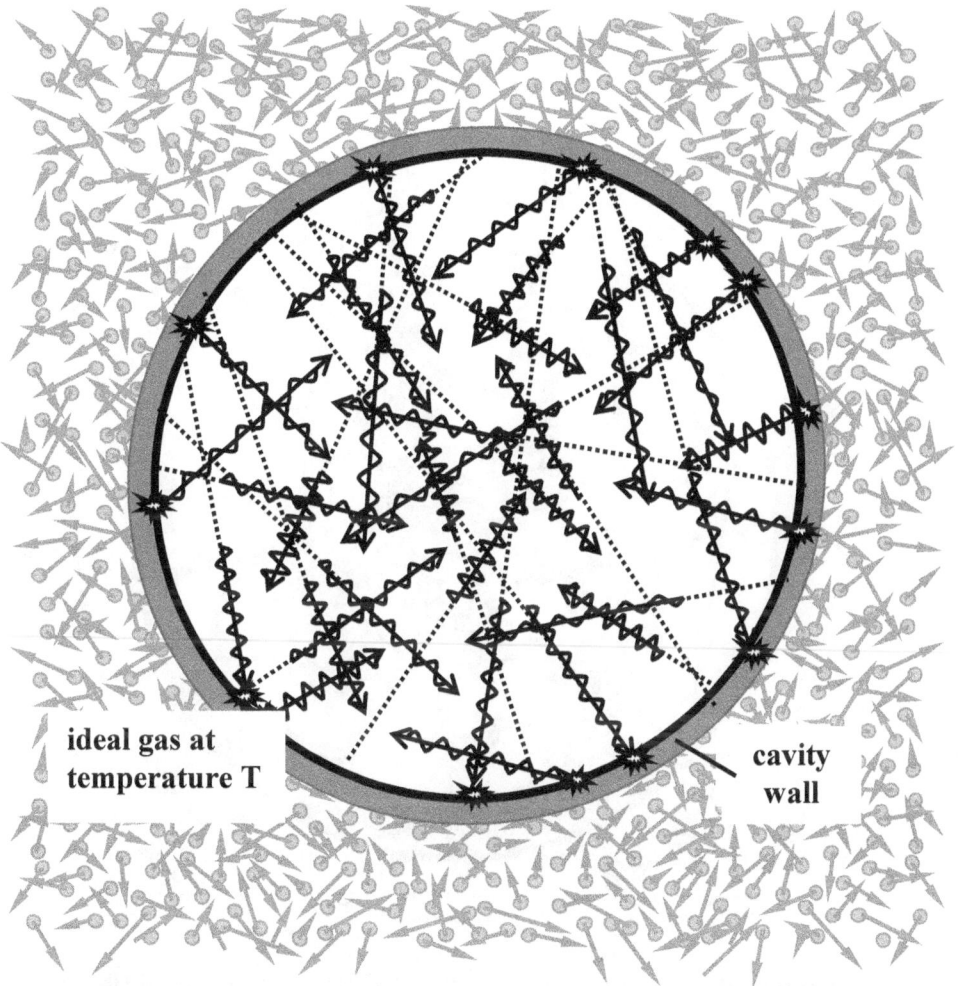

Figure 203: 'Cavity' embedded in an ideal gas at a fixed temperature

For cases such as that involved with the intergalactic plasma in which exchanges of energy between radiation and matter are dominated by scattering processes, conditions for thermal equilibrium pertain to those regions that are several optical thicknesses interior to surface boundaries. These conditions ensure that radiation and material particles will be in thermodynamic equilibrium throughout such interior regions. The optical thickness is a distance that radiation must penetrate in a medium before its intensity is reduced to 1/e of its original value. In intergalactic plasma, this is on the order of a Hubble distance H_o^{-1} as shown in figure 55.b on page 117. This involves energy exchanges with kinetic motions of particles in the medium.

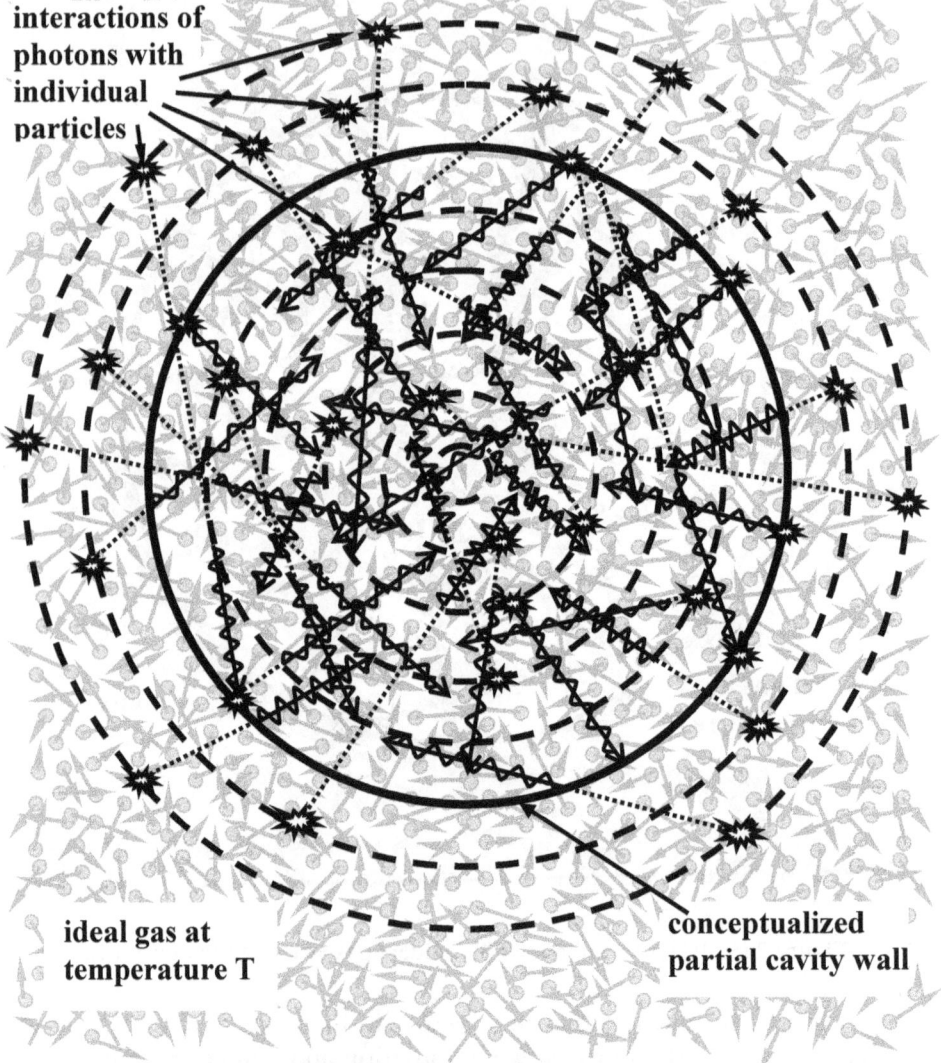

interactions of photons with individual particles

ideal gas at temperature T

conceptualized partial cavity wall

Figure 204: Radiation situation in an ideal gas

In Einstein's quantum theory of radiation (1917) he derived the Planck distribution from first principles using the Boltzmann energy distribution for molecules in an ideal gas. But it is significant that the density of material particles does not enter Planck's formula; the reason is that although 'heat' is currently perceived as tantamount to the movement of the constituent particles, the blackbody form of the distribution assumes an enclosed cavity as shown in figures 198, 199, and to the same effect in 204. If the conceptual cavity wall shown in figure 204 had been less than the optical depth of the medium from the boundary of the substance, then the blackbody form would be altered, which is why 'thin' plasmas radiate at much lower temperatures than their associated kinetic temperature.

Whereas particles can reach an equilibrium energy distribution denominated "Maxwell-Boltzmann distribution" by interacting among themselves, photons do not typically (except for rare situations we will not discuss) interact with other photons. So it is only by interacting with matter that photons can be thermalized to their equilibrium Planck blackbody distribution. Other than for thermonuclear reactions, the redistribution of energy in material substances does not alter the number of entities among which the energy will be distributed. This is not the case for electromagnetic energy, however. The number of photons of the various frequencies and the *total* number of photons will definitely be altered by this redistribution process.

Figures 203 and 204 illustrate two of the ways in which electromagnetic interactions achieve and maintain blackbody spectra, either by interacting with solid surfaces at a fixed temperature or by interacting with particles in an optically thick, gaseous substance that is in equilibrium at the given temperature. These interchanges ordinarily bring about the complete sharing of energy characterized by the phrase 'equipartition of energy'. In each figure it is individual exchanges of energy with particulate matter that produces the redistribution of electromagnetic energy to blackbody form. Usually the two instances depicted are extremely similar.

d. dependence on the density of matter

The preceding equations do not explicitly depend upon density of the material particles involved in the interactions that bring about thermodynamic equilibrium. This is notably because these equations pertain to surface brightness that incorporates a 'cavity surface' dependence rather than involving volumetric considerations directly. In the next chapter we will address more specifically the way in which surface constraints derive from volumetric considerations appropriate to situations of the thermodynamics of gases. This will be appropriate to what is illustrated in figure 204, but with redshifting occurring.

However, there are many implicit relationships that do apply. Certainly P V = n R T, with P the pressure, V the volume of gas, n the number of molecules in the volume, $R \equiv N_A k$ is the gas constant, and T temperature of the gas. This equation is the staple of thermodynamics and must apply. It provides the implied gas density dependence:

$$T(\rho_{gas}) = P / N_A k \, \rho_{gas}$$

Where the number density of the gas is defined as $\rho_{gas} \equiv n / V$. Only by artificially enforcing a constant pressure could one vary the temperature and density parameters independently.

There are other very general temperature-density relationships as well. In particular the energy density of radiation and the kinetic energy density of the gas must be equal at equilibrium according to the equipartition principle. Thus the kinetic energy density of a gas in equilibrium at temperature T will be:

$$E_{Tgas} = (3/2) \, k \, T \, \rho_{gas}$$

Thus, since $E_{Tgas} = E_{Trad}$ by the equipartition principle, we must have:

$$(3/2) \, k \, T \, \rho_{gas} = (2 \, \pi^5 \, k^4 / 15 \, h^3 \, c^3) \, T^4 \cong 7.56 \times 10^{-15} \, T^4$$

From which we obtain the generalized equipartition function:

$$T(\rho_{gas}) = (\, 0.027 \, \rho_{gas} \,)^{1/3}$$

These two expressions of constraints that link temperature and density in an extended thermal medium are illustrated as the solid lines in figure 205. The constant pressure instance of the thermodynamic equation must give the same results at the equilibrium temperature and density for which it applies naturally. Clearly the figure pertains to the situation of microwave background radiation realized in our universe.

The curve indicated as a dashed line in figure 205 is associated with the alternative accounting for observed microwave background radiation in the scattering cosmological model. The approach will be addressed in much more detail further on, but the need for explanation should be apparent, since the observed microwave background radiation does *not* emanate from current equilibrium conditions of a thermal medium capable of emitting blackbody radiation. For one thing, temperature and density of radiation does not seem to correspond to what we observe in our current universe. Unlike any blackbody radiation ever observed, its temperature does not directly reflect any of the conditions of its local or global surroundings. Both the average temperature and density of the universe as a whole differ markedly from that indicated by the observed spectrum.

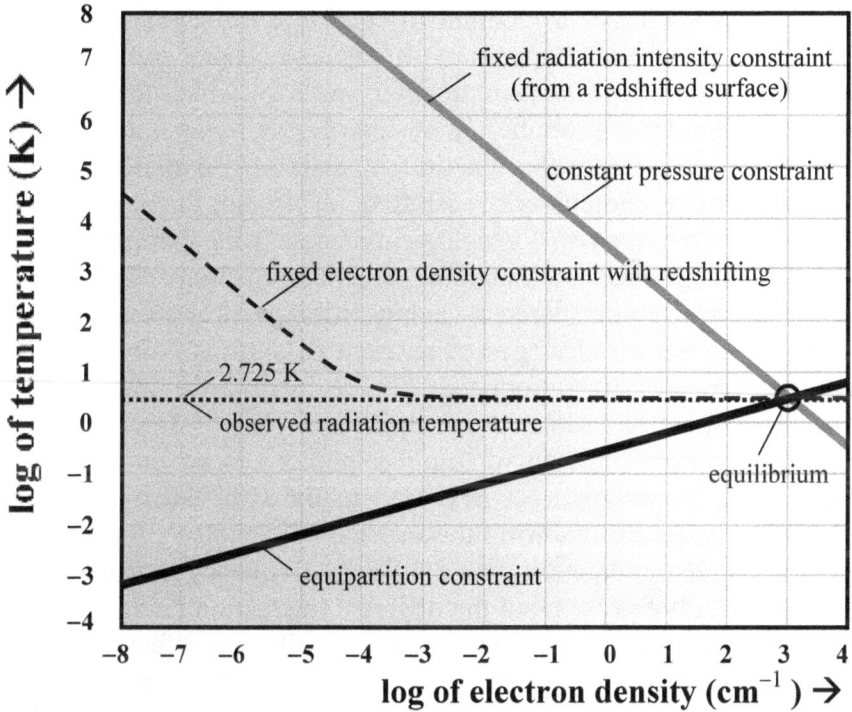

Figure 205: thermodynamic gas density constraints

Strategies for resolving why the background radiation is as it is require solutions that do not reflect current conditions. One method imposes equipartition constraints appropriate to the observed energy density in background radiation onto the material universe such that:

$$(3/2) \, k \, T_{gas} \, \rho_{gas} = 7.56 \times 10^{-15} \, T_{rad}^4 - 4.169 \times 10^{-13}$$, and thus:

$$T(\rho_{gas}) = 2.014 \times 10^3 / \rho_{gas}$$

This is precisely the forced thermodynamic pressure constraint equation described above and plotted in figure 205.

We will pick up this discussion again in the next chapter.

e. effect of redshift on a blackbody spectrum

It is a fact, although not generally known, that in a redshifting medium for which the optical depth is appreciable relative to the Hubble distance, H_o^{-1}, the stringent constraints of the equipartition of energy are somewhat relaxed. In all other cases this principle demands that all constituents of a medium in equilibrium, including radiation, share the same kinetic temperature. In the next chapter we will see that the temperature of the *conceptual* cavity wall for which the blackbody radiation distribution applies may involve a temperature that is significantly lower than the kinetic temperature of particles in a 'thin' substance where photons are redshifted. This fact has a tremendous impact on both of the two cosmological models being compared.

Although the two illustrated situations in figures 203 and 204 are not ordinarily considered as being different in any essential way, when photons in an ideal gas experience redshifting, subtle but very major, differences are introduced. These involve the interactions in concentric spheres, indicated by dashed circles in figure 204, not requiring the same amount of time for photons to cover the distance from a last interaction to a next one in the two cases. So that if redshifting is occurring during the interval of a photon's transit, it will arrive at its next interaction with a different frequency in that situation. This is true in both cases, but the different interaction histories implicit in the two cases result in unique redshift distribution effects of photons even within the confines of the cavity domain (or any cubic centimeter of the space). This may alter the distribution in ways that need to be investigated.

No one seems to have adequately addressed specifics of the subtle issues of the differences between emitted and observed spectra within continuously interacting and redshifting media. The standard models depend as intimately on such an explanation with regard to analogous problems as does the scattering model. Although there has been an attempt to simplify the situation into two discrete cases for which completely separate analyses would apply. The analyses would be straightforward indeed if a single hot surface were to have existed at only one point in time or particular redshift associated with a very definite redshift, distance, and/or time in the past. But those conditions do not hold to any very good approximation for any viable model.

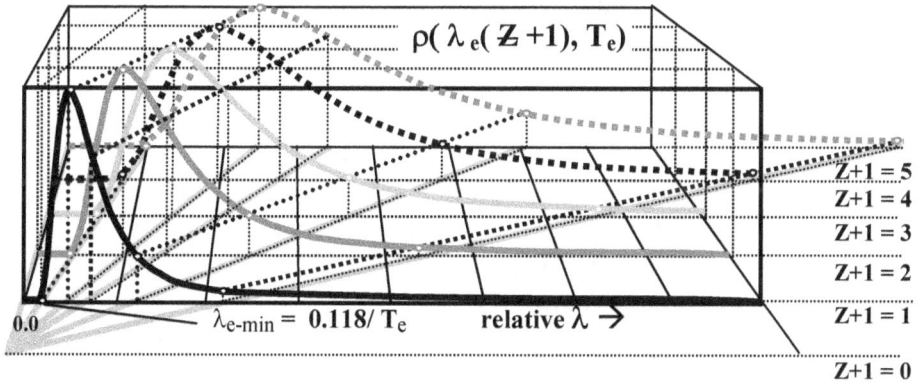

$\rho(\lambda_e(Z+1), T_e)$

$Z+1 = 5$
$Z+1 = 4$
$Z+1 = 3$
$Z+1 = 2$
$Z+1 = 1$

0.0 $\lambda_{e\text{-min}} = 0.118/ T_e$ relative $\lambda \rightarrow$

$Z+1 = 0$

Figure 206: The effect of redshift on blackbody spectral distribution

The various resultant distributions for a few redshifts are shown in figure 206. It is easily demonstrated that an emission distribution $\rho(\lambda_e, T)$ is modified to $\rho(\lambda_e(Z+1), T)$ in accounting for the redshifting of the spectral distribution. In these figures it can be seen also that for any particular value of λ, there is a readily determined maximum redshift from which blackbody radiation could be observed. This is because there is effectively a minimum emission wavelength, $\lambda_{e\text{-min}}$, at which the value $\rho(\lambda_e, T)$ drops to zero tremendously rapidly as was shown in figures 199 through 202. We'll discuss the effect of this in more detail farther on.

The projections of corresponding intensity values onto the $\rho = 0$ plane in figure 206 shows a convergence that would occur at $Z+1 = 0$, for which, of course, $Z = -1$. This domain obviously would correspond to a *blue* shift of the radiation rather than a red shift.

These diagrams (particularly those in figure 202) suggest some of the major differences between redshifted radiation from a high temperature medium and radiation from a correspondingly lower temperature, *non-*redshifted, thermal medium. The main difference is one involving the scale of the *intensity* at each corresponding wavelength of the radiation that does not affect the overall shape of the distribution itself. The *emission* temperature, T_e determines that scale height of the intensity on the wavelength distribution as we have seen. Redshifting does not alter this. It is very significant that in a redshifting environment the

407

temperature of the *observed* radiation does *not* determine the intensity of the radiation in any very straight-forward way. First of all, let us compare the related curves for $\rho(\lambda_e(Z+1),T)$ in figure 205 with corresponding curves for $\rho(\lambda_e, T / (Z+1))$ shown in figure 201. Obviously we have:

$$\rho(\lambda_e, T/(Z+1)) \, d\lambda_e = (Z+1)^{-4} \, \rho(\lambda_e(Z+1), T) \, d\lambda_e$$

Figure 202 illustrates this difference most dramatically on a log scale. The intensity of the distribution indicated on the right hand side of the previous equation exhibits the factor of $(Z + 1)^{-4}$. Clearly, the scale of the intensity does indeed say a great deal with regard to the redshift situation at which thermal radiation was emitted.

Let us look at the intensity effects of radiation emitted from a 'surface' at a single redshift shown in figures 199 and 200, as compared to cumulative effects of emission at all redshifts out to *extended* distances which must be considered with the scattering model that is under investigation. This analysis will highlight differences between major versions of the standard model and the scattering model.

As we have seen, blackbody radiation that is simply redshifted looks identical in *form* to blackbody radiation from the surface of a cooler object that has *not* been redshifted. That is, the *shape* of the curve is the same. *Significantly* however this does *not* include its *intensity* which provides clues to its unique origins.

f. Wien's law

Wien's law of blackbody spectra that involves the factor $\rho(\lambda,T)$ that is a function of the product λ times T guaranteeing similarity in form for both temperature T and wavelength λ. Wien's law of blackbody (or 'cavity') radiation is the following:

$$\rho(\lambda,T) = f(\lambda \bullet T) / \lambda^5$$

We use a heavy dot to emphasize the product. This expresses a product law functionality for temperature T and wavelength λ for the Planck factor $f(\lambda \bullet T)$ of the observed spectra.

Observed radiation wavelength will be given by $\lambda_o = (Z+1)\lambda_e$ in a redshifting environment, where λ_e is the emission wavelength discussed above. Wien's law assures us that whether radiation is from a surface at temperature T observed at a redshift of Z+1 or the surface is actually reduced to the temperature T/(Z+1) with no redshifting taking place at all, makes no difference at all as far as this factor is concerned:

$$f((\lambda_0/(Z+1)) \bullet T) = f(\lambda_0 \bullet (T/(Z+1)))$$

Wien's law concerning $f(\lambda \bullet T)$ was actually confirmed using Doppler shifted radiation reflected from a moving piston reassuring its applicability to redshifting environments.

Thus, observed redshifted radiation λ_0 will possess the same generic functional form of the distribution as non-redshifted radiation of temperature T / (Z+1) as clearly illustrated in figures 202 and 206. *However*, there is a very real difference in the scaled intensity, which will be reduced by the factor $(Z+1)^4$ in the one case in accordance with Stefan's Law as we have seen, once the expression for the distribution has been integrated over all possible wavelengths:

$$E_T = \int_0^\infty f(\lambda_0 \bullet T/(Z+1)) / \lambda_0^5 \, d\lambda_0 = \int_0^\infty \rho(\lambda_0, T/(Z+1)) \, d\lambda_0$$

$$= 7.56 \times 10^{-15} \, T^4 / (Z+1)^4$$

Here the total intensity of radiation from an emission 'surface' like that shown in figure 203 is reduced accordingly as a result of the operative redshifting. The temperature of the observed radiation determines both its form and intensity scale, but not whether or not redshifting had occurred or if the radiation just derived from a cooler surface.

Compare and contrast this situation with that for emission from areas throughout a continuously varying range of redshifts like that illustrated in figures 202 and 205. In this case we must integrate the energy density derived from each cubic centimeter of the entire space.

This integration to an indefinite extension, i. e., distances to which Z >> 1, results in the following intensity:

$$E'_T = 7.56 \times 10^{-15} \, T_e^4 \int_1^\infty \zeta^{-4} \, d\zeta = (1/3) \, E_T$$

In this case the effectual intensity is only diminished by a factor of one-third even though the radiation in this case will also be redshifted – so far we have not determined to what extent. The two implied intensities E_T and E'_T derived from two different alternative sets of assumptions may differ by as much as a factor of fifteen or twenty orders of magnitude when applied to cosmological theories of our universe.

So it should be clear why the standard cosmological model insists that the expanding 'surface' of scattering by a hot plasma gas must be a single 'closed' surface with no photon deriving in front of or behind it. Here it is inferred that the intensity of background radiation has resulted from thermodynamic properties of the surface of last scattering such that:

$$4.169 \times 10^{-13} \cong 7.56 \times 10^{-15} \, T^4_{sur} / (Z_{sur} + 1)^4$$

So that:

$$Z_{sur} = 0.367 \, T_{sur} - 1$$

From such considerations comes the conjecture that the 'surface' was at a redshift of about 1200 with a surface temperature of on the order of 3,270 K.

The Viability of that explanation depends on the applicability of there having been a single surface of last scattering; the explanation does not work otherwise. But, in fact, this required criterion is not met. The density at this point in the supposed history of the universe would not complete the associated 'cavity surface' and would thereafter still provide sufficient material that scattering would continue.

The scattering model hypothesizes something more in tune with the latter equation. But the closure criterion must be explored in more depth as we will in the next chapter where we discuss Olbers' paradox and the extent to which related concepts affect this situation.

g. considerations of an expanding thermal surface

More importantly, however, all of this about temperature of surfaces, presumably expanding at appreciable speeds relative to the observer so as to effect a Doppler (or expansionist 'stretching') redshift in standard cosmological models, in a real sense ignores the radiation itself. As we have shown, there *is* a major difference between the intensity of Doppler shifted radiation from expanding surfaces at temperature T redshifted to T/(Z+1) and redshifted radiation originating throughout a space filled with a scattering medium at temperature T.

The tremendous difference derives from the fact that Wien's law only applies to the Planck factor $f(\lambda \bullet T)$ of the emitted radiation and significantly, the complete distribution includes a Raleigh-Jeans factor of $1/\lambda^5$ as well. So although we can look at the Planck factor as independent of whether the temperature is actually T/(Z+1) or the radiation has been redshifted to $\lambda_o = (Z+1)\,\lambda_e$ so as to appear to be at that reduced temperature, the Raleigh-Jeans factor clearly discriminates between these two cases. Thus, we can determine with some certainly which situation has *actually* pertained. Significantly, however, the standard models requires the awkward splicing together of a two-step resolution to the microwave background radiation problem.

Notice in particular that in the integration step in obtaining the preceding expression that nothing had been specified with regard to the specifics of the distribution of the emitted radiation $\rho(\lambda_o,T)$, That dependency has been integrated out. The fact that the associated integration parameters are mutually independent assures us that the order in which the two implied integration steps are performed is irrelevant to the result. So we can, if we wish, perform the integration of the distribution over redshift first. If we do this, we obtain:

$$\int_0^\infty \rho(\lambda,T)\,/\,(Z+1)^4\,d(Z+1) = -\,\rho(\lambda,T)\,/\,3(Z+1)^3 \,\Big|_0^\infty = (1/3)\,\rho(\lambda,T)$$

Now consider the difference implied by placement of the Z+1 factor on the left and right-hand parameter in the following inequality.

$$\rho(\lambda,\ T/(Z+1))\ d\lambda \neq \rho(\lambda\,/(Z+1),\ T)\ d\lambda$$

Although the valid transference of redshift between temperature and wavelength applies to the Planck factor, it does *not* apply to the distribution as a whole as the expression clearly states. Therefore, integrated intensity of radiation over all wavelengths emitted at a single redshift-distance, such as a 'decoupling' surface or as depicted in figure 203 if you will, would differ considerably from what would have to have been emitted to realize the same observed temperature of radiation that had originated throughout an appreciable range of a redshifting environment. That is the essential difference between figures 203 and 198 that is pertinent to redshifting situations.

h. alternative consideration of a stationary state

As one looks out through a stationary state intergalactic medium as appropriate to the scattering model, the thermal radiation that is emitted will be subject to modeling by shells of uniform density as illustrated in figure 204. Such situations will be discussed in the next chapter in the context of Olbers' paradox. What is realized in such cases is the presence of contributions to observed background radiation coming from *each* incomplete shell rather than radiation from just a single complete shell.

The material substance in each shell would be statistically the same as that in any other and would emit radiation with the same spectral distribution. However, redshifting changes the situation as was shown in figures 202 and 206. It results in essentially the same form of blackbody curve with the same intensity applicable to each shell, but with a lower *effective* temperature emanating from each successive domain. At each wavelength, λ_0 there will be contributions from each of increasingly remote regions as indicated across the front of figure 207. This figure is derived from figure 206 by taking the cross section at the wavelength λ_0, where,

$$\rho(\lambda_0, Z_0, T) \, d\lambda_0 = (2 \, h \, c^2 \, (Z_0+1)^3 / \lambda_0^3 \,) / (e^{\,K(Z_0+1)/\lambda_0 T} - 1 \,) \, d\lambda_0$$

In observing radiation of the given wavelength λ_0 from an extended redshifting medium, one expects there to be increasingly large contributions from more distant regions – back to where the dramatic drop in blackbody radiation intensity is encountered at a distance of around where $Z+1 \approx 10 \, \lambda_0 \, T$. There is, of course, a weighting to be taken into account in integrating the effect from each distance. In the next chapter we will derive this factor involving the degree of cavity surface 'closure' realized at each 'shell' in this integration process.

'Emissivity' is the ratio of the power radiated from a surface of a substance to what would be realized by an associated blackbody of the same temperature. So although the observer is immersed in a vast medium with a temperature T – that of the equilibrium conditions of particles in the medium – only that very small portion of the total 'cavity surface' area realized by electron cross sections out to the distance for which radiation experiences no substantial redshift, will effectively radiate at that temperature. Particles in the next deeper region will appear to radiate at a slightly lower temperature, etc.. The effective surface area of each succeeding region will be reduced exponentially as we will see in the upcoming chapter where we discuss Olbers' paradox.

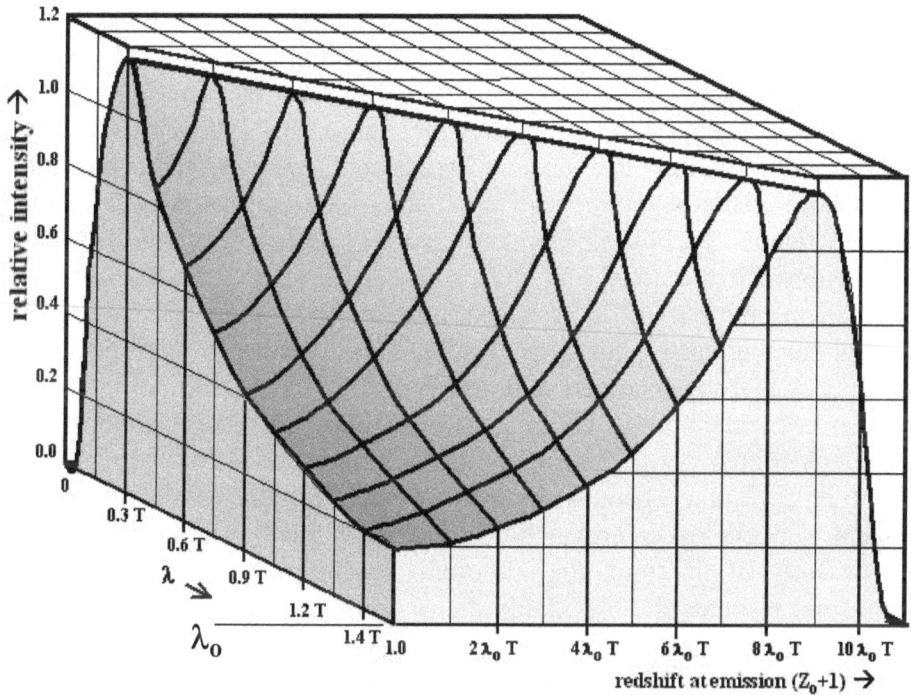

Figure 207:   The contributions to be integrated to obtain the intensity of radiation realized for a given wavelength in an extended redshifting medium

In consideration of the case of an intergalactic medium in a stationary state acting as a thermal medium, we must address radiation emitted from surface areas progressively deeper in the medium relative to the observation point. However, much beyond where $Z+1 \approx 10 \, \lambda_0 \, T$ there would be negligible contributions from the curvilinear parametric surface representing blackbody radiation in a redshifting environment. This was illustrated in figure 206 and more clearly in figure 207. So although one might integrate the parameter ζ (i. e., Z+1) from 1 to ∞, for all practical purposes integration would only be from 1 to a little beyond 10 λ_o T, since,

$$\rho\,(\lambda_o,\,T,\,Z+1) \cong 0.0, \text{ for } Z+1 >> 10\,\lambda_o\,T$$

This is illustrated quite clearly in the figure. However, other than as an interesting exercise, this does not very precisely reflect what happens either. In dense or 'optically thick' media (or where there is essentially no redshifting involved) there would be no seeing through to extended redshifts. It would just be as depicted on the left face in figure 207. That is what the standard model accounting for the microwave background radiation depends upon throughout the early phases of their explanation. But the degree to which that condition is satisfied must be demonstrated. Hubble's 'constant' as well as the electron density responsible for the scattering that is in turn also responsible for the thermalization process in that explanation vary quite dramatically throughout that early phase. Therefore, that explanation must provide assurances that the scattered radiation has not been redshifted between scattering events, but that claim is incompatible with the explanation.

Within any finite volume, unless it were very much larger than the Hubble volume of 'optically thin' plasma like that which we know currently occupies intergalactic space, the kinetic temperature of constituents of the medium, although generating thermal radiation, would not produce the blackbody distribution. Rather such a medium would generate radiation with the intensity of much lower temperature but still 'thermal' radiation appropriate to the kinetic temperature of the medium.

It is energy content that gets 'thermalized'. That is what forces a resultant blackbody radiation spectrum when an effectively complete 'cavity' surface characterizing an equilibrium situation in a sufficiently 'thick' medium is realized. With a diffuse medium the volume must be commensurably larger (in fact, *very* much larger) to effect such a 'surface'.

414

Chapter 21

The Relevance of Olbers' Paradox

Olbers paradox was first expressed hundreds of years ago to address perceived problems associated with cosmologies embracing an extensive universe full of hot, bright stars. In this chapter we will consider related issues including shells of partial thermal 'cavities' each filled with particles at a single temperature but with 'holes' through which one can observe progressively deeper into otherwise similar partial cavities from which increasingly redshifted radiation emanates.

There are two aspects to be considered here: One is the effect of extending the emission domain indefinitely (beyond the optical depth) and the other is the effect of redshifting radiation from the increasingly distant 'cavity' regions of the extended medium.

a. description of the paradox

The first aspect we will discuss involves considerations required to understand the traditional version of Olbers' paradox. We will focus on an aspect that is essential to both the proper understanding and resolution of this paradox with regard to an extended intergalactic medium. The resolution, because of the difference in the proper understanding of the problem from how the traditional formulation and resolution of Olbers' paradox has been treated, suggests effects that are other than might have been expected. It is well known, although still not *commonly* known, that redshift alone can *not* resolve Olbers' paradox. This paradox is illustrated graphically in figure 208.

It is an obvious fact that the night sky does not blaze as brightly as the surface of a star like our own sun. Olbers' paradox addresses the question of how this can be the case – particularly, if one considers the universe to be of virtually infinite extent and filled with a uniform density of luminous objects such as stars and galaxies of stars. It might certainly seem as though we are faced with resolving why we are not surrounded by extreme temperature radiation. We know that is not the case. Our night sky is actually quite dark. Why?

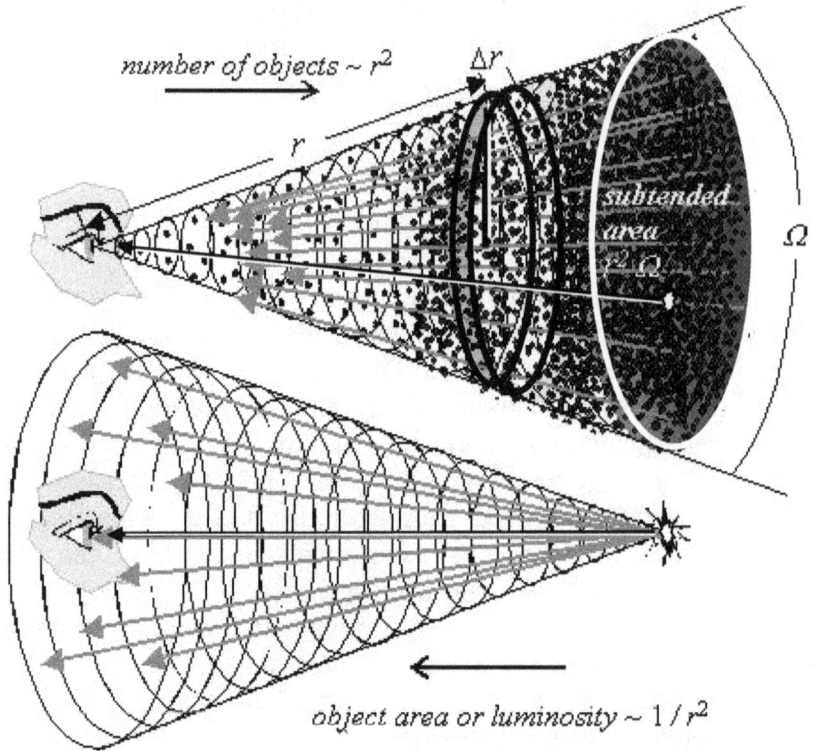

Figure 208: Cancellation of distance related relationships in observation

In typical expositions of Olbers' paradox the problem is presented as follows: A solid angle Ω subtends larger and larger areas out to increasing distances r as shown in figure 208 (top), while the intensity (number of photons to be observed) of this electromagnetic radiation originating at sources distributed within a spherical shell of width Δr at that distance is commensurably diminished (bottom). The increase in sources is *directly* proportional to distance squared while the decrease in intensity is *inversely* proportional to the square of

416

distance. Therefore, in an infinitely extended Euclidean universe, the intensity of radiation would simply accumulate linearly with increasing distance until finally every line of sight throughout the entire area of any solid angle would terminate at the surface of a star. Refer to figures 209 and 210 for an illustration of this fact.

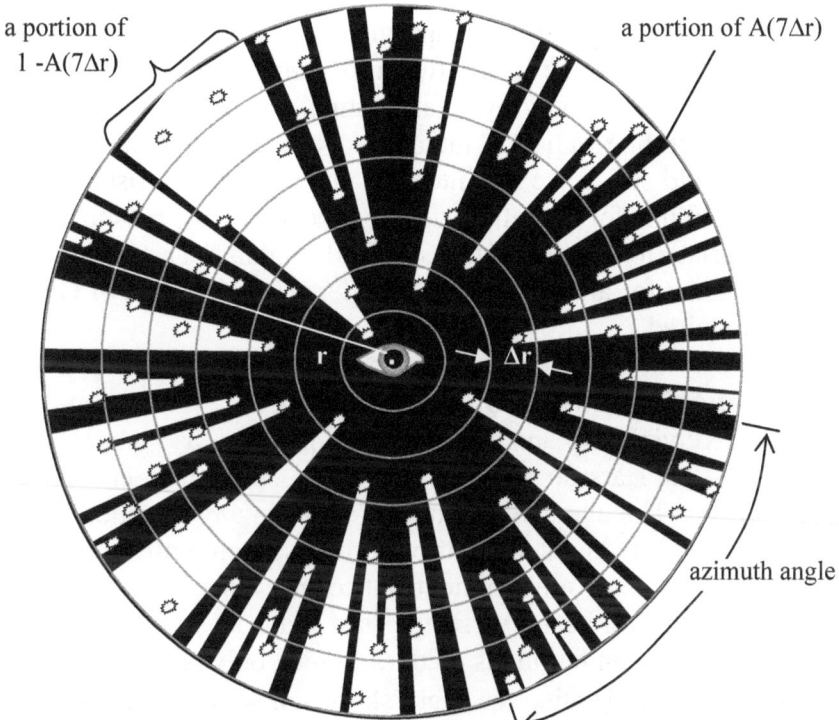

Figure 209: View through a slice of randomly distributed objects of uniform size

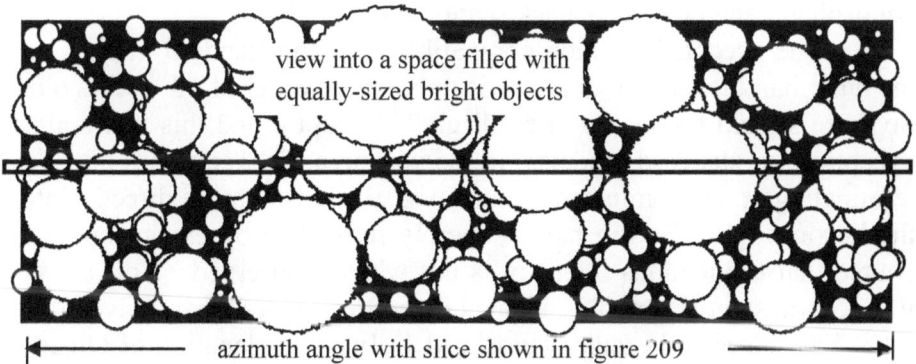

Figure 210: Surface brightness dominated by closer of uniformly sized objects

417

Integration of the intensity is similar to a simple addition whose result is proportional to the distance to which the summation takes place. As we look out into space, it would seem then that we should be looking directly into what would be equivalent to the sun's brightness in every direction if the universe were indeed infinite.

Some have argued that redshift itself would diminish the effect to insignificance. However, that particular argument has been known to be problematic for some time. Redshift would indeed diminish the intensity *somewhat*, but only by a fairly small fraction similar to what we saw by integrating radiation intensity over redshift in the previous chapter. It would still leave the intensity unbearable. Harrison (1991) provides a formula for radiation density reduction by redshifting in terms appropriate to several of the standard models. The basic result remains the same for all. Therefore, a different explanation of Olbers' paradox is required.

b. the proper resolution of the paradox

Lord Kelvin's explanation of Olbers' paradox, as elaborated by Harrison, was that stars do not burn forever. Usually they endure less than 10^{10} years. The average distance to a star in any direction would be on the order of 10^{23} light years if our universe were uniformly dense and similar to our local environment out to that distance and beyond. Thus, even in an infinite universe the vast majority of lines of sight would *not* terminate on luminous matter. In fact, only one in about 10^{13} = $10^{23} \div 10^{10}$ would, leaving the night sky brightness darker even than we know it to be. It is, of course, somewhat brighter because of the moon, the Milky Way, local stars, zodiacal light from the earth's atmosphere, and city lights back scattered off of clouds.

The accepted solution compatible with the standard models on the other hand is that the universe is only on the order of 10^{10} years old. So there should be no problem. Right? Except – and this is a major exception for standard models – in the vast areas between lines of sight terminating at an illuminated surface, we would be staring directly into the blazing light of a big bang – merely redshifted by expansion. Of course that tremendous intensity is not what is observed. Nor, as we will see, do standard models adequately address exactly why it is not.

To more fully understand Lord Kelvin's explanation of Olbers' paradox, let us consider our sun as representative of stars generally. As we have seen, its mass is 2 x 10^{33} gm and it is circumscribed by a sphere of radius 6.96 x 10^{10} cm. Its volume is, therefore, on the order of 1.5 x 10^{33} cm^3, for a density of 1.3 gm per cm^3. In order for the density of a universe made up of such entities to have a density of as little as on the order of 10^{-31} gm cm^{-3}, there could only be one such star occupying a volume of 2 x 10^{64} cm^3 on average, for a separation of approximately 5.4 x 10^{21} cm, which is about 5,100 light years. (In addition to Alpha Centauri at a little over four, there are six other stars within ten light years of the sun.) That is how Lord Kelvin's resolution works out, but let us pursue this line of argument a little further.

The cross sectional area of the sun is about 1.5 x 10^{22} cm^2. But consider the cross section of the occupancy sphere of a star, i. e., a sphere as defined above with radius about 2.7 x 10^{21} cm, for which the star is assumed to be the only occupant so as to effect the universal average density. The cross section of such an occupancy domain is on the order of 3 x 10^{42} cm^2. This is about 2 x 10^{20} times more cross section than the sun itself. This sparse distribution of stars within galaxies containing many billions of them is, of course, why entire galaxies can collide and pass completely through each other with only minimal impact to the individual stars of which they are comprised.

c. thermalization effectiveness of various forms of matter

Importantly, their extreme separations are why the stars and galaxies comprised of them do not contribute much to the scattering that produces thermodynamic balance in the universe. Nor, therefore, do they contribute much toward completing a 'cavity surface' essential to establishing a blackbody distribution of radiation, and most certainly in combination they do not result in the observed background spectrum.

The general topic of the amount of sky cover (or cavity surface area) afforded by various forms of matter, each possessing the same overall mass, is what is of interest to us here. This issue affects the determination of a complete cavity surface analogy in thermalization of energy in an expansive medium with no explicit boundaries.

Let us consider the average sky cover per individual electron in various configurations. Since electrons are primarily involved in the scattering of usual forms of radiation and since, in addition, electrons have much greater cross sections than do baryons, they are much more pertinent to establishment of a semblance of a cavity surface necessary for the associated thermalization process whereby the stable blackbody radiation distribution of energies is produced.

There are on the order of 10^{57} baryons contained in our sun, and therefore about that many electrons as well. The average cross section per electron in this configuration is therefore 1.5×10^{22} cm^2 / 10^{57} = 1.5×10^{-34} cm^2 per electron. In contrast, an individual electron's cross section is 6.65×10^{-25} cm^2 – over a billion times greater. However, since we are assuming on average only one such star per occupancy domain, the disparity in *overall* percentage of total cross section covered by electrons when in this configuration is even more extreme. The average distance for a line of sight to encounter one electron is on the order of 10^{23} light years as Lord Kelvin pointed out. Whereas, if the universe were made up of a uniformly dense hydrogenous plasma containing the same number of electrons, the line-of-sight distance to encounter an electron would be on the order of:

$$2 \times 10^{64} \text{ cm}^3 \div 6.65 \times 10^{-25} \text{ cm}^2 = 3 \times 10^{30} \text{ cm} = 3.2 \times 10^{12} \text{ light years}$$

Of course there are variations in electron density and scattering propensity encountered in observations at cosmological distances as we have seen. Each will be associated with unique total electron cross sections. Such analyses provide the key for resolving the responsibility for thermalization via scattering that ultimately reduces background radiation to microwaves. Clearly, without an intergalactic medium there would be virtually no blackbody spectrum whatsoever.

d. quantifying 'sky cover' for uniform density particles

Let us pursue this line of analysis more quantitatively then by assuming a uniform random distribution of density ρ per cubic centimeter of equal-sized spherical objects of cross sectional area πr_0^2, where r_0 is the radius of each object. Consider a solid angle that subtends shells of uniform thickness Δr as illustrated above. The proportion of the total surface area of just those objects subtended in each shell would then be:

$$\alpha = \pi \rho r_0^2 \Delta r$$

There is no explicit dependence on either radial distance or solid angle here because proportionate factors involving them cancel. However, some of the coverage in the nth shell (and indeed increasing percentages of it) will have been occluded by objects closer to the observer and thus will not be observed as clearly illustrated in figures 209 and 210.

So if we are interested in that portion of the observed field of view covered exclusively by objects in the shell at the distance r, we must subtract the amount we attribute to objects that are closer than r as suggested above. Thus, it is easier to solve the problem by working outward from the observer to more and more distant shells which we label 1, 2, 3, ... n. Then the proportion included in any solid angle not already occluded by objects included in preceding shells is:

$$a(0) = (1 - \alpha)^0 = 1$$
$$a(1) = (1 - \alpha)^1 = (1 - \pi \rho r_0^2 \Delta r)$$
$$a(2) = (1 - \alpha)^2 = (1 - \pi \rho r_0^2 \Delta r)^2$$

$$\cdots$$

$$a(n) = (1 - \alpha)^n = (1 - \pi \rho r_0^2 \Delta r)^n$$

Replacing Δr with its equivalent, r / n, and recognizing the usefulness of changing from a(n) to a(r), we obtain in the limit as $n \rightarrow \infty$,

$$a(r) = e^{-\pi \rho r_0^2 r}$$

This employs the definition of the exponential function:

$$e^{-x} \equiv \lim_{n \rightarrow \infty} (1 - x / n)^n,$$

Then from our derivation of a(r), the exponential approximation applies for large values of n, i. e., when $n \gg \pi \rho r_0^2 r$ so that $\Delta r = r / n \rightarrow 0$.

In order to make sure that there are no unaccounted objects occluded within a shell itself, the width Δr must be chosen such that the probability of occlusion within a shell is nil, which in the limit, will indeed be the case, of course. Curve 1 in figure 211 plots a(r). Curve 2, is its complement, total sky cover out to the distance r. In the figure as drawn, these curves are independent of the value of $\pi \rho_0 r_0^2$.

This discussion constitutes a proper approach to resolution of Olbers' paradox appropriate to its usual formulation. Brightness to any particular depth ultimately depends only upon the values of ρ_0 and r_0 and the distance corresponding to the lifetime of the bright objects.

There are, of course, additional thermodynamic issues concerning the ranges of temperatures of the various objects covering transitions between intensely hot stars and cold lumps or black holes, for example, that could be included in a fuller treatment of the problem.

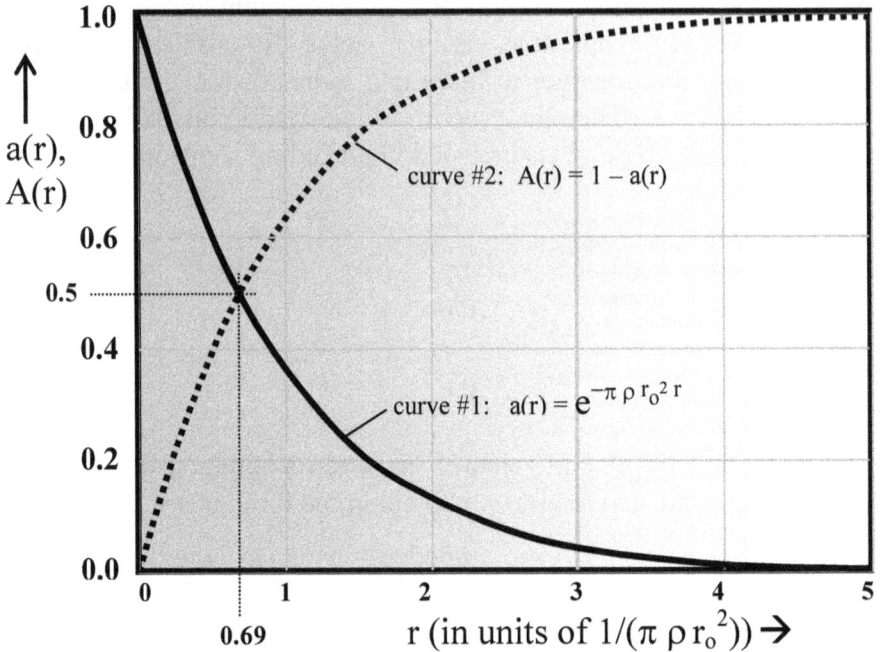

Figure 211: Sky coverage (curve #2) of uniformly sized objects

curve #2: $A(r) = 1 - a(r)$

curve #1: $a(r) = e^{-\pi \rho r_o^2 r}$

$a(r),$
$A(r)$

r (in units of $1/(\pi \rho r_o^2)$) $\rightarrow$

However, there are other aspects altogether that need to be taken into account in dealing with background radiation from an intergalactic medium both in the current model and also for the standard expansionary models. We will discuss those aspects more specifically in the next two chapters, but in this chapter we will play with a few concepts and numbers to determine to what extent Olber's paradox applies in general and in particular to intergalactic plasma.

e. sky cover as the closure criterion for a 'cavity surface'

Sky cover provides a basis for determining the percentage of 'closure' of a cavity surface. The rate of closure becomes a weighting factor pertinent to thermalization analyses addressed in the previous and upcoming chapters. It will prove of particular significance in redshifting environments.

First we define η, which is the exponential factor in the sky cover expression as derived above:

$$\eta \equiv \pi \, \rho_0 \, r_0^2.$$

Plasma electrons have considerably larger cross sections than do the various nuclides that also populate the intergalactic medium. If we used the classical electron radius, $r_e \cong 2.82 \times 10^{-13}$ cm, we would find: $\eta \cong 2.5 \times 10^{-25} \, \rho_e$. However, a more accurate value for the electron cross section is known to be 6.65×10^{-25} cm^2, so we use that value:

$$\eta = 6.65 \times 10^{-25} \, \rho_e.$$

Now, let us assess the distance $r_{1/2}$ for which the sky cover area, as shown in figures 209 and 210 would be 50 percent, i. e., $A(r) = 0.5$. So that:

$$e^{-\eta R_{1/2}} = \frac{1}{2}$$

Since the natural log of one half is equal to -0.69, we obtain:

$$R_{1/2} = \ln(e^{-\frac{1}{2}\eta}) = 0.69 / \eta \cong 1.038 \times 10^{24} / \rho_e.$$

If $\rho_e = 10^{-5}$, for example, then $R_{1/2} \cong 1.038 \times 10^{29}$. Multiplying this distance times the currently accepted value of Hubble's constant, $H_0 = 7.14 \times 10^{-29}$, we obtain the redshift at this half sky cover distance:

$$(Z_{1/2} + 1) = e^{(7.14 \times 10^{-29}) \times (1.038 \times 10^{+29})} = e^{7.408} = 1.649 \times 10^3$$

See figure 212 where the progression of both r and Z from emission to the observation of the thermal radiation are illustrated.

The remoteness of the mean distance to an electron participating in the analogy to a 'cavity surface' for the vast majority of observed thermal emissions implies, of course, that much of the thermal radiation from an indefinitely extended medium would be redshifted on the order of 10^3 to 10^4 for the case we have just calculated above. Thus, because of its extremely diffuse nature in a redshifting environment, we should expect the apparent temperature of radiation to be considerably reduced from that of its emitted spectrum. This is true for any redshifting mechanism, and in particular for the scattering model.

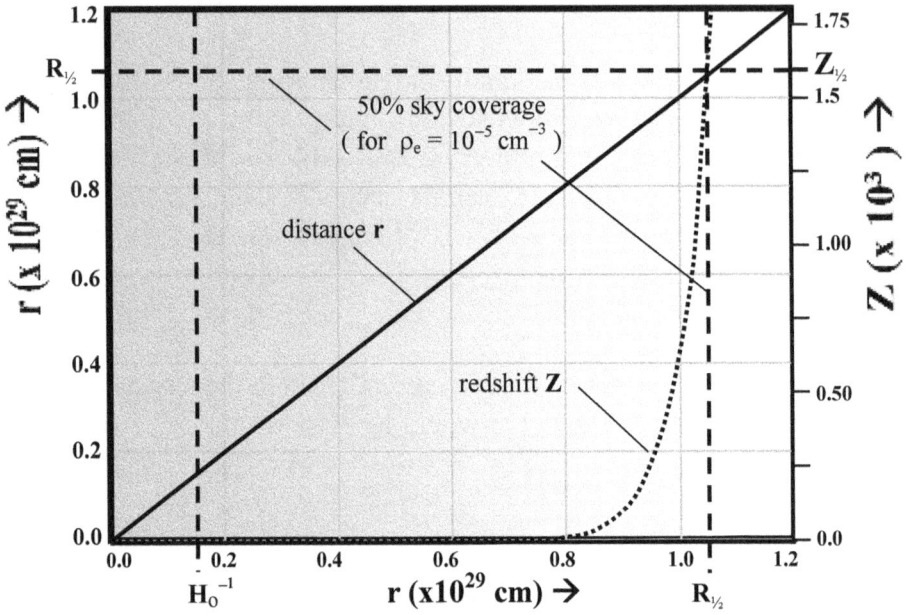

Figure 212: Distance and redshift to half sky cover of plasma electrons

To the extent that sky cover and redshift can be modeled as uniform continuous phenomena, the proportion of solid angle subtended by electrons out to a distance r would be given by:

$$A(r) = \int_{A(o)}^{A(r)} dA(\eta, r) = \eta \int_{o}^{r} e^{-\eta r'} dr' = 1 - e^{-\eta r}$$

From which the following differential equation results:

$$\frac{dA(\eta, r)}{dr} - \eta A(\eta, r) + \eta = 0$$

The function A(r) was plotted as curve #2 in figure 211. The fact that $A(\infty) = 1$ is as expected. For the 50% surface coverage calculation we tentatively assigned a value of $\rho_e = 10^{-5}$ cm^{-3}. And, in that case we had $\eta \cong 6.65 \times 10^{-30}$ cm^{-1}, so that $R_{1/2} \cong 1.038 \times 10^{+29}$ cm. This is in very basic agreement with the approach taken to the mean unobscured path calculation, i. e., the average length of a line of sight to encounter an electron in the intergalactic medium presented in figure 29 on page 51 where, however, we used the estimate $\rho_e = 10^{-6}$ cm^{-3}.

Clearly, the 50% sky cover situation is extremely sensitive to the value assigned to ρ_e. Using the estimate $\rho_e = 10^{-6}$ cm^{-3}, we obtain:

$$\eta \cong 2.88 \times 10^{-7} \; 6.65 \times 10^{-25} = 1.915 \times 10^{-31} \text{ cm}^{-1} \text{ and}$$

$$R_{\frac{1}{2}} \cong 3.6 \times 10^{30} \text{ cm}$$

The redshift associated with this value is, of course, very appreciable:

$$Z_{\frac{1}{2}} + 1 = e^{\,7.\,14 \times 10^{-29} \,\cdot\, 1.038 \times 10^{+30}} \cong 10^{32}$$

For all practical purposes this is an infinite value. A difference in electron density of a single order of magnitude precipitates this prodigious increase in redshift of the average line of sight distance to an electron 'surface' of more than ten orders of magnitude. However, the actual distance is less than 74 times the Hubble distance, still very appreciable, of course, but a more reasonable distance over which thermalization would occur in an extremely thin but extensive plasma.

f. applying the analysis to thermodynamic properties

We come now to the issue of using the weighting factor for integration in order to correctly assess effects that are dependent upon a redshift that occurs over the distances to partial surrounding surfaces to effect a total 'cavity surface' applicable to thermodynamic analyses as discussed in the previous chapter.

The temperature of virtually everything in the universe – the intergalactic medium in particular – is many orders of magnitude greater than that associated with the spectrum of the microwave background radiation. Not only that, the density of the universe – and again, that of the intergalactic medium in particular – is orders of magnitude less than what is directly implied by the background radiation temperature. It is certainly meaningful to inquire why this situation pertains? The standard cosmological model offers one reason; we provide a less complex alternative whose detailed explanation must await details concerning how the scattering model accounts for the origin of the ubiquitous microwave background radiation in chapter 23. Here we employ the analyses concerning sky cover to shed some light on related questions as preparation for that conclusion.

Our discussion has already illuminated some of the reasons why observed temperature of a blackbody radiation distribution emanating from an extended medium might be significantly less than would be calculated using only the kinetic temperature of the involved particles. Naturally, 'apparent' temperatures of encompassing 'surfaces' deeper and deeper in the medium from which 'cavity' radiation arises will be increasingly affected by cosmological redshift as we described in the previous chapter where it was determined that the temperature from a single expanding (or otherwise redshifted) surface would be reduced as the inverse first power of the redshift:

$$T_{rad}(r) = T_{sur}(r) / (Z(r) + 1)$$

For the scattering model in particular, over any *cosmological* distance the intergalactic medium is statistically uniform such that $T_{sur}(r) = T_e$, a constant. This implies:

$$T_{rad}(r) = T_e / (Z(r) + 1) = T_e \, e^{-H_o \, r},$$

Just as an example, using an estimate of electron density of $\rho_e = 10^{-5}$ cm^{-3} from which to calculate a half-sky cover for the thermal cavity and a kinetic temperature of the medium of 4.50 x 10^3 K, the apparent temperature of radiation originating at particles at the 'mean' distance of a surrounding 'surface' would be given by:

$$T_{rad}(r_{1/2}) = T_e / (Z_{1/2} + 1) \cong 4.50 \times 10^3 \text{ K} / 1.649 \times 10^3 \cong 2.725 \text{ K}$$

These assigned parameter values do determine a background radiation temperature of 2.725 K. However this is only because we specifically selected values of ρ_e and T_e to produce that result. Although such fudging of data may serve for some as explanation of the standard model, it does *not* legitimately accrue credibility on that account for several reasons. Not least of these reasons is that validity of the analysis depends upon the cavity 'surface' being *complete*, i. e., that it possess no closer obstacles or holes through which radiation of a different temperature (or none at all) could have arisen. This criterion is not met by the scenario to which it is applied in the standard model.

If a value of electron density had been chosen such as, $\rho_e = 10^{-7}$ cm^{-3}, an associated 'mean' radiation temperature would have been only insignificantly above absolute zero unless we were to assign a truly astronomical kinetic temperature at the surface. But a temperature of any such incomplete particular 'surface' is of very little significance in any case. What *is* important, however, is the composite effect of all the conceptual partial 'cavity surfaces' throughout the universe that produce closure as we have suggested repeatedly both here and in the previous chapter. A prorated average of the redshifted radiation temperatures, <T$_{rad}$(R)> for $0 \leq R \leq \infty$ from all partial radiating 'surfaces', is what is significant in attempts to predict background radiation in an extensive substance sufficiently diffuse for redshifting to be significant.

This can be assessed by integrating the effects of these radiation temperatures from surface areas throughout the extensive regions, each providing a portion of the total surface area of the cavity enclosure. In a medium of infinite extent, the total effect from beyond a distance, R can be obtained by integration using the following formula:

$$< T_{rad}(r>R) > = \int_R^\infty T_{rad}(r)\, dA(\eta, r) = \int_R^\infty \frac{T_e\, \eta\, e^{-\eta r}}{Z(r)+1}\, dr = \frac{T_e\, e^{-(\eta + H_o)R}}{1 + H_o/\eta}$$

Here we have addressed the closure problem with the weighting factor $A(\eta, r)$ that was derived in the previous sections of this chapter.

The generality of the three expressions for <T$_{rad}$(r>R)> above decreases from left to right. The first applies to any explanation of background radiation in a redshifting environment independent of the redshift mechanism. The next applies to the extent that the actual density and temperature are constant. However, since these parameters increase with lookback distance in the standard cosmological models, a somewhat different final formulation would be required. Nonetheless, such analyses are definitely required in that model that is deficient by not having incorporated a critical assessment of the impact of this phenomenon. The expression at far right incorporates the exponential-logarithmic redshift-distance relationship of the scattering model.

If we use this formula to determine the *overall average* radiation temperature (i. e., for r > 0) when looking out through the intergalactic medium from any observation point, we obtain:

$$\langle T_{rad}(r{>}0)\rangle = \frac{T_e\,\eta}{(H_o+\eta)} = \frac{T_e}{(1+H_o/\eta)}$$

So we have come finally to a determination of the relationship of the kinetic and radiational temperatures in a redshifting medium. Of course we know what this observed radiation temperature is for our universe. It is 2.725 K.

What should have been obvious, but because of bad habits seems to come as an extremely profound realization, is that background radiation does not impose its own temperature value on the rest of the universe. In fact, the ratio of kinetic and radiation temperatures in an extensive medium in equilibrium is given as:

$$T_{k/r} \equiv T_{kin}/\,T_{rad} = 1 + H_o/\eta$$

Here we have substituded $T_{kin} = T_e$ and $T_{rad} = \langle T_{rad}(r{>}0)\rangle$. By plugging in the values for Hubble's constant and the electron cross section in terms of electron density, we obtain:

$$T_{k/r} = 1 + 1.074 \times 10^{-4}/\,\rho_e$$

This formulation is illustrated in figure 213.

It is, of course, obvious that even for an extensive medium that is more dense than $\rho_e = 10^{-3}$ cm^{-3}, that the kinetic temperature of the particles in the medium and the temperature of radiation given off will, in fact, be identical as has typically been expected to be true in *every* case. But that is *not* the case for the intergalactic medium.

Certainly if the electron density distribution is similar to that depicted in figure 30 on page 53, we could have the average cavity surface temperature at about the ionization level of hydrogen with the density as low as the typically anticipated value in the further reaches away from the cores of galactic clusters.

g. parameter averaging appropriate to H_o and η

Viable points on the plot in figure 213 might seem to imply a basic incompatibility with the scattering model redshifting mechanism requirements defined earlier. Applied directly as operative averages of electron density and kinetic electron temperature of the intergalactic medium, these value do not accord well with the constraints we

428

established earlier for a plasma scattering mechanism if that is indeed responsible for cosmological redshift as proposed here.

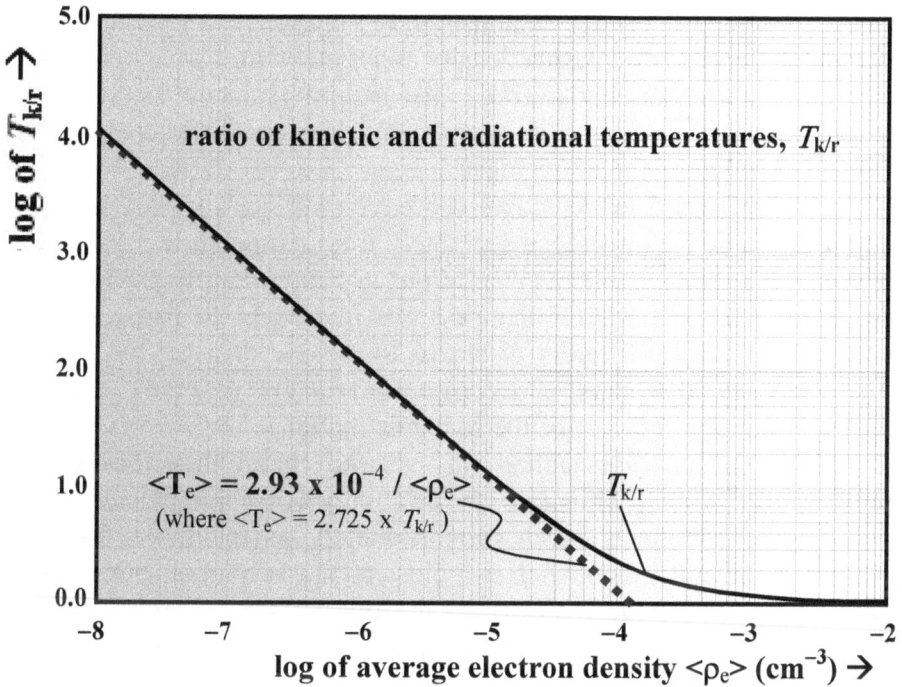

Figure 213: Log of the ratio of the average kinetic and radiational temperatures versus the average electron density of an extended thermal medium

We concluded that the explanation of cosmological redshift as a result of forward scattering required that $H_o \cong 1.719 \times 10^{-32} <T_e \rho_e>$. Therefore to effect an accurate fit with the Hubble constant, the average product of electron kinetic temperature and density of the plasma had to be $<T_e \rho_e> \cong 4.13 \times 10^3$ K cm^{-3}. This would certainly seem to be in conflict with what we have just found above to match the temperature of microwave background radiation. So we must resolve why there is no conflict in the disparate values predicted in the two situations.

The scattering model's redshift mechanism depends intimately upon a high degree of ionization and on a sufficiently high density to effect extinction at reasonably short intervals. But this is a statistically averaged condition as we have repeatedly illustrated. This averaging accommodates extensive variation, compatible with redshift survey

429

variations, while still matching the observed redshift phenomena over cosmological distances. The vast majority of the redshifting produced by this mechanism is on radiation passing through intermediate galaxy clusters. This was clearly demonstrated in chapter 17 in particular. In rich cluster cores both plasma density and temperature are much greater than their corresponding averages for the intergalactic medium as a whole. Where extremes exist averages of products differ considerably from the products of their averages.

With regard to sky cover (or 'cavity surface') we are naturally interested in density averaged separately from temperature. As we saw in sections b through d above, variations in matter density produce very different cross sections for the electrons involved. In particular they produce extremely different distances to closure of a 'cavity' surface. There we saw that stars, and compressed matter generally, although probably containing most of the baryonic matter in the universe, do not contribute in any substantial way to the thermodynamics of the universe as a whole. Thus also, the relatively dense intracluster plasma will not contribute as much proportionately as the more dispersed plasma in the outskirts of such rich clusters and in intergalactic regions. Distances to the large majority of the overall cavity surface will be determined by regions with the more dispersed electron densities.

Compatibility of H_0 and η, both of which are determined in large part by plasma density, is achieved by recognizing that they are based on unique averaging processes primarily involving different regions of the intergalactic medium. The former averages a product that places emphasis on regions of extremely large values of both temperature and density, the latter averages the distances to a cavity surface determined by the cross section contribution of electrons, and therefore, placing emphasis on the inverse of electron density. These differences quite naturally accrue values based primarily on the contributions from quite different regions.

To illustrate this more clearly, we substitute $T_{rad} = 2.725$ K and appropriate expressions for H_0 and η into the temperature dependence equation given above to obtain an expression relating individual averages with the average of the products for the intergalactic medium:

$$2.725 \text{ K} = \frac{<T_e> \eta}{(H_0 + \eta)} = \frac{6.65 \times 10^{-25} <T_e><\rho_e>}{1.719 \times 10^{-32} <T_e \rho_e> + 6.65 \times 10^{-25} <\rho_e>}$$

Again, <x> is used to indicate the average of x over its range of possible values.

The final term in the denominator will typically be considerably smaller than the first. This is because on the one hand, intracluster regions that impact redshift exhibit extremely high temperatures and densities which dominates that average. Secondly, the average of density outside of very small cluster cores is exceedingly small. This equation in essence specifies a ratio of two types of averaging, valid to the degree to which the assumptions of our analysis apply, as follows:

$$< T_e \, \rho_e > \cong 1.42 \text{ x } 10^7 < T_e > < \rho_e >$$

However, since we know that:

$$< T_e \, \rho_e > \cong 4.13 \text{ x } 10^3 \text{ K cm}^{-3}$$

We must have for the product of separately averaged parameters:

$$< T_e > < \rho_e > = 2.93 \text{ x } 10^{-4}$$

This is the dotted line plotted in figure 213 where $<T_e> = 2.725$ x $T_{k/r}$ and all of the preceding assumptions apply.

By such analyses it is clear that the temperature of background radiation does not impose the characteristic temperature (or even anything similar to that) of its blackbody profile onto the kinetic temperature of material aspects of an extended universe from which the radiation derives. While all of this provides some assurance that we are on the right track with the scattering model, it is still necessary to precisely determine these average values.

This determination will be made in conjunction with the clarification of what the origin of microwave background radiation must be, consistent with scattering model hypotheses. Thus, we will address that topic of the origin of background radiation separately for the standard and scattering cosmological models in the next two chapters.

Chapter 22

The Standard Model Explanation of Cosmological Effects

As we have seen, there are alternative ways to account for the redshifting of radiation from distant regions of space, for apparent 'evolutionary' effects, and even for popularized 'dark matter' effects. We will find that there are also alternative ways to account for the microwave background radiation and the abundances of the elements. But first let us consider the accepted means of this accomplishment.

a. the standard model scenario

In the initial seconds after a *big bang* is thought to mysteriously have occurred, the universe would have been extremely dense. After one second each cubic centimeter would have contained hundreds of kilograms of mass. Besides which, the universe would have been so small as to have been but a tiny dot within its Schwarzchild radius, constituting an enormous black hole by anyone's qualifications for such entities. And yet... rather than being gobbled into a singularity as theory dictates for such objects, it continued to expand explosively, having been tremendously assisted by Guth's hypothetical 'inflation' that defies current physical laws in expanding faster than the speed of light. Ultimately it escaped its Schwarzchild radius altogether.

The mass of the universe is envisioned as having initially been almost equally divided between matter and antimatter, which particles immediately proceeded to annihilate each other, producing prodigious amounts of high energy gamma radiation. This phase involved the destruction of a billion times as many material particles as are now left in the entire universe – a billion times the mass of our entire universe destroyed in just a very few seconds. Refer to figure 214.

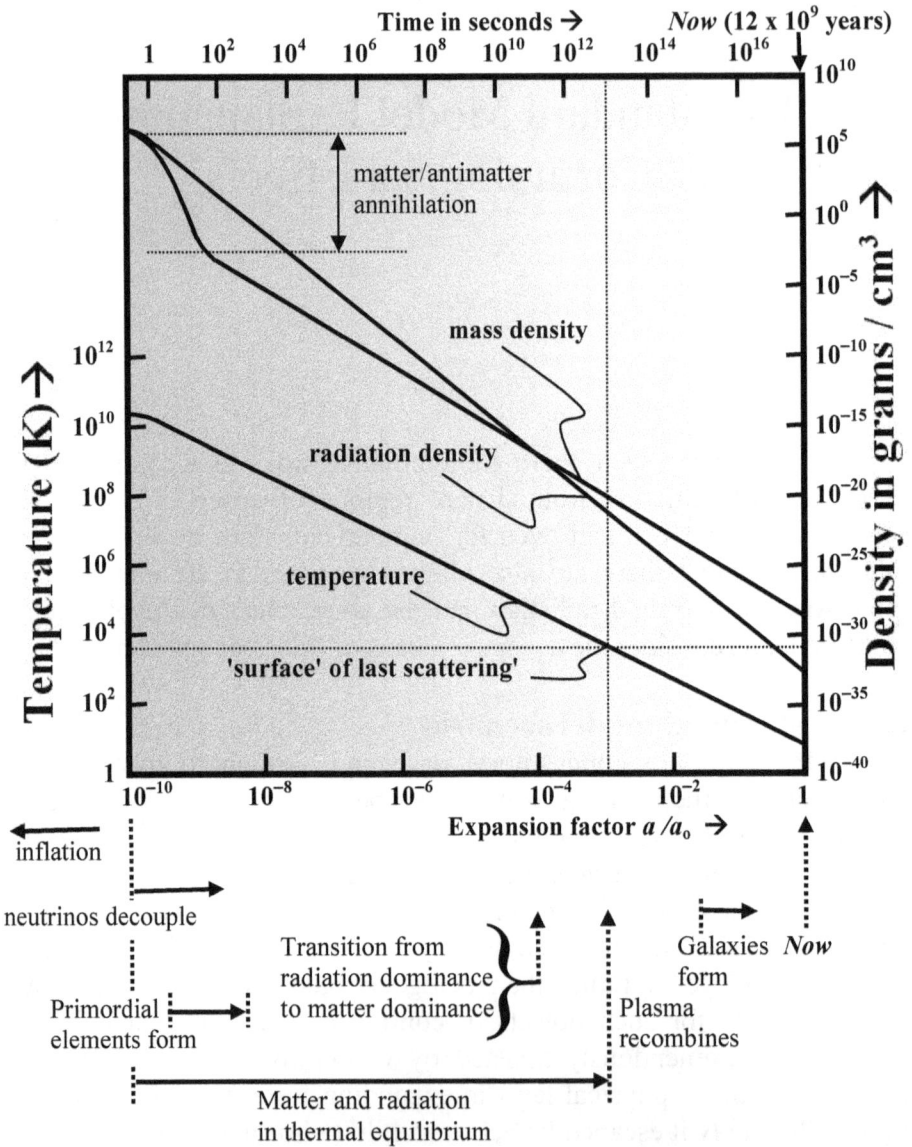

Figure 214: Reconstructed figure from Misner (1973) depicting key elements of the standard cosmological model

That's the big picture. But as we know, the devil is in the details, certainly in the details of how so gigantic a structure as an entire universe could spring from nothing. You'll notice that time only goes back to a little less than one second in the diagram of figure 214, to a redshift of 10^{10}. There is an obvious problem with what goes before that; it's known as the 'horizon problem'. Alan Guth (1981) addressed this problem by coming up with a scheme by which the universe would experienced phenomenal expansionary development. In Guth's account (1997), the solution to this problem was to 'grow' a region of the earliest phase of the pre-universe to eventually (after a lengthy 10^{-33} seconds) become the observable universe that can then be "described by the traditional Big Bang theory". Refer to figure 215 taken from Guth's account where the solid black line shows the growth curve for the inflation theory. This curve begins at a much smaller universe than accommodated by standard cosmological theory for the period. Uniformity throughout the universe is thought to have been established during the earliest pre-inflation stage. Then the universe is stretched to become large enough to provide compatibility with the standard model. Guth notes that ordinate values describing inflation are merely "illustrative, as the range of possibilities is very large".

So after an inflationary period and expulsion of anti-matter from the universe – that must sound more like a mythological account of a war between good an evil in a child's fairy tale than a modern day scientific account – radiation 'ruled'. This means in effect that the energy density of radiation throughout that era would have been greater than the density of the remaining matter. That remaining matter existed now in the form of subatomic particles and light element atomic nuclei, because the light elements were beginning to be produced by this time.

For the first few hours temperatures are thought to have been greater than 10^9 K such that nuclear fusion reactions took place producing deuterium, tritium, and light element nuclei such as isotopes of helium, and traces of lithium and other light elements. The *radiation* era settled in according to this standard cosmological model and is thought to have persisted for more than 10,000 years. The universe would have continued to expand 'adiabatically', meaning that rather than the energy densities, the radiation temperature and the kinetic temperature of material particles would have been forced to share the same value throughout this period. The radiation would have had a

Planck blackbody spectrum at each temperature as it dropped. It would have been continually re-thermalized by its interactions in scattering off of material particles with their Maxwell-Boltzmann distribution.

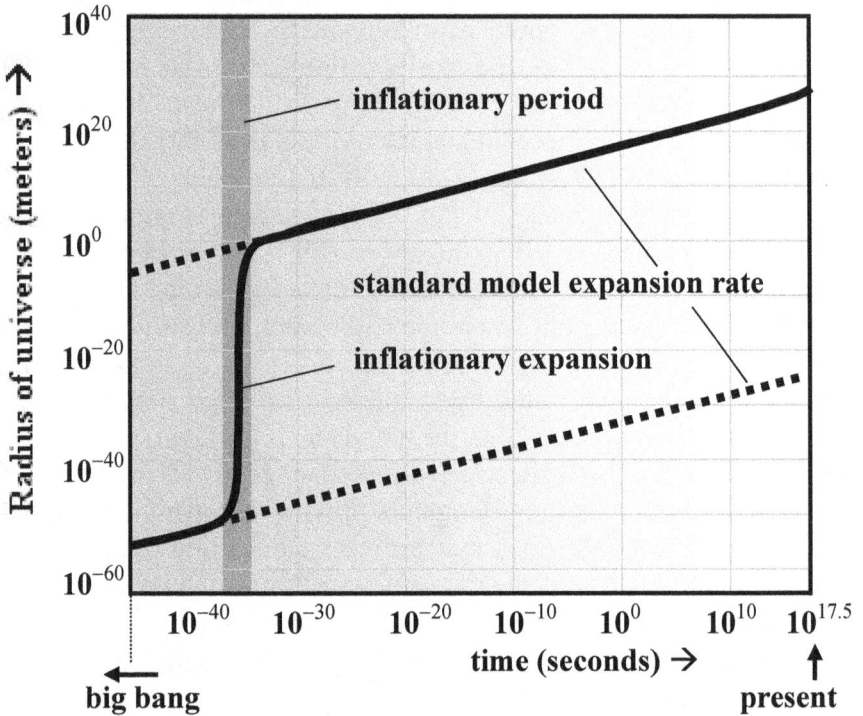

Figure 215: The period and impact of Alan Guth's inflation

Then according to this model, when the energy densities of matter and radiation had finally been equilibrated, the energy density in material particles would gradually have exceeded that in radiation. The universe is thought to have remained in a state of near equilibrium between matter and radiation for some time thereafter, the matter still being fully ionized with photons being scattered by free electrons and the electrons colliding with the heavier protons and light nuclei, ensuring constant thermalization of energy.

Perhaps the most illustrative diagram depicting the standard cosmological model, was provided by Misner, et al. (1973, page 764). This diagram was reproduced as figure 214. Of course that diagram was basically the same as that provided by Wagoner, et al. (1967).

Refer to figure 216 below which provides that diagram. A few details have changed, but even today the story is essentially the same.

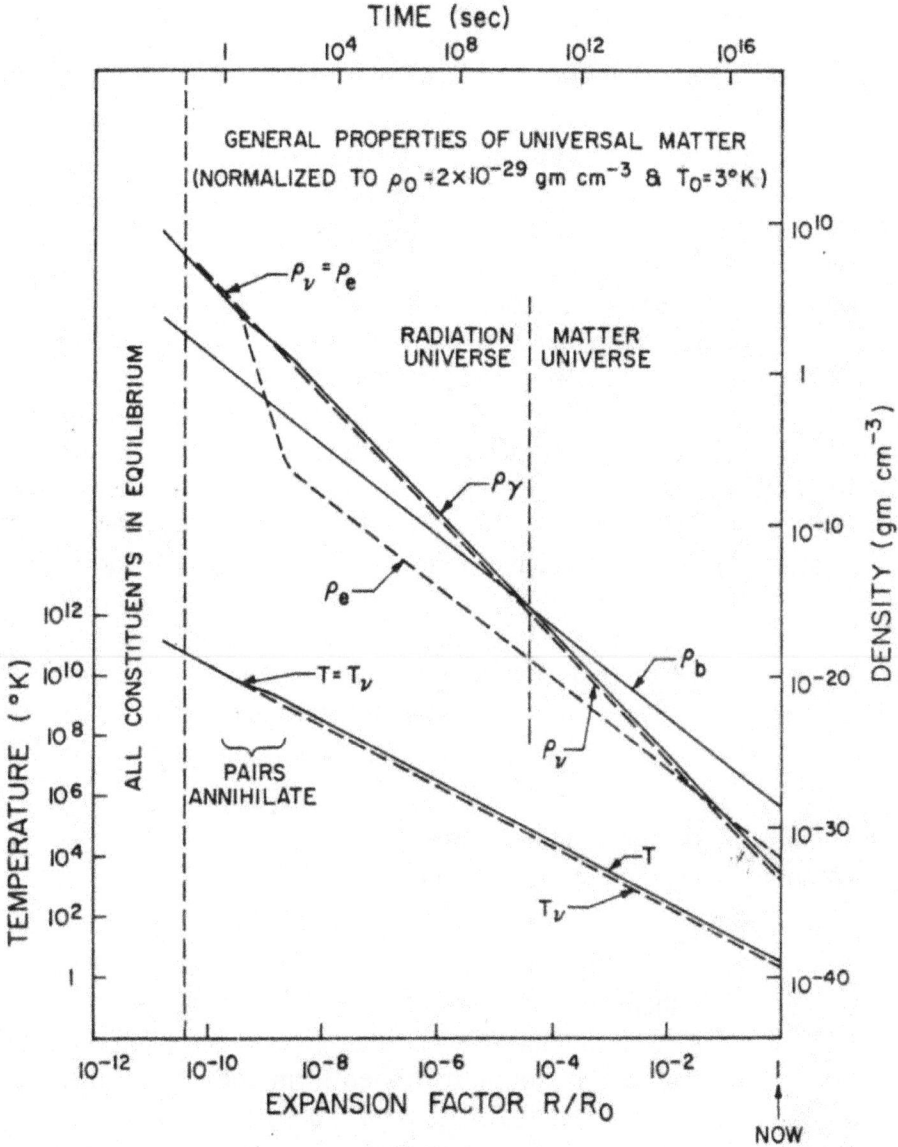

Figure 216: Diagram taken from Wagoner, et. al. (1967) depicting key elements of the standard cosmological model

From the early primordial phase there is assumed to have been on the order of 10^8 or 10^9 photons for every hadron (primarily baryons).

437

See Silk (1980, pp. 137-139), Misner et al. (1973, 765-766), and Birkinshaw (1999) for typical explanations. The rather awkward assumption of an initial mismatch between matter and anti-matter with the associated divergence of matter and radiation densities followed by a gradual convergence associated with the supposed annihilation of matter by antimatter to create a billion times more photons than remaining material particles is an essential part of the story. Why else would such a contorted narrative have been constructed?

Unconstrained expansion would ordinarily demand that the temperature of the material particles and photons diverge since the temperature of the radiation should drop proportionally with expansion and the temperature of material particles would drop as the square of the expansion factor. However, during *adiabatic* expansion, thermal equilibrium enforced by the scattering of the photons off the material particles would force temperatures rather than energy densities of the two to be equal. The energy densities eventually converge. Then about 250,000 to 350,000 years after the big bang ($\sim 10^{13}$ seconds), temperatures are assumed to have cooled to on the order of 10^4 to 10^3 K at which temperature neutral light element atoms would have become stable. This phase is customarily referred to as the '*recombination*'.

Misner, et al. (1973) maintain that the coincidence of equivalent densities at the ionization temperature happened because the baryons created at the big bang each possessed an amount of entropy, S_b of about 10^8, and that:

$$S_b \sim \frac{\text{number of photons in the universe}}{\text{number of baryons in the universe}}$$

From this and other aspects of the scenario, it is apparent that there are many aspects that are coincidental and have had *ad hoc* explanations created to explain them.

Following *recombination* matter would have been predominantly in a neutral atomic hydrogen form with the usual 24% helium by mass. It is further assumed to have become '*decoupled*' from the radiation that would formerly have been scattered by the more highly ionized matter. This assumes that matter would have been so exclusively clumped into neutral light element atomic and molecular forms that the electromagnetic radiation would have insufficient particles with which to scatter to force the natural tendency toward the equipartitioning of energy between matter and radiation. Thus, the

438

universe is assumed to no longer have been in thermal equilibrium. This is all part of the assumption that scattering interactions between radiation and material particles would no longer have appreciable effects in changing dynamics of a *separated* material and radiational status quo. However, that notion is seriously challenged in this volume with further explanations forthcoming in the next chapter.

This decoupling process would have been completed by 10^6 years after the big bang when supposedly no more than one electron in 10^5 would remain uncombined in atomic form. This would have occurred at a redshift of between 1600 and 1000. According to Silk, (1980):

"...once the electrons became bound, the universe became entirely transparent – the fog had lifted. From this point, the blackbody photons continued to cool but never again deviated in their motion. No longer were enough free electrons present in space to scatter the blackbody photons significantly. (The intergalactic gas atoms may subsequently, at a much later era, have become reionized into electrons and protons by a sudden injection of heat or ionizing radiation. If this happened before the density had fallen too far, further scattering could conceivably have occurred at a relatively recent era.) The radiation temperature continues to drop as the universe expands; at present, the blackbody temperature has dropped to 3 degrees above absolute zero."

Needless to say, this is logically unconvincing to the author despite the often cited Lyman-α forests of neutral hydrogen whose absorption we noted in chapter 7. Since currently there *is*, in fact, as against his "could conceivably have occurred', an intergalactic plasma medium as we have noted repeatedly. Silk's argument seems technically flawed in this regard as well. However, see the poster presentation of S. G. Djorgovski of Cal Tech origin in figure 217.

In the first place, however, readily absorptive atomic matter is much more opaque than a completely ionized diffuse plasma. So that rather than a "fog having lifted" by this scenario, a *fog* would actually have *settled in* unless by collapse into heavier units the remaining space is primarily still plasma or complete vacuum. Clearly, there would be a fuzzy boundary between ionized and neutral particles where scattering occurs just as shown in figure 103 on page 225 for a scattering model.

At decoupling, temperatures of the thermalized radiation would have been on the order of 10^4. This 'fact' would have been stamped on the intensity profile. Some references discuss employing a range of surfaces of from Z = 1,600 downward to 12,00 or 1,000, but integrating over these surfaces does not suffice either.

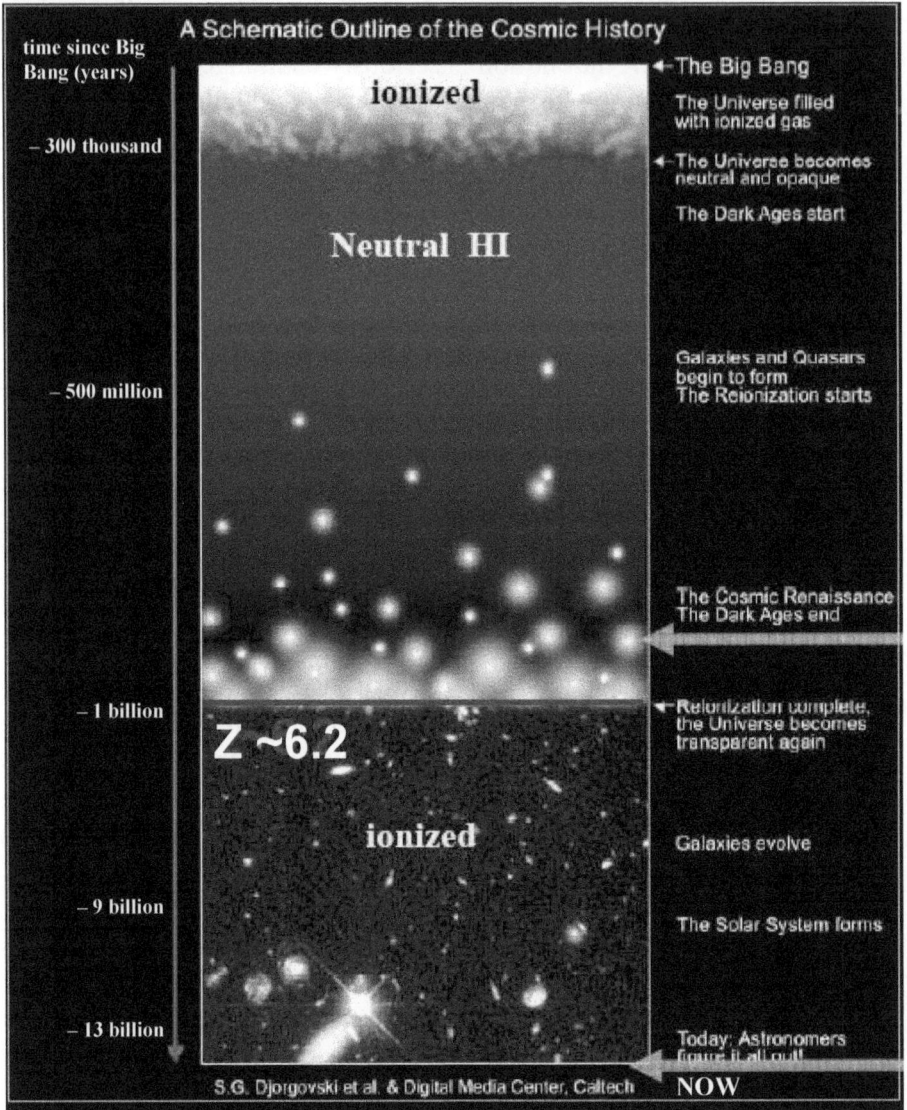

Figure 217: A schematic outline of cosmic history – the standard model

b. the standard model calculations

The numbers used in the foregoing discussion did not just come out of a hat. They are the result of very deliberate calculations. But unlike analogous calculations that we present for the scattering model, these calculated predictions begin with a single measured result and proceed backward to affix a value for when 'recoupling' had to have happened, what the temperature had to have been at that time, etc.. These values are not derived, nor observed values, they are inferred backward merely to force agreement with a single fact. It isn't as though decoupling is a direct implication of the standard model that can be predicted from the hypothesis of a big bang. No. The entire scenario has been retrofitted knowing that there is now a background blackbody spectrum at a temperature of 2.725 K. Here is the way that computation proceeds. In the earlier chapters we demonstrated the formula for the energy density present in a blackbody distribution of radiation. It is,

$$E_{Tbb} = (2 \pi^5 k^4 / 15 h^3 c^3) T_{bb}^4 \cong 7.56 \times 10^{-15} T_{bb}^4$$

For the temperature 2.725 K, this gives $E_{Tbb} = 4.17 \times 10^{-13}$ ergs per cm^3. This is linked to an expanding 'surface' inside this 'cavity' of our universe from which this radiation is conceived as last interacting with matter. The surface according to our diagram in figure 214 would have been at a temperature of about 3,500 K. Earlier we found the following formula appropriate to that happening:

$$E_{Tsur} = 7.56 \times 10^{-15} T_{sur}^4 / (Z_{sur}+1)^4$$

The subscript 'sur' in this formula refers to the parameter value at the referenced 'surface of last scattering'. Clearly, in order to make this work, the following must apply:

$$T_{sur} / (Z_{sur}+1) = T_{bb} = 2.725 \text{ K}$$

If $T_{sur} \cong 3,270$ K, then $Z_{sur} \cong 1,200$, corresponding to about one billion years after the big bang to assure feasibility.

But how does one get from a big bang down to the decoupling epoch? In the diagrams of figures 214 and 216, notice that only about one part in 10^9 of the entire matter in the universe remains after the annihilation of matter by antimatter. At this point the vast majority of the energy in the universe would have been vested in radiation – this would have had to be in the form of gamma radiation initially. Then since the conditions are said to have been in "thermal equilibrium", this radiation would have been scattered by the baryonic ionic soup forcing the radiation into the equilibrium blackbody distribution form of the same temperature as the remaining matter as we discussed earlier under theoretical considerations associated with blackbody radiation and illustrated in figure 205.

So, according to this standard model, as the universe expanded and cooled, it eventually became cool enough that the ionized plasma combined to form neutral hydrogen, helium and other light elements whose nuclei had to have been created earlier. At this point, scattering could supposedly no longer take place. That is the 'surface' considered to be that of a last scattering of the initially created energy associated with the annihilation scenario, after which the ionized state collapses into neutral combined states. Following radiation off of this final *surface*, the ambient temperature of that radiation would then have been redshifted to what we currently observe.

c. counter argument discussion

Of course the instantaneous origin and inflationary period that is said to have followed almost immediately afterward defy traditional laws of physics – in particular the laws of the conservation of energy and momentum. Something, and in particular a *something* as large as a universe, popping up out of nothing is a feat worthy of the gods of mythology, not modern day physics. Quantum fluctuations that have been studied and are understood fairly well are much more modest in scope – approaching a hundred orders of magnitude more modest – and do not ultimately defy conservation laws.

The required inflationary period institutes a temporary reprieve on velocity limits to the speed of light in order to get from a big bang to what might *realistically* have evolved into the universe we observe today. This requires an expansion of fifty orders of magnitude in a time interval so short as to be incomprehensible. So we have the instantaneous creation of a material universe containing a billion times the mass we observe today, followed by a virtually instantaneous expansion by fifty orders of magnitude. Such are the obstacles to credibility that must be hurdled to get where one might reasonably apply the accepted laws of physics, however capriciously, to this problem of the generation of the microwave background radiation in the arena of the standard cosmological model.

All analyses that presume to determine creation of the various light elements in the proper abundances assume continuous equilibrium conditions. This is true in particular for the standard model for which the equilibrium is conjectured to pertain even throughout these early phases. It is maintained that such a thermal equilibrium persisted throughout evolutionary stages up through the decoupling as shown in figures 214 and 216. This is, of course, notwithstanding the unilateral winding down of particle energies and increased wavelengths of the huge glut of radiation attributed to the annihilation of a billion times the amount of baryonic matter that remains in the universe. This total energy density dissipated as it expanded, with space itself envisioned as what expands.

At any rate, equilibrated thermal analyses are held to apply from the earliest times onward in all standard models. Furthermore, by any reasonable extinction distance estimate, background radiation would have continued to be forward scattered about once every thousand light years or so as visible light at about the time of decoupling. Even in the current epoch extinction occurs in the *current* intergalactic medium. How could all of this have been circumvented since the 'time of last scattering'? Of course the conjecture is that recombination was quickly 'completed' persisting only until radiation from resultant stellar structures could have accomplished reionization. However, the much shorter extinction intervals in a denser medium at about the decoupling epoch would be so short a period of time relative to the mean time to reach equilibrium (see chapter 3) that the radiation would have been continually in thermal equilibrium with the medium throughout this transitional period. This would have been true at each phase. So, even if we accept this explanation, we are left to ask why the blackbody spectrum is so precisely that of a single 'surface' temperature? This is particularly troubling under the assumption that thermalization processes had terminated whereby it could be brought back to blackbody form. Since radiation would *not* have been in equilibrium with its environment thereafter according to the standard model, how is that to be explained?

Of course current research is being expended into another area, that concerning why the spatial distribution of the radiation is not completely uniform to even more than the one part in ten-to-the-fifth.

That concern seems frivolous relative to the lack of concern for the distribution of 'surface-of-last scattering' temperatures.

The intergalactic medium is *not* (repeat, *not!*) now at a 2.725 K temperature by *anyone's* direct measurement – nor has it probably *ever* been much lower than that required for *decoupling*, the spectrum of background radiation notwithstanding. If it were *ever* to have been that cool in the last ten billion years or so, atomic absorption would have totally obliterated our view unless some tremendously effective 'sweeping' of atoms in intergalactic space could be confirmed to avoid the complete absorption we know does not take place. The Lyman-α forests of neutral hydrogen absorption discussed in chapter 7 were easily accounted with a uniform density of hydrogen clouds as far back as we can see. Nor are sufficient high-energy radiation sources observed in the intervening space to have completely re-ionized it.

So as justification for any model, inferences from continuous scattering through a primordial plasma *soup* (whether thick or thin) up to the current epoch must be included in any viable prediction of background radiation. This is so for any variant of any cosmological model; the advocates of the standard models have not addressed this convincingly.

To resolve these problems with regard to inevitable scattering would require temperature and/or redshift transition analyses analogous to those to be addressed in the next chapter for the current scattering model. The Sunyaev-Zel'd'ovich effect – in which the temperature of a foreground medium may alter the apparent temperature of background radiation – is an attempt to retrofit some tangentially related issues to the standard model in accounting for the ever so minor variations in the uniformity of the background radiation. But this effect would have to be integrated into an evolving process to adequately address the scope of the problem encountered by the standard models that there seems to be no attempt to accomplish. Even if such an approach were defined to address this as an evolutionary problem, it is this author's opinion that it would still be working from the wrong direction. Proper science addresses the implications of observations, not inferences from presumptions. Nowadays physics as well as cosmology is being couched in terms of inferences from conditions in a presumed big bang. But we can only understand causes after we have quite fully understood the effects.

We will show that blackbody radiation from an intergalactic plasma medium cannot be the one-shot effect of a single mythological instantaneous *epoch* in the universe as a developing phenomenon. Nor yet is it associated with a four dimensional surface at a single time/distance (redshift) in the past as has been anticipated with naïve hopefulness by standard models. For even these models assume continuity of temperatures evolving with redshift at least approximately as $1/(Z+1)^{\alpha}$ as illustrated in figure 214 and 216. The approach presupposes that a previous state of high temperature has now cooled to the extremely low temperature of current background microwave radiation, notwithstanding readily available knowledge of extreme current temperatures of the universe to the contrary and problems in accounting for the density of material substance.

Even in every version of the standard model there would be contributions from a continuous gradation of redshifts. There would have been a time when the plasma was somewhat hotter and denser with transitional states to when it was somewhat cooler and more diffuse. All would have to have contributed in some way, rather than a single very specific slice in the strata of temperature and density where that stops – one slice determining the number of photons to be 'frozen in', another providing a 'surface' to radiate appropriate to an earlier era. But all of that is the established explanation of the background radiation. It doesn't really work, does it?

All cosmological models require more conclusive transitional analyses than that. Certainly what is observed is blackbody radiation of a temperature of 2.725 K to within a few parts in 10^5. *But how does a medium transitioning in redshift effect a specific temperature and associated intensity for this radiation?* That is the question that needs to be asked and answered. That is the question that hasn't been asked and hasn't been answered. As we come to fully understand the question there is hope that we can obtain an answer to that and related questions. Only then can we honestly approach the problem of solving for the characteristics of the resulting radiation without fudging the answer.

Chapter 23

The Scattering Model Explanation
of Cosmological effects

Unlike the situation with the standard model, there is no linear story line for describing the scattering model. A simplistic cradle-to-grave narrative like one that might apply to a human being, who was born on a given date, graduated from university on another, performed his greatest achievements on other specific dates, and finally expired on another, does not apply. In this model there is no beginning and no end to the universe. So at the very least, topics concerning its beginning and its end are out of scope with regard to this model. So, also, there is no timeline to use as an outline for describing the model.

However, figure 218 is provided more or less as a worksheet with regard to which we can discuss those processes and events that occur without affixing them to specific times in a rigid scenario. Most of these on-going processes are envisioned as no different than could be described by an advocate of the standard model, but they are not assigned start and end dates. To be sure, there are differences that result from there being no sequential restrictions on their occurrence, even though in virtually every case there is a very direct analogy. We have numbered five aspects of cosmological phenomena that involve major differences from the standard model that will be described here.

The first of these, although not an integral part of the scattering model, and for which no theoretical basis is provided, is seen as necessary to perpetuate a scheme that seems self-sufficient in all other

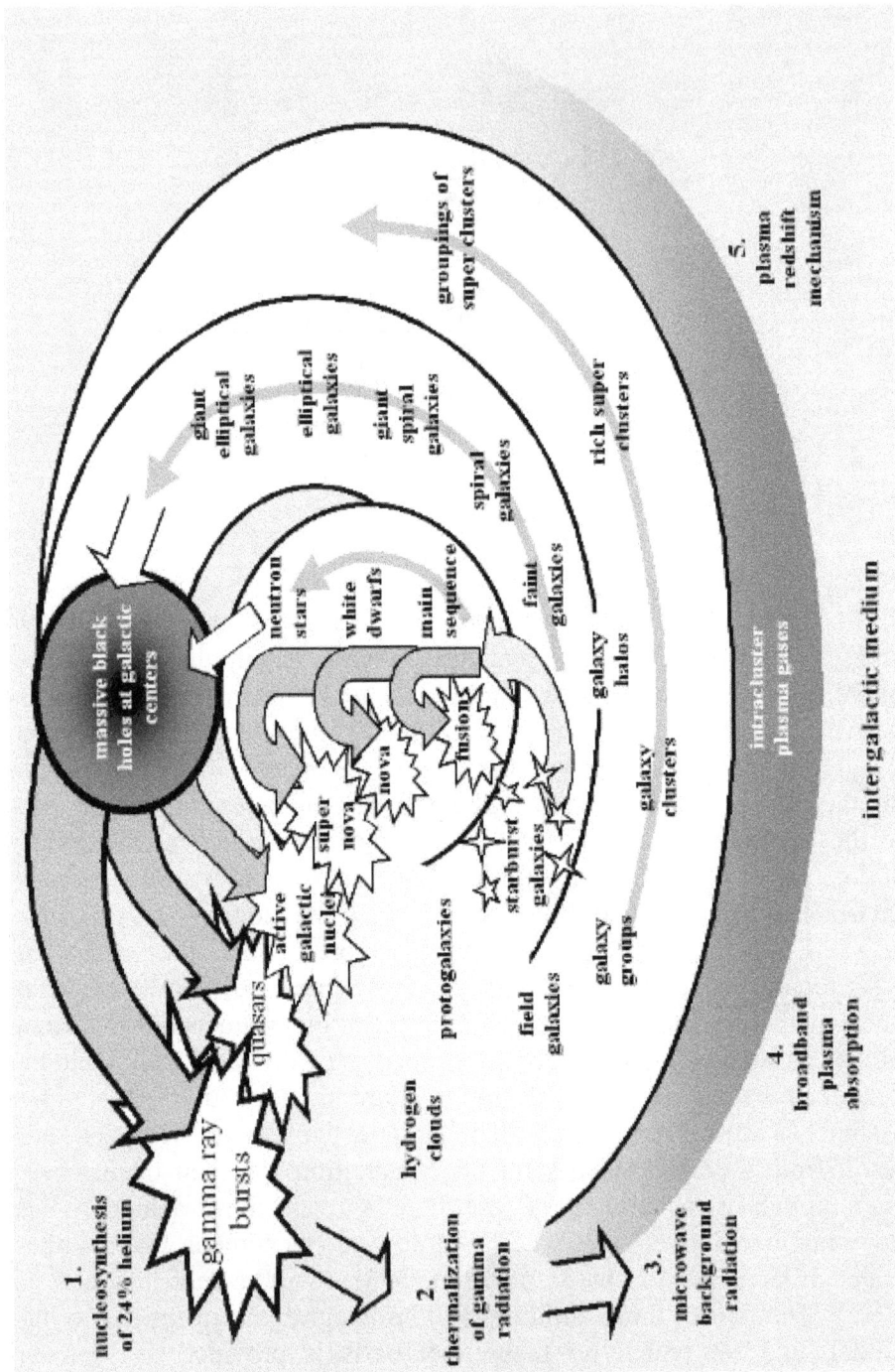

Figure 218: The universe as conceived according to the scattering model

regards. As an initial logical step in the process of the material development of the universe, it involves a process of emergence of matter from beneath shrouds associated with the Schwarzchild radii of black holes. This notion has been anathema until recently. Now some cosmologist entertain the idea of black holes 'popping out the other side' into another universe among an envisioned plethora of such structures. But why the idea of a black hole re-emerging from beneath its Schwarzchild radius into *this* universe should be abhorrent to those for whom the universe itself is glibly accepted as having done so, seems absurd to this author. Gamma ray bursts with virtually unfathomable releases of energy seem ideally suited to this role. Their energies and time release profiles map directly to essential features attributed to the 'big bang' by standard model cosmologists. We will not discuss this in any more depth lest we end up embarrassing ourselves by propounding an unexplainable inflation theory.

A next aspect where differences pertain is with regard to the nucleosynthesis of 24 percent helium by mass from the ubiquitous primordial hydrogenous plasma. This approach is taken almost verbatim from the work of Wagoner, Fowler, and Hoyle (1967). We will discuss the topic in some detail in this chapter but leave more technical aspects of the nuclear reactions to the next chapter.

Another major difference is treatment of the thermalization process whereby energy that began as gamma radiation becomes the microwave background radiation. Rather than adiabatic expansion as employed by the standard model which does not apply to the scattering model, our explanation of thermalization relies on a process incumbent upon scattering in any redshifting environment.

Yet another difference involves broadband plasma absorption incumbent upon any scattering process in a redshifted environment.

Finally the process whereby radiation transmitted through a plasma is redshifted without necessitating recessional motion of the source or expansion of space between the source and observer of the radiation must be acknowledged. This is the 'tired light' explanation that Hubble anticipated so long ago. Although this author does not quite understand why a mechanism associated with scattering should be assigned the denomination 'tired' when it is the resultant energy of light that is reduced (by tiring?) no differently than if it were 'tired out' in the same amount by an exhausting high velocity Doppler process.

Those are the processes that most distinguish the scattering model from the established explanation incorporated into the standard cosmological models. Despite extreme similarities, differences are indeed quite major as well. Unlike the standard cosmological models, the model hypothesized in this volume to account for Hubble's law and other phenomena properly classed as cosmological, it adheres to all time-honored physical laws including especially the conservation laws of energy and momentum. This feature applies also to its accounting of the origin of the energy invested in the microwave background and processes whereby that energy has been converted into its observed 2.725 K blackbody spectrum. It accounts for all the facts and accomplishes that without requiring creation from nothing followed by an inconceivably rapid inflation, or the annihilation of a billion times the mass of our current universe.

a. hydrogen-helium ratio – coincidence or explanation

According to the standard model the ultimate origin of the blackbody radiation currently present in the microwave background was the annihilation of more than 99.9999999% of the total matter created in an initial 'big bang', i. e., the energy of a billion universes just like ours went into creating it. So not only would our universe have been created from nothing, but a billion times that much more would have had to have been wasted just to account for a simple fact no one knew about until 1957. That glut of high energy gamma radiation is envisioned to have erupted about a minute after the initial creation. Figures 214 and 216 show the infusion of high-energy radiation resulting from particle annihilation thought to have later become the microwave background. That required a billion times more mass than would have been left in the universe immediately afterward.

Of course, cosmological facts have to have a reason for being as they are and so any complete model of cosmology must address those same facts. Thus, of course, the standard model had to broach these issues. The microwave background has to have come from somewhere and something – unless that too was to be posited as having miraculously sprung into existence directly with just the clap of hands. The scattering model too must, therefore, have answers to such reasonable questions – hopefully the reader will find that with this model there are much *more reasonable* answers.

The origin of the microwave background radiation according to the scattering model requires no such elaborate *ad hoc* explanation as fixing an initial ratio of matter to anti-matter for which there is no other requirement, evidence, nor major consequence. The amount of energy distributed throughout our universe as background radiation is extremely closely matched to the amount of energy that has gone into the production of the known percentage of helium and traces of other elements from a primordial hydrogenous plasma. There is no required violation of the conservation of energy or other time-honored laws of physics in supposing that the energy given off in producing that observable fact must still exist in some form or other in our universe today as an on-going fact. Where else could it have gone? These are two extremely significant universal numbers – the relative abundance of the first two elements and the energy density of background radiation. It would constitute a dereliction of scientific duty to ignore the obvious relationship between these two facts. This is especially the case when the two pertinent energy values happen to be in such complete agreement.

As an aside at this point, we must acknowledge that the standard model does account for release of this radiational energy. In that model, when the elements are said to have been created, the release of this same amount of energy would have been a miniscule (and therefore unaccountable) amount in comparison to the conjectured annihilation of a billion material universes that was required to account independently for background radiation. Thus required nucleosynthesis in that model is allocated to a time in the supposed history of the universe for which it would have been a comparatively insignificant amount of energy. That is how extreme conditions would have had to have been. But an explanation that allocates facts to a time at which they would have been insignificant, and therefore need not be accounted, does not legitimately enhance its credibility. That remains a fact of our universe whose accounting is indeed significant.

The *current* evidence overwhelmingly suggests that these two contemporaneously relevant numbers are intimately related. Occam's razor must certainly demand the simplest solution that relates the energy released in creating helium and the observed energy in the microwave background radiation. This obvious connection would have to be totally disregarded to accept the extravagant Rube Goldberg mechanism employed by the standard cosmological model instead.

A directly measurable amount of energy went into the universe with conversion of 24% (by mass) of primordial hydrogen into helium, and there is also an observable resulting radiational energy output. These two energy densities match precisely at the current time with no *ad hoc* assumptions having to be made. They can be envisioned as a perfectly balanced process. To ignore that obvious possibility, or to accept an alternative solution for which this coincidence is lost in the ineffectual expenditure of many orders of magnitude more energy than contained in all the matter in our universe to account for a simple fact would be fool hardy.

In Wagoner, Fowler, and Hoyle (1967, pp. 23-24) we find the following, part of which we quoted earlier:

"There has been no general disposition on the part of physicists and astronomers to question the cosmological significance of the measurements of Penzias and Wilson, and of those of Wilkinson and Roll (see, however, Kaufman 1965 and Layzer 1966). We do not wish to do so here in any very serious respect, but we do think it worthwhile pointing out the following remarkable coincidence. The average spatial density of galactic material is $\sim 3 - 7$ x 10^{-31} gm cm^{-3} (Oort 1958). Of this, about one third is probably helium, giving an average helium density of $\sim 10^{-31}$ gm cm^{-3}. Since the conversion of 1 gm of hydrogen to helium yields ~ 6 x 10^{18} ergs, the average energy production – if helium has come from hydrogen – has been ~ 6 x 10^{-13} ergs cm^{-3}. This energy density, *if thermalized*, would yield a temperature of just 3° K. Because in a cosmological expansion baryon density decreases as R^{-3} while the radiation density decreases as R^{-4}, the coincidence is an accident if the 3° K is a relic of a cosmological fireball. On this view the expansion factor R has increased since the fireball by a factor of 10^9 so that no such coincidence could have obtained over most of the expansion. It would be an accident of the present epoch. This is not the case if the observed radiation results from the thermalization of energy from the recent hydrogen to helium conversion in stars."

This is quoted here as contributing to the conjecture that the current – and the author tends to believe *stationary* – state of the universe can account not only for the distribution of the observed background blackbody radiation, but for its perpetuation as well. Note, however, that the alternative conjecture in the quotation above, i. e., that energy release and "thermalization of energy from recent hydrogen to helium conversion in stars" (emphasis added) is not required, and is, therefore, not integral to the scattering model proposed here.

The cosmological redshifting via forward scattering in this model is a side effect of the dynamic pressure of the intergalactic and especially intracluster plasma that exhibits the elemental abundance ratios. The energy profiles in gamma ray bursts, shown earlier as figure 16 on page 33, map directly to profiles assumed in the seminal paper by Wagoner, et al. (1967) that laid out the scenario for element creation for the standard model. We will address this and other issues germane to reinterpreting the quotation with regard to thermonuclear reaction properties of the current state of the universe.

We'll also discuss issues involved in what has seemed to many to be an ultimate disappearance of matter into black holes that might seem to demolish any prospect of a perpetuated balance. Some have assumed with Penrose's initial conclusion based in general relativity that these behemoths entail the ultimate doom of matter in singularities. More recently, however, recognizing that the universe itself must have emerged (or be emerging) from just such a singularity according to the standard cosmological model, alternative schemes have recently been proposed by standard model apologists. Quantum effects that must certainly pertain alter this picture (Barcelo et al., 2009). See also Smolin (2007) who discusses conjectures whereby matter might hop back into existence, albeit according to these conjectures *on the other side* as a schism, starting its own spacetime universe.

Such ad hoc mechanisms, including inflation, and the later universal acceleration, seem to arise upon the scene whenever the standard model is acknowledged as encountering difficulties. A veritable cottage industry has developed to accommodate 'multiverse' productions. But if such a supply-side universe re-cycling production is deemed feasible, why does it have to be to *the other side*? Is there not the possibility – no, even the likelihood – that matter swallowed by black holes might very well meet another previously-undiscovered quantum constraint whereby leviathan super-supernova repulsion ejects the neutron effluvium back onto a beach on *this side*? Back into the only universe we can ever know enough about to discuss rationally? Appendix C provides a detailed discussion of this topic.

b. determining the extent of the agreement

Let us follow up on more accurate current estimates of those quantities identified by Wagoner et al. (1967) in the passage quoted above as they apply to the scattering model predictions. Their comments pertained to the then-current estimates of mass density and elemental abundance percentages. Whatever the actual values happen to be, the scattering model must be compatible with them applying at all times in the past and foreseeable future.

In the next chapter we will address the specific thermonuclear reactions that are involved in this fusion sequence from hydrogen to the stable isotope of helium. There we will find that the difference in the rest mass energy of the nucleons required to produce one helium-4 nucleus minus the rest mass energy of the helium-4 isotope is:

$$\Delta m = 2\,m_p + 2\,m_n - m_{^4He} = 0.050603487 \times 10^{-24}\ \text{gm}$$

The subscripts p, n, and ^{4}He apply respectively to the proton, neutron, and alpha particle (^{4}He nucleus). These amounts apply to single nuclei. The total energy released in creating one gram of helium-4 is therefore:

$$E_{p \to He} = \Delta m \times c^2 \times 6.0225 \times 10^{23} \div 4$$

$$= 6.8571 \times 10^{+18}\ \text{ergs per gram of}\ ^4\text{He}$$

Notice that one *mole* of any substance involves Avagadro's number (i. e., 6.0225×10^{23}) constituent molecules – atomic nuclei in this case – with a total mass equal to the atomic number of the substance measured in grams. For helium with an atomic weight of four, this involves four grams of the substance. So to assess the energy loss per gram, this has had to be divided by the atomic weight of helium-4.

But to determine the *energy* density, $\rho_{p \to He}$, which is the amount of this energy per cubic centimeter, we must multiply the result above times the observed universal *mass* density of helium ρ_{He} (measured in units of grams per cubic centimeter) as follows:

$$\rho_{p \to He} = 6.8571 \times 10^{+18}\ \rho_{He}$$

Now for comparison with the quoted Wagoner et al. (1967) comment, notice that the energy per cubic centimeter in the microwave background radiation is slightly less than was estimated when that passage was written over forty years ago. It is now known that the radiation temperature is 2.725 K rather than 3.0 K.

When this ratio is raised to the fourth power as is appropriate in application of Stefan's law, it changes the estimate by a factor of:

$(2.725 / 3.0)^4 = 0.6807$

This reduces the Wagoner et al. rough estimate of the density of energy in the microwave background, $\rho_{\mu bb}$ to 4.08 x 10^{-13} ergs per cm^3 from the 6 x 10^{-13} ergs per cm^3 they estimated in what we quoted above. We now know that, in fact, it is precisely:

$$\rho_{\mu bb} \quad = \quad 4.169 \times 10^{-13} \text{ ergs per cm}^3$$

Their estimate was amazingly accurate.

But however amazing their work might have been at the time, their method is not exactly how the situation must be treated in lieu of the necessary considerations of the redshifting involved in the thermalization process.

c. the thermalization process

In the standard cosmological model treatment the situation involves emission from a single redshifted surface-of-last-scattering with no intermediate interactions with material substances so the only emergent energy density is at that redshift, with radiational energy reducing as $1/(Z_{sur}+1)$ from that surface onward. The scattering model embraces continuous thermalization of radiation over a continuously varying range of redshifts. So that in this case we must treat the energy density as being continuously redshifted, integrating the effects through *all* intermediate regions.

In chapters 20 and 21 covering theoretical issues with black body radiation we showed effects of continuously varying redshift that also reduces observed energy content per cubic centimeter. Although the reduction is not nearly as dramatic as that employed in the standard model. It involves integration of the inverse fourth power of redshift over the entire range of redshifts, rather than accepting only the one, to produce the following effect on the energy density of the system:

$$\rho_{\mu bb} = \rho_{p \to He} \int_0^\infty \zeta(r)^{-4} d\zeta(r) = \rho_{p \to He} / 3$$

Two-thirds of the original radiant energy density obviously is absorbed as increased kinetic energy of the medium. Therefore, we have that,

$$4.169 \times 10^{-13} = 6.8571 \times 10^{+18} \, \rho_{He} / 3$$

Thus, if as Wagoner et al. suggested as an alternative of interest and we have accepted as the likely reality, it is indeed the creation of the universally observed percentage of helium-4 from a primordial hydrogenous plasma that is responsible for the energy in the microwave background radiation, we should expect that:

$$\rho_{He} = 3 \, \rho_{\mu bb} / \rho_{p \to He} = 3 \times 4.176 \times 10^{-13} \div 6.8571 \times 10^{+18}$$

$$= 1.827 \times 10^{-31} \text{ grams of } {}^4He \text{ per cm}^3$$

Thus, we obtain as the overall baryonic density of the universe:

$$\rho_B = \rho_{He} / 0.24 = 7.613 \times 10^{-31} \text{ grams of baryonic matter per cm}^3$$

$$= 0.0931 \, \rho_o$$

This is about 10% of Einstein's critical density.

The following table indicates approximate numbers of the various constituent particles of the intergalactic medium that would exist in 3.375 cubic meter boxes appropriate to different regions in the medium. Boxes for two such regions are shown in figure 219. What is presented as typical of intracluster gas pertains to regions interior to cores of galactic clusters where densities are a thousand times higher.

per 3.375 cubic meters	universal average	typical of intracluster gases
contained baryons	1.42 protons, 0.194 neutrons (includes 0.06 helium-4 nuclei)	142 protons, 19 neutrons (includes about 10 helium-4 nuclei)
contained electrons	1.42 electron2	142 electrons

Let us see how well the overall estimate stands up with regard to the best estimates available today. In a recent *Discover* science magazine an article by Martin Rees and Priyamvada Natarajan states:

"Adding up the inferred gravitational effects in galaxies, galaxy clusters, and large-scale structures implies that the total amount of matter in the universe, including dark matter, comes to about 30 percent of critical density." This was March, 2009. Further on in the article they clarify the ratio of baryonic matter to what they assess as 'dark matter' as 0.37. This gives the most current estimate of the percentage of the projected baryonic matter to Einstein's critical density as 11%. This is based on current acceptance of the universe being 'flat'. *Flat*, of course, is meant to imply that the sum total of all forms of matter must equal the critical density, ρ_0, according to standard cosmological models. Of course we are attempting to justify estimates that are based on the standard model here, but that the estimate of baryonic mass density based exclusively on scattering model assumptions is in about as good of agreement with these most reliable recent estimates is reassuring nonetheless.

universal average   typical of intracluster gases

$$\rho_{uB} \cong 7.613 \times 10^{-31} \ gm \ cm^{-3} \qquad \rho_{cB} \cong 10^{-28} \ to \ 10^{-29} \ gm \ cm^{-3}$$

$$\rho_{ue} \cong 4.55 \times 10^{-7} \ cm^{-3} \qquad \rho_{ce} \cong 10^{-2} \ to \ 10^{-5} \ cm^{-3}$$

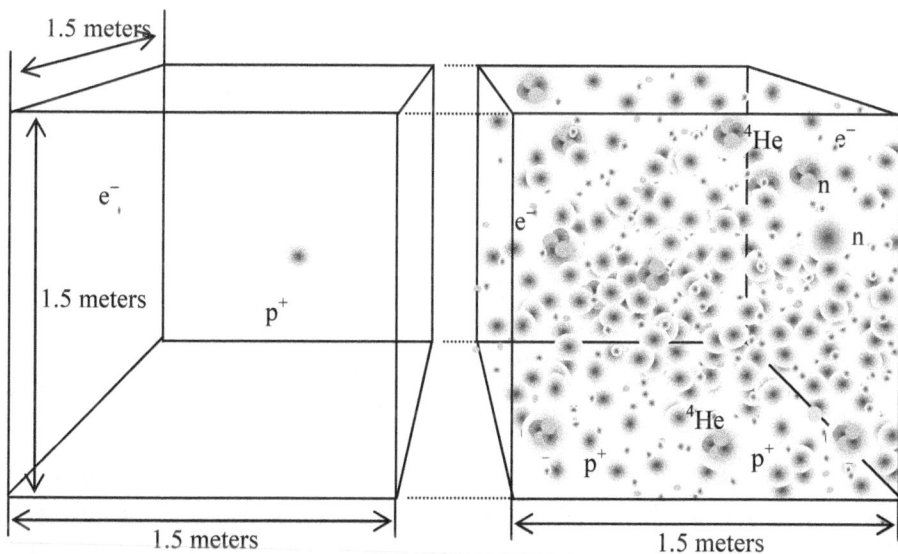

Figure 219: Universal average and typical intracluster gas densities

The value of ρ_{uB} determined above also determines the value for the universal average electron density, ρ_{ue}. According to the mix of material components of the universe as a whole with only traces of elements other than hydrogen and helium:

$$\rho_{ue} = 7.613 \times 10^{-31} / 1.673 \times 10^{-24} = 4.55 \times 10^{-7} \text{ per cm}^3$$

Of course this is not the average electron density of the intergalactic medium, nor certainly the electron density to be used in thermal analyses since an appreciable percentage of electrons are bound to nuclei and otherwise involved in compact matter in stars, et. cetera that do not contribute substantially to effecting a cavity surface. According to Bahcall (1999) plasma gases in and around clusters involve only about 0.07 times the mass of associated galaxies. A very similar conclusion is reached by Hicks et al. (2002) although for the cluster PKS 0745-191 they assess it at 0.18. More recently, however, Loewenstein (2003) attests to in excess of 80% of the baryons in the universe existing in the plasma gas state So that the average electron density of the intergalactic medium must be on the order of:

$$\langle\rho_e\rangle = \begin{cases} 3.2 \times 10^{-8} \text{ per cm}^3, & \text{if } \langle\rho_e\rangle = 0.07 \times \rho_{ue} \\ 3.64 \times 10^{-7} \text{ per cm}^3, & \text{if } \langle\rho_e\rangle = 0.80 \times \rho_{ue} \end{cases}$$

d. implied kinetic temperature in a redshifting medium

In earlier discussions of Olber's paradox, we determined that the kinetic temperature of an indefinitely extended medium in a redshifting environment would necessarily be appreciably higher than that associated with the blackbody radiation thermalized by such a medium in equilibrium. This was illustrated in particular in figure 213 on page 428. The reason is that all portions of an associated 'cavity surface' in an optically thin, i. e., sparse, medium that is required to effect blackbody (as against merely 'thermal') radiation would be redshifted in such a way as to effectively diminish the temperature of the associated radiation in attaining a blackbody form. In that chapter we learned that we must ignore as irrelevant electrons tied up in stars, a total of which account for a miniscule fraction of the effective thermal cavity surface.

We showed in that earlier chapter that we could determine an average kinetic temperature, $< T_e >$ of the intergalactic medium from the following: 1) The electron density of the medium that results in a total cross sectional areas that produces the effective cavity surface necessary for thermalization, 2) the value of Hubble's constant, and 3) the observed temperature of the microwave background radiation. The appropriate relationship was embodied in figure 213 where the impact of redshift occurring throughout an extensive medium of average kinetic temperature $< T_e >$ would be to reduce radiation temperature as a function of average electron density. Thus, for the intergalactic scattering medium the following formula applies:

$$< T_e > \ = \ 2.725 \ (\ 1 + 1.074 \times 10^{-4} \ / \ <\rho_e>)$$

This formulation together with the range of possible and expected average electron density values is illustrated in figure 220.

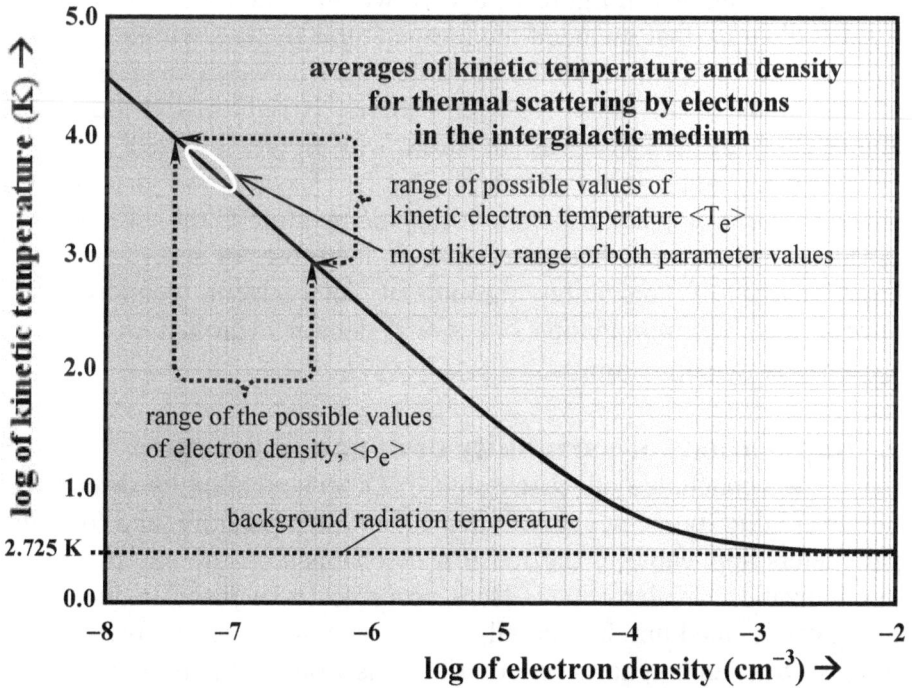

Figure 220: Kinetic temperature versus electron density of the medium as implied by the observed 2.725 K background radiation

Thus, having secured a range of estimates of $<\rho_e>$ we see that this results in the associated range of average temperature values:

$$< T_e > = \begin{cases} 9.0 \text{ x } 10^3 \text{ K}, & \text{if } <\rho_e> = 0.07 \text{ x } \rho_{ue} \\ \\ 0.8 \text{ x } 10^3 \text{ K}, & \text{if } <\rho_e> = 0.80 \text{ x } \rho_{ue} \end{cases}$$

The larger values seem the more realistic, for although 80% of the universe may indeed be in the plasma state, much of this is tied up in stellar structures that do not contribute much to the thermalization of the universe as a whole as we discussed in chapter 20. So 10 % of the mass of the universe being directly involved in its 'thermal cavity' surface brightness seems more realistic.

This entire temperature range, that derives in large part from the extremely hot gases deep within galaxy clusters that cool considerably in the outer fringes, seems to accord well with what is observed in the universe around us. In figure 1 page 18 we illustrated the Hertzsprung-Russell diagram that included the temperatures of stars within our own galaxy. The temperature range presented here is within about an order of magnitude of all stars. The temperature is right at the ionization temperature of hydrogen as illustrated in figure 31 page 55 and discussed in much more detail in chapter 7 where we discuss the formation of protogalaxies at about this temperature in the vast regions of intergalactic space. The prediction is, therefore, in accord with the high degree of ionization throughout intergalactic regions with temperatures accommodating occasional neutral hydrogen clouds that show up as Lyman-α forests.

e. implications of variability of dynamic pressure

We concluded our discussion of Olber's paradox by discussing the averaging processes involving kinetic temperature and density parameters applicable to the redshift mechanism on the one hand and thermalization on the other. Both processes take place in the same intergalactic medium but entail unique averaging requirements that bring to bear an emphasis on contributions from alternative regions of the medium. Hubble's constant resulting from the redshift mechanism involves an average of the product of the parameters $< T_e \, \rho_e >$ at each point along the path of electromagnetic radiation propagation. Whereas for thermalization purposes for which the preceding equations pertain, each of the parameters are averaged separately as $< T_e >$ and $<\rho_e>$ throughout all space. In the earlier treatment we demonstrated that for our analyses to apply, the following relationship must hold:

$$< T_e \, \rho_e > \cong 3.869 \times 10^7 < T_e > < \rho_e >$$

This relation emphasizes the degree to which the uniqueness applies to values obtained by averaging the product as against taking the product of the individual parameter averages.

This involves averages pertaining more explicitly to the cross section or volume density of electrons. It is inevitable, of course, that this difference in averages of temperature, density, and their products should come into play, since a universe without variability would be a dull place indeed. Where appreciable variability exists the product of parameter values like those of temperature and density averaged over the entire universe would not accurately reflect the average of the product of kinetic electron temperature and plasma density parameters throughout the intergalactic medium. Clearly the disparity between the product of individual averages and an average of the product of the individual parameters is very appreciable even within the intergalactic medium when the individual parameter values vary in concert as we have repeatedly illustrated. The product of these parameters varies by seven orders of magnitude through the interior of such clusters falling considerably lower outside these bounds.

The derivations of baryonic density value based on conversion of hydrogenous matter into helium that has been described above involves a straight forward averaging of all the matter in the universe. Thermalization of the radiation energy through its interactions with matter, primarily the electrons via scattering processes, involves primarily plasma because of its much greater cross section than is realized for the more compact baryonic material structures in the universe. We have thus recused matter that is thus otherwise occupied. But that too once met the same universal ratio.

The intergalactic medium obviously does not include people, planets, stars, galaxies, etc.. Nor does our designation "intergalactic medium" include neutral atomic or molecular material traces that also occur throughout intergalactic space. In the context of the scattering

461

model with regard to the processes of redshifting, absorption, and thermalization it is primarily the ionized plasma with which we are concerned.

The product of electron density and kinetic temperature as it appears in this volume refers to an average plasma dynamic pressure realized on any line of sight through intergalactic space, which supports a mechanism that produces the observed cosmological redshift. These lines of sight for which the average applies must be 'cosmological', i. e., extend for several hundred Mpc at a minimum as noted originally by Hubble. As is indicated by the above relation of dynamic pressure, the average value of this product will exhibit a considerably higher value than applicable to the separately averaged values that pertain separately if it is to produce the effect of Hubble's constant on lines of sight to cosmological distances. At any rate, the scattering model depends upon the average of that product having the value:

$$< T_{ige} \, \rho_{ige} > \cong 4.13 \times 10^3 \text{ K cm}^{-3}.$$

As we have shown by examining the extent of galaxy clusters, this average is indeed realized by a comprehensive averaging process that includes smaller intracluster regions of much higher dynamic pressure as well as those vast regions between clusters where a scarcity of matter contributes so substantially to a very low value of the overall universal average mass density.

We have demonstrated that the distance-redshift relation is not a uniform characteristic even at cosmological distances. Associated phenomena produce a very uneven distribution in redshift survey data. We know in particular that major amounts of the overall redshift of galaxies occur within rich cluster cores. The 'fingers of god' and pronounced ripple phenomena in redshift surveys discussed earlier illustrate this very dramatically. Even though the scattering model and the standard cosmological models differ with regard to explanations of these facts, the facts themselves are indisputable. The explanation of observational facts is an essential ingredient to success of a scientific model, and so these facts have required re-evaluation to emphasize the significance of using an average for the dynamic pressure that is central to the plasma scattering redshift mechanism.

462

f. arguments for perpetuation

Wagoner et al.'s (1967) tentative conjecture concerning the *origin* of microwave background radiation might seem to imply by the very word 'origin' that there must have been a time and place at which the ratio of hydrogen to helium that we now observe was established. Certainly observations have typically been taken by standard model advocates to imply that. However, the term 'origin' implies a logical relation no less significantly than a temporal one. It is in that logical sense that we refer to 'origins'.

The defunct "steady state" model argued for perpetuation, but still with creation ex nihilo to maintain the perpetuated mass density of the current universe as envisioned by Bondi, Gold, and Hoyle (See Sciama, 1959). It required the creation of but one neutron or proton-electron pair per cubic meter per 10^{10} years. The primary difference between that theory and the standard cosmological model was this assumed continuation of the creation ex nihilo process at a low level rather than a one-time cataclysmic event. Other than that difference in the mode of creation of the material universe, the steady state model was but another variation on a model dependent upon creation from nothing, expansion, and an associated Doppler interpretation of cosmological redshift.

Ultimately the failure to detect the implied smooth background of gamma radiation on the one hand or hard X-rays on the other doomed the theory. Its failure to provide an adequate explanation for the microwave background radiation was perhaps more legitimately major to its demise. Proponents postulated the universal presence of millimeter sized metallic dust grains that would absorb radiation produced by exceptionally luminous galaxies and reradiate it at the appropriate temperature (Arp et al., 1990). This contrivance served ultimately to more effectively refute the theory.

Defining a perpetuated process involving a logical rather than temporal sequence of events is the final challenge for the scattering model. The background radiation is completely compatible with its origin in the on-going creation of helium from hydrogen. But if that did not happen at one significant point in time but continuously, how has the ratio remained unchanged? And if the universe is not expanding and the ratio is not increasing, where does all this additional hydrogenous material come from, and why hasn't the percentage increased?

The answers involve the thermodynamic equilibrium associated with the 2.725 K radiation maintaining the status quo except for minor perturbations. The perturbations seem naturally to be two-fold, matter removed from the thermodynamic balance and matter returned back into this balance. The obvious culprits are black holes where matter with the effected ratio is removed from thermodynamic interaction, and gamma ray bursts that seem up to the challenge of recycling this matter back into its primordial state, which would then continue the cycle of thermonuclear reactions to re-establish the 24% ratio.

The current understanding of astrophysicists and cosmologists is that matter is being, and will continue to be, sucked out of the observable universe into black holes beneath whose Schwarzchild radii it is as though the matter that has been swallowed up no longer exists except for its collective gravitational influence without regard to any former hydrogen-helium ratio or other elemental distinctions. If there were no counter balance to this process, eventually, as surely as if the universe were to expand forever into oblivion, the universe we observe today would be no more.

Not least of the inhibitions against conjectures of a perpetuated state is the long history of crackpot inventions of perpetual motion machines. Surely a stationary state universe would constitute the ultimate in perpetual motion. So we must address the sense in which such notions are, in fact, solely in the domain of crackpot science and do not pertain in any direct way to the current discussion.

The second law of thermodynamics that applies to everything in our universe seems to enforce the escalation of change in the form of increasing entropy as a least common denominator that one might think would surely preclude an interesting stationary state. Every known process tends ineluctably to increase entropy. *There is no free lunch* – anywhere *in* the universe, at any time. But does this include the universe itself if the process whereby matter is dumped into black holes for which entropy is an on-going question, were excluded from the injunction? It is thought by most that black holes too must express the entropy that went into them. Various obscure theoretical means of assigning such a quantity to these 'hairless' critters have been propounded without observational confirmation. But what if, like the matter they swallow up, the information tied up as entropy in the

464

identity of the matter they swallow were also to disappear? And why wouldn't it, other than the fact that such leeway could not be exploited to improve steam engines? It would seem to constitute a much lesser violation of what is currently revered than throwing the conservation of energy overboard whenever a whim arises as is so often done by standard model cosmologists.

Hawking radiation from black holes derives from and/or produces a characteristic temperature of a black hole. This temperature can be calculated from its assigned entropy. The more massive a black hole becomes by accretion of more and more matter, the colder it will become. A black hole with the mass of the planet Mercury is propounded to have a kinetic temperature in equilibrium with the microwave background radiation. Supposedly more massive black holes will be colder than the background radiation, and gain energy faster than they give it up by Hawking radiation, becoming increasingly colder. What about our universe, shrouded as it has been by a Schwarzchild radius? What can be said about that? Questions. It would be singularly ridiculous to even attempt to answer them all as mere corollaries of the scattering model.

Gamma ray bursts – what are they? They are associated with thermal radiation on the order of 10^9 or 10^{10} K just as standard model theorists (albeit inconsistently) maintain our universe to have been shortly after its miraculous conception.

g. summarizing the scattering model

Wagoner et al.'s (1967) tentative conjecture concerning the origin of microwave background radiation is compatible with observed facts and with the implications of the author's scattering model. By anyone's cosmology the usual helium isotope, He-4, as well as traces of the isotopes of other elements have been created from high-energy hydrogenous plasma and in that multi-step fusion process energy has been (and is being) released. Admittedly that energy is in the form of extremely high-energy, low-density radiational and mechanical action. However, that energy has long since, and continuously since its release, become thermalized into a characteristic blackbody radiation spectrum that exhibits precisely the amount of energy content that is currently observed in the microwave background radiation.

Accepted laws of physics and observations of the cosmos suggest what has been propounded here as the 'scattering model':

1. There is no creation ex nihilo.

2. There is no need for faster than light expansion.

3. There is no need for expansion at all just to account for cosmological redshift.

4. There is no need to destroy a billion times the matter in the current universe to effect the microwave background radiation.

5. There is no need for acceleration and deceleration of expansion to match SNIA luminosity data.

6. There is no need to assume 82% of the matter in the universe cannot be observed, concerning which no explanation can be found.

7. There is no call for all the energy in the universe to be but 4% of some very mysterious and totally unaccounted 'vacuum energy'.

8. There is no reason to pretend that the material universe is 2.725 K when no observed part of it is that cool and the pretense is justified solely on the characteristic temperature of microwave background radiation exhibiting a 2.725 K blackbody spectrum. Observed radiation temperature does *not* constrain the emission temperature to that value in a redshifting environment.

In all these cases the scattering model provides a straightforward explanation for observations without demanding that we embrace what physics has repeatedly shown us to be impossibilities.

Chapter 24

Thermonuclear Reactions in the Intergalactic Plasma

One of the three types of phenomena that have been heralded as confirming an origin of our universe in a big bang has been the various atomic abundance percentages that seem to be universal. It is true, of course, that the number of thus-heralded percentages has waned considerably since Gamow (1952) following Alpher (1948) first disdained attempts to assign this function to stars and supernovas of stars past. Peebles (1993) and others have now acquiesced with regard to all light element production *not* being a direct result of a big bang. He states that, "It is now believed that the element production reactions essentially stop at helium, with trace amounts beyond that, and that the bulk of the heavier elements are produced in stars." This change in attitude came about largely because of the work of Wagoner, Fowler, and Hoyle (1967) and Wagoner (1969 and 1973).

a. hypotheses with regard to the hydrogen-helium ratio

There is, however, a consensus that at least the nuclides of hydrogen, deuterium, tritium, and helium necessitate something other than creation within stars. This is in large part because of agreement that there is a higher percentage of helium than could possibly result from stellar nucleosynthesis. Stars provide an environment that is too agreeable for proceeding further down the periodic table. There would, therefore, be much less helium than is observed if that were its primary source. There has seemed until recently to be no other rationale for the larger residual of helium after having synthesized the heavier elements than it's having originated in the echoes of a big bang. The work of Wagoner, Fowler, and Hoyle (1967) established details associated with that presumed origination.

Gradually a consensus on actual universal percentages seems to have been achieved. This percentage by mass of 24% helium has been used throughout this volume. But, see for example, Gamow (1952) where it was predicted to be 44%. Silk (1979) came up with 1 in 10 atoms or 29% by mass rather than the 1 in 13 (25%) that would have been closer to current estimates. Peebles (1993) showed that 22% was near the "maximum" theoretical value. Riess (1998) maintained that 23% was "the absolute minimum that would emerge from the fire ball." Riess (1998, p. 57) added that, "The helium abundance in the oldest objects is now very firmly pinned down to be 23 or 24 percent."

Gamow hypothesized a *frozen equilibrium* that is still in vogue although the 'freezing' is now assumed to have taken place at a slightly later epoch following a big bang so as to let the creation of the heavier elements occur within stars occurring at much later epochs. With the standard model, a continuous range of temperatures is assumed from which to choose a suitable epoch to which to assign the various origins of elemental matter. Primarily it is helium and deuterium which are of concern; a temperature on the order of 10^9 K seemed ideal to Gamow and more recently Peebles and others as the temperature of the crock pot for cooking primordial helium. (The reader is referred back to figures 214 and 216.) The standard model does require a somewhat rushed development to establish the eventual percentages.

Questioning the origin of such phenomena rather than accepting the determined ratio as one more of the givens of our universe might seem – illegitimately – to impose an evolutionary explanation more appropriate to creationist cosmologies. Nonetheless, providing an explanation for the apparent universality of these percentages seems meaningful. That the energy released by fusion of hydrogen into helium in the universally observed percentage is precisely the energy present in microwave background radiation is a related and most interesting fact for which there must certainly be an explanation.

468

b. significance of primordial hydrogenous plasma

Just as in molecular chemical reactions there is a two-way equilibrium of nuclear reactions that, at some temperature, density, and pressure, remains balanced, swinging gradually in one direction or the other with shifts in thermodynamic parameters values. There are equations that describe the conditions of this balance that we will discuss presently. Such conditions determine for a predominantly hydrogenous plasma, somewhat similar to that realized in intracluster plasma and in young stars, what relative abundance of all the nuclides will result once equilibrium is established. Gamow (1959, p. 70) states for example:

"...energy production in normal stars, such as our sun or Sirius, is due to a slow thermonuclear process in which hydrogen gradually transforms into helium. A temperature of about 20 million degrees which causes this nuclear reaction, is, however, not high enough to induce the reactions between heavier nuclei and to produce heavy elements in any appreciable amounts."

Of course what such descriptions do not address is what would accrue if gravity did not further collapse the star, raising the density and temperature to essentially destroy any semblance of balance in the initial nucleosynthesis process. Similarly, in big bang scenarios with extremely rapid expansionary progression, no *real* (lasting) equilibrium could occur. Helium will be created by nucleosynthesis and destroyed by collisions with energetic particles and radiation in all these cases.

However, what we address here are the conditions required of a plasma hot enough to sustain thermonuclear reactions that effect an equilibrium abundance percentage of 24% helium by mass and a much lesser universal percentage of deuterium, etc.. For this the plasma must possess an extreme temperature in excess of 10^9 K. As alternative to the scenario of an origin in the big bang we will consider the possibility of establishing the hydrogen-helium ratio as an integral aspect of intra cluster dynamics and gamma ray bursts that are receiving increasing attention as discussed in chapter 2. We will address these possibilities more directly at the end of this chapter. But what we are describing here applies to any theory of how this ratio is established.

At temperatures approaching 10^{10} K there could be no multiplet nuclei whatsoever whereas at 10^7 K there clearly would be. There is some minimal dependence on baryon density of course, with the onset of element production requiring a higher temperature for higher densities (Wagoner, 1973, p.349). At the higher temperatures, high-energy radiation destroys multiplets as rapidly as they are created. Reactions would still take place at lower temperatures, although at considerably reduced rates in such an environment because there will inevitably be some particles in the high-energy tail of even a fairly low kinetic energy distribution that would be capable of producing such reactions. Although the reaction equations would almost completely preclude reactions in one of the directions at lower temperatures.

Wagoner et al. (1967), are largely responsible for the theoretical justification for light element production from a big bang. They concluded that conditions that are somewhat similar to what we have determined to be required to replicate Hubble's constant in intracluster plasma in the scattering model may be responsible for light element abundances. Although the phrasing is somewhat complicated, what Wagoner (1969) points out is that whatever the state of the plasma at a given point in time, temperature, and density, what ensues with regard to internal processes is independent of any presumption of an earlier evolutionary state. In their words:

"The method of analysis which will be used is based on the fact that the abundances of the various elements and isotopes which are produced within bodies that reach sufficiently high temperatures (3×10^9 °K) are determined mainly by the physical conditions existing during subsequent evolution. This is because statistical equilibrium among the various nuclei erases any effects of previous evolution. Such temperatures also result in the production of neutrons, which will be seen to play a major role in the new type of synthesis process we shall investigate. On the other hand, maximum temperatures somewhat lower than these do not allow any significant further synthesis during the expansion time scales we shall be considering." (p.247)

Of course, the quasi-stability that was assumed at each step in the calculations by Wagoner, et al. (1967) might no longer constitute what they would consider a tentative "time scale" situation if it were to be sustained in a stationary state of the plasma. Thus, one might account for an equilibrium ratio of hydrogen to helium at a lower temperature. It should be clear that in at least some rich cluster cores temperatures are sufficient for this nucleosynthesis process to be taking place at a low level to reassert the universal ratio. (Note that that is *not* where the author presumes the bulk of this conversion process to have taken place since it would not re-establish primordial distributions of mass.)

c. equilibrium equations for an hydrogenous plasma

First in the sequence of thermonuclear processes (if we assume the existence of a hydrogenous plasma as a basis for any such processes as is usually done) is the production of neutrons. It is first in the sense of neutrons being required by subsequent processes, but as envisioned here, there is no temporal sequence to be inferred and no rush on the rates. We will not deal immediately with where the neutrons originate but later we will hypothesize that they are recycled matter swallowed first by black holes but ultimately spewed forth by gamma ray bursts.

Specific weak interactions involved in the creation of neutrons are the logical prerequisites to the creation of deuterium and more complex nuclei. The equilibrium equations for these weak interactions are the following:

$$e^- + p \leftrightarrow n + \nu,$$

$$\bar{\nu} + p \leftrightarrow n + e^+,$$

where e^- refers to the electron, e^+ to the positron, p the proton, n the neutron, ν the neutrino, and $\bar{\nu}$ the antineutrino. These reactions are what are called 'weak reactions' that convert baryons back and forth between protons and neutrons – the primary ingredients for further development. The equilibrium ratio of neutrons to protons is determined in large part by the decay rate of neutrons as dictated by the following interaction:

$$n \leftrightarrow p + e^- + \bar{\nu},$$

The decay rate of neutrons is approximately 1.1×10^{-3} per second so that a neutron survives on average only about 15 minutes after its creation unless it is merged into a stable nucleus.

In thermal equilibrium the amounts of each significant baryon product will be distributed in accordance with the Maxwell-Boltzmann distribution. Therefore, in a reaction equation the ratio of the baryon products on the left, N_{iL} and right hand, N_{iR} sides of a nuclear reaction equation identified by subscript i, will be given by:

$$N_{iL} / N_{iR} = e^{-Q_{iLR} / kT}.$$

In this equation,

$$Q_{iLR} \equiv (m_{iL} - m_{iR}) c^2$$

where m_{iL} and m_{iR} are left and right hand nucleon masses respectively in the reaction denominated by i. See Peebles (1993, p. 184). For converting protons to neutrons in a hydrogenous plasma in particular, we have that $Q_{iLR} = 1.2934$ MeV.

Clearly at $T = 10^{10}$ K (for which $kT \cong 0.86$ MeV) there would be a ratio of over 22 % neutrons to protons in the plasma, whereas at $T = 10^9$ K there would be less than one neutron per million protons if this were the only equation of pertinence to this situation. At lower temperatures there would be a commensurably lower percentage of neutrons. Each produced neutron would have a high likelihood, of course, of quite rapidly decaying, merging into, and/or being captured by other nuclei so that we should expect neutron densities in the plasma medium to be extremely small for lower temperatures and densities.

At high temperatures the neutron has an appreciable collision cross section relative to the disassociation probabilities of heavier nuclides. At lower temperatures there is more likelihood of nuclear absorption. All of these further reactions associated with neutron interactions would establish their own equilibrium ratios, ultimately forming a network with an interrelated baryonic product cycle that would determine a resultant mass percentage for each ensemble of nucleonic baryons in the plasma, as we will see in discussions of later sections of this chapter.

The deuteron is the next logical step in nucleosynthesis. It is very stable with regard to disassociation so that in the presence of a plethora of protons a percentage of whatever neutrons are created will eventually find their way into deuterons rather than decaying directly. This capture of neutrons by protons releasing a gamma ray in the process is described as follows:

$$n + p \leftrightarrow D + \gamma$$

where D is the deuteron (one-proton-one-neutron nucleus of deuterium) and γ denotes a gamma ray that removes the energy difference that would otherwise characterize left and right hand sides of the equation.

If there were no further capture of neutrons into heavier nuclei, equilibrium would result in essentially no free neutrons at temperatures beneath about 3.0×10^8 K. This is in large part because of their low production rate at this temperature, and in part because of radioactive decay of neutrons after about 15 minutes. However, the cross section for capture by a proton is sufficiently large that some of the few neutrons that are created would be directly incorporated into deuterons before that eventuality and some of the others would effect the disassociation of nuclei to effect equilibrium.

The binding energy of the deuterium nuclei is Q = 2.225 MeV; neutrons captured in this energy well do not decay and may accumulate as constituents of deuterium in addition to further accumulation after combining further into helium.

At high temperatures deuterium will suffer disassociation losses due to high-energy radiation. Such photons could derive from the high-energy end of a blackbody spectrum of lesser temperature as well as the gamma rays given off in the production of deuterium. In such a spectrum the number of photons in a given energy range may extend well beyond the mean of the distribution to very large E_ν. However, it is key to this analysis that in general the Planck distribution diminishes much faster at that extreme than does the Maxwell-Boltzmann distribution of particle energies that are primarily responsible for nucleosynthesis and the disassociation of deuterium. When the number of photons in the high-energy tail of the photon distribution above the energy E_0 = 2.225 MeV is less than the number of neutrons participating in deuterium formation, there will inevitably be too few photons to disassociate whatever deuterium is produced.

Since deuterium is recognized as the regulator of further nucleosynthesis of the elements, and the Wagoner, et al. model discussed here assumes equilibrium, we anticipate a medium whose temperature hovers around this reaction equilibrium value. Further developments could trigger a slight decline in temperature, but certainly diminish free baryons by that further development, that would effectively halt the process unless and until a further gradual leakage into more massive nuclei pushed the temperature upward again. Thus a naturally controlled process is anticipated.

We can establish the temperature at which this upper bound on temperature switches the balance in this process by assessing for each temperature at what energy E_0 this occurs and the relative abundance of photons and particles so characterized. This was done by Wagoner, Fowler, and Hoyle (1967) and by Wagoner (1969, 1973) and others who followed similar procedures. First one approximates the tail of the Planck photon distribution by the exponential $n(E_\nu)\ dE_\nu = 8\ \pi\ E_\nu^2 /$ $(hc)^3\ e^{-E_b\ /\ kt}\ dE_\nu$ and then obtains the integral of all photons with energies above E_0. This would then give

$$N_{tail}(E_\nu > E_0) = 8\pi/(hc)^3(kT)^3\ e^{-E_0/kt}\ [(E_0/kT)^2 + 2(E_0/kT)+2]$$

The determined number in this section of the distribution would then be divided by the total number density in the entire distribution for the given temperature to give a fraction of photons with energies above E_0 as,

$$f(E > E_0) \equiv N_{tail}\ /\ N_{total} \cong 0.42\ e^{-E_0/kt}\ [(E_0/kT)^2 + 2(E_0/kT) +2]$$

In the standard cosmological model analyses it is fashionable to then introduce the ratio of neutrons to photons within any volume of space as the determining factor on whether deuterium continues to be generated. This approach does not lend itself all that well to the current scattering model for which a lasting equilibrium obtains other than in those gamma ray bursts that occur continually throughout the universe. The standard model embraces the current ratio of photons to baryons as a universal situation that has not changed since a presumed annihilation of anti-baryons with baryons with the additional presumption that there must initially, therefore, have been a ratio of 10^{-9} baryons to the total of both baryons and anti-baryons. These are further presumed to be the same photons that after scattering are now distributed in accordance with a 2.725 K thermal blackbody radiation, but throughout the intervening epochs to have been distributed as blackbody radiation of proportionately higher temperature.

There is, however, another constraint that has more recently been levied on big bang nucleosynthesis scenarios. Walker et al. (1991) have pointed out that the ratio of baryon mass density to the 'critical mass' density, i. e., $\Omega_m=1$, cannot exceed 0.029. Current virial mass estimates make this criterion impossible to attain – one more incompatibility of Doppler interpretations in the standard cosmological model. In fact, Sarkar (1996) states with appropriate citations that, "some authors have gone so far as to question the consistency of the standard BBN [big bang nucleosynthesis] itself on this account."

Obviously, in our scattering model the temperature is the same *now* as it was *then*. There would *then* also have been the 2.725 K thermal blackbody radiation, which would *then*, as it does now, account for a predominant share of the greater number of photons with respect to baryons, but would not then or now, thereby contribute to thermonuclear reactions. What we need to know is what would be the number of *foreground* thermal photons originating in the immediate vicinity of any point in spacetime in the effective thermonuclear environment. Clearly the hotter the medium, the larger the vicinity for which photons could contribute to disassociation.

There is a critical temperature at which these reactions lead ultimately to a 'frozen-in' hydrogen-to-helium ratio. Escaping radiation in an extended homogeneous plasma does not contribute the kinds of problems encountered in experimental reactors, but similarities pertain.

Any high temperature hydrogenous plasma like rich intracluster gases or more particularly gamma ray burst environments will exhibit characteristics that lead to synthesis of ^{4}He. A series of reactions proceed from protons and neutrons via deuterium, tritium, and ^{3}He to effect this process. By comparing the energy released in the emitted radiation, it can be shown that the critical ignition temperature for the D-T reaction is between 10^7 and 10^8 K. For the other essential D-D and D-^{3}He reactions the required temperature is at least an order of magnitude higher. Refer to figure 221 where the cross sections of these reactions are shown as functions of the energy of bombarding particles. The figure was taken from Post (1993, p. 881).

Hydrogen fusion reaction rate data is shown in figure 222 as functions of temperature. The units of these rates are given to remove the dependence on density. To obtain a reaction rate per unit volume, the rate provided must be multiplied by the density of the two constituents involved in each reaction. These rates are strongly dependent on temperature. There are in addition, several reactions involving isotopes heavier than helium, but because these elements will be virtually nonexistent in intergalactic regions, we ignore them here.

At temperatures above 10 million degrees the reaction rate for producing deuterium from protons is very much less than any of the other reaction rates. This rate increases much more slowly with temperature than any of the other nucleon-nucleon rates, but it does increase by a factor of a thousand when the temperature increases from 10 million degrees to 100 million degrees. Also, although deuterium is created at a very slow rate, it is destroyed extremely rapidly, since its reaction rate with protons is the highest of the reactions in this fusion chain. Deuterium will be converted into helium-3 almost immediately.

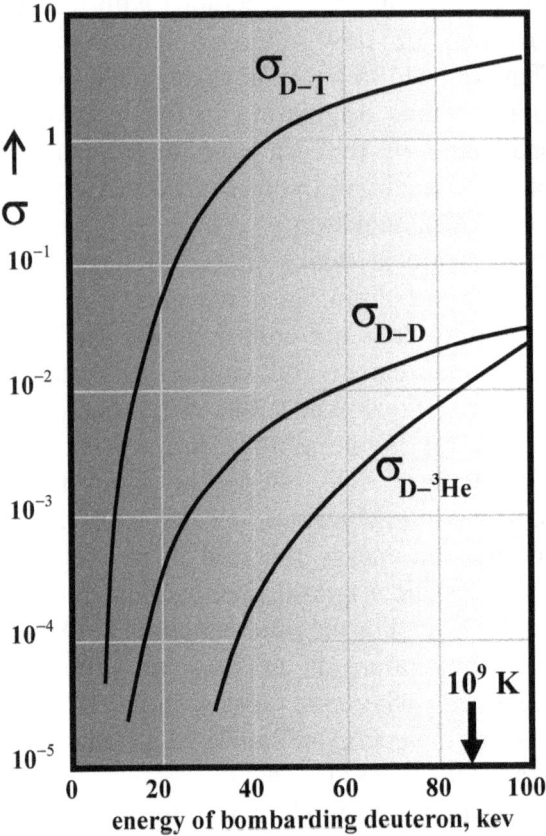

Figure 221: Cross section of D-T, D-D, and D-^{3}He reactions (in *barns*, 10^{-24} cm^2)

These critical temperatures are relatively independent of the ion density. Since the nuclear energy generated as well as the energy radiated away both vary with the square of the particle density, associated heating and cooling effects will cancel each other out. Typically, however, the critical temperatures cited above may not be quite sufficient because of other considerations. So the ambient temperatures must be in the range of 10^9 to 10^{10} K to guarantee that the reactions will be sustained. This is somewhat greater than characteristics observed in rich intracluster gases and prerequisite to cosmological redshifting to obtain an appropriate value of H_0 according to the scattering model as we have seen.

It was mentioned earlier that the current intragalactic gases in rich clusters and certainly gamma ray burst environments are similar to that conceived by the standard model as applying very shortly after the supposed origin of the universe. So for all intents and purposes big bang environments do indeed occur; they occur *now*, and always have been. Which is to say that primeval conditions are perpetuated in our universe rather than merely constituting its momentary origin.

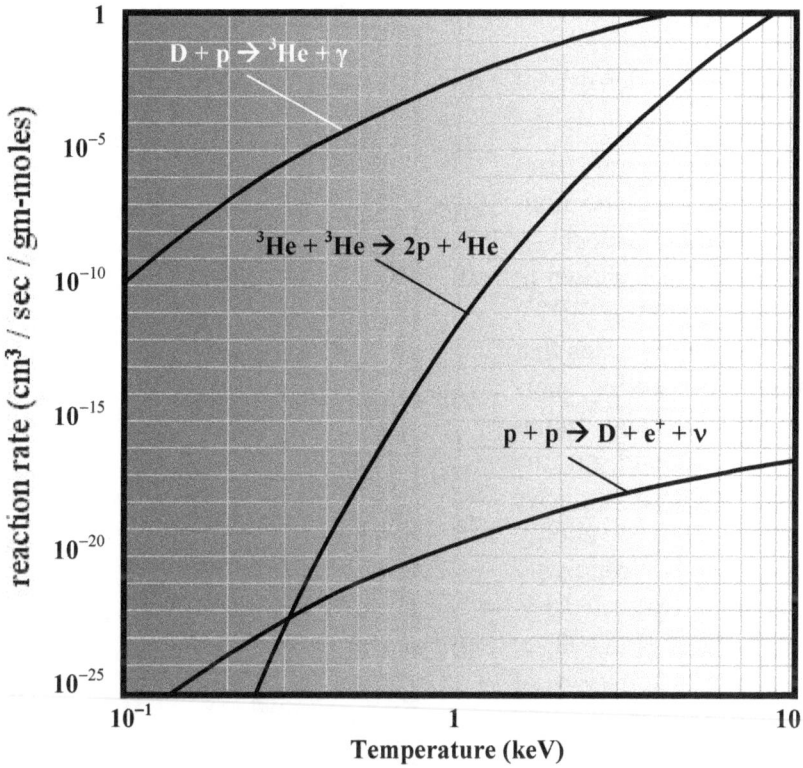

Figure 222: Proton fusion process reaction rates [from Lang (1980)]

d. proceeding as far as helium

The observed relative abundance of the light elements once propounded as requiring the big bang, in actuality requires no more of a bang than currently observed in intracluster gases and gamma ray bursts. These less extreme conditions in conjunction with a punctuated recycling of matter from gamma ray burst 'big bangs' provide adequate sustenance of elemental abundances with an equilibrium of expanding hydrogenous plasma sources and heavier material sinks, all of which contributes to microwave background radiation as already described.

At low temperatures and densities the efficiency of transmuting deuterium and the heavier tritium into heavier nuclei is slowed considerably so that residual amounts of deuterium will persist in accordance with observation. These reaction rates R_{xy} depend on the highly temperature (velocity) sensitive cross section of the reactions, σ shown in figure 221, and the density of the reactants as follows:

$$R_{\overline{xy}} = \rho_x \, \rho_y <\sigma \nu_{x-y}>$$

Here ρ_w is the density of the reactant w, ν_{w-z} is the relative velocity of reactants w and z. The non-density-related aspect of these reactions is shown in figure 223 per Post (1993, p. 881) who also notes that because the resulting energy production and radiation rates depend on the density squared as we discussed above, energy production is relatively unaffected by density. Also see Peebles (1993, p. 194) for a discussion of this and other related issues.

Rather than try to devise a scheme whereby enough helium is produced before the temperature precludes disassociation, we need to find the persistent temperature for which the production and destruction of helium are in permanent balance. The small probabilities of producing more helium would then be matched by the (also small) probability associated with neutron, proton or high energy photon absorption reverting back to deuterium and tritium on the one hand, or deuterium and the unstable isotope of helium on the other.

The thermonuclear processes that involve the further association of deuterium on into the stable helium-4 isotope are as follows:

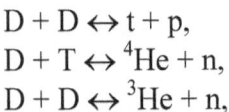

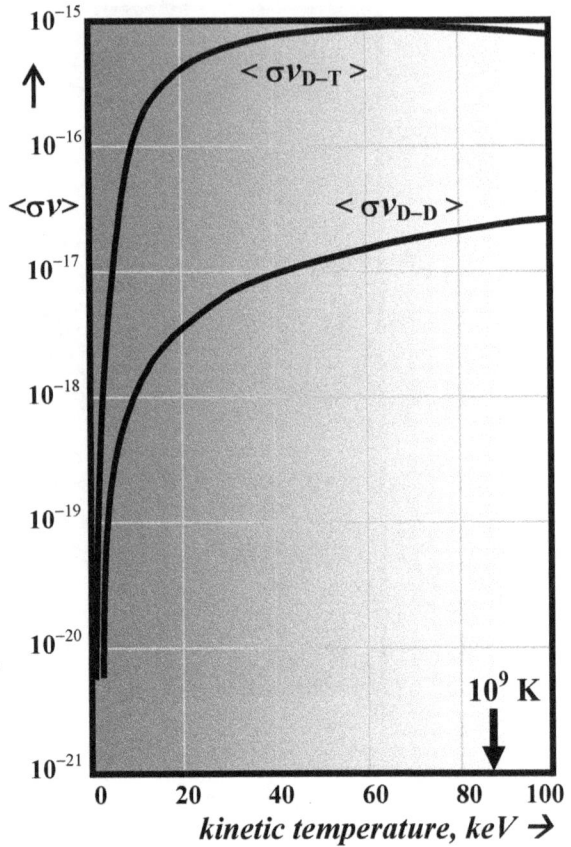

$$D + D \leftrightarrow t + p,$$
$$D + T \leftrightarrow {}^4He + n,$$
$$D + D \leftrightarrow {}^3He + n,$$

Figure 223: D-X reaction rate data

478

$$D + {}^{3}He \leftrightarrow {}^{4}He + p,$$

where T is a triton one-proton-two-neutron nuclide, ${}^{3}He$ is a two-proton-one-neutron isotope of helium, and ${}^{4}He$ is the more stable two-proton-two-neutron isotope of the helium atom nucleus. In figure 224 from Wagoner (1967, p. 18) these reactions have been arranged to show how they interactively define an equilibrium.

Figure 224 shows the first stages of the transmutation ladder, but the higher mass elements will not result at the usual temperatures we are talking about in gamma ray bursts and cluster gases. Although some such developments will occur in the latter environments, they occur primarily in stars, with only traces being created in the extra-stellar environment. This diagram is more or less a 'flow chart' of the computer generated results of the abundance percentages determined by Wagoner et al. (1967) and Wagoner (1969 and 1973).

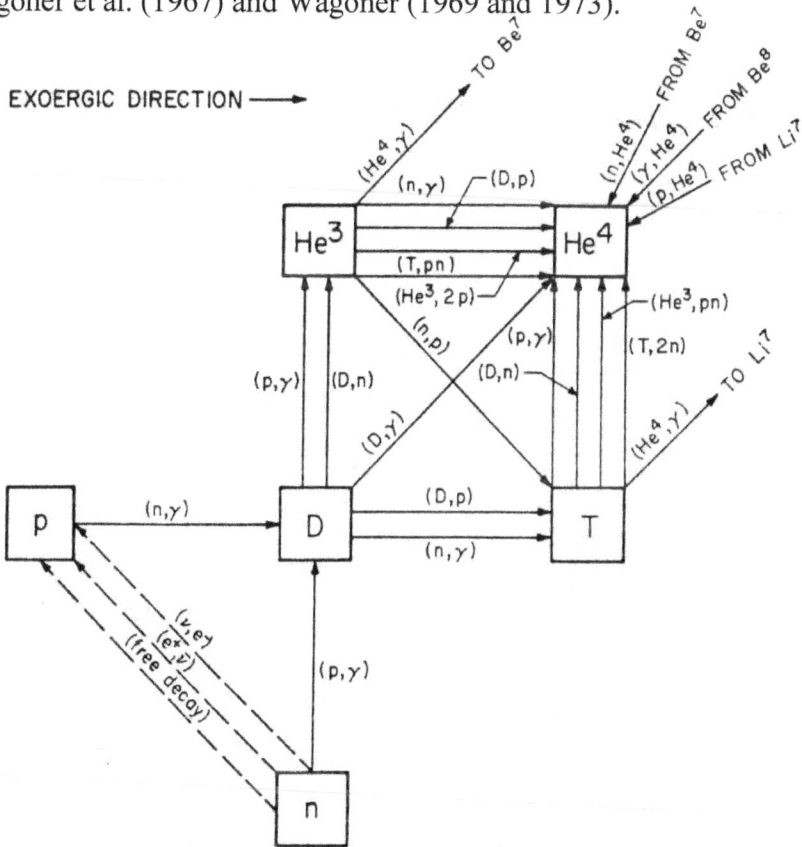

Figure 224: Interrelationships of thermonuclear reactions leading to ${}^{4}He$

In figure 225 a similar diagram is provided, but showing the two-way relationships of the associated reactions. Which way the reactions actually proceed depends on the specific energetics. In figure 224 the energy release (exoergic) reactions are emphasized.

The reactions involving the collision of a neutrino of either kind with a baryon have been eliminated from the diagram in figure 225 because of the extremely low probabilities associated with this phenomena. Therefore these are represented as one-way reactions. Also, the low availability of positrons has been used as reason to eliminate that reaction from the diagram altogether since both sides require products that are virtually unattainable in the environments of interest. The presumption has also been that the temperature will be such that nuclei beyond helium will be destroyed as rapidly as they are produced, without measurable accumulations.

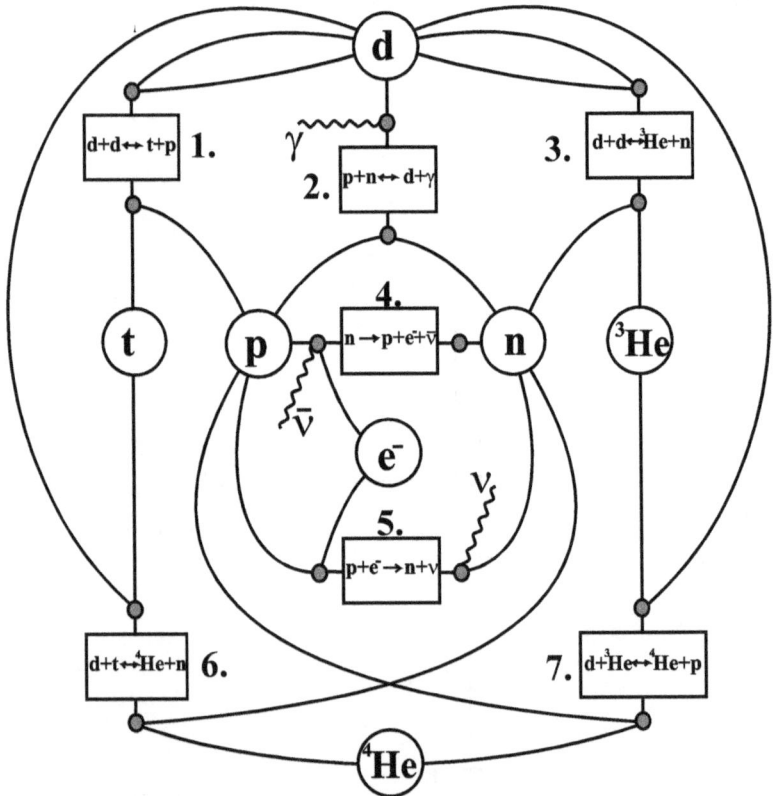

Figure 225: Thermonuclear stability diagram – excluding elements beyond helium

e. determining elemental abundance for two models

For a given temperature and initial density of a hydrogenic plasma, an equilibrium condition will eventually be reached for which there will be fixed percentages of each of the various nucleons. Applying Kirchoff's law to junctures and circuit analysis of the transmigration routes in the previously presented diagrams, it is possible to obtain a set of simultaneous equations for percentages of each of the various nucleons. Notice first, however, that from the Q values for each reaction, the percentage of incident and emergent products on each side of a reaction equation can be expressed as:

$r_{iL} = r_{iR} f_i(T)$, where

$$f_i(T) = e^{-Q_{iLR}/kT},$$

which is a constant for a given value of temperature T. We can drop the subscripts for left and right by substituting from the equation above and noting that $Q_i \equiv Q_{iLR}$. Since we are dealing exclusively with ratios, with eventual normalization to percentages of a total amount, we may as well define $r_{iR} = 1.0$ and $r_{iL} = f_i(T)$. The Q values for the seven reactions shown in figure 225 are identified below.

$Q_1 = 4.0$ MeV
$Q_2 = 2.227$ MeV
$Q_3 = 3.25$ MeV
$Q_4 = -1.2934$ MeV
$Q_5 = +1.2934$ MeV
$Q_6 = 17.5$ MeV
$Q_7 = 18.3$ MeV

The solutions to the associated equations yield the temperature profiles for the abundances of the various elements. This is how the analyses were performed for the standard model by Wagoner et al. (1967, p. 20). See figure 226 that was generated using assumptions appropriate to the big bang of the standard model. Clearly here the primary effects are realized at about 10^9 K. Here as in other variations of the analytical constraints, the process of generating these plots was to assume isolated conditions pertinent to each time, temperature, and density. But at each slot it is exclusively the constraints of that current situation that precipitates the equilibrium condition that is plotted.

Several versions of this plot were generated by Wagoner (1973, p. 347). Concerning this particular version for which all the curves are flattened out at temperatures less than 5×10^8 K except for neutrons, Wagoner states, "It is seen that the abundances 'freeze out' at temperatures $0.8 \geq T9 \geq 0.4$." Wagoner takes T9 to be temperature in units of 10^9 K. In the same paper, Wagoner also generated plots as functions of current mass density as shown in figure 227.

But Wagoner et al. (1967) also generated data for non big bang nucleosynthesis scenarios.

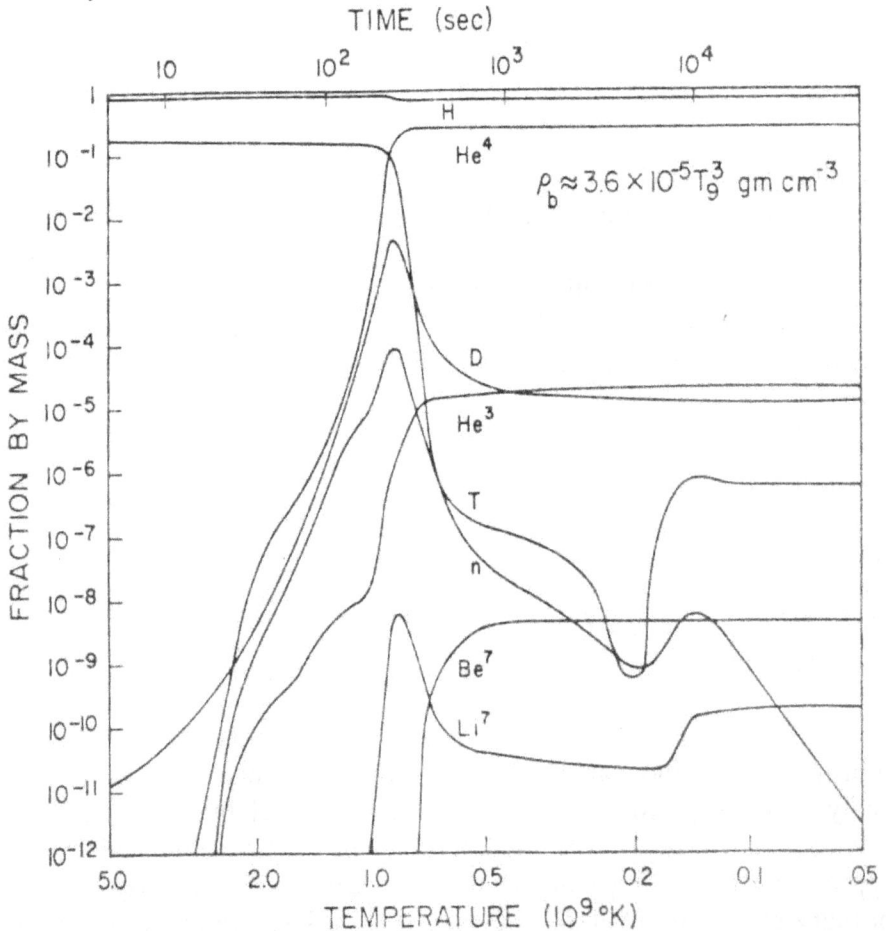

Figure 226: Nuclear abundance percentages that would be realized as functions of temperature, density, and time for the big bang model

482

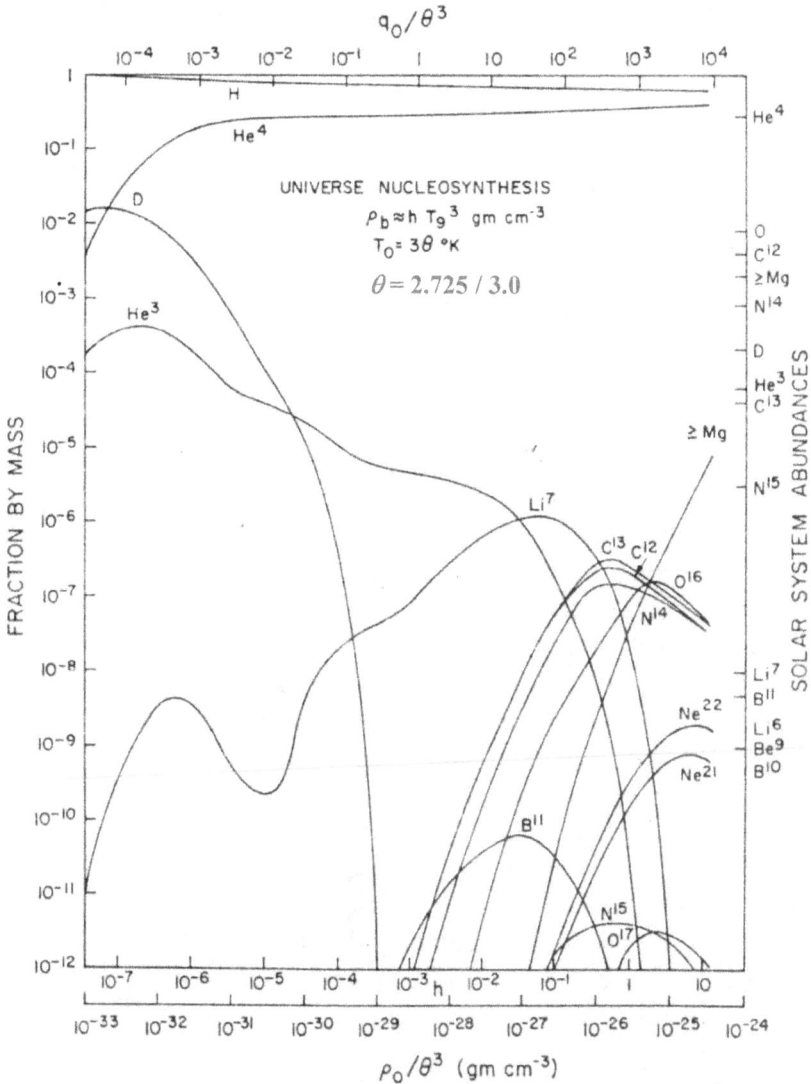

Figure 227:   Elemental abundances that would be realized as a function of
               density in a large high temperature exploding object

f. the other possibility – massive object explosions

In addition to the big bang model, Wagoner et al. (1967) ran the
same programmed analyses for exploding "massive objects" for which
the mass is assumed to be greater than on the order of 10^8 solar masses.
This is somewhat along the lines of what Alfven (1966) seemed to have
had in mind. The results are very similar to what we have presented

above as applicable to the big bang scenario. See figure 228 in which pertinent data is replicated from Wagoner et al. (1967, p. 34). The disparity in ^{3}He and D profiles from that shown in figure 226 may be explained in terms of these nucleons seeming not to have established universal ratios. See Sarkar (1996) where it is indicated that major differences in these ratios have been observed in various astronomical settings. It is clear that the mass for which figure 228 was specifically drawn corresponds to what might be anticipated of the eruption of massive black holes at the centers of huge galaxies. If these were to erupt as Appendix C suggests they might, there would be tremendous outbursts of gamma rays from the reaction of the freed neutrons as indicated by the reaction equation, $n + p \leftrightarrow D + \gamma$, which would initiate the processes indicated in figures 224 and 225.

Such massive gamma ray bursts are, of course, being observed regularly as of this writing. See for example the two panels in figure 16 page 36 taken from Zhang et al. (2006, p. 314).

Notice in particular that the time frame of these bursts corresponds to the critical phase between those shown in figures 226 and 226. Peaks in the gamma ray energy distributions take place at the time and energy of peak transmutational activity intermediate between the big bang and lesser explosion situations. So whether the rich intracluster plasma gas were to continuously maintain a hydrogen-helium balance or if it is established and maintained by refreshment from major sporadic eruptions spewing hydrogenous plasma back into more diffuse intergalactic regions, it is clear that the current universe provides ample capabilities of maintaining this ratio without recourse to a *big* bang. The recycled material that has gone down the chain of being to be spewed back, would not be all that different from what the standard cosmological model versions have envisioned – a punctuated equilibrium – rather than a one-time explosive origin.

What is significant is that an hydrogenous plasma in the temperature region around 10^9 K is where the hydrogen-helium ratio becomes established, and wherever and whenever such conditions arise, that ratio will tend to pertain. With regard to gamma ray burst phenomena, according to Band et al. (1996, p. 2), "In many bursts most of the energy is above 300 keV," i. e., well over 10^9 K.

In this region what is of primary importance according to Wagoner et al. (1967) is the availability of neutrons. In fact, they suggested that to begin the nucleosynthesis process (at least for the big

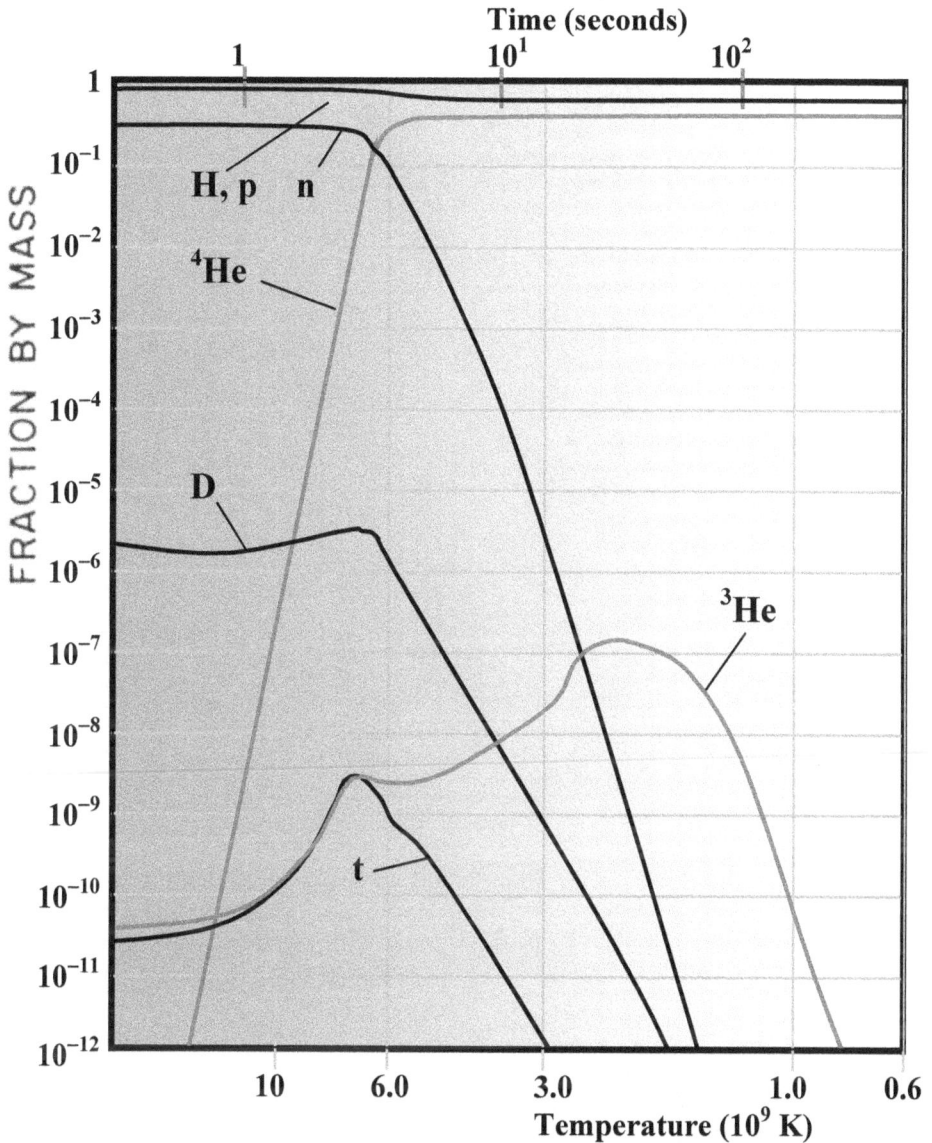

Figure 228: Elemental abundances realized as a function of temperature in a large high temperature exploding object

bang scenario) an approximate ratio of 50% neutrons is required. Since the last visible traces of matter dense enough to ultimately disappear beneath the shroud of an event horizon confined to a Schwarzchild radius, are emissions from neutron stars, certainly makes such situations a likely origin of such building blocks. So naturally if there is some physical process whereby black holes give up their bounty, a plethora of neutrons are to be expected. Gamma radiation is the ineluctable consequence.

Chapter 25

Matters of Gravity

"When an obscure Russian meteorologist named Alexander Friedmann proposed, in 1922, that the Universe might be expanding, Albert Einstein was sure that he was wrong. Five years earlier Einstein had published a static model of the Universe, and he was still convinced that it was correct. In a rare but dramatic blunder, Einstein bolstered his unfounded beliefs with an erroneous calculation, and fired off a note to the *Zeitschrift fur Physik* claiming that Friedmann's theory violated the conservation of energy. Eight months later, however, after a visit from a colleague of Friedmann's, Einstein admitted his mistake and published a retraction. The equations of general relativity do, he conceded, allow for the possibility of an expanding universe." Guth (1997)

Up until this chapter we have largely ignored what cosmologists have tended to see as the most essential aspects of cosmology – theory – Einstein's hypotheses with regard to space, time, and gravity in particular. These imponderables he integrated at the most basic level of reality. The generalization of his Special Theory of relativity he saw as requiring the incorporation of what had formerly been considered to be but another of the forces between objects within a Euclidean landscape. He removed gravity from its former status as a force transacted *through* space and time like any other, integrating it as an integral feature of the geometrical structure of a spacetime landscape.

We will not dally long in our discussion of complex theoretical considerations including latter day proliferations into string theory and 'multiverses', but we will briefly discuss Einstein's 'cosmological equation. We will also investigate the closure criterion' for the universe as he perceived it, and observe how this has given rise to all manner of conjectures concerning 'missing mass', etc..

We will begin our discussion by venturing into the topic of what Einstein considered to have been his 'greatest error' because it concerns the analogy with classical physics that motivated other aspects of his theoretical considerations. This will shed light on why cosmologists so regularly revive his acknowledged error as a *feature*, rather than a failing, of their own surmisings.

These topics inevitably lead to a discussion of 'dark matter', 'vacuum energy', and why it has seemed reasonable to cosmologists to believe that most of the matter in the universe is not visible. That is to say that they have become convinced that one cannot explain cosmic phenomena without embracing constructs that cannot be directly observed. Some conjectures involve illusive massive particles that, although unaffected by electromagnetic forces, would nonetheless affect and be affected by their surroundings through gravitational effects. Much of this has been by-passed as irrelevant to the work at hand in this volume, although, by providing an alternative resolution to the extreme redshift across galaxy clusters that does not require additional gravitational mass, that is not a problem for the scattering model. To fully resolve this we must also explain why so many cosmologists have been convinced that 'dark matter' must be a reality.

Concepts that preoccupy cosmologists concerning whether dark matter is 'hot' or 'cold' matters primarily to those convinced that there *is*, in fact, dark matter so we will not get into that. Of course, to dispense with such conjectures so off-handedly presupposes resolution by other means. We have concentrated on that alternative resolution, realizing that to a certain extent the intergalactic plasma medium is in itself 'dark matter' about which one is left to conjecture.

Finally we will address that most final of issues, black holes. It is, after all, these vortexes of concentrated matter that are perceived by so many as the penultimate doom of an evolving universe. To propound a stationary state universe with sinks into which matter can be totally removed from consideration would be irrational. We must, therefore, wrestle with these behemoths. So that is the agenda.

a. Einstein's "greatest error"

Einstein's General Theory gets into areas that attempt to explain the universe as a whole. In "cosmological considerations of the general theory of relativity" Einstein (1917) referred to Poisson's well-known equation that applies to gravitation. In particular:

$$\overrightarrow{\nabla}^2 \, \phi(\mathrm{r}) = 4 \, \pi \, \mathrm{K} \, \rho$$

where the second derivative of the gravitational potential energy $\phi(\mathrm{r})$ is equated to a constant times the mass density ρ as appropriate to inverse square law forces. He noted that there is an apparent incompatibility of this usual formulation and boundary conditions applicable to Newton's theory of gravity. It seemed to Einstein to imply that mass density must approach zero as the extent of the volume to which the equation applies becomes infinite, if the gravitational force were not to become infinite as well. This is certainly mathematically the case.

Clearly, the equation would seem to be incompatible with there being no net force, $\overrightarrow{\nabla}\phi = 0$ on matter in an extended uniformly dense universe as Newton was wont to accept as reality. This is a view which Einstein and others have disputed as being erroneous on Newton's part. Further discussion of this situation and how Einstein handled it is provided by Bonn (2008, pp. 130-150). Some of that discussion is duplicated here. It was no doubt to resolve just this quandary that Einstein introduced what he would later acknowledge as having been his greatest error. See Einstein (1952, p. 193), where he states:

"As I have shown in the previous paper, the general theory of relativity requires that the universe be spatially finite. But this view of the universe necessitated an extension of equations with the introduction of a new universal constant λ, standing in a fixed relation to the total mass of the universe (or, respectively, to the equilibrium density of matter). This is gravely detrimental to the formal beauty of the theory."

Here he audaciously presumes that one's methodologies and theoretical models might appropriately dictate requirements on the *actual* universe that one is attempting to model. This author considers that perspective to be a much more egregious error than what Einstein considered to have been his "greatest" in the above quotation. One must limit their theories and models to valid mathematical descriptions of *actual* phenomena from which to extract invariances and explanations. Theories are not specifications that must be followed by an unwilling universe. Dictums concerning nature must be accepted only to the extent that they *are* valid descriptions if we would have the entire universe acquiesce to such pronouncements. One easily falls prey to gibberish otherwise.

Einstein was concerned because solving Poisson's differential equation for the potential energy of a uniform distribution, resulted in:

$$\phi(r) = 2\,\pi\,K\,\rho_0\,r^2$$

which, of course, increases without limit as r becomes very large.

To resolve this problem, he conjectured that there must be some universal constant Λ, defined such that Poisson's equation could be replaced by the following:

$$\vec{\nabla}^2\,\phi + \Lambda\,\phi = 4\,\pi\,K\,\rho$$

The solution of this equation for a uniform density ρ_0 is,

$$\phi_0 = -\,4\,\pi\,K\,\rho_0\,/\,\Lambda$$

a constant everywhere. He proceeded to apply a similar kluge to higher dimensionality in his general theory as we will see. Later he would acknowledge this as his "greatest error". It is one that cosmologists continue unabashedly to precisely reincarnate to resolve mismatches between theory and observation.

The author attended a presentation by Philip Mannheim (2008) in which he described, among other topics, the major vagaries of the ill-begotten cosmological constant and how it fits into his own four-dimensional conformal theory of quantum gravity. He stated that not only was inclusion of the lambda term *not* an error, but that it would have been a serious error to have *omitted* it. After his presentation this author asked the presenter privately whether he felt that omitting lambda should be considered Poisson's greatest error instead of Einstein's? He laughed, of course, thought about it for a moment, and then acknowledged quite cheerfully that, yes, he would have to say that. Needless to say, this author does not agree.

The situation with lambda is one where we sometimes get so caught up in the mathematical symbolism that we forget to check for an isomorphic physical reality – the association that is the sole justification for any symbolic *representation* at all. Poisson's equation derives from

490

Gauss's integral theorem associated in turn with a divergence theorem discussed in detail in essays by Bonn (2008). This integral theorem illustrated at the left in figure 229 states that:

$$\iint \vec{\nabla}\phi \bullet d\vec{\sigma} = \iiint \rho \, dV$$

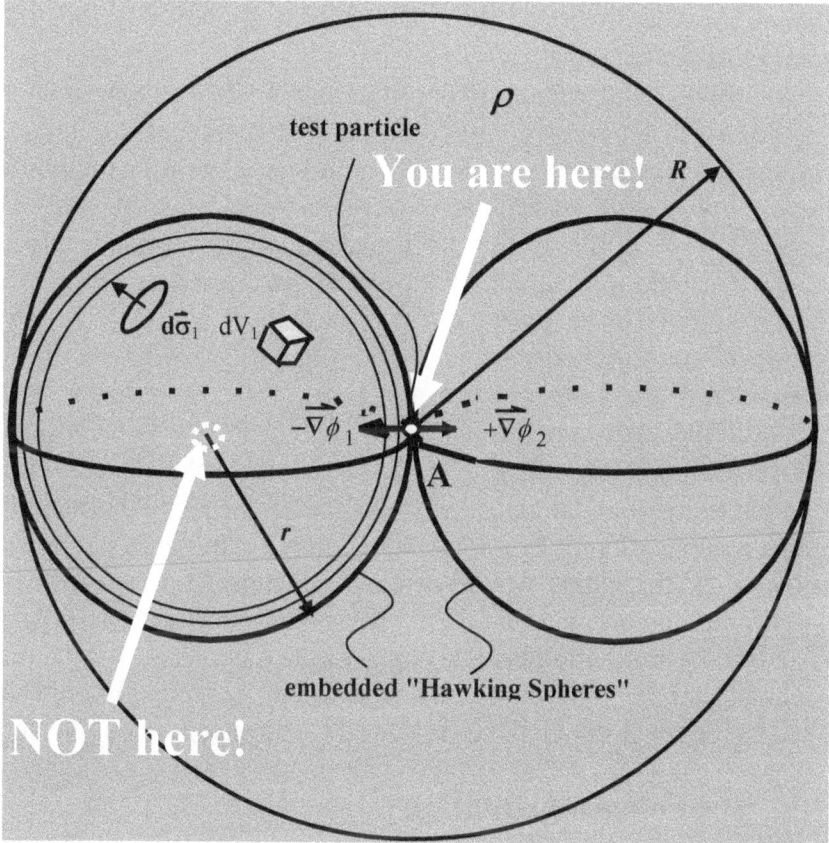

Figure 229: **Applying Gauss's integral theorem to embedded 'Hawking spheres'**

Here $d\vec{\sigma}$ is the outwardly-directed vector associated with an infinitesimal area on the sphere. The symbol dV represents the infinitesimal volume element within the sphere. The above equation expresses in mathematical terminology that the sum (integral) over an entire closed surface – such as the sphere on the left in figure 229 – of the outwardly-directed perpendicular component of the force field associated with the enclosed mass density distribution is equal to the total amount of mass enclosed by that surface. If the density is uniform throughout the enclosed sphere it corresponds to what Bonn (2008) refers to as a "Hawking sphere". This could be constructed of shells of equal thickness and uniform density to which Hawking (1988, p. 5) referred in siding with Einstein against Newton on the issue of whether an infinite homogeneous universe would necessarily collapse under its own weight. The illustrated shells are artifacts employed in integrating to an infinite limit.

Certainly the perpendicular component of the force field $F = -\vec{\nabla}\phi_1$ due to that portion of the uniform distribution in the left-hand 'Hawking sphere' shown in figure 229 is the same at every point on the sphere. However, in an infinite universe, similar relations apply with regard to $\vec{F} = -\vec{\nabla}\phi_2$ due to the mass distribution on the right, which is required if we are to maintain symmetry about a test particle on which the field is exerted at point A. Both values $\vec{\nabla}\phi_1$ and $\vec{\nabla}\phi_2$ can be determined using mutually exclusive portions of the mass distribution that maintains the proper symmetry about the test particle by this procedure, and their sum by the rules of field theory is therefore the legitimate solution at point A. So the total force field $-\vec{\nabla}\phi = -\vec{\nabla}\phi_1 + \vec{\nabla}\phi_2$ at the test particle at location A must be zero when we insist on the legitimate application of Poisson's equations to symmetric parts of this problem. And the *proper* way to extend such considerations to the limit of an infinite universe is to let R (not just r) go to infinity. This gets us away from the troubling necessity of our entire universe being either a gigantic black hole collapsing into a singularity or an equally grotesque, but otherwise required, big bang followed by an expanding universe.

b. the 'cosmological equation'

Einstein was motivated to generalize his work in an attempt to comprehend the universe as a whole, not satisfied with a 'special' theory that dealt exclusively with uniform relative motions that do not characterize much of the reality we observe. His 'world model' of the universe presupposed a finite, static spacetime large enough so that the galaxies, and even clusters of galaxies would constitute insignificant ripples in a uniform mass distribution. To avoid an 'edge' problem to his universe required some alterations. The traditional Euclidean concepts of geometry had to be extended so that three-space could be accommodated as a finite 'surface' within the overall scheme of things.

This required a 'metric tensor' to characterize spacetime, into which the concept of gravity itself was incorporated. He developed a tensor differential equation to characterize this model, for which in the limit of small enough volumes, it reverted to something very similar to the usual Poisson equation with Newtonian gravitational force:

$$\vec{\nabla}^2 \phi = 4 \pi K (\rho + 3p),$$

Dynamic pressure p is included in a stress-energy equivalent. And here again we see the problem that confronted Einstein and that precipitated his error. He needed lambda so his universe would not collapse. Thus, his cosmological equation as cosmologists accept it is:

$$R_{ij} - \tfrac{1}{2} g_{ij} R - \Lambda g_{ij} = 8 \pi G T_{ij},$$

A double subscript indicates the construct is a tensor quantity; R_{ij} and R are functions of the metric tensor g_{ij} and its derivatives, and T_{ij} is the stress-energy tensor that includes the dynamic pressure. We will leave it to others to tell us what this equation implies to them, and concern ourselves primarily with whether those implications are realized.

c. the effects of pressure

Certainly gravitational collapse into stars and galaxies occurs. Over (and under) densities will (do) certainly occur for various reasons. In regions of overdensity gravitational effects will produce contraction into gravitationally bound systems.

But gravity is not the only 'force' to be considered with regard to the resultant behavior of distributions of matter throughout an otherwise uniformly extended universe. There are thermodynamic considerations to be taken into account as well. Any volume of matter at a temperature above absolute zero experiences an outward pressure that would, if it were constrained by a spherical membrane such as in a balloon, for example, force continued outward expansion of that volume in accordance with the following traditional thermodynamic formula:

$$p V = n k T$$

where p is thermodynamic pressure, V is the volume within the surface, n is the number of particles of gas within the volume, k is Boltzmann's constant, and T is the temperature of the gas within the volume. This produces a force on each unit of the surface area as shown in figure 230; the associated force f_p would be experienced in an outward direction due to the thermodynamic pressure, p, where r_J indicates the radius of the enclosed overdensity volume.

This force would be countered in such a hypothetical situation by the gravitational force operative on the matter contained within the volume as assumed by Einstein's analysis of an inward gravitational force. This would reduce the volume, raise the kinetic temperature, thereby increasing outward pressure.

The *Jeans criterion* for collapse takes both forces into account in assessing the conditions throughout the volume for overall stability. The resulting criterion for an over density in an ideal gas with no external forces, is

$$r_J > \sqrt{\gamma \, k \, T / m_p \, G \, \rho_J}$$

The symbols k, T, and G are as defined previously. Here r_J is the *Jeans length* beyond which gravitational collapse would be inevitable, γ is the adiabatic expansion factor (approximately unity), m_p is the average molecular mass, and ρ_J is the mass overdensity.

$$\rho_J \equiv \rho_M - \rho_u$$

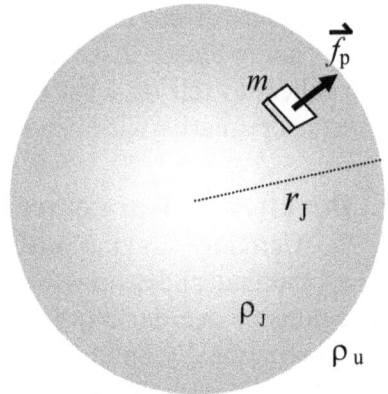

Figure 230: Thermodynamic forces active in the structure-producing processes of the universe

which is merely the amount of density in excess of the immediate surroundings.

Thus, we could define regions of interplay between the forces gravity and thermodynamic considerations.

$$r_J \cong 1.5 \times 10^8 \ \sqrt{T / \rho_J},$$

At a minimum, it must be obvious to the reader that there are various forces that have acted in concert in determining how our universe has come to its observed current conditions. One might theorize various models to derive conditions we perceive as essential to the universe we observe, but such models are of little significance relative to factual observation.

What is observed is a pattern of structures for which any conceivable aggregation process from a uniform distribution of galaxies would take upwards of ten times longer than the alleged age of the universe according to any version of the standard model.

d. uniformity of matter in the universe

In any case, Einstein was persuaded that the universe must indeed be homogeneous and very uniform at distances large with respect to our own galaxy and it's immediate environs such that a uniform density seemed a reasonable assumption. That assumption seems even more valid now that many millions of galaxies have already been observed and mapped. See figure 231 from Maddox et al. (1990) and refer again to diagrams provided as figures 5 through 7 in chapter 2 above. The two-dimensional map of the sky provided in figure 231 covers a region 100° by 50° around the South Galactic Pole. Automatic Plate Measuring (APM3) that provided this galaxy survey provided the positions, magnitudes, sizes, and shapes for about 3 million galaxies. Each pixel covers a small patch of the sky that is 0.1° on a side. The image is shaded according to the number of galaxies per pixel area. The pixels are brightest where there are the most galaxies. Clusters, containing hundreds of galaxies are seen as merely a bright patch. Larger elongated bright patches are 'superclusters' and 'filaments'. Small empty 'holes' are excluded viewing regions around bright foreground stars in our own galaxy, nearby dwarf galaxies, and globular clusters. Clearly the galaxy distribution by angle seems quite uniform on the sky.

But Einstein was also convinced by his interpretation of Poisson's equation that the universe must be finite to keep the velocities of distant galaxies within bounds. That is a conviction we must question – not for reasons of Hubble's hypothesis, which ultimately persuaded him to disavow his arbitrary insertion of Λ, but for physical and mathematical reasons we have just discussed.

e. expanding universe hypothesis

Hubble's hypothesis did seem to have come to Einstein's rescue with regard to the universal constant Λ such that, assuming an extreme initial velocity of the matter at remote distances, one could suppose that gravitation was indeed operative at these extreme ranges in bringing the velocities of distant galaxies into check.

The APM Galaxy Survey
Maddox et al

Figure 231: The manifest uniformity of the universe at large enough scales

That would seem to put us at a central non-Copernican position in spacetime. However, in the four-dimensional geometrical approach of the general theory, our place in three-space would still be equivalent to any other. Our place in time is quite another matter. We would occupy a very unique place in the history of the universe as Hawking (1988) noted by the title of his popular best seller, "A brief history of time – from the big bang to black holes".

At any rate it was Hubble's hypothesis of expansion of the universe that effectively did away with any need for Λ in Einstein's mind, especially in a finite universe. So he acknowledged that it had all been a bad mistake – that he should have let his equations guide him without fear that the universe might not follow. He recognized that Hubble's constant provided a means for assessing gravitational values of cosmological significance including the average density and radius of the entire universe. His cosmology was conceived somewhat as shown in cartoon form in figure 232.

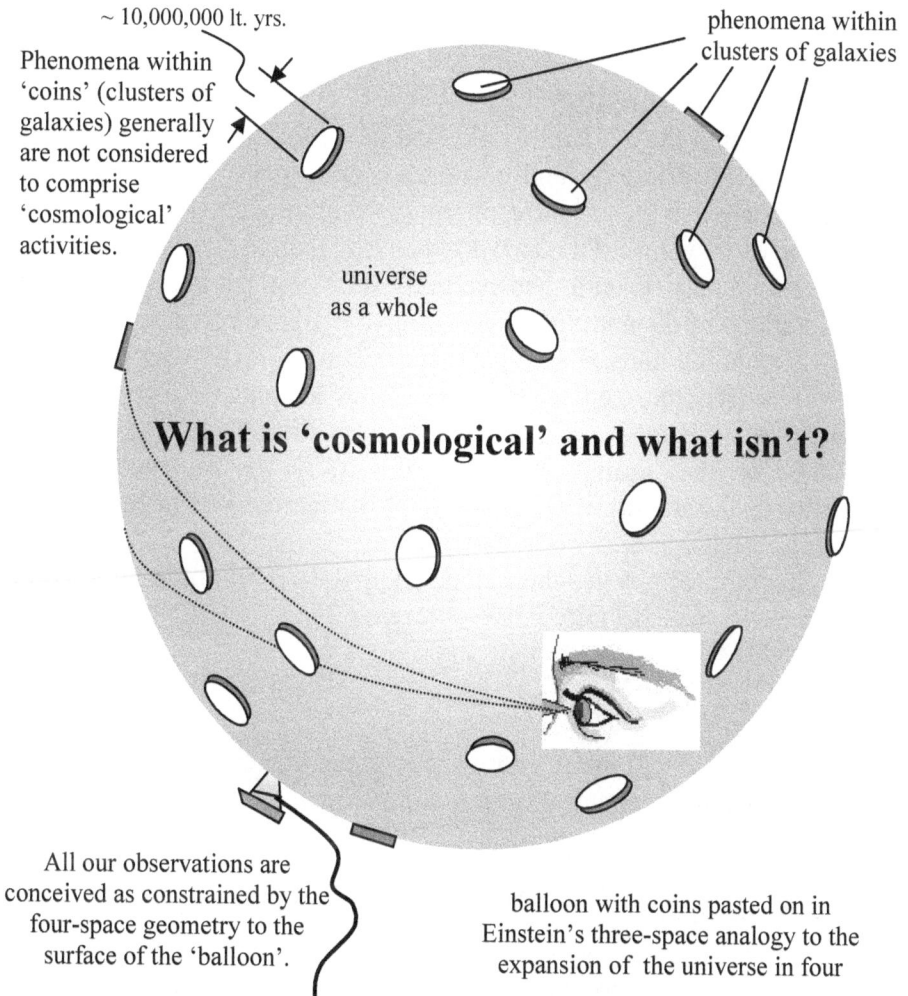

~ 10,000,000 lt. yrs.

Phenomena within 'coins' (clusters of galaxies) generally are not considered to comprise 'cosmological' activities.

phenomena within clusters of galaxies

universe as a whole

What is 'cosmological' and what isn't?

All our observations are conceived as constrained by the four-space geometry to the surface of the 'balloon'.

balloon with coins pasted on in Einstein's three-space analogy to the expansion of the universe in four

Figure 232: A visualization of Einstein's conception of a four-space universe

Clearly Einstein's "greatest error" was purosely incorporated to avoid gravitational collapse. It retains this role in models that have resurrected it. Inflation and the recent discovery that at great distances 'expansion rate' seems actually to be increasing rather than decelerating has emboldened its reincarnation and in some applications turned it into a variable rather than a constant. See for example, Bothun (1998) who says, "In the cosmological equations Λ appears as a long range repulsive term and acts like a source of negative pressure," thus adding another $10 - 20$ % to the presumed age of the universe.

f. the 'critical density'

Although the *critical density* and its derivation are cornerstones of what general relativity and current cosmology are all about, it is a simple concept with a correspondingly simplistic, non-relativistic, derivation of its value. The derivation begins with the classical concept of 'escape velocity' from a massive body such as earth and proceeds to considerations of distant objects receding at extreme velocities as a part of the expanding universe hypothesis. The collective mass of the universe is hypothesized as the retro force keeping expansion from getting out of hand. Of course this derivation does not get into rationale for the strange initial condition, its cause, nor yet the criticality of the tuning of the model that is required just to realize this condition. That was addressed by Guth.

In classical physics the kinetic energy, T, of an object of mass m that is moving with velocity v is,

$$T = \tfrac{1}{2}\, m\, v^2$$

The gravitational potential energy, V, of an object of mass m at a distance, r, from the center of gravity of a spherical mass distribution of total mass M is,

$$V = -\,G\,M\,m\,/\,r.$$

Here G is Newton's gravitational constant we defined earlier.

An object will escape the gravitational field of the distributed mass if its kinetic energy exceeds the absolute value of the gravitational potential energy by which it is bound, such that:

$$\tfrac{1}{2}\, m\, v^2 \geq G\,M\,m\,/\,r.$$

498

Kinetic energy will be converted into gravitational potential energy as it is slowed down in proceeding further from the center, satisfying the energy conservation law. If the two forms of energy happen to be equal then the object would come to a stop at a very great distance with essentially zero velocity and zero potential energy.

It is virtually the same calculation for stars and dust circulating about a galaxy or galaxies in a cluster whose escape would be from the attraction of all the other galaxies and intragalactic gases. And it is the same equations that would be applied to a finite (Hawking sphere) universe that is in question. Will they stop, turn around, or fall back to swirl with the other galaxies until finally they dissipate their energies and collapse into a gigantic black hole?

If there is just enough material in the universe to stop the galaxies then perhaps the universe will go on forever expanding ever more slowly – never escaping and never collapsing. Einstein preferred that solution for obvious reasons. And the universe seems to have acquiesced amazingly well, although apparently not quite. How it could be so close – and yet so far – is one of the difficulties facing standard model cosmologists

According to Hubble's law the approximate velocity of a distant galaxy is proportional to its distance, $v = c\,H_o\,r$, so the kinetic energy of a galaxy can be written $\frac{1}{2}\,m\,(H_o\,r)^2$. The mass of all the material inside a sphere of radius r is given by $M = 4/3\,\pi\,r^3\,\rho$, where ρ is the average density of the universe. Substituting these two expressions into the inequality provided above produces the inequality:

$$\tfrac{1}{2}\,m\,(H_o\,r)^2 \geq G\,(4/3\,\pi\,r^3\,\rho)\,m\,/\,r.$$

By simplifying and rearranging to solve for the critical situation for which equality applies with $\rho = \rho_o$, we obtain:

$$\rho_o = 3\,c^2\,H_o^2\,/\,8\,\pi\,G$$

This 'critical density' depends only upon universal constants. It is approximately 8×10^{-30} gm cm^{-3}.

Interestingly, from the usual standard model understanding that the radius of the universe is equal to the Hubble distance, $r_u = 1\,/\,H_o$, we can determine the 'critical' Schwarzschild radius r_s of the universe itself, since $v \rightarrow c$ as $r_u \rightarrow r_s$, as follows:

$$r_s \quad = 2\ G\ M_u\ /\ c^2 = 2\ G\ (4/3\ \pi\ r_u^3\ \rho_u)\ /\ c^2$$

$$= 8\ \pi\ G\ \rho_u\ /\ 3\ c^2\ H_o^3\ = 1.74\ x\ 10^{56}\ \rho_u$$

So that, if $\rho_u = \rho_o$, then the universe would be neatly tucked into its own black hole. But, of course, $\rho_u = \rho_o$ is by no means confirmed, and in fact $\rho_u < \rho_o$ seems to be the case. Since according to the standard cosmological model the same amount of mass has existed in smaller and smaller confines in the past, this means that the entire universe would have to have emerged from the confines of a gigantic black hole in the very recent past by cosmological standards.

g. the missing matter

Estimates of the mass of stars and galaxies comprised of them have been obtained using methods described previously in chapter 16. By adding the masses of all clusters and individual galaxies in observed regions and dividing by the volume of space involved in the survey one obtains an estimate of ρ_u. As larger and larger regions of space are included in such surveys the mean density of baryonic mass has approached a figure more like $5\ x\ 10^{-31}$ than a value significantly greater than 10^{-30} gm cm^{-3} as Einstein would have preferred. Certainly there is some fairly large degree of uncertainty or 'wiggle room' in this value because it is based on a series of estimations that do not do too well on accounting for dispersed plasma, but the degree to which there is a shortfall is quite appreciable. This gives rise to many heated discussions of 'missing mass' that inevitably devolve into discussions of 'dark matter' and the even more mysterious 'vacuum energy'.

By any accounting the observations imply an 'actual' density of the universe that is a relatively small fraction of Einstein's 'critical density'. This in turn should imply that the universe will not collapse back onto itself according to those same theoretical considerations. There are discrepancies in behavior from what is predicted by standard models that have promoted the various 'dark matter' theories, of course, which some believe ups that percentage a little closer to 100 percent. At any rate Einstein's 'greatest error' continues its ill-begotten success, suggesting to those who should know better that a mysterious 'vacuum energy' might save the day.

If the critical mass density is not realized – as it evidently is not, other than in mysterious ways – we have succeeded in escaping from the biggest of all possible black holes, a supposedly impossible feat.

So what's to doubt?

h.   inherent problems in the theory

Let's just list a few of the objections to the certitude given this bit of cosmic mysticism that constitutes the theoretical underpinnings of the standard cosmological model. These objections are not necessarily listed in the order of the significance the author places on them:

1) There are the inconsistency problems in observed data – stars in our own galaxy that are older than, or nearly as old as, the supposed age of the universe, too early giant elliptical galaxies, and other too early structures throughout the universe, which are continuously being excused away as even earlier representatives are found.

2) Supposed requirements for 'dark matter' to support the virial theorem calculations with regard to galaxy and galactic cluster dynamics.

3) Evidence for acceleration (rather than deceleration) of expansion on which the whole calculation is based – revitalizing Einstein's 'mistake'.

4) A general willingness to entertain Einstein's admitted "greatest error" or any other alteration of time-honored principles, laws of physics, or universal constants just to make calculations work.

5) Theoretical inconsistency with black hole theory since the universe by these calculations has been a black hole for most of its existence but is now apparently emerging from that ultimate lethality, contradicting notions put forward by the same theorists in the context of black holes being inescapable.

6) The current understanding that gravitational forces are transmitted via gravitons in analogy to photons transmitting electromagnetic forces, suggests that these must also be limited to speed-of-light

travel and involve wavelengths and frequencies proportional to the momentum and energy transmitted. This would certainly be associated with redshifting in accordance with Hubble's hypothesis with an associated diminution of both with distance if the effect is geometry-dependent. So that to presume unabated inverse square gravitational forces or its equivalent in the general theory (like the otherwise inverse square luminosity relationship) at such extreme distances seems unwarranted in the case of gravitation also.

7) The critical density calculation is based on an arcane model of the universe as discussed above with regard to an inappropriate application of the divergence theorem to all space and Poisson's law to the universe as a whole.

The first six of these are more or less nitpicking. The seventh addresses underlying assumptions of the theory with respect to incorrectly applying gravitation to cosmology, and virtually all current theoretical thinking in cosmology. There is no reason to believe the underlying assumption should be considered valid for the universe as a whole. Presumed validity in this domain is based on precedence in other domains and the reputations of those who have previously made the assumption, perhaps most notably Einstein and Hawking.

Chapter 26

Cosmogony and Other Flights of Fancy

Just as astrology pretends to legitimacy via ties to astronomy, pretenses often come to play in otherwise legitimate discussions of cosmology that involve where the universe came from or – worse yet – *why?* Or *by whom?* Scientific study of the cosmos can easily gives rise to such categorical errors and unscientific, i. e., totally irrefutable and meaningless, blather concerning origins of the universe as a whole.

Whether the universe sits on the back of a giant tortoise that in turn sits on the back of another all the way down in infinite regress or if, at the bottom there is an elephant (read "big bang" if you like as shown in figure 233) can be fun stuff for students in bull sessions when they are drinking. But like discussions of whether the universe was created by Jaweh in the twinkling of an eye or existed forever a-causally, such word games don't tend to be very intellectually productive. Inebriates don't tend to comprehend that properties of a set just might not apply to the set to which the set in question belongs.

a. theories of everything and other flawed logic

Theories Of Everything (TOEs) and Grand Unified Theories (GUTs) are on the rise. Standard models that incorporate the big bang are such, but by today's standards, pretty minor as these things go. Once one defines how an egg can pop into existence from nothing, or alternatively how a chicken can fly in from nowhere, there is no more 'chicken and egg' problem, but science has not been served. If God said, "Let there be light and there was light" a mere six thousand years ago, or a major 'quantum fluctuation' occurred fifteen billion years ago from which all else derives, the case is pretty much closed. Isn't it?

Once one has a theory of everything, explanation tends to become a mere deduction from that underlying premise. Quite honestly, that is just how our minds prefer to operate. We each have a model of reality into which our observations seem to fit compatibly until, and unless, one day it all fails. At the point at which our conjectures fail, if we are scientists, we alter them rather than just shoring them up and providing excuses. We make new conjectures and test them to see whether they too can be refuted. That is how legitimate science is performed.

This author is convinced of the *advantages* of not attempting to explain *everything* just because we may happen to be able to explain a lot. This distinguishes his approach from elaborate constructions involving multiverses and employing string theory on the one hand and theories such as Witt's (2007) "null physics" that attempts to even explain *why* a universe exists on the other.

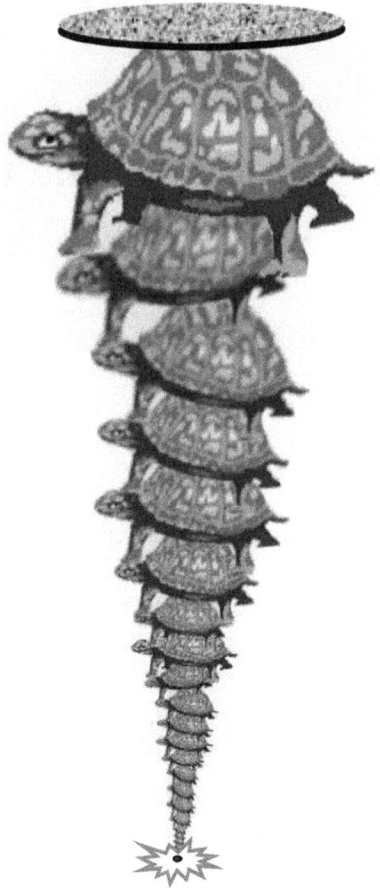

Figure 233: Theories of everything tend to divert attention away from science

The hallmark of current approaches to cosmology seems to be a blatant disregard for the hard-earned basics of science itself. These basics include a disenthralled perspective with respect for earlier discoveries of science – conservation laws of energy and momentum, well-known physical limitations including a light velocity proscription, and laboriously established universal constants. That Einstein's admitted "greatest mistake" is so readily accepted as a *feature* rather than a *flaw* in his theory is characteristic of conduct in this field.

504

All major theories of cosmology seem to involve apologies for the spontaneous emergence of the entire universe from nothing – well, a 'quantum fluctuation', if you will. The most illustrious competitor of the current 'standard' theory was the discredited 'Steady State' model of the universe in which the creation of matter from nothing was institutionalized as an on-going process.

"Nothing comes from nothing; nothing ever could" encapsulates a wealth of scientific maturity that seems to have been forgotten. Religious traditions that originated before natural laws were understood continue to play havoc with such healthy skepticism, preferring mythological origins that do violence to the laws of physics. And yet – to mimic notable critics of the early critics of relativity – many scientists, having learned such quasi science at an early age are not appalled by such illogical ideas. Much of current thinking in the realm of cosmogony maps quite directly to just such naive sentimental presumptions.

Current cosmology tends to hide behind the mathematics of elaborate theories that have not been subjected to rigorous testing for refutation. "Elegant" seems to have replaced "repeatedly observed" as an adjective to describe the degree to which conjectures should be seriously considered as explaining the nature of reality. Wherever observations are deemed impossible, theories rest as though confirmed. Alan Guth's flirtatious idea of 'inflation' (1981) was delivered with such eloquence that those who heard him lecture in 1983 apparently considered him "inspired". He audaciously suggested that the major problems in the standard cosmological model would be solved if the universe had inflated faster than the speed of light in the tiniest fraction of a second after its birth. Twenty-five years later no one has been able to justify Guth's fantasy, and yet it stands as though a fact and he is now advocating "eternal inflation". The logic behind his contribution seems to have been the following: Since *we know* the universe began with a big bang, there must have been an inflationary period of expansion, otherwise that hypothesis wouldn't work. Figure 215, page 436, shows 55 orders of magnitude of fudge factor to make it work. Riess et al. (1998, p. 169-170) state:

"Guth's original proposal for driving, and then ending, the inflation ran into various snags. Indeed, the mechanisms are still speculative, because they depend on physics at ultrahigh energies that is almost completely unknown...

"The idea of inflation is now more than 15 years old. There is still no consensus on the link between any specific unified theory and the mechanics of inflation."

That is science on its ear. What Guth (and Riess) might, more ingenuously, have said was that the *only* way the currently preferred theory can be salvaged is if the time-honored laws of physics are all wrong. Other scientists should have insisted on such ingenuousness. But instead, the quotation of Riess et al. goes on to state,

"...There is an iteration between well-defined (albeit speculative) theories and data that can constrain them: to that extent, inflationary theories are squarely within the frame of serious science."

Read that last sentence again if you will. As this author reads it, it says: "A 'well-formulated (albeit speculative)' theory such as the big bang can legitimately be supported by another unsupported theory that manipulates the data used by the former as its confirmation." In the parlance of everyday language that is tantamount to legitimizing the fabrication of truth.

For yet another perspective of the value of Guth's contribution, consider the following statement by Peebles (1993, pp. 392-393):

"...The great influence of inflation on the directions of research in theoretical cosmology has led people to term it the new paradigm for this subject. As usually understood in physical science, however, a paradigm is a pattern for research that one has reason to believe really is a useful approximation to the physical world, because the pattern has passed nontrivial experimental/observational tests. Since that has not yet happened in inflation, and there is not even a generally accepted and definite inflation model, we will continue to term inflation a scenario. It is notable, though, and perhaps significant, that in the decade since the concept was discovered and the homogeneity puzzle made very visible, nobody has proposed a reasonably definite alternative resolution to the puzzle. Unless and until that happens, or the concept somehow can be shown to be untenable, we must expect that inflation will continue to occupy a central place in the exploration of concepts in theoretical cosmology."

That this quote comes from a revered name in cosmological circles is truly disappointing. The logic baffles the mind like daunting caveats for promoted drugs in TV commercials. A scientist should not defer rejection of a conjecture that defies all known physics pending a better alternative. That rejects the only legitimate method of refuting conjectures. Science is not about the lesser of two evils.

b. mixing mythology and theology with science

There is also the issue of how quite naturally our rich heritage of mythology sometimes affects our modes of expression. A common symbolism provides a ready means of communication that is at once very powerful, but also very hazardous to clear thinking.

When Einstein stated, "God does not play dice," it was not presumed to have had theological significance, i. e., no one takes it literally but rather arrives at the intended notion of Einstein's understanding of how the natural world works. He was repelled by the very idea of inherent uncertainty with regard to where things are, as against the limitations of our abilities to determine where they might be. Similarly, when he states, "God would have done it that way," we understand his profound sense of logical necessity. Brent Tully's coining of the term "fingers of god", in reference to the spoke-like strings of galaxies that seem (as indeed they are) to be pointing directly at us from the depths of space, is no more or less than a picturesque reference to a scene like that on the ceiling of the Sistine chapel in which we seem to be blessed as though by God himself as at the center of everything. No doubt Tully saw the humor in it. One hopes he did.

But sometimes religious heritage becomes more intellectually obtrusive than such examples. In 2006 George Smoot was awarded the Nobel Prize in physics jointly with John Mather. The latter was heralded as "the true driving force" behind COBE – NASA's Cosmic Background Explorer satellite. His co-recipient George Smoot is the one, you may recall, who stated as a conclusion to his related work in mapping the tiny variations of the cosmic background radiation data that, "If you are a religious person it's like seeing the face of God."

The author remembers the comment as first appearing without the initial clause and was truly appalled. Besides the analogy seeming to imply that the nose on Smoot's God is no more than one part in 10^4 higher than his eye sockets or chin dimple, the statement is scientifically idiotic. Did he mean that "a religious person" could take his proclamation of the big bang confirmed as confirming also the veracity of mythological creation of the universe by the God of their choice? It smacks of religious zealotry no matter how one interprets it. At the very least it smacks of sanctimonious deference to demagoguery that attributes presumptively whatever is discovered as somehow having warranted unjustifiable religious faith. This author finds the use of scientific discoveries to justify religious claims of victory over heretical science repugnant.

Smoot's enthusiasm is somewhat understandable of course. Certainly it is not alien among investigators in any field. But it is hard to imagine what result would not have given him such glee if random variations of one part in 10^4 or 10^5 would suffice. Later in concert with D. Scott (2004) they make the appeal, "The most important outcome of the newer experimental results is that the standard cosmological paradigm is in good shape." Really? "Good shape"? Is that what science is about?

Other otherwise-respected researchers, most of whom would claim no religious affiliation whatsoever, exhibit that same ostensible theological preference in the analogies they use to describe their findings as justifying former beliefs while vociferously fighting the teaching of creationism in elementary schools.

c. exploitation of the 'anthropic principle'

"...I find myself even more puzzled than when I began as to why the Anthropic Principle has such strong support by so many otherwise good scientists. Having carefully considered the arguments, and engaged several proponents who I deeply respect in conversation and correspondence, I come to the conclusions I have explained here. The logic seems to me incontrovertible, and it leads to the conclusion that not only is the Anthropic Principle not science, its role may be negative. To the extent that the Anthropic Principle is espoused to justify continued interest in unfalsifiable theories, it may play a destructive role in the progress of science.

"If I am mistaken about any of this, I hope someone will set me straight."

Smolin (2007)

Clearly the standard model of cosmology requires some sort of perturbation to the time-honored 'cosmological principle' whereby the guiding principle of science since Copernicus has been to disenthrall ourselves of our peculiar situation in the universe. This principle has insisted that what we determine to be *true* (and *significant*) about the universe must be independent of particular perspectives whether spatial, temporal, or ideological. Conditional facts associated with what we believe or that we orbit a typical star in a typical galaxy at a particular time must be considered irrelevant. What *is* relevant to the nature of the universe itself involves only those invariances that pertain no matter what the platform from which observations are made. That we happen to be here rather than ten billion light years away in any direction – and whether our observations were made a few minutes ago or ten billion years in the past or future must be considered irrelevant.

However, according to the standard cosmological model, any observations made from anywhere in the universe ten billion years ago would have revealed a very different universe than the one we know today. Adherents still retain the sense that the universe would look essentially the same from anywhere at a particular time in the universe's history, but that *history*, and the sense in which the universe can even be supposed to *have* a historical scenario is what is at issue. Is the universe – as the very name implies – all there is, ever has been, and always will be? Or is there more? Before? After? In parallel? These are questions all men have asked about their personal existence from time immemorial with answers that have typically been framed as irrefutable hypotheses concerning realities that cannot be observed, and that therefore cannot be directly refuted. We typically view the hypotheses of others as, "Well, if you want to believe that, fine. But it makes no sense to me." So, putting aside the questionable value of such personal perspectives, rhetorical questions, and irrefutable responses generally given to such questions, let us consider whether any such endeavor to intellectually circumnavigate the universe as a whole makes any sense whatsoever.

Philosophies of physics have come and gone as rapidly as the ever-eclipsing theories of physics. Logical positivism, and the ideas of empiricism more generally, insisted that we accept as *real* only that which could be directly observed. Metaphysical intuitions of how the universe *is* ('in reality') beneath its mere *appearance* became anathema. However, current hyperbole concerning a 'multiverse' that provides a setting for the jewel we know as 'our universe' is gaining acceptance as a means of generalizing cosmology in ways that observation has denied us with regard to our modest universal environs. Perhaps unwittingly this emerging philosophy of idealism embraces the 'immaterialism' of Bishop Berkeley. The mere fact that there are *notions* of other universes seems to too many to endow those notions with all those "certain inalienable rights" formerly thought to pertain only to what was actually observed in scientific realms.

'Our *particular* universe' in this subjective scheme of things *is* the way it is in large part just because *we* are here to see it. We would not have evolved to be able to observe any other kind of universe, or so that story goes, so naturally our particular situation must be more significant than formerly envisioned. We and our universe have more or less brought each other into existence. In its various forms, this is essentially the 'anthropic principle'. It is in form very like that of professor Pangloss in Voltaire's *Candide* who maintained that we live in "the best of all possible worlds".

In Smolin's fine essay (2007) that discusses the problems with use of the 'anthropic principle' in scientific work, he begins as follows:

"I have chosen a deliberatively provocative title [Scientific alternatives to the anthropic principle] in order to communicate a sense of frustration I've felt for many years about how otherwise sensible people, some of whom are among the scientists I most respect and admire, espouse an approach to cosmological problems that is easily seen to be unscientific. I am referring of course to the anthropic principle. By calling it unscientific I mean something very specific, which is that it fails to have a necessary property to be considered a scientific hypothesis. This is that it be *falsifiable*. According to Popper, a theory is falsifiable if one can derive from it unambiguous predictions for doable experiments such that, were contrary results seen, at least one premise of the theory would have been proven not to apply to nature."

Further on in his introduction he proceeds, as follows:

"In recent discussions, the version of the anthropic principle that is usually put forward by its proponents as a scientific idea is based on two premises.

"A There exists (in the same sense that our chairs, tables and our universe exists) a very large ensemble of 'universes', μ which are completely or almost completely causally disjoint regions of spacetime, within which the parameters of the standard models of physics and cosmology differ. To the extent that they are causally disjoint, we have no ability to make observations in other universes than our own. The parameters of the standard models of particle physics and cosmology vary over the ensemble of universes.

"B The distribution of parameters in μ is random (in some measure) and the parameters that govern our universe are rare. This is the form of the Anthropic Principle most invoked in discussions related to inflationary cosmology and string theory, and it is the one I will critique here."

Neither proposition A nor B could ever be disproved by any means whatsoever. Many cosmological claims derived from these two unprovable premises and therefore those resulting claims are invalid to the extent that they do, in fact, rely on A or B for their veracity. Most such arguments that pretend to rely on them actually do not.

510

There are alternative formulations of A and B which Smolin also demolishes as unverifiable on the basis of irrefutability. Among them are arguments that state: "Our theory has many solutions, S_i. One of them, S_1 gives rise to a prediction X. If X is found that will confirm the combination of our theory and the particular solution S_1. But if X is not found belief in the theory is not diminished, for there are a large number of solutions that don't predict X."

He proceeds to point out how such arguments lead ultimately to a "situation in which the scientific community is indefinitely split into groups that disagree on the likelihood that the theory is true, with no possibility for resolution by rational argument." Does that not sound precisely like the situation with the standard cosmological model?

Smolin suggests that string theory may be at that point now. That is worth noting, but the current author believes that it is especially the case with proponents of the standard model of cosmology generally. As we have seen, one version or another of the model seems to be able to match almost any cosmological observation, although typically involving alternative versions enshrining different values of key parameters or the introduction of entirely new parameters to obtain a fit. That is a feature of unfalsifiable theories. The possibility of some complicated set of values matching observations is not what good science should be about. Smooth piecewise polynomial curves called splines can always be found to fit any number of observed data points, but that does not endow the resulting curves with epistemological significance with regard to explaining the associated data.

There is a version of the anthropic principle for which Smolin is more sympathetic and believes there to be some evidence:

"The anthropic observation: Our universe is much more complex than most universes with the same laws but different values of the parameters of those laws. In particular, it has a complex astrophysics, including galaxies and long lived stars, and a complex chemistry, including carbon chemistry. These necessary conditions for life are present in our universe as a consequence of the complexity which is made possible by the special values of the parameters."

He tenders that his acquiescence to this tenet is because he accepts what he calls "Cosmological natural selection". Conditions that accommodate the emergence of life he sees as contributing also to "reproduction of the universe itself." We'll ignore the words "natural selection" and "reproduction" for the moment. What leads him to his conclusion derived from what he sees as the significant achievements in the standard model of fundamental particles as well as the standard model of cosmology that were developed contemporaneously in the mid seventies of the last century. 'Standardization' in both cases involved a unifying principle that realized symmetries by describing all forces in terms of gauge fields. The second principle in that standardization provided a description of how the symmetries can be broken when the force fields are coupled to matter fields.

In the case of the standard model of cosmology, he states that this second principle "leads generally to the existence of a non-zero vacuum energy, which can both drive an early inflation of the universe and act today, accelerating the expansion." Plionis (2001) takes on the task of tying down some of these parameters for the "concordance" version of the standard model. The problem is, as he points out, that it "requires a large number of parameters to describe" and there is no defined method for determining their values. They must all be "hand tuned". He identifies the standard model of cosmology as having fifteen such parameters, bemoaning the fact that for these standard theories *no parameter has ever been explained by the theory.* So here we are, right back at square one, splining curves to fit observations and presuming to thereby understand what we observe.

But back to "anthropic observation" and why Smolin sees it as making some sort of sense. His argument is as follows:

"Were the neutron heavier by only one percent, the proton light by the same amount, the electron twice as massive, its electric charge twenty percent stronger, the neutrino as massive as the electron etc. there would be no stable nuclei at all. There would be no stars, no chemistry. The universe would be just hydrogen gas. The anthropic observation stated in the introduction is one way to state the complexity problem."

Okay, so this author sort of gets that as a way to express, "Gee, it's complicated!" But isn't it a bit sophomoric? This author remembers thoughts of, "What if like charges attracted each other rather than repelling each other?" In that case immediately after its having been forcibly created the universe would have split in opposite directions with no possibility of *anything* of meaning. Anything! "What if there were no Pauli exclusion principle?" "What if there were no attractive force of gravity?" What if?

512

Oh. Did a form of the word "creation" sneak in there? Oh well. As long as it's *create*-and-select rather than *design*-and-create science will buy it nowadays. Or maybe it would anyway if it were only with regard to worlds we cannot see.

Have we not been allowing ourselves to ponder totally unscientific teleological gibberish! Did fish evolve fins so they could swim better, or did the fins just evolve and thereafter got used to swim better and assist survival? What kind of thinkers are we? There is a kind of mindless wonder that is amazed at how photons 'know' which path to follow in order to always and without exception take the fastest path between the source of radiation and its observation no matter how contorted that path might be as noted by Fermat. But the 'law' with his name on it is merely shorthand nomenclature for a localized scattering process that ends up having that rather more sensational effect. Einstein is purported to have said that there are two kinds of people – those for whom *everything* is a miracle, and those for whom *nothing* is. Alas, this author finds himself among the latter.

Certainly this author has contemplated the nature of the universe 'in the round' and wondered why it is the way it is. But just as in Einstein's view of the two types of people, in Fourier analysis there are two ways to look at all things. There are two views of cosmological or any other phenomena. One is what the author considers to be a naïve perspective that demands to know the sequence of events starting with creation that leads up to the current moment. But there is another the author accepts as more meaningful that demands to know the logical interrelationships that perpetuate the situation.

Intelligence tests have traditionally had mathematical and spatial puzzles for which a subject was asked to find a missing, or the next, term in a series. Some of these problems form a group (in the mathematical sense) for which the expected term 'closes' the group. There is a unifying consistency with *all* the other terms that form a ring rather than merely filling a gap or 'missing link' in an on-going sequence. This kind of puzzle is particularly satisfying to solve. Those are the kinds that the physical universe gives us.

The author sees himself as obviously having evolved, but sees that as sort of an irrelevant consequence of this universe being the way it is, our planet having formed, etc.. That his distant and recent ancestors had survived on account of having a little better legs and brains than the next species or specimen, and could therefore outrun that bear or avoid situations altogether for which extreme speed or strength would have been required, he sees as a blessing indeed. Because now here he is – and yes, it is because the universe is the way it is and his ancestors were the way they were, and some of them were extremely lucky to boot. So what? What relevance does any of that have to a multiverse, the universe, or anybody, even himself? That is all merged into the rather mundane fact of his being here now. So here he sits at square one, not having to run like hell or wrack his brains just to survive. A reasonably long period of survival is pretty trivial for him really in this little sheltered nook where he sits.

So he has a brain that evolved based on entirely different selection criteria than those he chooses to employ. Furthermore, rather than studying biology or genetics, he wants to know how the world works at the highest (or is it *lowest*?) level and why it has the appearance that it does. Did it all happen so he could do this? Did the fins on fish evolve so they could swim, so they might become legs for later generations? It would be a serious misunderstanding to think so.

Meanwhile back at the ranch here on planet earth, Smolin analyzes Hoyle's successful but suspect argument from the anthropic principle that since life has evolved and requires carbon which could not have been created in a big bang, that it must have been created in stars. Hoyle's result was found as a fact of nature, but the argument is fallacious nonetheless. There is carbon; that is the essential fact here, the consequences of that fact have no bearing on how it might have gotten here or what might have transpired on account of that separate eventually deciphered fact. We too easily are confused about what is *implied* and what has been *inferred*. There is a tremendous difference.

In an analogous way cosmologists have argued that since the microwave background exists, it must have come from some process. And because life probably doesn't depend upon it, there was no tendency to get caught up in anthropic principles. However, in fact, big bang proponents associated that happening with the big bang rather than acknowledging the direct relationship between an observed energy of the precise amount in an observed source of energy. This is a different kind of typical *post hoc ergo propter hoc* fallacy. Presupposition inserts itself to justify designing causes.

Smolin points out that such fallacies are rampant with regard to selection criteria for universes. It was, of course, Einstein who opened Pandora's box with regard to the appropriateness of designer universes.

514

As suggested in the previous chapter, unbeknownst to him at that time, or apparently at any time thereafter, this was actually the greatest of his errors. So not only has his acknowledged "greatest error" been resurrected as a major cornerstone of cosmology, but his specifying of requirements for the creation of universes that induced him to make it in that same inadvisable paragraph has become the obsession of cosmologists of our time. Once one has accepted that we live in a "multiverse" of which we can know only parts of one but reason freely concerning the nature of all others, and the likelihood of our own as a member of that hypothetical set, we are ready for straight jackets.

The argument goes: Since there *is* a multiverse and we live in a universe that supports life, therefore, etc., etc.. If there is a universal consequence of life having evolved somewhere in this universe, it has nothing to do with teapots orbiting mars, Jesus of Nazareth having been resurrected on planet earth, or the vagaries of unknowable universes. Our brains have evolved further than this. In all such arguments, the prediction not having been found would not refute the false premise for anyone who argues in this way. That intelligent life exists, and that our universe exists are not subject to debate. These are facts that add no relevance to any argument other than to illegitimately load dire consequences on those who would deny them.

However, Smolin finds an argument from multiverses that he does find satisfying with regard to allowing falsification. It is the following: Let a property of a universe be posited that is logically independent of the existence of life in that universe. He qualifies this further stating that *in general* the probability of a universe having this property must be small. In other words it is physically and logically conceivable (even likely) that a universe might exist without it. Now according to some theory of universes that has been put forward for test, there is a strong correlation between the capability to support life and the particular property. If the property has not yet been observed in our universe, then he claims that it would be a "genuine prediction" of the theory of universes that has been put forward. Discovery of the predicted property that was vulnerable to falsification would be significant in proving that our universe is one of *them* – having supposedly thereby established the existence of 'them'. He concludes that in this case one "can now proceed to do real science with such a multiverse theory" and "draw a very important conclusion from this".

It is the supposed small probability and its previously not having been observed on the one hand, and its strong correlation on the other that seems to be so persuasive to Smolin. But why? Whatever it is that leads one to believe that universes might be one way or another derives from what we know of our own. Can one not ponder "What if?" questions without positing the existence of all the alternatives.

In Smolin's arguments he takes a particular theory of a universe such as ours having been born from a black hole "bouncing" into another spacetime from that in which it had become an unsightly blotch. Of course this is a theory on the heels of the previous theory that merely plunged the content of black holes into singularity. So hopefully some insightful person will follow that theory with a theory that, "Oh, yeah! Maybe black holes *can* 'bounce' violently – back into *this* universe. Maybe that is what gamma ray bursts are all about." But Smolin is looking at the other side of the doughnut hole, or holes, where universes are spawned with only vague similarities to their parent universes, not just replenished, but with variations possible on which something like natural selection of *universal* constants can operate as though they were genes. But what are they fighting for? Survival? Is our entire universe in some king of colossal struggle with parallel universes just to survive? Are universes greedy and selfish like Dawkins's genes? This seems like Wall Street anthropomorphism. So the analogies don't work for this author.

But any discussion of what comes out the other side of a black hole that "has no hair", i. e., is unknowable beneath its shroud, reminds this author of discussions of life after death. There seem always to be those skeptics – the author admits to playing this role on occasion – who ask, "Where is it? What is it like? Would I be like I am now?" There used to be the other side of the moon that *worked* for some zealots until photos showed that there is no paradise there... or hell either. But mostly they just express sadness for one's naivete in asking such mundane questions about paradise. "It will all be different there!"

Yeah... brand new parameter values for physical constants.

The credence that Smolin gives to probability arguments causes this author to think of tabulating the results of flipping a coin a hundred times. Whether the result is all heads, all tails, or any combination of heads and tails, would such a test have achieved anything amazing? Either a miracle would have happened if they're all heads, for which one would acknowledge that there had been only one chance in ten-to-the-thirtieth, which is greater than the number of centimeters in the girth of the universe according to the standard cosmological model. Or on the other hand the result is just a bunch of both heads and tails that one sees as of no real consequence. That there is an identical probability seems to have no relevance, life just goes on as usual. This author insists that its life as usual no matter which combination results; there is no hocus pocus by which we can lasso a universe and jerk it out of a herd by flipping coins or thinking about it.

Probability arguments after the fact are typically invalid – *everything* is unlikely to a degree of virtual impossibility. So what? Having observed a fact of unfathomable unlikelihood – perhaps the existence of the platypus – doesn't alter the membership status of our universe among multiverses no matter what your theory of universes happens to be. No argument or hypothesis concerning what might be observed in the universe can bring into existence a host of universes no matter how unlikely one assesses that fact to be. Again, from Smolin:

"Thus, so long as we prefer a science based on what can be rationally argued from shared evidence, there is an ethical imperative to examine only hypotheses that lead to falsifiable theories. If none are available, our job must be to invent some. So long as there are falsifiable – and not yet falsified – theories that account for the phenomena in question, the history of science teaches us to prefer them to their non-falsifiable rivals. The simple reason is that once a non-falsifiable theory is preferred to falsifiable alternatives, the process of science stops and further increases in knowledge are ruled out. There are many occasions in the history of science when this might have happened; we know more than people who espoused Ptolemaic astronomy, or Lysenko's biology, or Mach and others who dismissed atoms as forever unobservable, because at least some scientists preferred to go on examining falsifiable theories."

How could it be said better?

d. more mundane errors in reasoning

There are, of course, many less egregious flaws of reasoning to which we humans are all susceptible. Naturally astronomers and cosmologists are not excepted. The inimitable Richard Feynman is cited by Sarkar (1999) as having stated in 1962 that,

"When a physicist reads a paper by a typical astronomer, he finds an unfamiliar style in the treatment of uncertainties and errors... The authors are apparently unwilling to state precisely the odds that their number is correct although they have pointed out very carefully the many sources of error, and although it is quite clear that the error is a considerable fraction of the number. The evil is that often other cosmologists or astrophysicists take this number without regard to the possible error, treating it as an astronomical observation as accurate as the period of a planet." This is quoted from Feynman (1995).

Certainly his was a wry wit, but surely he would not have been joking in this regard.

In looking at the stated uncertainties in Lubin and Sandage's data presented on pages 253 and 254 with regard to surface brightness measures one finds uncertainties whose determinations involve a wide range of sources including instrumentation, mathematical model that compares distant galaxies to a fiducial set, and mode of presumed evolution. One must suppose that their determination of the uncertainty was rigorous. Yet they discard data as "unreliable" (and this author agrees with their decision) but the uncertainties provided for the data that is discarded are essentially the same as for the accepted data. It would seem that uncertainty values should be more useful than that.

Statistics, as applied to model validity, tend to affirm the degree to which one or another curve of a generalized functional form fits the data at hand – *not* whether the form itself is correct mind you, but whether, *if it is correct*, the chosen parameters provide the *best* fit. Refer for example to the diagram provided in figure 120 on page 265 where Peebles (1993) provides a best linear fit that obviously misses the more essential curvature of the data. Similarly in figure 162 on page 337 originally from Lubin and Bahcall (1993) the original authors plotted several second order curves along with the velocity scatter data, illustrating which of the curves provided the best fit. However, they did not show the much better first order fit because they had no basis to support that first order relationship. These are but two examples that the author sees as typical of presented data.

Kochanek (1996, p. 36) acknowledged that, "Astronomy is, of course, replete with examples where hidden systematic errors lead to incorrect conclusions even though all possible attention was given to statistical uncertainties."

These classes of errors are unavoidable. Naturally we can only compare and evaluate those options that have occurred to us.

Chapter Z+1

Conclusions

"The machines that are first invented to perform any particular movement are always the most complex, and succeeding artists generally discover that with fewer wheels, with fewer principles of motion than had originally been employed, the same effects may more easily be produced. The first philosophical systems, in the same manner, are always the most complex, and a particular connecting chain, or principle, is generally thought necessary to unite every two seemingly disjointed appearances; but it often happens that one great connecting principle is afterward found to be sufficient to bind together all the discordant phenomena that occur in a whole species of things." – Adam Smith, *Essay on the Principles which Lead and Direct Philosophical Enquiries, as Illustrated by the History of Astronomy*

The scattering model of cosmology that has been presented here attributes many of what have been considered 'cosmological' effects to the ramifications of the scattering of electromagnetic radiation by an intergalactic plasma medium. Thus, we attribute a host of observed cosmological effects that have baffled generations of cosmologists to the mode of observation through a diffuse scattering medium between and among galaxies using only previously acknowledged physical laws.

Of course this is a conjecture to be tested for refutation. That is how legitimate science works. It is necessary also that the applications of scattering theory be compatible with tested theories that explain what it cannot, and that it not ignore any data that might *ipso facto* refute the approach. The author is confident that this is indeed the case. Later efforts can amalgamate what we get right ignoring what is wrong.

519

a. brief summary of results of the scattering model

The author is proud to have presented for readers' consideration only hypotheses for which mechanisms are based squarely on existing physical laws that have been repeatedly demonstrated or at a minimum have been extrapolated to their most obvious conclusion. Instead of frivolous claims of what *might* have happened in the distant past, he has shown that the established mechanisms of forward scattering when applied to high temperature plasma such as that currently present in rich intracluster gases produces a redshifting of forward scattered radiations. The densities and temperatures of plasmas in intergalactic regions are completely compatible with this hypothesis. Pronounced spoke-like streaks associated with galaxy clusters and spherical wave-like density variations on redshift surveys collaborate the veracity of explanations of the plasma redshifting mechanism presented here.

These results depend exclusively on traditional physics. The theory is, therefore, conservative in its approach to accounting for cosmological redshift, redshift dependence of luminosity and surface brightness data, comoving number densities of galaxies, and other phenomena. It allows for the time required for the observed structures to have developed, and does not invoke unknown physical concepts such as inflation theories to fit vaguely anticipated deductions of a supposed early universe. Nor does this scattering model require that the universe be comprised of predominately mysterious forms of exotic matter or obscurely defined 'energy' to account for observations in our current universe. It does not bemoan 'missing matter'. It seems hardly coincidental that the characteristics of the intergalactic plasma should so precisely account for such extremely diverse observations. The standard cosmological models have required alternative inconsistent parameter values and hitherto unknown types of matter and physical concepts to account for them with varying success.

When the associated scattering phenomena are studied in detail, it becomes apparent that they account for a broadband absorption that combines with the implicit cosmological redshift to produce observed luminosity and surface brightness relationships with redshift. Other cosmographic relations including angular effects at great distance and comoving number densities by which models can be tested also support the same conclusion, with the agreement at least as good and typically better than for any of the versions of the standard cosmological model.

That is heartening, but any theory of the cosmos must account for the microwave background radiation and light element percentages, most notably the universal helium-to-hydrogen ratio. According to the author's thesis, these significant phenomena are associated with the thermodynamic balance of a stationary state universe. The associated explanation is integral to the scattering model. Energy released primarily as gamma radiation in producing 24% helium by mass from a hydrogenous plasma base is precisely the energy density that is found in the microwave background radiation that has been 'thermalized'. It goes without saying that the thermalization of electromagnetic radiation requires scattering of the radiation by matter.

Normally radiation temperature and kinetic temperature of the associated material particles in thermal equilibrium would be the same in any system that produces black body radiation. However, what has seemed 'normal' of the thermodynamics studied in laboratories has never involved a 'cavity' filled with a redshifting medium as is pertinent to cosmology. In the standard model it is assumed that radiation was continuously in equilibrium with originally-dense plasma that expanded adiabatically until, at a redshift of about 1,200 it is argued to have become so cool and diffuse as to no longer support scattering. That point in time at which it was no longer considered to have interacted with matter is termed the 'surface of last scattering'. These two phases had necessarily to be handled differently in the expansion model.

It is necessary to incorporate the fact that in addition to radiation being scattered from particles characterized by a given temperature, there will inevitably be redshifted photons in the mix that have been scattered off of more distant particles that would thereby seem to have been scattered by a considerably lower temperature 'surface'. This problem should definitely have been addressed by proponents of the standard model with regard to the first phase thermalization in that model. At a minimum it would have to be shown that the plasma was sufficiently dense in each era that this phenomenon did not apply. But it would have to have applied. Nor is this awkward problem adequately addressed for the transition between phases.

In the scattering model the unique nature of the equilibrium in a redshifting medium is of paramount significance. When one does solve that problem, one discovers that the temperature of the radiation and the average kinetic temperature of material particles from which scattering occurs may differ significantly one from the other. It is a demonstrated fact that the microwave background radiation is at 2.725 K; its energy density is 4.176 x 10^{-13} ergs per cubic centimeter. The density of baryonic matter for which a 24% conversion from hydrogenous plasma to helium would produce the very same energy density of radiation is 7.6 x 10^{-31} grams per cubic centimeter. This value is between the best current estimates of the universal baryon density. At that mass density, the equilibrium kinetic temperature that is determined using the redshifted blackbody analysis is between 10^3 and 10^4 K, the uncertainty depending upon the percentage of baryonic mass is vested in the intergalactic plasma. That number has been variously estimated as between about ten and eighty percent. The temperature range is a realistic assessment of temperatures that characterize the bulk of the material universe observed all around us in planets, dust, stars, galaxies, and the intergalactic medium. It avoids a naïve pretense that the bulk of the universe shares the temperature of the microwave background radiation.

These temperature and density values that characterize ongoing thermalization processes are too low for the redshifting mechanism that is hypothesized in this volume to account for the Hubble constant that contributes to thermalization. However, it is in the rich cores of galaxy clusters that the much higher *average value of the product of the temperature and density* is achieved. Here observed temperatures are as high as 10^9 K and electron densities are as high as 10^{-1} cm^{-3}, but the separation of cluster cores on any line of sight is upwards of a hundred Mpc. Thus, these intracluster plasma gases produce both the spoke-like features of cluster galaxies often called the "fingers of god" and the wave-like lumpiness seen in redshift surveys. Thus, the scattering model provides mutually consistent parameter values that collectively account both qualitatively and quantitatively for cosmological features.

Since the effects of scattering discussed in this paper assume an equilibrium situation of the universe as a whole, the theory does not accommodate evolutionary effects of the universe itself, although individual galaxy and galaxy cluster level developments naturally occur in any viable cosmological model. The standard models do not excel in accounting for apparent evolutionary effects, but such developments are at least compatible with the standard cosmological model. There are a number of observations that have suggested to many researchers that their explanations *require* evolutionary effects, and obviously these must be accommodated otherwise by the scattering model. Not least of such cited phenomena is microwave background radiation, comoving number densities of galaxies, the 'blueness', morphology, and sizes of some types of distant galaxies, light element abundance percentages, Lyman-α forest data, etc.. Therefore, it has been necessary for the author to provide at least cursory alternative explanations of all of these diverse phenomena in order for his approach to be taken seriously. Resolutions follow as natural concomitants of the scattering model.

Operating always in the spirit of Peebles' suggestion, quoted as an introduction to chapter 1 with regard to it being "sensible and prudent that people should continue to think about alternatives to the standard model, because evidence is not all that abundant", the author is convinced that his scattering model has excelled in this endeavor. Having addressed many, although by no means all, peripherally related phenomena, the author finds his results most gratifying. The microwave background radiation formerly considered to have been the exclusive claim of standard models has been shown to result from an indefinitely extended universe in the state in which we observe it today. There is actually much better agreement than previously accounted by standard models inasmuch as the resolution ties in other pertinent data considered awkward or irrelevant by the standard model. The explanation is actually much more complete as far as having relied on a more thorough analysis of blackbody equilibrium conditions applicable to a redshifting medium, which would actually refute claims made for the standard model scenario.

The implied values of temperature and density that account for redshift have been shown to be completely compatible with observed background radiation as well as with the observed variability in redshift surveys of galaxies. It accounts for the uniform X-ray background emanating from galaxy cluster cores. Light element thermonuclear production is in accord with the approach taken here with an attractive hypothesis based upon the range of parameter values observed in gamma ray bursts that suggests a recycling of baryonic matter from black holes that seem to 'bounce' back into *this* universe rather than into an alternative one as some proponents of multiverse metaphysics have intoned. Ultimately it must be observation that is the arbiter.

The scattering model provides more convincing explanations to a wide range of heretofore-unexplained phenomena. Naturally, it has no problems with 'a too early appearance of galaxies'; the distribution of galaxies fits a uniform pattern once standard model assumptions are backed out of the data. It eliminates quandaries of recent observations by the Hubble space telescope that indicate that the ages of certain stars within our own galaxy based on their metalicity may actually exceed the attributed age of the universe predicted by many standard model versions using parameter values required to match other observations. It resolves dilemmas that have given rise to mysterious 'dark matter' and 'vacuum energies' without necessitating these exigencies.

Importantly, the velocity dispersion of the galaxies, particularly in rich clusters, is readily accounted as resulting from combinations of the denser plasma medium producing a more rapid redshifting through clusters rather than by presuming mysterious forms of matter. This merely involves the same redshifting mechanism that is responsible for cosmological redshift, only more so in the higher dynamic pressure environment. Rotational anomalies of individual galaxies are explained now in similar terms of increased plasma densities of spherical 'halos' interior to and extending into remote regions of the spirals arms and beyond. These halos seem to extend beyond individual galaxies, merging into the extremely hot intracluster gases that have thereby suggested to some researchers that the inferred 'dark matter' is virtually all associated with such extended halos (Bahcall, 1999). The observed temperatures and densities of this plasma gas produce a redshift rate well in excess of H_o across the extent of these cluster cores that mimics what has been attributed to a virial 'velocity scatter'.

b. extending analytical results

Analytical results reported for what the author believes to be the first time in this volume have been produced by the use of a few new analytical methods that the author also believes never to have been applied before. He is convinced that it is important that these same approaches be applied to analyses of any and all cosmological models to determine legitimacy of their associated predictions. Previous research has been remiss in assuming there would be no associated effect in these areas without having performed appropriate analytical computations in situations where the impact has been shown to be major for *any* cosmological model.

Basically there are three such analytical areas that have been used in support of the current investigation:

1. **Absorption effects applicable to the dispersion associated with the propagation of electromagnetic radiation through a plasma.**

Absorption is the immediate consequence of transmission of electromagnetic radiation through any scattering medium. In both the standard model and the scattering model the Lyman-α forests provide examples for which such analyses are required, and have been performed, to determine the implicit absorption effects of the column density of neutral hydrogen between astronomical objects and their observation. However, there is also a significant plasma component present in intergalactic regions through which we observe the distant cosmos and the absorption effects of a plasma differ considerably from those of neutral substances.

These effects must necessarily be (and have been) analyzed for the scattering model. Significantly, it is even more essential to the standard cosmological model because in that conjecture there is a much hotter and more dense plasma hypothesized as present over much of the time period since a big bang, as well as a unique transition hypothesis for which analyses must also be performed.

Whether this absorption is characterized as broadband as is the case for the scattering model, or absorption with unique wavelength functionality depends upon the absorption coefficient determined to apply to the medium at each point in time as well as the redshift functionality appropriate to the model. But in any case, the analysis must be performed to determine the amount of luminosity loss that must be attributed to associated absorption processes. It cannot legitimately just be ignored as has been done.

In the standard models there is a luminosity diminution factor attributed to time dilation whose functionality is precisely that of a broadband absorption like that which has been determined to apply to the scattering model. But if time dilation is claimed as the cause of this reduction in luminosity in the standard model, an associated assertion must be made, and verified, that at no phase of the scenario would there also be *any* observable plasma absorption. That in the propagation of electromagnetic radiation though billions of light years of intergalactic plasma none would be absorbed seems highly unlikely. In any case, that is another conjecture of the standard model that needs to be explicitly addressed rather than ignored.

2. Convergent diffraction effect associated with forward scattering processes through a plasma medium.

Forward scattering is involved in the imaging of objects viewed through any intermediate medium. In our atmosphere at sea level photons are replaced by virtually identical forward-scattered photons at sub-centimeter intervals on every light transmission. Other than quite unrelated noncoherent scattering and absorption processes, this does not affect our ability to see 'objects' per se out to distances even somewhat beyond the optical depth of the medium that is also determined by scattering processes. This phenomenon of forward scattering is certainly pertinent to all cosmological observations and their interpretations.

Of course forward scattering has typically involved nothing other than mundane optical physics to little spectacular effect. However, the intergalactic medium is hardly typical of media that have been studied in the laboratory. In addition, those who have been the significant contributors to forward scattering theory have specifically excluded media for which relativistic velocities of charged particles are involved. Thus, the well-known wavelength invariance that has been claimed to apply to this process is not directly applicable to high temperature plasma media without a determination of the magnitude of associated effects. With regard to the intergalactic medium, that investigation is essential.

The author has performed this determination for his scattering model and found that the relativistic aberration and transverse Doppler effects collaborate to produce an effective diffraction angle for which conservation of energy and momentum imply a lengthening of wavelength at each 'extinction' in this forward scattering process. Significantly, the analyses also show that the relativistic aberration angle of the diffraction does not preclude forward scattering. In effect, its only impact is to lengthen wavelength, which, when combined with a wavelength dependence of extinction intervals, produces Doppler-like redshifts.

This physical phenomenon is not unique to the author's scattering model; it is a physical effect associated with forward scattering in a plasma generally. Therefore, this effect must apply in the context of the standard model as well. At a minimum this must impact the 'dark matter' controversy applicable to domains for which plasma densities and temperatures are appreciable. The effects of forward scattering through any such medium must be addressed.

3. Implications of redshift on the thermalization processes that occur throughout a scattering medium.

It seems singularly amazing to the author that no one seems to have ever even considered the most immediate impact of redshift on the equilibrium conditions essential to blackbody radiation. But that does indeed seem to be the case.

Of course standard model advocates have addressed the redshift impact simplistically as appropriate to the two separate phases that characterize that model. The first involves plasma scattering for which the unstated assumption is that there is sufficient density that no redshift occurs between scattering events. The second applies Wien's law to a receding cavity wall associated with 'decoupling'. That is essentially the depth of the redshift-related analyses that have been performed to 'predict' the eventual state of background radiation. Of course that is not *pre*-diction, but *post*-diction. The associated redshift and temperature of that 'wall' of last scattering the two phases has been retro-fitted to the observation. There is a stark contrast between that gross over simplification and the complexity of the actual problem.

In the scattering model for which the universe is assumed to be in an essentially stationary state, there is definitely redshifting that takes place between diffuse plasma scattering events in particular, but scattering for thermalization involves all the absorption and re-emissions and non-coherent scattering as well, so that essentially all the objects we see are included in thermalization. When one takes these facts into account, the kinetic temperature of the matter with which the radiation is thermalized by scattering is no longer constrained to the value of associated blackbody radiation as in the adiabatic expansion phase assumed by the standard cosmological model. Nor is that constraint valid once the impact of intermediate redshifting is included in the scope of the analyses.

So it is manifestly clear that a cosmological model that does not take these factors into account as the standard model does not, has not adequately addressed the implications of the propounded model.

c. deeper questions

Of course, one must address cosmogony and even eschatology – if one might be so bold as to apply the theological term that many have unwittingly assumed without specific allusion to have legitimacy in an otherwise entirely scientific debate – concerning asymptotic trends. Topics that frequently arise in such discussions involve the beginning and end of the universe as we know it. We have purposely given short shrift to such 'talking points' in the previous chapter as unworthy of detailed scientific discussion. Many of those aspects that have seemed to place our entire universe into a realm for which it might be appropriate to ask such questions have effectively been removed from consideration by the scattering model to which they do not apply.

The primordial origin of the elements has been addressed as an on-going equilibrium situation, whether continuously in cluster plasma or in lesser 'bangs' involving gamma ray bursts rather than with an initial 'big' bang. Similarly the microwave background that is so directly associated with the origin of the light elements has been removed from requiring a primitive one-shot origination. However, there is the issue of the gravitational effects in over densities that create galaxies and ultimately huge 'super' cluster structures throughout the universe. This seems on the face of it to be the kind of one-way development that in itself might totally deny perpetuity. The death of stars and entire galaxies that plummet ultimately beneath Schwarzchild radii, if that is the process, would seem also to violate the principle assumed by the author's assumption of an overall enduring equilibrium.

The author does not claim to know the answers to all, or even many, such questions. However, he is convinced of current ignorance of possible stages in the collapse and violent eruption of matter in the realm of extremely high-energy physics, an arena in which Alan Guth has slain dragons to much applause. The breadth of our ignorance in this area certainly extends beyond the gradual disappearance of matter beneath Schwarzschild radii into black holes of which we know primarily that few differences can ever be discriminated. Perhaps the exciting new ion collider technology will enlighten us.

There is certainly insufficient reason to believe that the structure of neutrons envisioned as collapsing into black holes would continue to collapse to singularity. That neutrons, whose *comfort zones* are protected by forces so many times more powerful than an inverse square law, should succumb to the lesser force and do a swan dive into the oblivion of a spacetime singularity is not a foregone conclusion to this author's way of thinking. Barcelo et al. (2009) share this opinion. In their recent article in *Scientific American* they state:

"In particular, the old calculations assume that collapse proceeds very rapidly, taking about the same time as would be needed for material at the star's surface to free-fall to the star's center. We found that for a slower collapse, quantum effects may produce a new kind of very compact object that does not have an event horizon and is thus much less problematic." p. 44

This proceeds along the line of reasoning described in Appendix C. The demise of these compact objects – much more massive than white dwarf stars that erupt as supernovae – would doubtless result in more spectacular displays. The author considers there to be a real possibility that at some level of massiveness, explosive new forces would recycle primordial baryonic matter back into the visible universe with a fresh influx of neutrons from which nucleosynthesis proceeds as suggested in Appendix C. Thus, throughout our universe, we might expect at regular, even if infrequent, intervals to be witness to such gargantuan explosions.

Increasingly at extreme distances in space we witness the tremendous release of energy associated with gamma ray bursts. It is true that some in the community tend to consider the observed intensity of the released energy to correspond to a narrow streaming so as to minimize the overwhelming amount of energy implied otherwise. Other than reducing the scope to comprehensible magnitudes, while commensurably increasing their frequency of occurrence, there seems to be little to recommend that conjecture as fact. The amount of energy released if these were truly full blown explosions rather than focused emissions that happen to be directed right at us like 'fingers of god', is completely compatible with the reverse of black hole creation processes that would spew forth the inchoate substance of the universe. From such occurrences on-going maintenance of the hydrogen-helium ratio would continue, and importantly, a major redistribution of baryonic matter back out into the intergalactic medium would be the result.

So it seems altogether likely to this author that one day it will be generally acknowledged that these eruptions we are witness to from the extreme depths of space represent the final stage of the black hole process in which they 'bounce back' into *this* universe. The outpouring of neutrons from such cataclysmic events would then decay into, and react with, surrounding hydrogenous plasma, in essence refreshing the intergalactic medium in a continuing process of nucleosynthesis of helium from hydrogen. Black holes may reach some super-supernova level of explosive power that we can not as yet comprehend. These eruptions might well be heralded by a slow and gradual quantum cloud-like expansion at first out past a Schwarzschild radius barrier, if there is one, as the author has suggested as a possibility in Appendix C.

You may wish to contrast and compare the plausibility of this epiphany with hypotheses of creation ex nihilo followed by a necessary inflation that has been glibly incorporated in all versions of the standard cosmological model. In these models also, whether acknowledged or not, the universe itself is envisioned as having emerged from a single gigantic black hole, while the same proponents of this scenario deny that possibility to the more usual magnitude of black hole.

Having cleared Schwarzschild radius limitations allowed by quantum fermion distribution probabilities of the neutrons themselves, material could be thrust back into the universe in highly energetic states that would redistribute matter – perhaps creating and/or filling voids like those that have been observed. But that is just speculation that the author has not wanted to associate directly with his more scientific investigations.   Nothing in the author's scattering model actually depends on that possibility other than his and his readers' natural expectation of a global stability in the universe.

This is very unlike Guth's inflation (1981) idea that proposed that major problems with origination in a big bang could be solved by violating physical laws.   Whatever problems there are with Guth's conjecture, it is at the very least a completely necessary adjunct to a model that would be invalidated without it.   Certainly any scientific hypothesis should properly be evaluated within "the scope of the scientific domain of its origin" but mystical violations to known physics should never be tolerated as legitimate science.   Science must once again proceed systematically by observation to determine that which may one day be deduced rather than the other way around.

d. looking back at strongly held misinterpretations

There is also the issue of the 'elegance' of Einstein's general relativity (his "greatest error" notwithstanding) and the forces that would seem to require expansion or eventual collapse of the entire universe. Valid scientific theories, although sometimes initially understood based on metaphysical reasoning, must be placed on a more solid footing as the relationship between measurable effects before they are legitimately accepted as descriptive of reality.

The author supposes with ready evidence that Hubble and many other able scientists were unwillingly misled by observed effects of the larger universe and succumbed to extravagant conjectures reluctantly. See, for example, Assis et al. (2008) who document Hubble's life long reluctance to accept his own hypothesis.

So also must the universe be fooled! The 'effects' of presumed expansion do indeed appear as though they were caused by Doppler redshifts. These effects, to the extent to that they are emulated in precise detail, must affect gravitational as well as electromagnetic systems. Gravitational energies, if their propagation is limited by light speed transmission, must also be redshifted, and all associated effects substantially reduced. De Sitter's solution to Einstein's equations for an empty universe whose distance-redshift relation matches functionality of the scattering model in many ways may actually be just another way of looking at the very same situation. In effect many of the legitimate gravitational arguments that have been put forward may well, therefore, apply even according to the scattering model. So that all these issues that involve gravitation, which might seem to be major distinctions between models, are very possibly moot.

One would do well to consider the relative uncertainties of the data that has resulted in various current theories of cosmology. The scientific revolutionaries Arp, Burbidge, and Hoyle (1990) put a special perspective on this matter in their statement:

"Cosmology is unique in science in that it is a very large intellectual edifice based on very few facts. The strong tendency is to replace a need for more facts by conformity, which is accorded the dubious role of supplying the element of certainty in people's minds that properly should only belong to science with far more extensive observational support."

And again, Oldershaw states (1990):

"A curious dynamic tension has been arising in the field of cosmology. Some widely held theoretical assumptions are coming into increasing conflict with observational results, and yet those assumptions continue to receive strong support."

He provides notorious examples that are still outstanding. Acceptance of the big bang is certainly a reasonable disposition in the opinion of virtually every cosmologist with any considerable credentials, but at the same time it is well known that none of the particular models that are propounded under that aegis have been 'confirmed'. That includes the 'concordance' model. The data conflicts with every respected model with regard to one observation or another. So it is hardly scientifically embarrassing to back an alternative in this arena.

It is admittedly an awkward time for cosmology. Thomas Kuhn, who has chronicled the changing of the guard with regard to the overthrow of major theories in science, states (1962):

"In fact, however, step by step their deep divergences and incoherencies emerge increasingly within the scientific community, but people do not see them until finally the confusion becomes so great that the situation breaks down."

It is inevitable that physical theories should be continually replaced, but a completely smooth evolution does not occur. This is partly because of the "incommensurabilities" that Khun identified as associated with alternative theoretical paradigms. This is as inevitable as change itself, of course. These *incommensurabilities* are never more apparent than when one must *back out* the data from theory-ridden constructs to support an alternative, as we have had to do in some cases.

Currently cosmological data is no more gathered in a spirit of falsifying standard cosmological models than are concerts held to humiliate composers. There is a spirit of reconfirmation of time-honored beliefs in virtually every peer-reviewed paper on any such topic published nowadays. In fact papers on topics that are far afield from cosmology tap into the vogue of acceptability of a big bang.

The spirit of independent hard-nosed critical analysis that is so befitting the sciences has suffered severe setbacks in the last fifty years. The open-minded search for knowledge has been an inspiration to the young ever since the ever-so-dark ages tended ineluctably toward enlightenment as explained so masterfully by Gribbon (2002) in describing what being a scientist has been about. But that is hardly what is happening now in cosmology. What does the 29th co-author of a ten page technical article, who was charged by someone of authority over him or her to tabulate reams of data in support of some minor assumption made with regard to some minor aspect of the 'standard' propounded by someone else of authority, really know about what being a *scientist* is all about?

As a final note it is worth acknowledging that the big bang has not been a popular concept even with many of its advocates who have merely been convinced by its ability to account in one way or another for diverse observations but have found it intellectually stifling nonetheless. In this regard the standard cosmological model has broken with a long-standing scientific tradition. Scientists have always tried to disenthrall themselves of their peculiar time and place in the scheme of things, assuming that their perspectives were not unique – that their work could have been done equally well by anyone at any place or time in the universe. Replacing that healthy-minded perspective by dillusionary attempts to situate the entire universe in that diminutive role in a sophomoric conjecture concerning 'multiverses' really doesn't have the same intellectual impact, now does it? There is no possible observation of the 'scheme of things' outside of the universe that could excuse that as being a *scientific* activity.

This *perfect cosmological principle* that has guided science for centuries was lost with acceptance of a *big bang*. Four dimensions that some pretend to visualize, didn't quite salvage the legitimacy of the argument for expansion in all directions away from us. Nor have the concepts of space and time merged in the way predicted by Minkowski a century ago; we let go of either concept at our own peril.

What about time? Does it really have a "brief history" with an actual "beginning" as Hawking (1988) thought to be a possibility?

Before acceptance of universal expansion and the inevitability of a big bang, the cosmological principle was touted as a philosophical position that was virtually synonymous with science itself. It is a major intellectual property loss that that is no longer so. Do the merits of what we have been left with warrant having abandoned it? The author thinks not.

That major *loss* is indeed a major *difference* between the standard cosmological model and the scattering model hypothesis this author has presented. The approach that has been described here brings us back to a more restrained and disenthralled approach to determining the nature of reality. It eschews flights of fantasy and deductions from elaborately conceived, elegantly expressed, but essentially irrefutable (and therefore scientifically meaningless) arguments based solely on mathematics with no experimental component. Naively embraced artifacts are religiously supported in many cases by zealots whose hopes remain high that outlandish claims will ultimately be confirmed as justifying an unjustifiable faith. They are *not* confirmed, nor can they be, and the fact that they remain unrefuted is only because they *cannot be* refuted; they are irrefutable by their very nature. They are unscientific by their very nature. 'Confirmation' is not what science is about. One is 'confirmed' in one's religion; science is about refutation.

We can do better than the 'standard' cosmological model.

Appendix A

Electromagnetic Theory of Radiation

In this presentation of Maxwell's equations and their solutions, which pertain to electromagnetic radiation, we address issues that are of particular interest in with regard to relativistic electrons encountered in a hot plasma such as the intergalactic medium. The discussion accommodates theoretical considerations of absorption theory which would have to be extended somewhat to address relativistic interactions involving paired emission and absorption of electromagnetic radiation. This natural pairing of specific emission and absorption events relates closely to work by Lewis (1926), Wheeler and Feynman (1945), Cramer (1980 and 1986), and others who have shown that propagation of light may require an explicit pre-association of such emission and absorption events between material agents. Whereas observational aspects associated with this description are identical to those of more usual presentations of electromagnetic theory in the cases involving relatively stationary emitters and absorbers, observational aspects of relativistic aberration and Doppler effects associated with relatively moving observers are characterized quite naturally in this approach as well.

Terminology
This appendix will shy away from much in the way of difficult mathematical prerequisites. However, equations will be presented wherever appropriate because there is much that can be inferred from an understanding of the symmetries of the equations and descriptions of the implied operations even by someone for whom the

equations themselves may seem obtuse. Descriptions will be explicit – graphic where possible – but attempts have been made to avoid the more difficult aspects of the associated mathematics. Some minimal understanding of vector products and *divergence* and *curl* differential operations on a vector is essential to an understanding of the vector approach to electromagnetic field theory, of course. These definitions in Cartesian coordinates are as follows:

Inner or *dot* product: $U \bullet V \equiv U_x V_x + U_y V_y + U_z V_z$

Outer (*cross*) product: $U \times V \equiv i\,(U_y V_z - U_z V_y) + j\,(U_z V_x - U_x V_z) + k\,(U_x V_y - U_y V_x)$

Gradient: $\nabla \alpha \equiv i\,\partial \alpha /\partial x + j\,\partial \alpha /\partial y + k\,\partial \alpha /\partial z$

Divergence: $\nabla \bullet U \equiv \partial U_x /\partial x + \partial U_y /\partial y + \partial U_z /\partial z$

Curl: $\nabla \times U \equiv i(\partial U_z /\partial y - \partial U_y /\partial z) + j(\partial U_x /\partial z - \partial U_z /\partial x) + k(\partial U_y /\partial x - \partial U_x /\partial y)$

In the above definitions, U and V are vector fields; α, U_i's, and V_i's are scalars. The scalar U_x is the component of the vector U along the x axis. The *right-hand rule* (see figure A.1) states that if you use the fingers on your right hand to indicate the direction of rotation of U into V, then the extended thumb will be in the direction of the vector *cross* product. In these definitions, U is a vector function of x, y, z, t. The *basis* vectors *i, j, k* are unit vectors in the directions of the x, y, z axes, respectively. The vector $\partial a(s)/\partial s$ is the partial derivative (the "slope" or rate of change) of the function $a(s)$ with respect to the independent variable s. Scalars $\partial U_i /\partial s$ are the partial derivatives of scalar components of the vector U with respect to the independent variable s. The electric field E is, for example, the gradient of a scalar potential field. Note: Determining the *divergence* and *curl* of a vector is sufficient to determine the vector itself to within a vector constant throughout the region for which the relations apply.

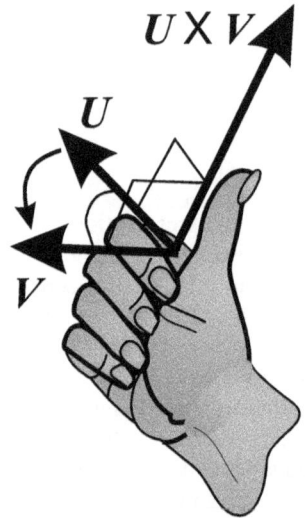

Figure A1: The right-hand vector cross product rule

Electromagnetic theory

In the interaction theory of radiation being discussed here, all of the overwhelming evidence of experimental confirmation of the theoretical origins of electromagnetic theory remain unchallenged and have intentionally not been altered. Maxwell's differential equations consolidate these results and are, therefore, accepted without change. They are:

536

Cosmological Effects of Scattering in the Intergalactic Medium

1)   Coulomb's law: *macroscopic* field – inhomogeneous equation

$$\nabla \bullet D = \rho$$

2)   Absence of monopoles: *microscopic* field – homogeneous equation

$$\nabla \bullet B = 0$$

3)   Faraday's law: *microscopic* fields – homogeneous equation

$$\nabla \times E = - \partial B/\partial t$$

4)   Amphere/Maxwell's law: *macroscopic* fields – inhomogeneous equation

$$\nabla \times H = J + \partial D/\partial t$$

In these equations in *rationalized mks* units experimentally known vector functions D, H, E, B, and J are related one to another.[5] The scalar function ρ is the *charge density* throughout the region for which the equations pertain. The vector quantity J is a characterization of the amount and direction of *conduction current* throughout the region. Boundary conditions of the region to which the equations are to pertain may further constrain the relationships among the various vector field quantities. The relationships define a nearly symmetric cycle; if ρ and J vanish throughout the region, all the equations become homogeneous differential equations of identical form and the symmetry is obviously complete. Since we will be dealing with the propagation of light in a vacuum between encounters, this symmetry will be assumed throughout the remainder of this article. Of the four remaining vector field quantities, two involve fields associated with electrical effects and two involve fields associated with magnetic effects. Two *constitutive* relation equations define and relate dual *microscopic* and *macroscopic* electric and magnetic fields as follows:

5)   electrical: $D = \varepsilon E$ *macroscopic* field relation to *microscopic* field

6)   magnetic: $H = \mu^{-1}B$ *macroscopic* field relation to *microscopic* field

where ε is the *permittivity* and μ the *permeability* of the medium. Both these quantities are typically scalars, but in certain media there are anisotropic distortion

---

[5] The quantum theory of light does not substantially alter the results of Maxwell's approach that was historically significant to the development of relativity and so we will go with that more intuitive approach. This is in accordance with decisions by Wheeler and Feynman, and Cramer cited above in their similarly motivated analyses. The fashionable geometrical approach using generic differentiation of an electromagnetic field strength tensor to represent these equations, while economical in terminology, de-emphasizes the integrated nature of emission and absorption processes envisioned here, since typically the tensor has been deployed with exclusively microscopic fields.

effects that can be characterized by a tensor representation of these quantities. These two equations reflect the fact that only one of the quantities (called the *microscopic* field – on the right) in each field category will be associated directly with emission; it is independent of the structural characteristics of interacting media throughout the region of consideration. The other two are *induced* in part by the *microscopic* fields and are called the *macroscopic* fields; these terms have more to do with *externality* of origination than with the *size* in electromagnetic theory. In a vacuum, the scalar constitutive coefficients are typically identified as μ_0 and ε_0, whose values depend upon the system of units chosen. The speed of propagation of a wave function that satisfies Maxwell's equations will be seen to be determined by these quantities and in particular for propagation in a vacuum, that instantaneous speed will be:

7)   speed of light in vacuum: $c \equiv (\mu_0 \, \varepsilon_0)^{-\frac{1}{2}}$

In addition to Maxwell's equations, one must acknowledge the role of the *Lorentz force* on isolated charges as of extreme relevance to electrodynamics where there is relative motion of the charge in *microscopic* electromagnetic fields. It is given by:

8)   Lorentz force: $L = q \, (\, E + v \times B \,)$ *microscopic* fields

where q is the scalar quantity of a specific charge that is in motion and v is the vector velocity of the charge relative to a test charge of unit magnitude experiencing the force. Thus the instantaneous electromotive force on a unit charge depends on magnetic as well as the usual electric forces in that case.

Deriving and solving radiation wave equations
Derivation of the wave equations from Maxwell's equations is problematical in several regards. Although there are two *microscopic* (2 and 3) and two *macroscopic* (1 and 4) equations, substitutions using *constitutive* relations (5 and 6) must be used to obtain the wave equations. The implications of the original four field equations, which seem clear, can easily be lost in the process of solution. For example, by these substitutions, solutions can be obtained for the *microscopic* fields E and B with the resulting equations looking *as though* they should be interpreted as the fluctuating electric and magnetic fields of an emitter *independent* of the medium or the ultimate absorber of the radiation. Here only the speed of propagation appears to be affected by the medium:

9) $\nabla^2 E = - \mu\varepsilon \, \partial^2 E / \partial t^2$

10) $\nabla^2 B = - \mu\varepsilon \, \partial^2 B / \partial t^2$

The definition of $\nabla^2 U$ can be elaborated from the definitions above for the dot product of a gradient operator: $\nabla^2 U \equiv \nabla \bullet \nabla U$. The wave equations themselves derive from the vector identity $\nabla \times (\nabla \times U) \equiv \nabla(\nabla \bullet U) - \nabla^2 U$ and by substitutions from constitutive relations into Maxwell's equations. The wave equations 9) and 10) each derive directly from Maxwell's equations 3) and 4) in addition to either 1) or 2) with

538

constitutive relation substitutions occurring twice in the process. So these are hardly isolated conditions applicable solely to an emitter.

These equations describe propagational wave phenomena. In general solutions will be complex quantities, only the real parts of which are of any interest experimentally. Solutions shown in figure A2 are of the form:

11) $\qquad E = E_0 \, e^{\pm i (\kappa \cdot \mathbf{r} - i \, \omega t)}$

12) $\qquad B = B_0 \, e^{\pm i (\kappa \cdot \mathbf{r} - i \, \omega t)}$

E_0 and B_0 are constant vectors for plane polarized waves. Substitution back into Maxwell's *divergence* equations results in further constraints on E and B such that both must be perpendicular to the direction of propagation given by the *wave vector* κ, whose magnitude is given by $\kappa = (\mu \, \varepsilon)^{\frac{1}{2}} \omega$, where ω is the *angular frequency* of the radiation. This constraint is the basis of the notable *transverse* wave nature of light. Substituting into Maxwell's *curl* equations places additional constraints on E and B such that they must always be in phase and of equal in magnitude in addition to being at right angles to each other. By superposition of linearly independent solutions with uniquely paired E_0 and B_0 values, one obtains the more general elliptical polarization solutions – plane and circular polarization being the special cases shown in figure A2.

Are there preferred solutions to Maxwell's equations?

It is apparent that Maxwell's equations may be used to determine valid solutions for all four of the fields. But which wave equations (if any) *inherently couple* as a single transverse wave? In other words, do E and B, E and H, D and H, or D and B constitute the most meaningful description of the radiation we associate with these equations? With such a plethora of possibilities, which (if any) of these solutions should be preferred?

Radiation energy density and energy flow (as electromagnetic *momentum*) equations both involve equally coupled *microscopic* and *macroscopic* fields for each as follows:

13) $\qquad$ energy density: $\qquad u = \frac{1}{2} (\, E \bullet D + B \bullet H \,)$

14) $\qquad$ energy flow: $\qquad P = E \times H$

More than any other single equation, the latter Poynting vector equation symbolizes the *transverse* nature of electromagnetic radiation (refer to the *right hand rule* above for an intuitive feel for this quantity) that distinguishes it from *longitudinal* vibrations characteristic of sound propagation. Furthermore, *this* equation clearly indicates equal participation by *macroscopic* fields associated within the medium and/or absorption. With only an emitting and an absorbing atom under consideration, E would clearly be associated with the emitter, H with the absorber. Thus, energy and momentum considerations would seem to suggest that E and H occupy preeminent positions, as the fields most naturally characterizing radiative energy transfer.

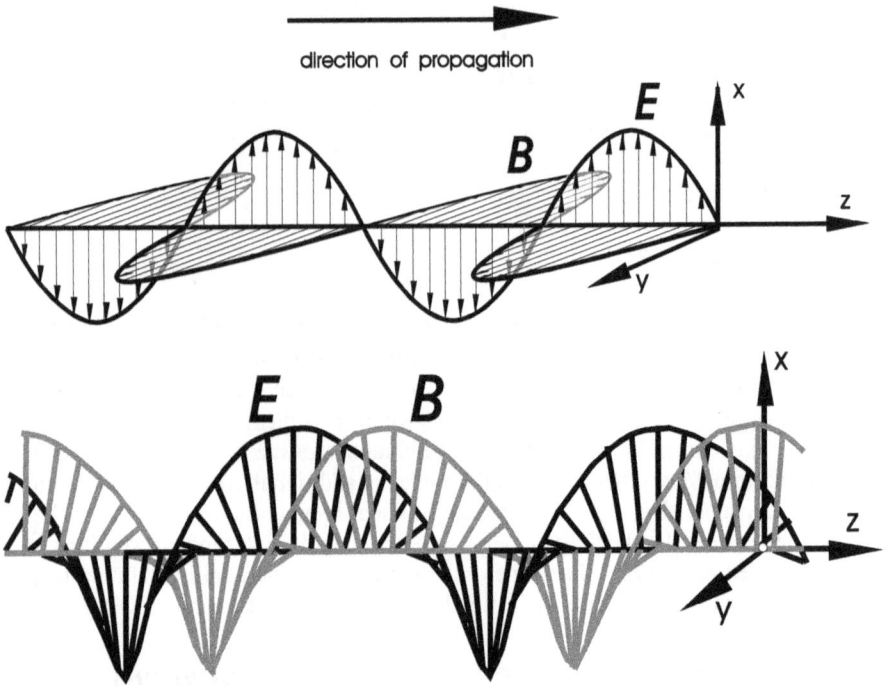

Figure A2: **Plane and circularly polarized solutions of Maxwell's homogeneous differential equations**

Proponents of absorption theory have advocated an equal role for absorption to the one usually associated exclusively with emission. They have pointed out that, in addition to field alternatives, there are two sets of valid solutions to *whichever* set of wave equations are selected. One of these alternatives – identified as the *retarded potential* solution (associated with propagation from the emitter toward the absorber) – has been the traditionally selected solution to Maxwell's equations. The other allowed solution identified as the *advanced potential* solution (associated with propagation from the absorber toward the emitter) was subsequently proposed as being equally legitimate by Wheeler and Feynman (1945). Naturally the *retarded* solution was exclusively in vogue until absorption theory was seriously considered, the *advanced* solution having always seemed to correspond to the *non-physical* situations of a signal *arriving* at the moment that emission occurs as though by divine intervention. More recently Cramer has proposed a similar reinstatement to vitalize a "transaction interpretation" of quantum mechanics. He demonstrates the role of the two waves as illustrated in figure A3 taken from his presentations (1986, p. 659).

Here there is an *arithmetic* assignment of plus and minus signs to be associated with advanced and retarded waves, but nothing that could be considered a *physical* assignment specific to the roles of emission and absorption so clearly integral to this whole process. None of these early investigators addressed the more obviously physical allocation of fields specific to material entities associated with the emission and absorption of the radiation. The assignments fit naturally into this scheme.

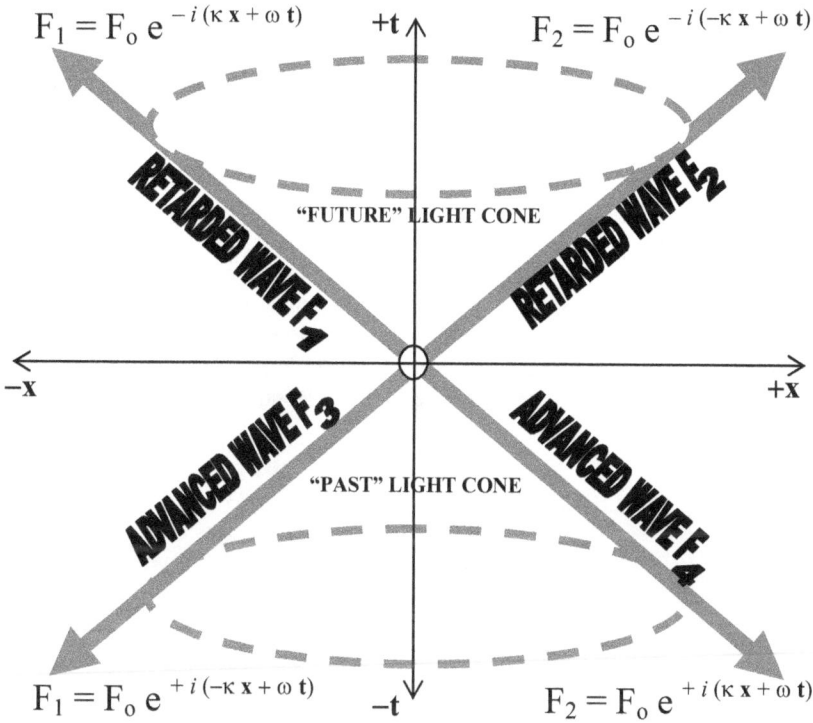

$$F_1 = F_o\, e^{-i\,(\kappa\, x\, +\, \omega\, t)} \qquad +t$$

$$F_2 = F_o\, e^{-i\,(-\kappa\, x\, +\, \omega\, t)}$$

RETARDED WAVE F_1

"FUTURE" LIGHT CONE

RETARDED WAVE F_2

$-x$

$+x$

ADVANCED WAVE F_3

"PAST" LIGHT CONE

ADVANCED WAVE F_4

$$F_1 = F_o\, e^{+i\,(-\kappa\, x\, +\, \omega\, t)} \qquad -t$$

$$F_2 = F_o\, e^{+i\,(\kappa\, x\, +\, \omega\, t)}$$

Figure A3: **Minkowski spacetime diagram showing the propagation of advanced and retarded waves from an emission locus at (x,t)=(0,0)**

But the conclusion that redundant sets of solutions are involved equally in the transaction is a conclusion that absorption theorists have long maintained, advocating acceptance of both the plus and minus signs in the exponential expression of the wave solutions provided in the equations 11) and 12). This author is convinced that the respective microscopic and macroscopic physical fields should also be acknowledged as being uniquely associated with these four solutions as well rather than merely including solutions with an arbitrary alternation of arithmetic sign in an attempt to restore physically meaningful interpretations to the two solutions. There is obviously much more to it than that.

This reluctance to make distinctions between the frame of reference of the fields is no doubt an outgrowth of the frame independence that has resulted from Einstein's *law of the transmission of light* for which it should make no difference in which frame the source of the emission and the absorber of the radiation happen to reside. Thus, the early investigators did not allocate *macroscopic* fields associated specifically with *absorption* or the *microscopic* ones with *emission* as seems only reasonable to this author. Nor did they attempt to exploit complimentary symmetries among the fields, which would seem so natural to that endeavor. If we had solved Maxwell's equations for H and D instead of E and B, for example, we might in effect have solved for what could be called an *absorber wave equation* as against an *emitter wave equation*. For reasons cited above and others beyond the scope of the current effort, the author believes neither of these to be precisely valid designations, however. There is in either case an interaction between the microscopic and macroscopic fields to be taken into account. Perhaps we are at least discovering why *four*, seemingly redundant, rather than just *two* such field vectors have been required to fully determine electromagnetic transactions even in a vacuum.

Of course, when dealing with a relatively stationary emitter and absorber there would be no measurable difference, but in dynamic situations epistemological differences abound. These differences derive from directional distortions associated with relativistic aberration. But again, further discussion of this topic is beyond the bounds of the current Appendix and may be found in Bonn (2008).

Appendix B

Refraction in Spherically Symmetric Electron Densities

The determination of the deflection angles experienced due to a variable index of refraction is complicated by the fact that traditional calculus of variation solution of the brachistochrone ("minimum time") trajectory for light does not provide an analytic result. This necessitates a numerical approach to solution.

Snell's law discussed in chapter 5 provides a basis for developing a recursive algorithm from which to obtain such a solution. Figure B1 illustrates the situational progress of a photon along a path that accommodates propagation from points **A** to **B** in minimum time. The incrementally changing angles are determined by Snell's law at each juncture determined by a finite integration interval of angle, $d\phi$.

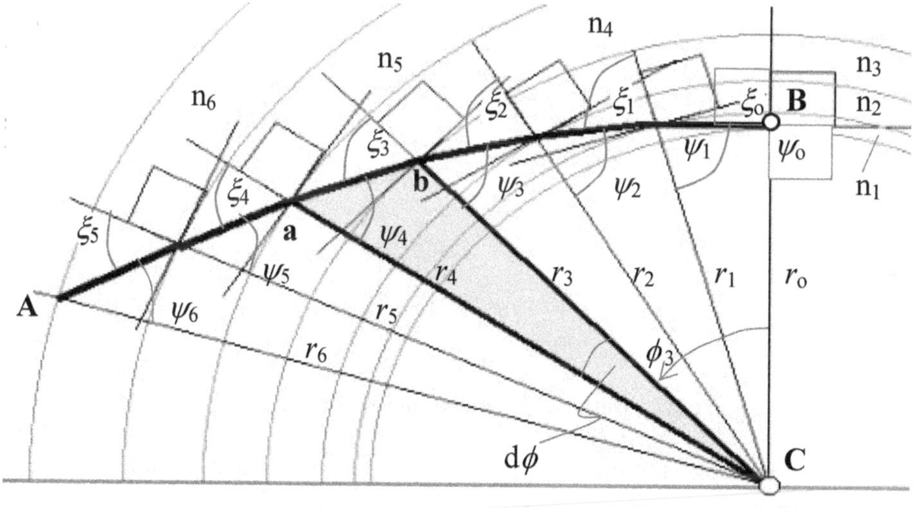

Figure B1: Geometry of refraction in a spherically symmetry situation

Clearly the path from **A** to **B** will be the same as the path from **B** to **A**. So we will proceed as though following a photon backwards from **B** to **A** so as to more easily accommodate an indefinite extension of the extremity **A** at a large distance from the center of symmetry, **C**. The path is taken as horizontal with regard to the electron distribution at **B**. This is arbitrary and any initial angle may be chosen, but it should be noted that every path will have a location of closest approach to **C** for which these conditions will apply.

In this figure the indices of refraction at each successive radius are characterized as n_i, i=1,2,3,... Each successive value of the index of refraction can be computed from the designated symmetric function of radius, $n_{i+1} = N_{ref}(r_i)$. Similarly the angles ξ_i and ψ_i are angles appropriate to Snell's law for which:

$$n_{i+1} \sin \xi_i = n_i \sin \psi_i$$

Exploiting these relationships for successive triangles **abC** in the figure, generalized so as to describe *any* of the corresponding triangles along the path, we obtain a recursive relation from which to fully define the entire curve **A** to **B**. We can readily establish values for three of the six characteristic parameters of this triangle. We use traditional geometry to fully determine the triangle based on a plane geometrical angle-side-angle argument as included succinctly beside figure B2 below.

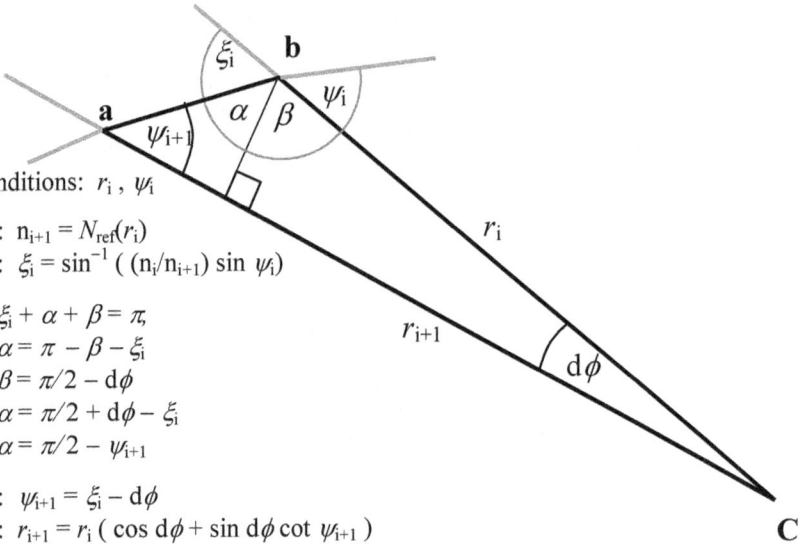

Initial conditions: r_i, ψ_i

Compute: $n_{i+1} = N_{ref}(r_i)$
Compute: $\xi_i = \sin^{-1} ((n_i/n_{i+1}) \sin \psi_i)$

Since: $\xi_i + \alpha + \beta = \pi$,
$\quad\quad\ \alpha = \pi - \beta - \xi_i$
But: $\quad \beta = \pi/2 - d\phi$
Thus: $\quad \alpha = \pi/2 + d\phi - \xi_i$
Also: $\quad \alpha = \pi/2 - \psi_{i+1}$

Compute: $\psi_{i+1} = \xi_i - d\phi$
Compute: $r_{i+1} = r_i (\cos d\phi + \sin d\phi \cot \psi_{i+1})$

Final conditions: r_{i+1}, ψ_{i+1}

Figure B2: Detail of geometry for recursive triangle

To solve for the total refraction angle, we continue this procedure until the angle ψ_i is sufficiently close to zero. This is accomplished by defining $d\phi$ sufficiently small

544

as in the integral calculus. Ultimately at great distances for which $r_o / r_i \ll 1$, it will be the angle ϕ_i that provides the assessment of the resultant angle of deflection.

Below we provide a JavaScript program that implements this algorithm. It is set up to determine the refraction angle to be expected of starlight that passes just above the limb of the sun during an eclipse. The final deflection angle is converted to seconds of arc.

```
<html><head><title>Refraction Program</title>
<head>
  </body>
  <script type="text/javascript">

    Math.cot = function (x) { return 1/Math.tan(x)}

    var φ=0;                         // (integration parameter)
    var Δφ=0.000001;                 // (Adjust this for desired accuracy.)
    var cdf=Math.cos(Δφ);
    var sdf=Math.sin(Δφ);
    var r=6.955*Math.pow(10,5);      // (initial radius -- radius of sun here)
    var no=nref(r);                  // (initial index of refraction)
    var ψ=0.5*Math.PI;               // (initial observation angle.)
    var conv=180/Math.PI;            // (conversion factor -- radians to degrees)
    y=conv*ψ;
    document.body.innerHTML += (r+", "+ψ+", "+no+"br>");
    if( !(ψ > Δφ) ) {
          break;
        } else {
            n=nref(r);
            x=Math.asin((no/n)*Math.sin(ψ));
            no= n;
            ψ=ξ–dφ;
            r=r*(cdf+sdf*Math.cot(ψ));
            φ=φ+Δφ;
            phe=conv*φ;
        }
    Dphe=3600*(90–phe);             // (final deflection angle in arc-seconds)
    document.body.innerHTML += (+Dphe+"<br>"); });

    function Nref(r) { // (index of refraction function based on electron density data)
        rS=6.955*Math.pow(10,5);    // (radius of sun in km)
        return 1+2.020*Math.pow(10,–8)*Math.pow(10,–2*Math.sqrt(r/rS–1)); } });

  </script>
  </body>
<html>
```

The determination of an appropriate integration interval $\Delta\phi$ follows procedures that employ the fundamental theorem of the calculus, which is that the interval is selected as sufficiently small that the error in the result is within tolerance. In figure B3 we have shown the impact of the size of $\Delta\phi$ on the quality of the result that was used in determining the maximum deflection due to refraction in the earth's atmosphere.

deflection
angle (degrees)

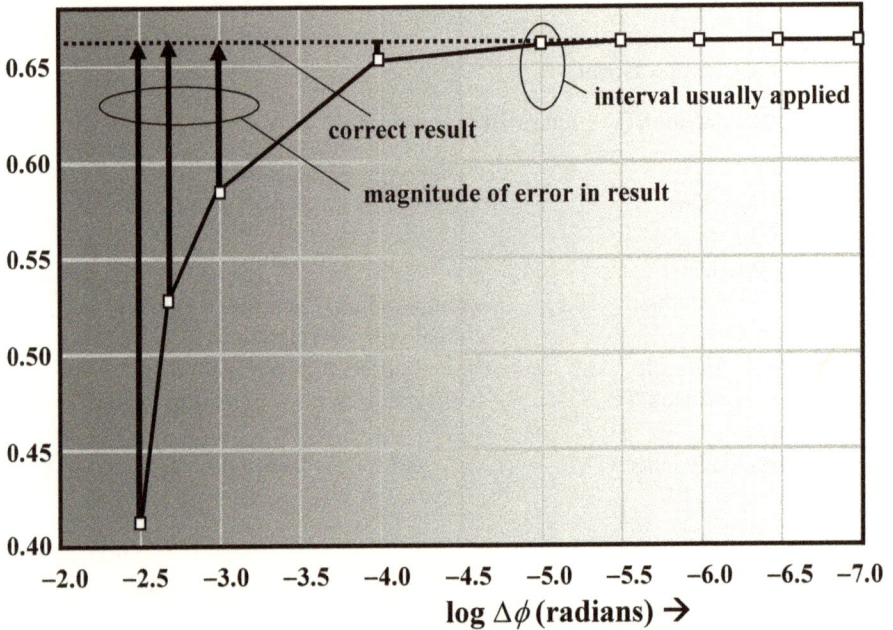

Figure B3: Determination of integration interval

546

Appendix C

Reopening the Book on Black Holes[1]

The ugly specter of a black hole is somehow quite enchanting to physicists in this new millenium, in part I suppose this is because they mirror conditions perceived by many as pertinent to our ultimate womb and doom – a narcissistic perspective that has seemed to beckon physicists for well over forty years now. That the geniuses of Hawking and Penrose have been greeted with such enthusiasm is due in large part to priorities they have assigned to these elusive objects of their unique insights – insights involving the inner workings of what have been perceived as seething vortexes of matter. But the most salient features of black holes can easily be understood by virtually anyone – even those with minimal backgrounds in the sciences. Black holes had been anticipated hundreds of years ago by a member of the clergy who stated in his paper presented to the Royal Society back in 1783 that escape velocities from an extremely massive object could exceed the speed of light under prescribed conditions. Thus, a lowly holy man augured prophetically that "all light emitted from such a body would be made to return towards it."[2]

For a particle of mass m to escape from a massive body of mass M, the kinetic energy imparted to it must involve a velocity larger than the 'escape velocity' v_s in order to overcome the negative gravitational potential energy such that:

$$\tfrac{1}{2}\,m\,v_s^2 \geq G\,M\,m\,/\,r,$$

where G is the gravitational constant 6.7×10^{-8} erg-cm/gm^2, r the distance of m from the center of gravity of the object of mass M when it possesses the velocity v_s. Since the upper limit on achievable velocities is that of light, we have:

$$r_s = 2\,G\,M\,/\,c^2.$$

---

[1] This essay is reproduced from ***Aberrations of Relativity*** by the current author. (Bonn, 2008)

[2] Although John Mitchell was indeed a member of the clergy he was also a polymath of no mean talent who had given up a post as professor of geology in Cambridge in 1764. (Gribbon, 2002, p. 293.)

where, c is the speed of light, 3.0×10^{10} cm/sec, and r_s the *Schwarzschild radius* to the 'event horizon' from within which even photons of light could not escape. This formula derives from classical analyses as shown, but is compatible with Einstein's gravitational model. Thus, if an object were sufficiently dense, it would be invisible. That is, if it were smaller than its Schwarzschild radius r_s, it could not be observed other than by external effects of matter being dragged to its doom and a minor associated effervescence. Let us ignore for now the ability to 'observe' it by means of its gravitational 'field'; i. e., how do these fields escape if electromagnetic ones cannot? How fast do gravitons move? Etc..

Thus, Newton's formulation of gravity in which forces act through the center of mass of an object reduces the complexity of calculating the Schwarzschild radius of an event horizon from beneath which no light can escape to mere child's play. The minimum mass that is required by evolving stellar masses if they would attain unto this status is similarly easy to determine as we will see. It is about two solar masses. We now know also from Hawking's and Penrose's extensive work that there are no particular subtleties with respect to black holes; they must all be 'standard' inasmuch as distinguishing characteristics outside their 'event horizons' can only be their unique mass and angular momentum – net charge not being much of a possibility. (Thus, "Black holes have no hair" is every bit as sophisticated as, but certainly no more so than, the statement, "There is no free lunch.")

But despite such dispassionate determinations of their simplicity there is still a tremendous amount of conjecture pertaining to internal structures – or lack thereof – with popularized conceptions promoted by those who should know better dictating an associated spacetime singularity. That general relativity, whose equations cannot even be solved for trivial planar cases, implies that spacetime may be "pinched off" in the vicinity of a black hole is a factal of which I will deny myself other than an amused awareness (for reasons to be discussed in more detail below). From the outside, however dark, a black hole is just an object. There persists this notion that having once sunk beneath its Schwarzschild radius all its mass would have been swallowed into a single mathematical point never to return, although we have been told by the same individuals that our current universe emerged (or is just about to emerge) from beneath just such a shroud. It's hard for me to distinguish just what should be believed before breakfast. From such fanciful theorizings come fantasies of "worm holes," Einstein-Rosen Bridges, "quantum foam," and time machines. Notwithstanding these absurd (Oh, did I say "absurd?") presumptions, Hawking has shown that given 10^{85} years (regrettably somewhat less than a picturesque googol) black holes would eventually effervesce back into visible matter. And as usual, I'm skeptical – not of the effectuality of his effervescence which seems reasonable mind you, but of a need for it in this case.

We are all aware of the frequent news flashes claiming repeatedly to have confirmed the existence of black holes. It is claimed that there are giant black holes at the centers of many distant galaxies and even our own Milky Way. The galaxy M87 is thought to possess a black hole at its center with a gravitational pull three billion times that of our sun. These "messy eaters" have become the engines of choice for the prodigious energies generated by quasars, etc. Statistical estimates place the number of black holes resulting from collapsed neutron stars at as many as 100 million in our

Milky Way galaxy alone. With respect to the news flashes, there is considerable reason to believe that black holes do indeed exist. But on logical grounds I currently have very serious doubts – outside the scope of mathematical games played with general relativity – about their being associated with singularities in spacetime as popularly envisioned. Let us consider that notion.

There is, of course, the minimum mass requirement for astronomical objects that proceed down the thermonuclear ash ladder based on thermodynamic pressures and simple gravitational collapse considerations. There are observationally confirmed stopping off places in the collapse of matter into its densest states. In a penultimate state, an entire massive star may be comprised of a single nuclear blob of juxtaposed protons and neutrons surrounded by an atmosphere of electrons. This structure is known as a "white Dwarf." Quantum solutions for such high Z (proton count) "Hartree atoms" would provide an extremely wide range of orbits for degenerate (as in Pauli exclusion principle) electrons. The inner shells would be constrained well within even their own Schwarzschild radii while the outer shells would be virtually free of gravitational attachments altogether. Such stars are thought to be particularly stable because electron degeneracy that precludes the particles occupying the same angular-momentum-space-spin attributes, would preclude their being packed more tightly such that they would then have to share mutually exclusive allotments as in the shell structures of their more mundane atomic counterparts. Neutron stars are those that fall through this rung on the downward spiral staircase by virtue of exceeding the Chandrasekar threshold of 1.4 solar masses. Exceeding this limit suffices to allow gravity-induced pressures to exceed electron degeneracy forces by increasing temperatures such that thermonuclear reactions that merge electrons and protons into neutrons occur, so that the star plunges to the next rung on the ladder. If the stellar mass is less than about 2.0 solar masses the surface of the neutron star will remain above its Schwarzschild radius. Such neutron stars are now well-known as "pulsars." Those that have been observed have radii of about ten kilometers just safely larger than their Schwarzschild radius of approximately five kilometers. However, stars more massive than this threshold, will eventually disappear. Their collapse is envisioned by many, however, as hounding them like Bill Clinton's tireless detractors even beyond their new-found obscurity. But how can that happen when the mass density must now be determined by neutron degeneracy? It is conventionally thought that processes similar to those whereby electron degeneracy is overcome by gravitational pressures would eventually force neutron stars also to succumb. But this would not occur as soon as the neutron star sank beneath its event horizon – these two phenomena are certainly not directly coupled.

For modeling purposes calculations of gravitational collapse phenomena can be simplified by unrealistic assumptions involving constant densities such that any macroscopic region of a neutron star would have the same density. As compaction proceeds in search of a new compressed equilibrium under such (unrealistic) assumptions, the object would more or less continuously reach higher and higher densities. This process is perceived as proceeding "beyond" the neutron star stage once a black hole is created with an associated abandonment of the conservation of baryons as the trapped heat from the increasing pressure cannot be released. Assumptions appropriate to a neutrino-quark gas are what are inferred and in this form the indivisibility of major atomic components is seen as having finally been lost. In this case the density profile is intuited to proceed down the path to singularity. Collapse would force density toward infinity more rapidly than the radius tends toward zero. The tremendous gravity would turn surface mountains into submicroscopic ripples, smoothness, then oblivion. One might argue thus that for matter comprised of point particles distributed evenly as in a gas in a spherical gravitational well there is no reason why, if degeneracy gives way to the ineluctable pressures of gravity, sufficient matter should not collapse indefinitely. So singularity might seem to be inevitable such that black holes would become point particles of extremely large mass – the big bang happening in reverse! Such fantasies of thought engage even the brightest notwithstanding the established facts to the effect that whether black holes collapse to singular points or hover forever just beneath their event horizons could never be scientifically distinguished unless there were some possible consequence that could be observed – that there isn't. But singularities are the stuff of dreams for string theorists who anticipate so many large point particles they don't know what to do with them all. That the truth might forever be shrouded from falsifiability by experimental and even theoretical means has never been an obstacle to such theorists; it may even subconsciously be acknowledged as an advantage. But let's just consider the simplified model of matter involving uniform distributions of infinitesimally small point particles. How legitimate is it?

It is true that the divergence theorem legitimizes the assumption of all symmetric mass distributions acting *as though* (but certainly not *as in actual fact*) operating through a *single point* at the center of mass of the distributed body for gravitational consideration. It is also true that the Schrödinger equation that nailed down the behavior of electronic matter did assume *point particles*, but that treatment used little more than broad analogies. It turns out that solution of these equations involving the very same *point particles* results in their inevitably being *smeared out* as mere probability clouds with absolutely no credentials for existence at a single point at any particular time. The validation of these solutions by experiment is legend. But despite success in the laboratory, the derivation of the equation itself and the assumptions that went into it remain entangled in hocus-pocus. Notice also in this regard that although it assumed that attractive forces of the nucleus act through a single point this is *only* in the sense of the divergence theorem, and that in cases with more than a single proton it obviously cannot actually *be* a single point other than as the abstracted center of mass. So... so much for those lame arguments. If particles are, in fact, as most theorists maintain, *point* particles, one might ask why protons and neutrons do not ultimately just collapse into their own gravitational potential wells. Their Schwarzschild radii are on the order of $r_s = 5 \times 10^{-19}$ cm, but that is one hell of a lot bigger than a *point* particle and would provide a very dangerous environment for a particle that dashes about violently within strict confines! It would be like a man in an Edgar Allan Poe nightmare with a manhole-sized abyss in the middle of his dark cell – simply a matter of time. The answer to this dilemma is simple if one accepts data from the real world. The theoretically and experimentally inferred radii of their associated clouds exceed 10^{-13} cm. They are alas, despite theoretical arguments to the contrary, like neutron stars of less than several solar masses, everyday planets, people, baseballs, and M&M's, just too damn big to fit within any such confinement as their own event horizon!

Mass and charge are concepts that are not all that well defined other than with respect to their effects on apples and cat's fur, and I will not make conjectures here other than in that same time-honored tradition. In figure C1 there is a set of curves representing the density of nuclear charge as a function of radius for a few garden variety atomic nuclei as determined by electron scattering methods appropriate to this endeavor. You will notice that all these nuclei are too big to fit into their Schwarzschild radii and I would wager that there is little danger of component quarks falling into theirs either. It is inherently reasonable to assume there are nearly identical distributions of mass and charge in such cases. There is, of course, the slight increase in the percentage of the uncharged neutrons relative to protons with increasing atomic number, but otherwise the curves in figure C1 are much more like what one should expect for mass distribution of elementary nuclear particles than for the soup model described above. But again, when dealing with units of miles or kilometers such fuzziness about the edges would have been on the order of 10^{-20} smaller – in fact the mere "ripples" of which we spoke earlier.

But before we talk too glibly of singularities, for which such fuzziness becomes huge, let's consider effects of such fuzziness on the ultimate collapse of matter into the abyss of its own black hole.

When electron degeneracy breaks down in the collapse into a neutron star and in proceedings thereafter (if there is, in fact, a thereafter), is it reasonable to assume that the generic aspect of a probability distribution associated with the building blocks of matter would be drastically altered also? And if the structure were to be so altered, who is to say it would be to a distribution along the lines of a simplistic soupy model? Does it seem reasonable to anyone capable of coherent thought on the subject that Quantum organization would be abandoned at this point? Would God have thrown up his hands at that point and said, "Oh, I never thought about that?" I don't think so. Be aware that no one knows correct answers to

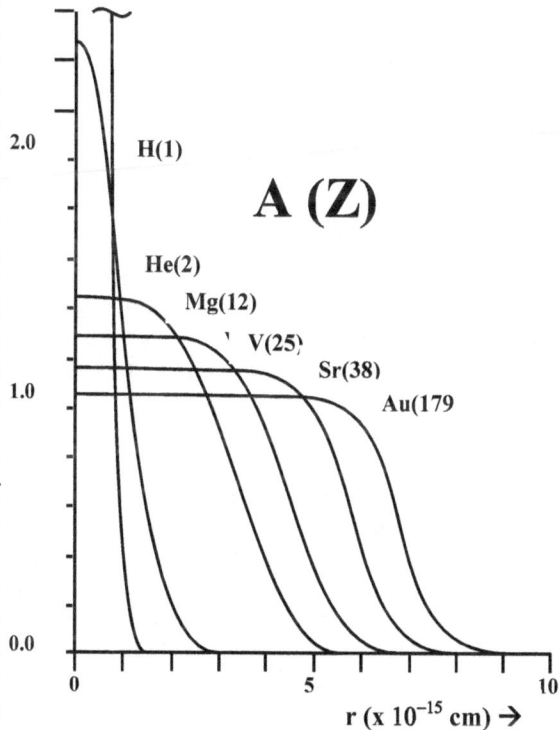

Figure C1: Nuclear charge densities

such metaphysical questions since we have no snap shots from the supposed *bang*, some time after which neutrino degeneracy is praised, but I don't think that matter in black holes would turn to soup. Occam's Razor would surely take a swipe at that assumption and I see no reason to fight such a weapon myself. There is a continuous record of soupy models of matter having repeatedly been replaced by previously unsuspected models involving a more organized structure as heady endeavors provided additional information about phenomena associated with submicroscopic matter. In particular, there would have been every reason to believe that a stable hydrogen atom would prove to be an utter impossibility. But nature has vehemently insisted on particle indivisibility that precluded an electron soup from spiraling into a proton soup and their two charges dissipating in a sayonara swan song as they disappeared altogether into however romantic a unity in an electromagnetic vortex. The forces were there for exactly that eventuality, but... it turns out that there are *other* forces than electromagnetism and gravitation that have precluded that. How could tiny nuclei contain multiple protons whose inverse square repulsion would skyrocket these juxtaposed objects to the opposite ends of the universe? But of course the nuclear attractive and repellent forces involving lower levels of fundamental particles enforce comfort distances using forces of much higher order than an inverse square relationship to preclude such disasters. No one could have anticipated the nature of these additional forces until sufficient data was available. Now there's a concept! All the high powered deductive reasoning on then current models was laughably insufficient to scale these peaks of knowledge. It has been our scientific heritage that by employing inductive methods we do systematically scale such peaks, and ultimately smile down on our former ignorance. But there seems currently to be little inclination to such humility on that account or patience for just plain "finding out!"

It should be noted that nuclear forces although symmetric do not involve inverse square relationships and that, therefore, the divergence theorem that is so essential in the context of black holes no longer even applies in that domain. Certainly as a neutron star becomes more massive by accretion, more significant gravitational forces become increasingly pertinent to any quantum solution. However, it seems a bit rash to predict that a tiny force, that in domains for which we have actual data pertaining to it being smaller by a factor of less than 10^{-40} than another, should prematurely be declared the victor based on interpolations from an ultimate dearth of data. Never mind the fact that G. W. Bush achieved as much in Florida – that was third world politics not heady science.

To assume that an inverse square law attractive force could suck objects into a singularity in the real world when those same objects repel each other by much more extreme forces is a bit...well...extreme! Much more likely it seems to me is the possibility that increasingly massive stars would go quietly to that good night behind the curtains of their event horizons. As a neutron star's mass attained several solar masses, whether initially or eventually through gradual accretion, whatever associated increase in volume it achieved by adding particulate matter would be dwarfed by more dramatic cubic increases in the volume increases due to its increased Schwarzschild radius. So it would seem reasonable to assume that the object might indeed eventually sink beneath its event horizon. But it seems unlikely without further evidence that it would proceed from such a gradual demise directly to the hidden singularity too often propounded as a necessary consequence. Why would it? No one now, nor will anyone *ever*, have empirical evidence of what happens beneath an event horizon other than that of our segment of the universe, because alternative inner workings of black holes must forever remain moot points in accordance with the findings of Hawking and Penrose. But one thing seems certain and that is that there is so far no adequate justification to conclude that they must proceed in one fell swoop to a mathematical point rather than the externally equivalent alternative! As mentioned, their radii and all other features are fixed independent of their internal workings so why is it scientific to presume such an impossible situation when all possibility of evidence for that eventuality is foregone? This gets back to the meaning of the divergence theorem and the equivalence of any symmetric distribution to one in which all mass is concentrated at a point: That equivalence applies to inverse square law forces and even in that case does not confuse anyone with regard to our sun, earth, and moon possibly thereby being merely mathematical points assigned the given masses. Why is this so-related point so hard to understand?

The neutron star rung in the matter ladder may ultimately arrest collapse altogether – perhaps it's the basement floor itself or the trampoline beneath the trapeze of being! In some cases such an object's surface may actually indeed immerse into and beneath an event horizon, but the internal workings of the associated object itself need not undergo transmogrification on that account. It is my guess that it will remain the embodiment of the very same generic rung on the ladder notwithstanding its understandable new shyness. It is obvious that we know too little about neutron stars other than pulsar radiations we attribute to them. What is the structure of a neutron star – whether it involves 1.0, 1.4 or 5.0 solar masses? Whatever it is, it must involve a lump of neutrons whose organization is determined by quantum considerations pertinent to a fermi gas trapped in a tremendous gravitational well. Complimentarity suggests that classical expressions for energy of a neutron added to such an object of radius r must bear some resemblance to the corresponding quantum mechanically determined value. So $E \approx 4/3 \, \pi \, G \, m_n \, \rho \, r^2$, where $m_n=1.67 \times 10^{-24}$ gm is the mass of a neutron with density 1.67×10^{15} gm cm^{-3}, which is not much more dense than typical neutron stars as one might expect. But now let's consider how a distribution of fermions is affected by increasing temperatures that would accompany additional gravitational pressure. As is typical of quantum solutions, the distribution becomes much broader by skipping energy levels and hopping into extended orbits as implied in figure C2. Only at the temperature of absolute zero Kelvin would such a gas be completely compacted within its minimum radius determined by E_f (the highest compacted energy level). At 10,000 K the distribution would be totally out of any bounds we could associate with complete compaction in any way similar to a soupy model restricted within an event horizon let alone presume it to have collapsed to a mathematical point. At hundreds of millions of degrees – reasonable temperatures for such ensembles – associated neutrons would exist throughout a vast cloud much larger than the event horizon. Nor would this involve impossibilities of faster than light travel; in quantum solutions there is no sense in which probabilities of being here or a light-year away involve the concept of 'escape velocity'. And since a high-energy neutron has a definite propensity for disintegrating and/or interacting with other matter no matter where it is found in the vicissitudes of its 'travels', this scenario involves something totally *other* than being 'confined to a black hole'. These real world considerations are why the contents of such objects cannot be dismissed like debris shoved down a garbage disposal. High-energy neutrons light years away from the center of the neutron star or black hole would disassociate atoms, create deuterium in collision with plasma protons, and ultimately create helium and traces of heavier elements far removed from the hole itself. In short, this would ape big bang behavior. The pertinent question is, "How could this *not* happen?"

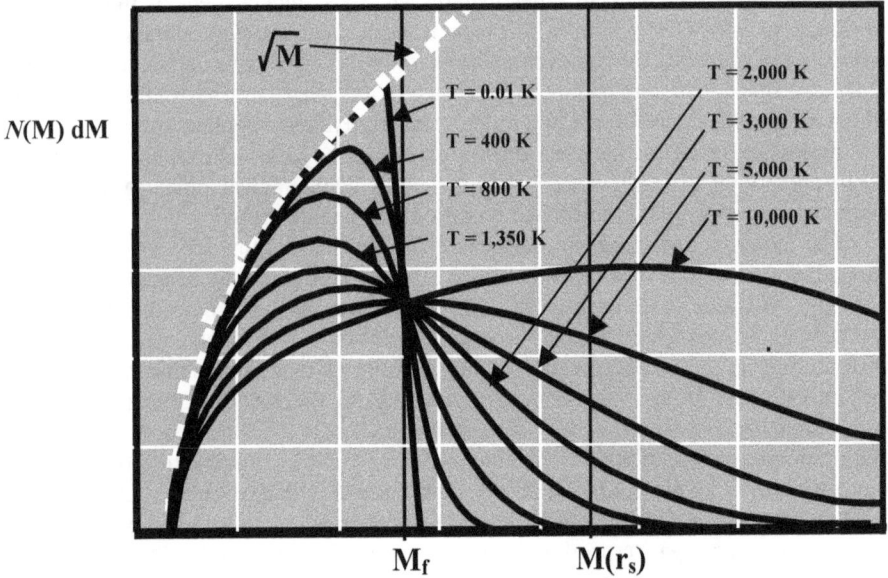

Figure C2: Significance of fuzziness in the mass distribution in a 'fermion gas' of neutrons as would be realized in a collapsed neutron star

Being compressed to a Schwarzschild radius is *not* like reaching *Mach one* or *the boiling point*! There is no qualitative new torture awaiting matter at this coincidental (as against universal) threshold as popular thought insists. (For example, scientists are having one hell of a time determining whether our entire universe is beneath or has somehow crawled out from underneath such a shroud. If it made such a tremendous difference, why could we not tell? And if our entire universe escaped its own event horizon as data increasingly suggests to most that it must have a long time ago now according to the standard cosmological model, *how* did it get out?) Internal phenomena might very well reach a state (even if one anticipates some method of circumventing fermi gas restrictions) in which it becomes sufficiently energetic whereby internal eruptions (the next up the Richter scale from supernova) associated with quantum distribution phenomena occur. We may already have observed this at the centers of active galaxies – quasars or gamma ray bursts – about which we have had plenty of Jungian *inflationary* dreams concerning primordial origins. There is no reason to presume that such once-obscured matter might not reappear as a result of internal reorganizations that swells it first back beyond its Schwarzschild radius in a process that might afterwards explode the entire now visible contents back into luminous interaction. Such a process could free all of the trapped matter with no violation of any physical law – freeing the hot neutrons in one gigantic (although not *that*!) big bang from which the rest of all we know about the universe proceeds. There is nothing magical here. This would not involve the spewing forth of iron, gold, Europium, Americanium, or the various other heavy elements of a supernova, but the basic building blocks that have naively been assumed as only *initial* primordial prerequisites of the universe. "Cosmocentrism" propounded by Frank Luger (2000) may be actualized by such rising phoenixes – not everywhere all at once, but all black holes at some point in their maturity so as to maintain an infinite and eternal equilibrium between these sources and sinks of all material existence. It is enough to titillate and frustrate the fantasies of creationists of all ages and scientific persuasions.

I wish I could flap my lips to produce the mellifluous sounds of a Carl Sagan on one of those old Public Broadcasting System *Nova* programs my children used to deplore when I say the following because it expresses the awe-inspiring religious sense in which I feel it. Anyway, getting away from this epiphany, and whether with eloquence or a more characteristic bombasity, here goes: "Nothing says that a book, a mind, or even a black hole, having once been closed, cannot be re-opened."

Afterward:

There would seem to be some level of hypocrisy for those propounding the origin of the universe from what they consider to be a singularity with a "big bang" when these same individuals insist on the penultimate death of the material universe into just such singularities. For example, Ed Seidel (NCSA and University of Illinois) states with regard to what he considers to be cosmic "decency laws" that what happens beneath Event horizons must in essence forever remain no one's business such that:

"All singularities within the universe must therefore be 'clothed.'"

"But inside what? The event horizon, of course! Cosmic censorship is thus enforced. Not so, however, for that ultimate cosmic singularity that gave rise to the Big Bang."

That is not the introduction to an explanation, but the end of one. And this ultimately is the hypocritical lie to be told – where we find that what is good for the goose in *not*, in fact, good for the gander!

I was recently accosted by an individual who claimed that the universe could not possibly exist in a stationary state because of the multiple levels of fundamental particles, and indeed the 'standard' models of fundamental particles and cosmology have been very purposely, but the author believes illegitimately, linked. I asked the accoster just how he conceived that such a logical structure could imply a temporal origin to the universe. I was told in essence that many, if not indeed *most*, of these particles would have no role if it were not for the big bang where they could conceivably have had some play. It was as though my critic had perceived the universe as a staged production being somehow *directed*; and why would a playwright write a play with specified actors for some of whom there were no parts written. A theatre group that hired actors for which there were no roles would be a madhouse. In such case there should as likely be roles for which there were no actors.

I understood his point. I could tell from whence he came.

However, what did he not understand about the similarity presented by the possibility that black holes might ultimately spew forth matter back into the useful universe just as what is envisioned as having happened with an even bigger bang?

Certainly the high-energy conditions under which these lower levels of fundamental particles have been discovered are realized inside black holes. So just maybe these neutron lumps transform to heavier but similarly structured matter as a next rung on the ladder of material being that retards the ultimate collapse – until it also reaches its own analogy to a supernova. Who knows?

There is a lot we do not know about gamma ray bursts other than that they seem to occur even at the extremities of the visible universe and to be associated with optical galaxies. Maybe these are the evidence of black holes erupting.

Appendix D
Frequently encountered constants

a. physical constants applicable to universe as a whole

c (speed of light in a vacuum) = 2.9979×10^{10} cm/sec.

h (Planck's constant) = 6.626×10^{-27} erg sec

k (Boltzmann's constant) = 1.380×10^{-16} ergs/degree kelvin

σ (Stefan-Boltzmann constant) = 2.268×10^{-4} erg-cm^{-2}-deg^{-4}

N_A (Avagadro's number) = 6.0225×10^{23}

G (Newton's gravitational constant) = 6.67428×10^{-8} cm^3 / gm sec^2

H_o (Hubble's, constant) $\cong 7.14 \times 10^{-29}$ cm^{-1}

$$\cong 67.4 \text{ km sec}^{-1} \text{ Mpc}^{-1} / c$$

ρ_o (Einstein's critical density) $\cong 8.0 \times 10^{-30}$ gm cm^{-3}

ρ_{mu} (mass density of the universe) $\cong 5.49 \times 10^{-31}$ gm cm^{-3}

$\rho_{\mu bb}$ (microwave background density) = 4.176×10^{-13} ergs cm^{-3}

$T_{\mu bb}$ (microwave background temperature) = 2.725 K

b. constant properties of objects in the universe

e (electronic charge) = 4.80×10^{-10} stat coulombs

m_e (electron rest mass) = 9.109×10^{-28} gm

r_e (classical electron radius) = $e^2/ m_e c^2 \cong 2.82 \times 10^{-13}$ cm

m_p (proton mass) = $1.672621637 \times 10^{-24}$ gm

m_n (neutron mass) = $1.67492729 \times 10^{-24}$ gm

$M_\odot$ (mass of the sun) $\cong 2.0 \times 10^{33}$ gm

$L_\odot$ (luminosity of the sun) $\cong 4.0 \times 10^{33}$ ergs/sec

$R_\odot$ (radius of the sun) $\cong 6.9550 \times 10^{10}$ cm

$M_{\odot MW}$ (mass of Milky Way) $\sim 6 \times 10^{11}$ $M_\odot = 1.2 \times 10^{45}$ gm

L_{MW} (luminosity of Milky Way) $\sim 2 \times 10^{10}$ $L_\odot = 8 \times 10^{43}$ ergs/sec

c. unit conversion constants

1 light year (distance) = 9.4606×10^{17} cm

1 parsec (distance) = 3.26 light-years

1 Mpc (distance) = 10^6 parsecs

1 Angstrom = 10^{-8} cm

1 eV (electron volt) = 1.602×10^{-12} ergs $\rightarrow$ 1.161×10^4 K

1 MeV (energy) = 10^3 KeV = 10^6 eV

Bibliography

1.   G. Aldering, W. M. Wood-Vasey , B. C. Lee, S. Loken, P. Nugent, S. Perlmutter, J. Siegrist, L. Wang, P. Antilogus, P. Astier, D. Hardin, R. Pain, Y. Copin, G. Smadja, E. Gangler, A. Castera, G. Adam, R. Bacon, J-P. Lemonnier, A. Pecontal, E. Pecontal, R. Kessler, "The Nearby Supernova Factory", Proc. SPIE, 4835 146 (2002)

2.   H. Alfvén, Worlds-Antiworlds, W. H. Freeman and Co., San Francisco, p. 13, (1966)

3.   R. A. Alpher and R. Herman, "On the Relative Abundance of the Elements," Physical Review 74 (1948), 1577.

4.   Arp, H. C; Burbidge, G.; and Hoyle, F., "The extragalactic universe: an alternative view," Nature, Vol. 346, pp. 807-812, (30 August 1990)

5.   L. E. Ashmore, "Hydrogen cloud separation as direct evidence of the dynamics of the universe", CC2 Proceeding, p.3 September 7-11, (2008)

6.   A. K. T. Assis, M. C. D. Neves, S. S. L. Soares, "Hubble's Cosmology: From a finite Expanding Universe to a Static Endless One", Crisis in Cosmology Conference–2 (September 11, 2008)

7.   Neta A. Bahcall, "Clusters and Superclusters of Galaxies", Formation of Structure in the Universe, Ed. A. Dekel and J. Ostriker, Cambridge Press (1999)

8.   David Band, Raul Jimenez, and Tsvi Piran, "Host Galaxies As Gamma-Ray Burst Distance Indicators", LA-UR-01-0655 (October, 96) Submitted to: http://lib-www.lanl.gov/la-pubs/00367072.pdf

9.   Carlos Barcelo, Stefano Liberati, Sebastiano Sonego, and Mat Visser, "Black Stars, Not Holes", Scientific American, Vol. 301, Number 4, October (2009)

10.   X. Barcons, A. C. Fabian, and M. J. Rees, "The physical state of the intergalactic medium," Nature, Vol. 350, pp. 685-687 (25 April 1991)

11. M. R. Becker, T. A. McKay, B. Koester, R. H. Wechsler, E. Rozo, A. Evrard, D. Johnson, E. Sheldon, J. Annis, E. Lau, R. Nichol, C. Miller, "The Mean Scatter of the Velocity Dispersion – Optical Richness Relation for maxBCG Galaxy Clusters", The Astrophysics Journal, 669, 905 (2007)

12. Mark Birkinshaw, "The Sunyaev-Zel'dovich Effect", Physics Reports, 310, 97 (1999)

13. Andrea Biviano, "Our Best Friend, The Coma Cluster (A Historical Review)", Published in Untangling Coma Berenices: A New Vision of an Old Cluster, Proceedings of the meeting held in Marseilles (France), Eds.: Mazure, A., Casoli F., Durret F. , Gerbal D., Word Scientific Publishing Co Pte Ltd. (June 17-20, 1997)

14. Andrew W. Blain, Ian Smail, R. J. Ivison, J.-P. Kneib, and David T. Frayer, "Submillimeter Galaxies", Physics Reports, Volume 369, Issue 2, p. 111-176 (2002)

15. M. R. Blanton, J. Dalcanton, D. Eisenstein, J. Loveday, M A. Strauss, M. Subbarao, D. H. Weinberg, J. E. Anderson, Jr., J.S Annis, N. A. Bahcall, M. BernardiI, J. Brinkmann, R. J. Brunner, S. Burles, L. Carey, F. J. Castander, A. J. Connolly, I. Csabai, M. Doi, D. Finkbeiner, S. Friedmann, J. A. Frieman, M. Fukugita, J. E. Gunn, S. Hennessy, R. B. Hindsley, D. W. Hogg, T. Ichikawa, Z. Ivezic, S. Kent, G. R. Knapp, D. Q. Lamb, R. F. Leger, D. C. Long, R. H. Lupton, T. A. McKay, A. Meiksin, A. Merelli, J. A. Munn, V. Narayanan, M. Newcomb, R. C. Nichol, S. Okamura, R. Owen, J. R. Pier, A. Pope, M. Postmann, T. Quinn, C. M. Rockosi, D. J. Schlegel, D. P. Schneider, K. Shimasaku, W. A. Siegmund, S. Smee, Y. Snir, C.S Stoughton, C. Stubbs, A. S. Szalay, G P. Szokoly, A. R. Thakar, C. Tremonti, D. L. Tucker, A. Uomoto, D. Vanden Berk, M. S. Vogeley, P. Waddell, B. Yanny, N. Yasuda, and D. G. Yo, "The Luminosity Function of Galaxies in SDSS Commissioning Data", The Astronomical Journal, 121:2358-2380 (May, 2001)

16. S. Blondin, T. M. Davis, K. Krisciunas, B. P. Schmidt, J. Sollerman, W. M. Wood-Vasey, A. C. Becker, P. Challis, A. Clocchiatti, G. Damke, A. V. Filippenko, R. J. Foley, P. M. Garnavich, S. W. Jha, R. P. Kirshner, B. Leibundgut, W. Li, T. Matheson, G. Miknaitis, G. Narayan, G. Pignata, A. Rest, A. G. Riess, J. M. Silverman, R. C. Smith, J. Spyromilio, M. Stritzinger, C. W. Stubbs, N. B. Suntzeff, J. L. Tonry, B. E. Tucker, and A. Zenteno, "Time Dilation in Type Ia Supernova Spectra at High Redshift", The Astrophysics Journal, 682, 724-736 (2008)

17. Bonn, R. F., The Aberrations of Relativity, ISBN 978-0-6151-9781-4, Vaughan Publishing, Seattle (2008)

18. M. Born and E. Wolf, Principles of Optics, 6th Ed., Pergamon Press, New York, NY, pp. 93-96 (1980)

560

19.	Boroson, T. A., Meyers, K. A., Morris, S. L., and Perrson, S. E., "The appearance of a new redshift system Markarian 231", <u>The Astrophysics Journal</u>, 370, L19-L21 (1991)

20.	Gregory Bothun, <u>Modern Cosmological Observations and Problems</u>, Taylor and Francis Publishing group (1998)

21.	K. Brecher, "Is the Speed of Light Independent of the Velocity of the Source?" <u>Phys. Rev. Letters,</u> Vol. 39, N0. 17, pp. 1051-1054 (24 October 1977)

22.	Michael J. I. Brown, Arjun Dey, Buell T. Jannuzi, Tod R. Lauer, Glenn P. Tiede, and Valerie J. Mikles, "Red Galaxy Clustering in the NOAO Deep Wide-Field Survey", <u>The Astrophysics Journal</u>, 597 225-2 (2003)

23.	Brune, R. A., Jr., Cobb, C. L., Dewitt, B. S., Dewitt-Morette, C., Evans, D. S., Floyd, J. E., "Gravitational deflection of light: solar eclipse of 30 June 1973 – Description of procedures and final result", <u>The Astronomical Journal</u>, Vol. 81, p. 452 - 454 (1976)

24.	Ari Brynjolfsson, "Redshift of photons penetrating a hot plasma", (2005) arXiv:astro-ph/0401420v3

25.	D. N. Burrows and J. A. Mendenhall, "Soft X-ray shadowing by the Draco cloud," <u>Nature,</u> Vol. 351, pp 629-631 (20 June 1991)

26.	John E. Carlstrom, Gilbert P. Holder, and Erik D. Reese, "Cosmology With The Sunyaev-Zel'dovich Effect," <u>Annu. Rev. Astron. Astrophys</u>, 40:643–80 (2002)

27.	Campbell, W. W. and Trumpler, R. J., "Testing the theory of relativity", <u>A Source Book in Astronomy</u>, Harlow Shapley editor, Harvard Press, Cambridge (1960)

28.	Timothy Clifton and Pedro G. Ferreira, "Does Dark Energy Really Exist?", <u>Scientific American,</u> April (2009)

29.	Matthew Colless, "2dFGRS Image Gallery", (June 30, 2001 image) http://vo.iucaa.ernet.in/2df/Gallery/index.html

30.	Matthew Colless, "First results from the 2dF Galaxy Redshift Survey", <u>Phil. Trans. R. Soc. Lond. A</u> (1999)

31.	Christopher J. Conselice, Norman A. Grogin, Shardha Jogee, Ray A. Lucas, Tomas Dahlen, Duilia de Mello, Jonathan P. Gardner, Bahram Mobasher, and Swara Ravindranath, "Observing the Formation of the Hubble Sequence in the Great Observatories Origins Deep Survey", <u>The Astrophysics Journal</u>, 600L.139C (2004)

32. Christopher J. Conselice, "The Galaxy Structure-Redshift Relation", Invited Review to appear in "Penetrating Bars Through Masks of Cosmic Dust: The Hubble Tuning Fork Strikes A New Note", ed. D. Block et al. (2004) astro-ph/0407463 (July 2004)

33. Cowen, "Safe from a heavenly doom: gamma-ray bursts not a threat to Earth", <u>Science News</u>, pp 88-89 (2006)

34. L. L. Cowie and S. C. Perrenod, "Origin and distribution of gas within rich clusters of galaxies: The evolution of cluster x-ray sources over cosmological time scales", <u>The Astrophysics Journal</u>, 219, 254 (1978)

35. J. G. Cramer, "Generalized Absorber Theory and the Einstein-Podolsky-Rosen Paradox," <u>Phys. Rev. D</u>, 22, 2, 362-376 (1980).

36. J. G. Cramer, "The Transactional Interpretation of Quantum Mechanics," <u>Rev. Mod. Phy.</u>, 58,3, 647-687 (1986),

37. Robert H. Dicke, Robert Beringer, Robert L. Kyhl, and A. B. Vane, "Atmospheric Absorption Measurements with a Microwave Radiometer" (1946) <u>Phys. Rev. 70</u>, 340–348

38. M. Dickenson, "Typical Lyman-break galaxy spectra vs. redshift," GOODS Legacy team (and STScI), site updated (January 2009) http://ssc.spitzer.caltech.edu/documents/compendium/xgalsci/lybreakgal.gif

39. R. W. Ditchburn, <u>Light,</u> Blackie & Son, Limited, London and Glasgow (1963)

40. N. Drory, R. Bender, J. Snigula, G. Feulner, U. Hopp, and C. Maraston, <u>The Astrophysics Journal</u>, 562:L111-L114, Dec. 2001.

41. A. Einstein, "Cosmological Considerations of the General Theory of Relativity," <u>On the Shoulders of Giants</u> – edited by Stephen Hawking, Running Press, London ,pp 1248- 1257 (1917, 2002)

42. A. Einstein, <u>Relativity – The Special and The General Theory, 15th Ed.</u>, Crown, New York (1952)

43. A. Einstein, "On the Quantum Theory of Radiation", (1917), <u>Sources of Quantum Mechanics</u>, Dover, New York (1967)

44. Richard Ellis, "Faint Blue Galaxies", <u>Annual Revue of Astronomy and Astrophysics</u>, 35: 389-443 (1997)

45. Ruth Esser and Dimitar Sasselov, "On the Disagreement   between Atmospheric and Coronal Electron Densities", The Astrophysical Journal, 521:L145–L148 (August 20, 1999)

46. A. C. Fabian and X. Barcons, "The Origin of Tthe X-Ray Background," Annu. Rev. Astron. Astrophys, 30: 429-456 (1992)

47. Alan J. Fenn, Donald H. Temme, William P. Delaney, and William E. Courtney, "The Development of Phased-Array Radar Technology", Linclon Laboratory Journal, Vol. 12, No. 2 (2000)

48. R. P. Feynman, Feynman Lectures on Gravitation, Ed. F. B. Mornings, and W. G. Wagner, Addison Wesley (1995)

49. Fischer, Philippe; McKay, Timothy A.; Sheldon, Erin; Connolly, Andrew; Stebbins, Albert; Frieman, Joshua A.; Jain, Bhuvnesh; Joffre, Michael; Johnston, David; Bernstein, Gary; Annis, James; Bahcall, Neta A.; Brinkmann, J.; Carr, Michael A.; Csabai, István; Gunn, James E.; Hennessy, G. S.; Hindsley, Robert B.; Hull, Charles; Ivezić, Željko; Knapp, G. R.; Limmongkol, Siriluk; Lupton, Robert H.; Munn, Jeffrey A.; Nash, Thomas; Newberg, Heidi Jo; Owen, Russell; Pier, Jeffrey R.; Rockosi, Constance M.; Schneider, Donald P.; Smith, J. Allyn; Stoughton, Chris; Szalay, Alexander S.; Szokoly, Gyula P.; Thakar, Aniruddha R.; Vogeley, Michael S.; Waddell, Patrick; Weinberg, David H.; York, Donald G.; The SDSS Collaboration, "Weak Lensing with Sloan Digital Sky Survey Commissioning Data: The Galaxy-Mass Correlation Function to 1 H^{-1} Mpc", The Astronomical Journal, Vol. 120, Issue 3, pp 1198-1208 (2000)

50. D. J. Fixsen, "The Cosmic Microwave Background Spectrum from the Full COBE FIRAS Data Set," The Astrophysical Journal, (May 1996)

51. M. Fukugita, C. J. Hogan, and P. J. E. Peebles, "The cosmic distance scale and the Hubble constant," Nature, Vol. 366, pp 309-312 (25 November 1993)

52. G. Gamow, "The Origin of Elements and the Separation of Galaxies," Physical Review 74 (1948), 505.

53. G. Gamow, "The evolution of the universe", Nature 162 (1948), 680.

54. M. J. Geller and P. J. E. Peebles, "Test of the Expanding Universe Postulate," The Astrophysical Journal, Vol. 174, pp. 1-5 (15 May 1972)

55. G. Ghirlanda , G. Ghisellini, and C. Firmani, "Gamma-ray bursts as standard candles to constrain the cosmological parameters", New Journal of Physics, 8, 123 (2006)

56. Philippe Gondoin, V. Kirschner, A. Santovincenzo, A Lyngvi, A. Short, U. Telljohann, G. Chirulli, P. de Pascale, D. de Wilde, D.M. Di Cara, P. Fabry, A. Figgess, L. Gaspar Venancio, D. Hagelschuer, P. Holsters, A. Jeanes, M. Khan, S. Mangunson, A. Mestreau-Garreau, C. Monteleone, J-L Pellon-Bailon, P. Ponzio, H. Rozemeijer, M. Tuti, P. Villar, S-F. Wu, S. Zimmermann, "ESA Study of a Wide Field Imager for Supernovae Surveys and Dark Energy Characterization", proceedings of the 6th International Conference on Space Optics, (2006)

57. John Gribbon, The Scientist, Random House, New York (2002)

58. Gruber, D. E., Matteson, J. L., Peterson, L. E., & Jung, G. V., "The Cosmic X-Ray Background Radiation", The Astrophysical Journal, 520 124 (1999)

59. J. E. Gunn and B. A. Petersen, (1972), The Astrophysical Journal, 142, 1633 (1965) [See also Gunn and Gould concerning initial estimate of upper limit on intergalactic medium density from infall limits, (Sarazin 1986 reference).]

60. A. H. Guth, "Inflationary universe: A possible solution to the horizon and flatness problems", Phys. Rev. D, Vol. 23, pp. 347-356 (January 1981)

61. A. Guth, "Was Inflation the 'Bang' of the Big Bang?", The Beamline, 27, 14 (1997).

62. M. Guhathakurta, A. Fludra, S. E. Gibson, D. Biesecker, R. Fisher, "Physical properties of a coronal hole from a coronal diagnostic spectrometer, Mauna Loa Coronagraph, and LASCO observations during the Whole Sun Month", Journal of Geophysical Research, 104 (A5), 9801–9808 (October, 1998)

63. Luigi Guzzo, "Large-Scale Structure at the Turn of the Millennium," Invited review at the 19th Texas Symposium on Relativistic Astrophysics and Cosmology, eds. J. Paul, T. Montmerle, and E. Aubourg; astro-ph/9911115 (November 1999)

64. N. Haramein, M. Hyson, E. A. Rauscher, "Scale Unification: A Universal Scaling Law for Organized Matter," Proceedings of The Unified Theories Conference in Cs Varga, I. Dienes & R.L. Amoroso (eds.) (2008) http://www.theresonanceproject.org/pdf/scalinglaw_paper.pdf

65. Edward Harrison, "The redshift-distance and velocity-distance laws", The Astrophysical Journal, Part 1 (ISSN 0004-637X), vol. 403, no. 1, p. 28-31. (January 1993)

66. J.G. Hartnett and K. Hirano, "Galaxy redshift abundance periodicity from Fourier analysis of number counts N(z) using SDSS and 2dF GRS galaxy surveys", The Astrophysical Journal, Vol. (11 September 2008)

67.   Stephen Hawking, <u>A Brief History of Time – from the Big Bang to Black Holes</u>, Bantam, New York (1988)

68.   A. K. Hicks, M. W. Wise, J. C. Houck, and C. R. Canizares, "CHANDRA X-RAY Spectroscopy and Imaging of the Galaxy Cluster PKS 0745-191", <u>The Astrophysical Journal</u>, 580:763–773 (December , 2002)

69.   David W. Hogg, "Distance measures in cosmology", <u>The Astrophysical Journal</u>, Vol. (December 2000)

70.   David W. Hogg, Ivan K. Baldry, Michael R. Blanton, and Daniel J. Eisenstein, "The K correction", <u>The Astrophysical Journal</u>, arXiv:astro-ph/0210394v1 (2002)

71.   Wayne Hu, "Lecture I: Thermal History and Acoustic Kinematics", Tenerife, November (2007)
http://background.uchicago.edu/~whu/Presentations/canaries1.pdf

72.   John P. Huchra, web site: http://cfa-www.harvard.edu/~huchra/hubble/ (2008)

73.   J. D. Jackson, <u>Classical Electrodynamics</u>, John Wiley & Sons, New York and London (1962)

74.   Tess Jaffe, "X-Ray Background, everpresent noise", <u>Astronomy Today</u> (2008). http://www.astronomytoday.com/cosmology/xray.html

75.   D. F. V. James, M. P. Savedoff, and E. Wolf, "Shifts of Spectral Lines produced by Scattering from Fluctuating Random Media," <u>The Astrophysical Journal</u>, Vol. 359, pp. 67-71 (August 1990)

76.   T. Jarret, "'Finger of God' Radial Velocity Artifacts", <u>2MASS Galaxy Redshift Catalog (XSCz)</u>, (last updated Dec 11, 2006)
http://spider.ipac.caltech.edu/staff/jarrett/XSCz/index.html

77.   Jensen, J. W., "Supernova 2006gy and the Copernicus Principle: Modern Cosmology Meets Goliath, Crisis in Cosmology Conference-2 (Sept., 2008)

78.   Kahn, S. M., "Draco's silver lining," <u>Nature</u>, Vol. 351, pp 607-608 (20 June 1991)

79.   Kaiser, N., "Clustering in real space and in redshift space", <u>Monthly Notices of the Royal Astronomical Society</u>, 227, 1-21 (1987)

80.   Kellermann, K. I., "The cosmological deceleration parameter estimated from the angular-size/redshift relation for compact radio sources," Nature, Vol. 351, pp 607-608 (20 June 1991)

81.   Chris Kochanek, "Is there a Cosmological Constant?" The Astrophysical Journal, Vol. 466, 638 (1996)

82.   Kochanek C. S., Pahre M. A., Falco E. E., Huchra J. P., Mader J., Jarrett T. H., Chester T., Cutri R., Schneider S. E., "The K-band galaxy luminosity function", The Astrophysical Journal, 560, 566 (2001)

83.   M. Kowalski, D. Rubin, G. Aldering, R. J. Agostinho, A. Amadon, R. Amanullah, C. Balland, K. Barbary, G. Blanc, P. J. Challis, A. Conley, N. V. Connolly, R. Covarrubias, K. S. Dawson, S. E. Deustua, R. Ellis, S. Fabbro, V. Fadeyev, X. Fan, B. Farris, G. Folatelli, B. L. Frye, G. Garavini, E. L. Gates, L. Germany, G. Goldhaber, B. Goldman, A. Goobar, D. E. Groom, J. Haissinski, D. Hardin, I. Hook, S. Kent, A. G. Kim, R. A. Knop, C. Lidman, E. V. Linder, J. Mendez, J. Meyers, G. J. Miller, M. Moniez, A. M. Mourao, H. Newberg, S. Nobili, P. E. Nugent, R. Pain, O. Perdereau, S. Perlmutter, M. M. Phillips, V. Prasad, R. Quimby, N. Regnault, J. Rich, E. P. Rubenstein, P. Ruiz-Lapuente, F. D. Santos, B. E. Schaefer, R. A. Schommer, R. C. Smith, A. M. Soderberg, A. L. Spadafora, L.-G. Strolger, M. Strovink, N. B. Suntzeff, N. Suzuki, R. C. Thomas, N. A. Walton, L. Wang, W. M. Wood-Vasey, "Improved Cosmological Constraints from New, Old and Combined Supernova Datasets", accepted for publication in The Astrophysical Journal, (2008)

84.   Edward W. Kolb, "Cosmology and the Unexpected", Searching for the 'totally unexpected' in the LHC era, two lectures presented at the International School of Subnuclear Physics, Erice, Italy (2007) arXiv:0709.3102v1 [astro-ph]

85.   Thomas S. Kuhn, The Structure of Scientific Revolutions, 3rd Ed., Univ. of Chicago Press, Chicago (1962, 1970, 1996)

86.   Kenneth R. Lang, Astrophysical Formulae: A Compendium for the Physicist and the Astrophysicist - 2nd edition, New York: Springer-Verlag (1980)

87.   Lehner, N.; Savage, B. D.; Richter, P.; Sembach, K. R.; Tripp, T. M.; Wakker, B. P., "Physical Properties, Baryon Content, and Evolution of the Lyα Forest: New Insights from High-Resolution Observations at z <~ 0.4", The Astrophysical Journal, Volume 658, Issue 2, pp. 680-709 (2007)

88.   Eric Lerner, The Big Bang Never Happened, Times Books, New York (1991)

89. G. N. Lewis, "The Nature of Light," Proceedings. N. A. S., 12, 22-29 (1926).

90. Loewenstein, M., "Chemical Composition of the Intracluster Medium", Carnegie Observatories Astrophysics Series, Vol. 4: Origin and Evolution of the Elements, arXiv.org > astro-ph > arXiv:astro-ph/0310557v1, (October 2003)

91. Loveday, J., "The K-band Luminosity Function of Nearby Field Galaxies", Monthly Notices of the Royal Astronomical Society, Volume 312, Issue 3, pp. 557-566 (2000)

92. Lowenthal, J. D.; Koo, D. C.; Guzman, R.; Gallego, J.; Phillips, A. C.; Faber, S. M.;Vogt, N. P.; Illingworth, G. D.; Gronwall, C.; "Keck spectroscopy of redshift z~3 galaxies in the Hubble deep field", The Astrophysical Journal, 481, 673 (1997)

93. Lubin, Lori M. and Bahcall, Neta A., "The relation between velocity dispersion and temperature in clusters – Limiting the velocity bias", The Astrophysical Journal, Part 2 – Letters (ISSN 0004-637X), vol. 415, no. 1, p. L17-L20 (1993)

94. Lubin, L. M., Postman, M., Oke, J. B., Ratnatunga, K. U., Gunn, J. E., Hoessell, J. G., & Schneider, D. P. , The Astronomical Journal, 116, 584 (1998)

95. Lubin, L. M., & Sandage, "The Tolman Surface brightness Test for the Reality of the Expansion, II", The Astronomical Journal, 121, 2289 (2001a)

96. Lubin, L. M., & Sandage, A., "The Tolman Surface brightness Test for the Reality of the Expansion, III", The Astronomical Journal, 122, 1071 (2001b)

97. Lubin, L. M., & Sandage, A., A Measurement of the Tolman Signal and the Luminosity Evolution of Early-Type Galaxies – The Tolman Surface brightness Test for the Reality of the Expansion, IV", 122:1084-1103, (2001 September)

98. Luger, Frank, "Antropocentrism vs. Cosmocentrism – Groping Toward a Paradigm Shift", Commensal – Newsletter of the Philosophical Discussion Group Of British Mensa, issue 102 (August 2000)

99. Maddox, S. J.; Efstathiou, G.; Sutherland, W. J., "The APM Galaxy Survey - Part Two - Photometric Corrections," R.A.S. Monthly Notices, V.246, NO. 3/OCT1, P. 433, (1990)

100.   Maloney, Phil and Leiden, Sterrewacht, "The Extragalactic Radiation Field and Sharp Edges to HI Disks in Galaxies," Proceeding NASA Conference, Wyoming (1991)

101.   Mannheim, Philip, "Does the Cosmological Constant Problem Presage a Paradigm Shift in Gravitational Theory?" Proceedings CCC2 – conference on cosmology (2008)

102.   Marshall, F. E., Boldt, E.A., Holt, S. S., Miller, R. B., Mushotzky, R. F., et al., "The diffuse X-ray background spectrum from 3 to 50 keV," The Astrophysical Journal, 235: 4-10 (1980)

103.   J. C. Mather, "Measurement of the Cosmic Microwave Background Spectrum by the COBE FIRAS Instrument," The Astrophysical Journal, Vol. 361, pp. 134-136 (January 1993)

104.   J. C. Mather, "A Preliminary Measurement of the Cosmic Microwave Background Spectrum by the Cosmic Background Explorer (COBE) Satellite," The Astrophysical Journal, Vol. 354, pp. L37- L40 (May 1990)

105.   J. C. Mather, et al., "Measurement of the Cosmic Microwave Background Spectrum by the COBE FIRAS Instrument," The Astrophysical Journal, 420, 439 (1994)

106.   J. Melbourne, A. C. Phillips, J. Harker, G. Novak, D. C. Koo, and S. M. Faber, "Radius-Dependent Luminosity Evolution Of Blue Galaxies In Goods-N", The Astrophysical Journal, 660:81 Y 96 (2007)

107.   Felipe Menanteau, Holland C. Ford, Vero´nica Motta, Narciso Benl´tez, Andre´ R. Martel, John P. Blakeslee, and Leopoldo Infante, "The Morphological Demographics of Galaxies in the Advanced Camera Surveys of Hubble Ultra Deep Parallel Fields", The Astronomical Journal, 131:208– 215, (2006 January)

108.   Mordehai Milgrom, ""A Modification of the newtonian Dynamics as a Possible Alternative to the Hidden Mass Hypothesis", The Astrophysical Journal, Vol. 270, pp. 365-370 (July 1983)

109.   Mordehai Milgrom, ""Does Dark Matter Exist?" A Scientific American Special report reprint, Scientific American (2009)

110.   C. W. Misner, K. S. Thorne and J. A. Wheeler, "Part VI: The Universe," Gravitation, W. H. Freeman and Co., NY, p. 703-816, (1973)

111.   Monier, Eric M.; Turnshek, David A.; Hazard, Cyril, "Hubble Space Telescope Faint Object Spectrograph Observations of a Unique Grouping of

568

Five QSOS: The Sizes and Shapes of Low-z LYalpha Forest Absorbers", The Astrophysical Journal, Volume 522, Issue 2, pp. 627-646. (1999)

112. R. L. Oldershaw, "Cosmology theory compromised," Nature, Vol. 346, p 800 (19 July 1990)

113. A. N. Parmar, M. Guainazzi, T. Oosterbroek, A. Orr, F. Favata, D. Lumb, and A. Malizia "The low-energy cosmic X-ray background spectrum observed by the Beppo SAX LECS", Astronomy and Astrophysics, 345, 611–617 (1999)

114. P. J. E. Peebles, Principles of Physical Cosmology, Princeton University Press; Princeton, New Jersey (1993)

115. P. J. E. Peebles and J. Silk, "A cosmic book of phenomena," Nature, Vol. 346, pp 233-239 (19 July 1990)

116. S. Perelmuter, S., G. Aldering, G. Goldhaber, R. A. Knop, P. Nugent, P. G. Castro1, S. Deustua, S. Fabbro1, A. Goobar, D. E. Groom, I. M. Hook, A. G. Kim, M. Y. Kim, J. C. Lee, N. J. Nunes, R. Pain, C. R. Pennypacker, R. Quimby, C. Lidman, R. S. Ellis, M. Irwin, R. G. McMahon, P. Ruiz-Lapuente, N. Walton, B. Schaefer, B. J. Boyle7, A. V. Filippenko8, T. Matheson8, A. S. Fruchter9, N. Panagia, H. J. M. Newberg, W. J. Couch, "Measurements of Ω and Λ from 42 High-Redshift Supernovae", (Supernova Cosmology Project), The Astrophysical Journal, 517, 565 (1999)

117. Petrosian, V., The Astrophysical Journal, 209, L1 (1976)

118. M. Plionis, "The Quest for the Cosmological Parameters", Proceedings of the 1st Aegean Cosmology Summer School (2001). astro-ph/0205166

119. R. F. Post, "Nuclear Fusion", McGraw-Hill Encyclopedia of Physics – 2nd Ed., pp. 879-881 (1993)

120. Postman, M., & Lauer, T., The Astrophysical Journal, 440, 28 (1995)

121. T. H. Reiprich, "The galaxy cluster X-ray luminosity–gravitational mass relation in the light of the WMAP 3rd year data," Astronomy & Astrophysics (2006)

122. P. M. Ricker and P. Mézáros, "Contributions of Starburst Galaxies and Reflection Dominated Active Galactic Nuclei to the Cosmic X-Ray Background," The Astrophysical Journal, 418: 49-54 (1993)

123. Paul Ricker, "The Cosmic X-Ray Background Radiation,"

http://www.astro.uiuc.edu/~pmricker/research/xrb/#References, last update 05/04/07.

124.   Martin Rees and Priyamvada Natarajan, "A Field Guide to the Invisible Universe", Discover: Space / Telescopes, March (2009)

125.   Adam G. Riess, Alexei V. Filippenko, Peter Challis, Alejandro Clocchiatti, Alan Diercks, Peter M. Garnavich, Ron L. Gilliland, Craig J Hogan, Saurabh Jha, Robert P. Kirshner, B. Leibundgut, M. M. Phillips, David Reiss, Brian P. Schmidt, Robert A Schommer, Nicholas B. Suntzeff, and John Tonry, "Observational Evidence from Supernovae for an Accelerating Universe and a Cosmological Constant", The Astronomical Journal, 116:1009-1038 (September 1998)

126.   Adam G. Riess, Louis-Gregory Strolger, John Tonry, Stefano Casertano, Henry C. Ferguson, Bahram Mobasher, Peter Challis, Alexei V. Filippenko, Saurabh Jha, Weidong Li, Ryan Chornock, Robert P. Kirshner, Bruno Leibundgut, Mark Dickinson, Mario Livio, Mauro Giavalisco, Charles C. Steidel, Narciso Benitez, Zlatan Tsvetanov, "Type Ia Supernova Discoveries at z>1 From the Hubble Space Telescope: Evidence for Past Deceleration and Constraints on Dark Energy Evolution", Astrophysics Journal 607 665-687 (2004)

127.   D. H. Roberts, J. Lehar, J. N. Hewitt and B. F. Burke, "The Hubble constant from VLA measurements of the time delay in the double quasar 0957+561," Nature, Vol. 352, pp 43-45 (4 July 1991)

128.   P. Ruiz-LaPuente, "Cosmology With Supernovae! *The Road from Galactic Supernovae to the edge of the Universe*", Invited Review given at JENAM 2002, Porto 4-6 September, Astrophysics and Space Science, 290: 43–59, 2004.

129.   Sarazin, C. L., "X-ray emission from clusters of galaxies," Review of Modern Physics, Vol. 58, 1-115 (1986)

130.   Sarkar, Subin, "Primordial Nucleosynthesis and Dark Matter", astro-phy/9611232, 27 Nov (1996)

131.   Sciama, D. W., The Unity of the Universe, Anchor Books, Doubleday & Co., New York (1959)

132.   Science news, (January 12, 2006)

133.   D. Scott and G. F. Smoot, "Cosmic microwave background", Physics Letters B592, S. Eidelman et al., June 16 (2004) 23:32   Also available at: http://pdg.lbl.gov/

134. Silk, J., and Sunyaev, R. A., "Ionising flux of cosmic background radiation," <u>Nature,</u> 260, 508, (1976)

135. J. Singh, <u>Great Ideas and Theories of Modern Cosmology,</u> 2nd Ed., Dover Publications, New York (1970)

136. Smith, L., "Hydrogen clouds colder than expected", <u>Science News</u> 138, 253 (1990)

137. Lee Smolin, "Scientific alternatives to the anthropic principle", <u>Universe or Multiverse</u>, ed. by Bernard Carr, Cambridge University Press, ISBN-13: 9780521848411 (June 2007)

138. Yoshiaki Sofue and Vera Rubin, "Rotation Curves of Spiral Galaxies," <u>Annu. Rev. Astron. Astrophys,</u> 39: 137-174 (2001)

139. Johann Soldner, "On the Deflection of a Light Ray from It Straight Motion Due to the Attraction of a World Body Which it Passes Closely". <u>Berliner Astronomisches Jahrbuch</u> (1804) (from a secondary account)

140. Steidel, C.; Giavalisco, M.; Pettini, M.; Dickinson, M.; Adelberger, K., "Spectroscopic Confirmation of a Population of Normal Star-Forming Galaxies at Redshifts z>3", <u>The Astrophysical Journal,</u> 462, L17 (1996)

141. Steidel, C. C., Giavalisco, M., Pettini, M., Dickinson, M. E., and Adelberger, K., "The Distribution of Lya-Emitting Galaxies at z=2.3", <u>The Astrophysical Journal,</u> 462, L17 (1996)

142. Sunyaev, R. A., "The Interaction of the Metagalactic Ultraviolet Background Radiation with Galaxies and the Limit on the Density of the Intergalactic Gas," <u>Astrophysical Letters</u> 3, 33 (1969)

143. Sunyaev, R. A., and Y. B. Zel'dovich, <u>Comm. Astr. Space Phys.,</u> 4, 173 (1969)

144. Sunyaev, R. A., and Y. B. Zel'dovich, Astr. Space Sci. 7, 3 (1970)

145. David Sutton and Sarah Bridle, "Quantifying the Mass Distribution of a Galaxy Cluster by Weak Gravitational Lensing," <u>MSci Astrophysics</u> (2006)

146. Tolman, R. C., <u>Proc. Nat. Acad. Sci.,</u> 16, 5111 (1930)

147. J. A. Tyson, "The Optical Extragalactic Background Radiation", appeared in in "Extragalactic Background Radiation", <u>Space Telescope Science Institute Symposium Series 7</u>, eds. D. Calzetti, M. Livio and P. Madau (1995)

148. G. L. Verschuur, and K. I. Kellermann, <u>Galactic and Extragalactic Radio Astronomy, 2nd Ed.</u>, Springer-Verlag, New York (1988)

149. R. V. Wagoner, William A. Fowler, and F. Hoyle, "On the Synthesis of Elements at Very High Temperatures", <u>The Astrophysical Journal</u>, Vol. 148 (1967)

150. R. V. Wagoner, "Synthesis of the Elements within Objects Exploding from Very High Temperatures", <u>The Astrophysical J. Supl. Ser.</u>, Vol. 162 (1969)

151. R. V. Wagoner, "Big Bang Nucleosynthesis Revisited", <u>The Astrophysical J.</u>, Vol. 179 (1973)

152. Walker, T. P., Steigman, G., Kang, H., Schramm, D. M., and Olive, K. A., "Standard big bang nucleosynthesis", <u>The Astrophysical Journal</u>, 376, 51 (1991)

153. J. A. Wheeler and R. P. Feynman, "Interactions with the Absorber as the Mechanism of Radiation," <u>Rev. Mod. Phys.</u>, 17, 157-181 (1945).

154. Terence Witt, <u>Our Undiscovered Universe</u>, ISBN-10: 0-9785931-3-8 (2007)

155. E. Wolf and D. N. Pattanayak, "General Form and New Interpretation of the Ewald-Oseen Extinction Theorem," <u>Optics Communications,</u> Vol. 6, No. 3, pp. 217-220 (November 1972)

156. E. Wolf, "Non-cosmological redshifts of spectral lines," <u>Nature,</u> Vol. 326, No. 6111, pp. 363-365 (26 March 1987)

157. E. Wolf, J. T. Foley, and F. Gori, "Frequency shifts of spectral lines produced by scattering from spatially random medium," <u>J. Optical Soc. Amer. A,</u> Vol. 6, No. 8, pp. 1142-1149 (August 1989)

158. E. Wolf, "Invariance of the Spectrum of Light on Propagation," <u>Phys. Rev. Letters,</u> Vol. 56, No. 13, pp. 1370-1372 (31 March 1986)

159. E. Wolf, "Recent Work on the Ewald-Oseen Extinction Theorem," <u>Atomic and Molecular Optics, Proc. of the 1971 Rochester Symp.,</u> J. H. Eberly, Ed., Univ. of Rochester, NY, pp. 55-67 (March 1971)

160. E. L. Wright, "Cosmic Tutorial," from <u>Ned Wright's home page</u>, (October, 2007) http://www.astro.ucla.edu/~wright/cosmo_01.htm

161. E. L. Wright, "Measuring the Curvature of the Universe by Measuring the Curvature of the Hubble Diagram", from <u>Ned Wright's home page</u>, (May, 2008) http://www.astro.ucla.edu/~wright/sne_cosmology.html

162.   Wu, X.; Hamilton, T.; Helfand, D. J.; Wang, Q.; "The Intensity And Spectrum Of The Diffuse X-Ray Background," The Astrophysical Journal, 379: 564-75 (1991)

163.   Yost and Daunt (2008), "Astronomy 162 – Web Syllabus", Dept. Physics & Astronomy, University of Tennessee, (current in 2008) http://csep10.phys.utk.edu/astr162/index.html

164.   Ya. B. Zel'dovich, "The theory of the expanding universe as originated by A. A. Friedmann" [English translation], Sov. Phys. – Uspekhi. 6, 475-494 (1964)

165.   Zel'dovich, Ya.B. & R.A. Sunyaev, "The Interaction of Matter and Radiation in a Hot-Model Universe," Astrophysics and Space Science 4, 301-16 (1969)

166.   Zhi-Bin Zhang, Jia-Gan Deng, Rui-Jing Lu, and Hai-Feng Gao, "Relative Spectral Lag: a New Redshift Indicator of Gamma-ray Bursts", Chin. J. Astron. Astrophys. Vol. 6 (2006), No. 3, 312–322 (2006)

Index of (First-Appearing) Authors

Aldering	32, 559, 566, 569
Alfvén	12, 559
Alpher	387, 467, 559
Arp	8, 9, 55, 462, 531, 559
Ashmore	148, 559
Assis	12, 531, 559
Bahcall	52, 54, 55, 57, 332-334, 337, 347, 358-359, 365, 367, 379, 458, 518, 524, 559, 560, 561, 567
Band	484, 559
Barcelo	43, 453, 528, 559
Barcons	47-48, 386, 559, 563
Becker	333, 365, 560
Birkinshaw	203, 438, 560
Biviano	326, 560
Blain	45, 560
Blanton	298, 560
Blondin	13, 269, 282-283, 285-287, 560
Bonn	xxi, 104, 257, 279, 337, 489, 491-492, 542, 560
Born	xxviii, 156, 163, 278, 560
Boroson	135, 561
Bothun	498, 561
Brecher	156, 160, 561
Brown	292, 295-296, 561
Brune	104, 106, 561
Brynjolfsson	225, 561
Burrows	55, 385, 561
Carlstrom	203, 263, 561
Campbell	102, 561
Clifton	278, 561
Colless	39, 304, 364, 561
Conselice	38, 273, 314-315, 562
Cowen	254, 562
Cowie	52, 54, 108, 333, 346, 449, 562
Cramer	535, 540, 562
Dicke	103, 387, 562
Dickenson	140, 562
Ditchburn	77, 150, 562
Drory	296, 298, 307, 309, 311, 562
Einstein	ix, xviii, xxiii, xxviii, xxxi-xxxiii, xxxvii-xxxviii, 4, 10, 13, 14, 49, 53, 92, 102-105, 191, 222, 227-228, 236-238, 241, 257, 314, 321,

326-329, 370, 402, 456-457, 487-490, 492-497, 499-502, 504, 507, 512-513, 530-531, 541, 548, 557, 562

Ellis 45, 255, 310, 313, 562, 566, 569
Esser 104-105, 563
Fabian 386, 559, 563
Fenn 152, 563
Feynman 517-518, 535, 540, 563, 572
Fischer 329, 379, 563
Fixsen 388, 563
Fukugita 219, 560, 563
Gamow 387, 467-469, 563
Geller 12-13, 366, 563
Ghirlanda 32, 269-270, 563
Gondoin 266, 564
Gribbon 532, 564
Gruber 382, 564
Gunn xiii, 48, 138-139, 145, 560, 563, 564, 567
Guth xviii, xxxv, 433, 435-436, 487, 498, 505-506, 528, 530, 564
Guzzo 374, 564
Haramein 341-342, 564
Harrison 234, 416, 564
Hartnett 365, 374-375, 564
Hawking xviii, 463-464, 491-492, 496, 499, 502, 532, 547-548, 553, 562, 565
Hicks 52, 54, 334, 337, 339, 343, 346-347, 356, 565
Hogg 235, 238, 243, 264, 560, 565
Hu 393, 565
Huchra 220, 364, 565-566
Jackson 77, 565
Jaffe 386, 565
James 162, 565
Jarret 355, 358, 365, 565-566
Jensen 280, 565
Kahn 55, 565
Kaiser 365, 565
Kellermann 135-136, 261-264, 383-384, 566, 571
Kochanek 236, 238, 294, 297-298, 300-301, 318, 329, 379, 518, 566
Kowalski 267-268, 566
Kuhn 14, 530, 566
Lang 477, 566
Kolb 329, 566
Lehner 54, 148, 566
Lerner 64, 566
Lewis 533, 567
Loewenstein 52, 319, 458, 567
Loveday 298, 300, 560, 567

Lowenthal 301, 567
Lubin xxx, 271-274, 334, 337, 518, 567
Luger 555, 567
Maddox 495, 567
Maloney 55, 568
Mannheim 488, 568
Marshall 367, 382-383, 386, 568
Mather 385-386, 388, 507, 568
Melbourne 38, 568
Menanteau 272, 568
Milgrom 366-367, 568
Misner xviii, 48, 103, 200, 241, 434, 436, 438, 568
Monier 148, 568
Oldershaw 531, 569
Parmar 382, 569
Peebles 1, 12, 14, 235, 238, 241, 262, 265, 385, 467-468, 472, 478, 506, 518, 523, 563, 569
Perelmuter 266, 569
Petrosian 272, 569
Plionis 28, 512, 569
Post 56, 383, 475, 478, 569
Postman 271-272, 560, 567, 569
Reiprich 52, 333, 569
Ricker 382, 384, 569
Rees 456, 559, 561, 570
Riess 8, 62, 236, 259, 266, 277-278, 281, 468, 505-506, 560, 570
Roberts 219, 570
Ruiz-LaPuente 29-30, 280, 286, 570
Sandage xxx, 271-274, 518, 567
Sarazin 48, 57, 60, 564, 570
Sarkar 474, 484, 517, 570
Sciama 462, 570
Scott 508, 570
Silk 48, 55, 384, 438-439, 468, 569, 570
Singh 171, 570
Smith 135, 570
Smolin 32, 453, 508, 510-512, 514-517, 571
Sofue 339-341, 379-380, 571
Smoot 507-508, 570
Soldner 326, 571
Steidel 301, 570, 571
Sunyaev 48, 54-55, 202-203, 263, 444, 560-561, 570-571, 573
Sutton 329, 571
Tolman vii, xxx, 255-257, 270-274, 567, 571
Tyson 386, 571

Verschuur	383-384, 571
Wagoner	xviii, xxxvii, 9, 436-437, 449, 452-456, 461, 464-468, 470, 473-474, 479, 481-483, 572
Walker	474, 572
Wheeler	241, 535, 540, 568, 572
Witt	504, 572
Wolf	xxviii, 48, 156, 162-163, 165-166, 192, 560, 565, 572
Wright	388, 572
Wu	382, 573
Yost	394, 573
Zel'dovich	12, 200-203, 263, 560-561, 571, 573
Zhang	484, 573

A Related Book by Raymond Bonn

"Observation" (as against a mere Lorentz mapping of the fou
coordinate aspects of objects in relative motion is more
complex than generally acknowledged. One must incorporat
the effects of Penrose's "transformation of the field of vision

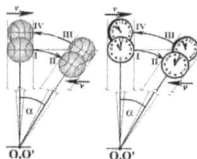

Visual observations made by two coincident observers involv
four possibilities rather than the two usually considered.

ISBN 978-0-6151-9781-4

Aberrations of Relativity

R. F. Bonn

Aberrations
of Relativity

Raymond F. Bonn

The Aberrations of Relativity, 202 pages, by Raymond F. Bonn

This is a collection of essays and articles that place emphasis on the most observable aspects of relative motion (aberration effects) to ultimately define an "observational relativity." Readers will gain insights into all aspects of relativity using the many informative diagrams and illustrations. These will prove invaluable whether the author's occasional alternative reinterpretations are accepted or not. It is written for the intelligent (perhaps quite intelligent) layman. Very little in the way of advanced mathematics and physics is required to fully comprehend the arguments, however.

*This is the book you may have read about, which took Ray Bonn to New York as described in **Not Julie**, a novel by R. F. Vaughan.*

ISBN 978-0-6151-9781-4

http://www.lulu.com/content/paperback-book/aberrations-of-relativity/572819
http://www.lulu.com/content/hardcover-book/aberations-of-relativity/945246
An electronic download is also available.

www.ingramcontent.com/pod-product-compliance
Lightning Source LLC
Chambersburg PA
CBHW021542210326
41599CB00010B/284